T0137400

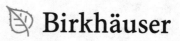

Advances in Mathematical Fluid Mechanics

Series editors

Giovanni P. Galdi, Pittsburgh, USA
John G. Heywood, Vancouver, Canada
Rolf Rannacher, Heidelberg, Germany

Advances in Mathematical Fluid Mechanics is a forum for the publication of high quality monographs, or collections of works, on the mathematical theory of fluid mechanics, with special regards to the Navier-Stokes equations. Its mathematical aims and scope are similar to those of the *Journal of Mathematical Fluid Mechanics*. In particular, mathematical aspects of computational methods and of applications to science and engineering are welcome as an important part of the theory. So also are works in related areas of mathematics that have a direct bearing on fluid mechanics.

More information about this series at http://www.springer.com/series/5032

Sergey Shklyaev • Alexander Nepomnyashchy

Longwave Instabilities and Patterns in Fluids

 Birkhäuser

Sergey Shklyaev
Institute of Continuous Media Mechanics
UB RAS
Perm, Russia

Alexander Nepomnyashchy
Technion
Department of Mathematics
Haifa, Israel

ISSN 2297-0320   ISSN 2297-0339 (electronic)
Advances in Mathematical Fluid Mechanics
ISBN 978-1-4939-8529-6   ISBN 978-1-4939-7590-7 (eBook)
https://doi.org/10.1007/978-1-4939-7590-7

Mathematics Subject Classification: 76E06, 76E17, 76E30, 35B36, 35Q35

This book is published under the trade name Birkhäuser, www.birkhauser-science.com
The registered company is Springer Science+Business Media, LLC
The registered company address is: 233 Spring Street, New York, NY 10013, U.S.A.

In memory of Prof. Alexander Golovin

Preface

The longwave instabilities are inherent to a variety of systems in fluid dynamics (from dynamics of atmosphere and ocean to motion in microfluidic devices), geophysics (mantle dynamics, surges, avalanches), electrodynamics (electronic beams, plasma), biophysics (wetting and lubrication in biophysical applications, such as tear dynamics and flows in the respiratory system), and many others.

The characteristic feature of all these systems is the existence of two significantly different length scales, a "short scale" (e.g., thickness of a thin liquid film, a reaction zone, a phase transition front) and a "long scale" of patterns generated by instabilities. In many cases, such a difference makes it possible to apply asymptotic methods (e.g., averaging over the "fast" coordinate or other multiscale approaches), which allow to significantly simplify the problem and reduce its dimension.

The abovementioned difference in length scales gives rise to several clearly distinguished time scales. Some appropriate mathematical tools (e.g., averaging technique or disregarding fast and/or slow processes) are efficient in many cases, and they drastically simplify the problem under consideration. For instance, within the lubrication approximation in fluid mechanics, one can safely neglect such processes as transversal transfer of momentum and other physical quantities (heat, mass, etc.), because they immediately adjust to the instantaneous distribution of corresponding forces and thermodynamic forces. In contrast, the relaxation of longitudinal fluxes is a slow process, which has to be a focus of attention. Such a simplification makes the lubrication approximation a very powerful and frequently used approach (see useful surveys on that subject by A. Oron, S.H. Davis, and S.G. Bankoff, Rev. Mod. Phys. **69**, 931 (1997), and R.V. Craster and O.K. Matar, Rev. Mod. Phys. **81**, 1131 (2009)). It should be emphasized that the abovementioned lubrication approximation, which usually leads to the derivation of a generalized Cahn–Hilliard equation, is a specific example of a more general approach, which results in a much wider class of reduced equations. This approach is the subject of the present book.

Historically, *shortwave instabilities* characterized (near their thresholds) by a well-defined critical wavelength have been extensively investigated during the last decades (see S. Fauve, Pattern forming instabilities, in *Hydrodynamics and Nonlinear Instabilities*, edited by C. Godrèche and P. Manneville (Cambridge University Press, 1998); L.M. Pismen, *Patterns and Interfaces in Dissipative Dynamics* (Springer, 2006); R. Hoyle, *Pattern Formation, An Introduction to Methods* (Cambridge University Press, 2006); and M. Cross and H. Greenside, *Pattern Formation and Dynamics in Nonequilibrium Systems* (Cambridge University Press, 2009)). The analysis of the nonlinear development of that kind of instabilities is simplified by the existence of a generic nonlinear equation, the so-called Ginzburg–Landau equation, which is valid near the stability threshold independently of the physical nature of the problem under consideration. An extensive review on that equation has been written by I.S. Aranson and L. Kramer, Rev. Mod. Phys. **74**, 99 (2002).

In contrast, the longwave instabilities are much more diverse. A number of prominent equations (Fisher–Kolmogorov equation, Allen–Cahn equation, Cahn–Hilliard equation, Kuramoto–Sivashinsky equation, and many others) have been obtained and investigated in different physical contexts. Moreover, during the last two decades, there appeared a lot of new papers devoted to nonlinear longwave regimes. In particular, an outstanding progress has been made in studying oscillatory longwave patterns. Furthermore, the number of applications, where the longwave perturbations are crucial, grows fast. Such modern fields as micro- and nanofluidics, materials science, chemical engineering, and biophysics can be mentioned to name but a few.

On one hand, this diversity of applications leads to a situation when the same results are rediscovered in different contexts. Therefore, we believe that a book containing a description of main advances in that field from a general point of view would be sought for. On the other hand, in our opinion, the scientific community would need a handbook with receipts, methods, and approaches collected in order to use this experience in new investigations. We hope that both these factors would make the suggested book potentially useful.

In our book, which is addressed to both students and experts, we describe the techniques of the derivation of longwave amplitude equations, as well as the analysis of numerous nonlinear equations which govern the development of longwave instabilities. The book is based on the following general principle: we consider *physical problems* that lead to definite mathematical problems, rather than mathematical problems per se. Each equation is introduced in a context of a *particular physical problem*, and it is derived by means of multiscale asymptotic methods. Thus, we present a sequence of paradigmatic physical problems, rather than a sequence of equations. Most of the basic examples belong to the field of fluid mechanics, though applications in other fields of science are also indicated. Longwave patterns created by buoyancy and surface-tension-driven convection are most often considered in order to present a physical interpretation of equations. The reason of doing that is twofold. First, the fluid dynamics and convection as its subdomain are intuitively clear for most readers independently of their background. The patterns can be easily imagined and described. Moreover, most of the known nonlinear amplitude equations can be derived in applications to fluid dynamics. Second, the buoyancy and surface-tension-driven convection is the authors' field of expertise. However, we believe that the focus on pattern formation in convection does not make the book too specific in that subject; the main ideas and results can be easily transferred to other fields.

The book consists of nine chapters.

Chapter 1 presents basic physical examples and basic classification of the types of instabilities and corresponding nonlinear amplitude equations, which are the subject of the book.

The physical problem of convection, discussed in Chapters 2–4, gives the opportunity to introduce basic amplitude equations. We start with the problems of the cylindrical geometry in Chapter 2 that allows a simple derivation of one-dimensional equations for a real amplitude function, which is a good starting point for the demonstration of the main approach and presentation of transparent physical results that make it possible to discuss such basic notions as Lyapunov functional, stability of stationary patterns, wavenumber selection, defect dynamics, influence of inhomogeneities, pinning of defects, influence of distant boundaries on the patterns, and higher-order corrections to the amplitude equation derived at the leading order of asymptotic expansions. In Chapter 3, we revisit the classical problem of buoyancy convection in a horizontal layer heated from below, which is a paradigmatic example of the cellular pattern formation, but from a different point of view: we discuss the case where the instability is a longwave one; hence, it is not characterized by a selected critical wavenumber. Also, the surface-tension-driven convection is considered.

While the Rayleigh–Bénard convection in a pure liquid serves as a paradigmatic example of stationary patterns generated by monotonic instability, convection in binary liquids (Chapter 4) provides a basic example of an oscillatory instability generating wave patterns. Therefore, within this chapter, we mainly focus on time-periodic patterns; the stationary convection is discussed only when it is qualitatively different from that in a pure liquid. The description of longwave oscillatory patterns is strongly different from both longwave stationary patterns and shortwave oscillatory patterns. Partial differential amplitude equations, which are local in space, cannot be derived in this case, and a different approach, based on the analysis of resonances, is developed.

In Chapter 5, we consider instabilities of the simplest kind of flows, the parallel flows, which are characterized by a number of new effects; specifically, the problem is anisotropic. From a large set of problems, we have selected two problems, where the progress in understanding the nonlinear development of instabilities is significant, namely, the flow of a liquid film over an inclined plane and the famous Kolmogorov flow, which provides an understanding of the generation of large-scale structures by instabilities of short-scale structures.

Within Chapter 6, we discuss instabilities of propagating fronts. The main subject of Chapter 7 is the description of longwave modulations of spatially periodic, stationary, and oscillatory patterns. We consider intrinsic longwave instabilities of short-scale patterns, which are crucial for the pattern selection and development of the spatiotemporal chaos. In Chapter 8, we consider the ways to control the long-wave instabilities by variation of parameters. Both open loop and closed loop (feedback) approaches are discussed. Chapter 9 contains a review of directions for a further development in the field. Some technical stuff is put into the appendices.

In our opinion, this book will be useful for experts in fluid mechanics, heat and mass transfer theory, nonlinear dynamics, and applied mathematics as well as for physicists and chemical engineers interested in the investigation of the nonlinear evolution and pattern formation caused by longwave instabilities. The book can be used also by graduate students.

The scientific results included in this book have been obtained due to collaboration and fruitful discussions with our colleagues and friends: S.I. Abarzhi, A.A. Alabuzhev, D.E. Bar, A. Bayliss, S.H. Davis, I.S. Fayzrakhmanova, E.A. Glasman, A.A. Golovin, V. Gubareva, Y. Kanevsky, M. Khenner, B.J. Matkowsky, A. Mikishev, M. Morozov, B.I. Myznikova, A. Oron, A. Pikovsky, L.M. Pismen, A. Podolny, B.I. Rubinstein, A.E. Samoilova, T. Savina, I.B. Simanovskii, B.L. Smorodin, A.V. Straube, M.G. Velarde, V.A. Volpert, and M.A. Zaks. The help of I.S. Fayzrakhmanova in the manuscript preparation is highly appreciated. The book would not be written without the inspiring suggestion of U. Frisch.

Contents

Chapter 1
Introduction

1.1 The Phenomenon of Pattern Formation

In the last decades, the spontaneous formation of spatially nonhomogeneous patterns was an object of extensive investigations.

A spatially nonuniform state can correspond to the minimum of the free energy functional of a system in a thermodynamic equilibrium. Examples include hexagonal Abrikosov vortex lattices in superconductors [1], magnetic stripe phases in ferromagnetic garnet films [2], and spatially periodic phases of diblock copolymers [3]. The pattern formation is however especially widespread in physical, chemical, and biological systems which are far from the thermodynamic equilibrium state.

The paradigmatic example of that phenomenon is the Rayleigh–Bénard convection producing roll-like (stripe) or hexagonal patterns. The same kinds of patterns are observed on different spatial scales. For instance, spatially regular surface structures in epitaxial solid films [4] and hexagonal arrays of nano-pores in aluminum oxide produced by anodization [5] have a very small typical scale. Self-assembly of spatially regular nanoscale structures is used in several areas of nanotechnology [6, 7]. The typical size of hexagonal solar granules is about 700 km [8].

Also, the same patterns appear in completely different natural systems. For instance, stripe patterns are observed in human fingerprints, on zebra's skin, and in the visual cortex [9]. Hexagonal patterns result from the propagation of laser beams through a nonlinear medium [10] and in systems with chemically reacting and diffusing species [11]. Square patterns [12] and even quasiperiodic patterns [13] are also observed.

Moreover, there exist nonstationary, wavy patterns. The paradigmatic example is the Belousov–Zhabotinsky reaction demonstrating rotating spirals. Traveling [14] and standing waves [15], rolls with alternating directions [16], traveling squares [17], and other kinds of patterns are possible.

© Springer Science+Business Media, LLC 2017
S. Shklyaev, A. Nepomnyashchy, *Longwave Instabilities and Patterns in Fluids*,
Advances in Mathematical Fluid Mechanics,
https://doi.org/10.1007/978-1-4939-7590-7_1

A systematic description of the phenomenon of pattern formation and numerous examples can be found in the review paper by Cross and Hohenberg [18] and monographs [19–21].

In many cases, the spontaneous pattern formation has been described by means of a universal equation named *complex Ginzburg–Landau equation* (CGLE)

$$\frac{\partial a}{\partial t} = \sigma_0 a + \sigma_2 \frac{\partial^2 a}{\partial x^2} - \kappa |a|^2 a \tag{1.1}$$

where $a(x,t)$ is the so-called order parameter or amplitude function and σ_0, σ_2, and κ are, generally speaking, complex coefficients (see Section 1.3). This equation and its multidimensional and multicomponent generalizations were used in order to explain such phenomena as stability of patterns, pattern selection, wavenumber selection, dynamics of defects, spatiotemporal chaos, and so on. Equation (1.1) can be derived near the instability threshold by means of asymptotic methods, starting from original nonlinear equations which describe different physical systems. It can be done for both *short-wavelength* instabilities and some types of *long-wavelength* instabilities (the definitions are given below). In the case of a short-wavelength instability, the CGLE is an equation for envelope function, and it describes the propagation of modulation waves [22–24]. In the case of long-wavelength instabilities, this equation governs directly the evolution of the local amplitude of an unstable mode [25]. Properties of solutions of the CGLE have been summarized in the review paper by Aranson and Kramer [26].

However, there are important kinds of instabilities that cannot be described by the CGLE. Specifically, that equation is not valid for long-wavelength instabilities in presence of a *conservation law* preventing the growth of spatially homogeneous disturbances or another kind of invariance leading to the existence of a one-parametric family of base solutions. In that case, the situation is much more intricate, and there is a variety of equations describing the nonlinear evolution of large-scale disturbances generated by the instability.

In this chapter, we present a basic classification of the types of instabilities and corresponding nonlinear amplitude equations, which are the subject of the book. First, we define the two main classes of instabilities, shortwave and longwave instabilities (Section 1.2). In Section 1.3 we discuss the case of a shortwave instability, which is governed by the CGLE. Though the description of the general derivation technique for the CGLE can be found in many books [27–31], for the sake of reader's convenience, we reproduce an outline of such a derivation. In Section 1.4 we give a list of some most widespread equations, together with heuristic arguments motivating their structure. The derivation of those equations is postponed to chapters where specific physical problems are considered.

1.2 Shortwave and Longwave Instabilities

Let us consider a physical system described by a set of real variables $U(\mathbf{r},t) = \{U_n(\mathbf{r},t)\}, n = 1, \dots N$, and governed by a certain system of nonlinear partial differential equations. For the sake of simplicity, we assume that the system is *two-dimensional* ($\mathbf{r} = (x, y)$) and situated in a region D which is infinitely extended in the *unconstrained* direction x and finite in the *constrained* direction y:

$$D: -\infty < x < +\infty, \; y_- < y < y_+. \tag{1.2}$$

Also, we assume that the external conditions are t- and x-independent and depend on the only parameter R. Thus, we can write the equations of motion and boundary conditions in a symbolic form:

$$f\left(y, \frac{\partial}{\partial x}, \frac{\partial}{\partial y}, \frac{\partial}{\partial t}; U; R\right) = 0 \;\; \text{in} \; D, \tag{1.3}$$

$$g_\pm\left(\frac{\partial}{\partial x}, \frac{\partial}{\partial y}, \frac{\partial}{\partial t}; U; R\right) = 0 \;\; \text{at} \; y = y_\pm. \tag{1.4}$$

For the sake of simplicity, later on we restrict ourselves to the case

$$g_\pm = U(\mathbf{r}, t) - U_\pm^\gamma, \tag{1.5}$$

where $U_\pm^\gamma$ are known. Finally, we suppose that for each R the set of equations (1.3)–(1.5) has a solution,

$$U = U_0(y; R), \tag{1.6}$$

which does not depend on t and x. This solution will be named *base state* or *base flow* depending on its physical meaning. There are many examples of problems where our assumptions are satisfied, e.g., heat convection, instabilities of parallel flows, diffusion–reaction systems, etc.

Typically, the base state is *unstable* in a certain region of the parameter R, and this instability is usually connected with the bifurcations of new solutions characterized by a more complex spatiotemporal structure (see Appendix B). The investigation of the variety of solutions to the nonlinear problem (1.3)–(1.5) starts with the *linear stability theory*.

We construct solutions in the form

$$U = U_0 + \tilde{U}, \tag{1.7}$$

where $\tilde{U}$ is a *small disturbance* of the base state U_0. Substituting (1.7) into (1.3)–(1.5) and neglecting all the terms nonlinear in $\tilde{U}$, we obtain a *linearized* system of equations

$$L\left(y; \frac{\partial}{\partial x}, \frac{\partial}{\partial y}, \frac{\partial}{\partial t}; R\right) \tilde{U}(\mathbf{r}, t) = 0 \; \text{in} \; D, \tag{1.8}$$

$$\tilde{U}(\mathbf{r}, t) = 0 \; \text{at} \; y = y_\pm, \tag{1.9}$$

where L is the real linear operator (Fréchet derivative). We can consider complex solutions to equations (1.8) and (1.9). Certainly, $\tilde{U}$ and $\tilde{U}^*$ are solutions to equations

(1.8) and (1.9) simultaneously, and the real physical disturbance is $\text{Re}\tilde{U} = (\tilde{U} + \tilde{U}^*)/2$.

Usually, it is sufficient to consider the so-called *normal disturbances* compatible with translational invariance of the problem (1.8) and (1.9) with respect to both x and t:

$$\tilde{U}(\mathbf{r},t) = \tilde{u}(y;k)\exp(\sigma t + ikx). \tag{1.10}$$

In an infinite region D, only real values of the *wavenumber k* have a physical meaning. The *growth rate σ* is generally a complex number. For normal disturbances, the set of evolution equations (1.8) with boundary conditions (1.9) is reduced to an *eigenvalue boundary problem*

$$L\left(y;ik,\frac{\partial}{\partial y},\sigma;R\right)\tilde{u}(y;k) = 0 \text{ at } y_- < y < y_+, \tag{1.11}$$

$$\tilde{u}(y;k) = 0 \text{ at } y = y_\pm, \tag{1.12}$$

which determines usually a countable set of branches of the *dispersion relation* between the growth rates σ_n and the wavenumber k:

$$\sigma = \sigma_n(k,R), \ n = 1,2,\ldots \tag{1.13}$$

and the corresponding eigenfunctions (*eigenmodes*):

$$\tilde{u} = \tilde{u}_n(y;k,R), \ n = 1,2,\ldots \tag{1.14}$$

It is obvious that together with an eigenmode $(\tilde{u}_n(k,R),\sigma_n(k,R))$, there exists an eigenmode

$$\tilde{u}_{n'}(k,R) = \tilde{u}_n^*(-k,R), \ \sigma_{n'}(k,R) = \sigma_n^*(-k,R). \tag{1.15}$$

These two eigenmodes can coincide, but it is not necessary. In many cases, the completeness of the set of functions u_n can be proved. That allows to construct a general solution of problem (1.11), (1.12) as a superposition of normal disturbances.

Later on, we assume that the instability is generated by a single eigenmode with the growth rate $\sigma_1(k,R) \equiv \sigma(k,R) = \sigma_r(k,R) + i\sigma_i(k,R)$. If $\sigma_r(k) > 0$ for some k, the solution given by (1.6) is *linearly unstable*. In that case, if $\sigma_i = 0$, we have a *stationary* (monotonic) instability; otherwise, there is an *oscillatory* (wave) instability.

Let us assume that the base state is stable at $R \le R_c$ (in the *subcritical* region) and unstable at $R > R_c$ (in the *supercritical* region). At $R = R_c$, the maximum of the function $\sigma_r(k,R_c)$ can be situated at $k = k_c \ne 0$ or at $k = 0$. Accordingly, there are two main possibilities (see Figure 1.1): (1) the instability arises near a certain wavenumber $k_c = O(1)$ (*short-wavelength* instability); and (2) the instability domain is localized around $k_c = 0$ (*long-wavelength* instability). The latter type of instabilities is the subject of the present book.

In the case of a short-wavelength instability, the growth rate can be expanded near the point (k_c,R_c) into the Taylor series. It is obvious that in this point

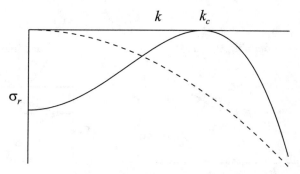

Fig. 1.1 Variation of the real part of the growth rate σ_r with the wavenumber k at $R = R_c$. The solid line and the dashed line correspond to the short-wavelength instability (type I) and the long-wavelength instability (type III), respectively

$$\sigma_r = 0, \quad \frac{\partial \sigma_r}{\partial k} = 0, \quad \frac{\partial^2 \sigma_r}{\partial k^2} < 0, \quad \frac{\partial \sigma_r}{\partial R} > 0, \tag{1.16}$$

and the Taylor expansion has the form

$$\sigma_r = \left(\frac{\partial \sigma_r}{\partial R}\right)_c (R - R_c) + \frac{1}{2}\left(\frac{\partial^2 \sigma_r}{\partial k^2}\right)_c (k - k_c)^2 + \dots, \tag{1.17}$$

$$\sigma_i = (\sigma_i)_c + \left(\frac{\partial \sigma_i}{\partial k}\right)_c (k - k_c) + \left(\frac{\partial \sigma_i}{\partial R}\right)_c (R - R_c)$$

$$+ \frac{1}{2}\left(\frac{\partial^2 \sigma_i}{\partial k^2}\right)_c (k - k_c)^2 + \dots, \tag{1.18}$$

(subscript c means $k = k_c$, $R = R_c$). The instability region is bounded by the *neutral* (marginal) *stability curve* $\sigma_r(k, R) = 0$ (see Figure 1.2), which is described near the threshold by the equation

$$R = R_c + R_2(k - k_c)^2 + \dots, \tag{1.19}$$

where

$$R_2 = -\frac{1}{2}\frac{(\partial^2 \sigma_r / \partial k^2)_c}{(\partial \sigma_r / \partial R)_c}. \tag{1.20}$$

The width of the instability interval at small $R - R_c$ is

$$\delta k = 2\left(\frac{R - R_c}{R_2}\right)^{1/2}. \tag{1.21}$$

In the review paper of Cross and Hohenberg [18], the short-wavelength instability is called *type I instability*.

In the case of a long-wavelength instability, there are several different possibilities. First, the growth rate can be described near the instability threshold by the same

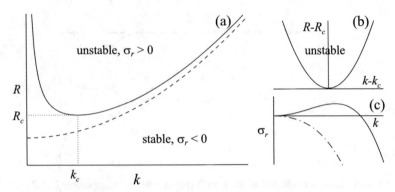

Fig. 1.2 (a) Neutral stability curves for short-wavelength (type I, solid line) and long-wavelength (types II and III, dashed line) instabilities; (b) a zoomed fragment of panel (a), where (1.19) works; (c) variation of the real part of the growth rate σ_r for type II instability, according to (1.25); solid (dashed) line corresponds to $R > R_c$ ($R < R_c$)

formulas, (1.17) and (1.18), but with $k_c = 0$ (*type III instability*, according to Cross and Hohenberg [18]). In that case, the spatially homogeneous disturbance has the largest real part of the growth rate. In the case of oscillations ($\sigma_i \neq 0$), because of the symmetry property, given by (1.15), there are *two* unstable, mutually complex conjugate modes with complex conjugate growth rates at $k = 0$.

However, there is a wide class of problems where the growth of a spatially homogeneous disturbance is ruled out by a *conservation law*. For instance, in the case of a wave instability of an incompressible fluid flow with a free surface, the spontaneous *spatially homogeneous* change of the depth of the fluid layer is impossible because of the conservation of the fluid volume. A spatially homogeneous disturbance introduced into the system does not change in time; that means that there exists a mode with

$$\sigma(0, R) = 0, \tag{1.22}$$

which is called *Goldstone mode*.

Another kind of problems, where a spatially homogeneous disturbance does not change in time, is the instability of a plane front of phase transition or chemical reaction propagating with a constant velocity, where the base state is a traveling wave, $U = U_0(\xi)$, $\xi = x - vt$. In that case, the origin of the neutral mode is the translational symmetry of the problem: there is a one-parameter family of solutions, $U = U_0(\xi + C)$, where C is an arbitrary constant.

At small but nonzero k, the Goldstone mode can produce an instability. Assuming that the unstable mode is *unique*, we conclude that for such a mode

$$\sigma(k, R) = \sigma^*(-k, R) \tag{1.23}$$

(see (1.15)), thus

$$\left(\frac{\partial \sigma_r}{\partial k}\right)_{k=0} = 0. \tag{1.24}$$

At the instability threshold, $R = R_c$ not $(\sigma_r)_{k=0}$, but $\left(\partial^2 \sigma_r / \partial k^2\right)_{k=0}$ changes its sign. Near the threshold we find:

$$\sigma_r = \frac{1}{2}\left(\frac{\partial^3 \sigma_r}{\partial k^2 \partial R}\right)_c k^2 (R - R_c) + \frac{1}{24}\left(\frac{\partial^4 \sigma_r}{\partial k^4}\right)_c k^4 + \ldots, \tag{1.25}$$

$$\sigma_i = \left(\frac{\partial \sigma_i}{\partial k}\right)_c k + \left(\frac{\partial^2 \sigma_i}{\partial k \partial R}\right)_c k (R - R_c) + \frac{1}{6}\left(\frac{\partial^3 \sigma_i}{\partial k^3}\right)_c k^3 + \ldots. \tag{1.26}$$

If the base state is stable at $R < R_c$ and unstable at $R > R_c$, then $(\partial^3 \sigma_r / \partial k^2 \partial R)_c > 0$; see Figure 1.2(c). It is obvious that $(\partial^4 \sigma_r / \partial k^4)_c < 0$, otherwise $k_c \neq 0$. In the supercritical region $R > R_c$, the parameter σ_r is proportional to k^2. This phenomenon is called *type II instability* and referred to as "*negative viscosity*" [32, 33].

Some additional possibilities should be mentioned. If there are *two* Goldstone modes connected by relation (1.15), we find instead of (1.23):

$$\sigma_{\pm}(k, R) = \sigma_{\mp}^*(-k, R) \tag{1.27}$$

(we use $\pm$ instead of n, n'). Thus, instead of (1.24), we obtain

$$\left(\frac{\partial \sigma_{+r}}{\partial k}\right)_{k=0} = -\left(\frac{\partial \sigma_{-r}}{\partial k}\right)_{k=0}. \tag{1.28}$$

That means that one of the modes is unstable for $k > 0$ and another one for $k < 0$. Such a situation takes place, for instance, in the case of anisotropic kinetic alpha effect [34].

Also, there are some physical problems where the growth rate is *not analytical* at $k = 0$. Recall that for gravity waves in an *infinitely deep* basin of a *nonviscous* liquid, the dispersion relation that connects the wave frequency ω with the wavenumber k contains a nonanalytical term: $\omega^2(k) = g|k|$, where g is the gravity acceleration [35]. The origin of the nonanalytical expression $|k|$ is the decay condition for disturbances infinitely far from the interface, as the vertical coordinate $y \to -\infty$, which leads to the dependence $\exp(|k|y)$ for eigenfunctions. Due to a similar reason, the nonanalytical dependence of the growth rate

$$\sigma = \sigma_1|k| - \sigma_2 k^2 + o(k^2) \tag{1.29}$$

is obtained sometimes for *dissipative* systems if the region of the motion is not bounded in y-direction, i.e., $y_- \to -\infty$, $y_+ \to \infty$. For instance, relation (1.29) is characteristic for the instabilities of reaction fronts [36, 37].

Another example of a nonanalytical dependence of the growth rate is given in [38], where $\sigma(k \to 0)$ depends on the ratio of the components of the wavevector k_x/k_y with the maximum at a certain value of that ratio.

1.3 The Complex Ginzburg–Landau Equation

The next step in the investigation of patterns generated by an instability is the construction of small-amplitude solutions to nonlinear problem (1.3)–(1.5). In the present section, we discuss the case of a shortwave instability, $k_c \neq 0$. Fixing the wavenumber k, one can perform the bifurcation analysis for a spatially periodic solution satisfying the condition $U(x + 2\pi/k) = U(x)$ near each point of the neutral stability curve $R = R(k)$ defined by the condition $\sigma_r(k, R) = 0$. The translational invariance of the problem is the crucial point in the analysis of bifurcations of spatially periodic solutions, specifically, in the case of a degenerate eigenvalue [39].

However, it is not obvious beforehand that any spatially periodic patterns will be really observed as a result of the instability. Indeed, for each $R > R_c$, there is an *infinite* set of modes with $(k_c - \delta k/2 < k < k_c + \delta k/2)$ growing on the background of the base state (δk is defined by (1.21)). Thus, the limit regimes could be quite diverse, including patterns with defects or even spatiotemporal chaos. The crucial point simplifying the description of patterns near the threshold is the following circumstance. If $R - R_c$ is small, the base state is unstable to *slowly* growing disturbances with wavenumbers from a *narrow* interval near k_c. The transverse spatial structure of all these disturbances is practically of the same type. Because of that, the nonlinear evolution leading to the final nonlinear state (also called "secondary flow" in view of numerous applications in fluid dynamics) is relatively simple and can be described by means of few functions slowly changing in space and in time that are called *amplitude functions* or *order parameters*. The reader can find an introduction to that approach in [40].

At the lowest order of an asymptotic expansion, the evolution of the order parameters is governed by some nonlinear partial differential equations, which are essentially simpler than the equations describing the original physical problem. In the case of a short-wavelength instability, one obtains a *universal* equation named *complex Ginzburg–Landau equation* (CGLE). The values of a few coefficients, which enter that equation, depend on the parameters of the original problem. In the context of hydrodynamic instabilities, that equation was first derived by Stewartson and Stuart [24]. The detailed derivation of CGLE in the general case can be found, e.g., in [30].

Let us describe briefly the technique of the derivation of the CGLE in the case of a nondegenerate oscillatory short-wavelength instability mode. For the sake of simplicity, we restrict ourselves to the case where equation (1.3) has only quadratic nonlinearities, and the nonlinear equation for the deviation $\tilde{U} = U - U_0$ can be presented in the form:

$$L\tilde{U} + N(\tilde{U}, \tilde{U}) = 0, \tag{1.30}$$

where L is the linear operator introduced in (1.8), and

$$L = -S\frac{\partial}{\partial t} + F\left(y, \frac{\partial}{\partial x}, \frac{\partial}{\partial y}; R\right), \tag{1.31}$$

where S is a constant matrix, F is a linear operator, and the form $N(U,V)$ is a finite sum of bilinear terms,

$$N(U,V) = \sum_i G_i\left(y, \frac{\partial}{\partial x}, \frac{\partial}{\partial y}\right) U \cdot H_i\left(y, \frac{\partial}{\partial x}, \frac{\partial}{\partial y}\right) V.$$

The boundary conditions are assumed to be homogeneous (see (1.9)).

Assuming that near the threshold the emerging secondary flow is close to the base state, we construct the solution of the problem (1.30), (1.9) in the form of the expansion

$$\tilde{U} = \sum_{n=1}^{\infty} \varepsilon^n U^{(n)}, \tag{1.32}$$

where $\varepsilon = \varepsilon(R)$ is a small parameter. The dependence between ε and R is presented also in the form of a series:

$$R - R_c = \sum_{n=1}^{\infty} \varepsilon^n R^{(n)}. \tag{1.33}$$

The secondary flows near the instability threshold are characterized by several strongly different temporal and spatial scales. For instance, the characteristic growth time $(\sigma_r)^{-1}$ of oscillatory disturbances is large as compared with the period of oscillations $2\pi/|\sigma_i|$ and with the decay time of stable modes. The spatial scale $(\delta k)^{-1}$ of the envelope of a wave train, which contains a superposition of disturbances whose wavenumbers are taken from the instability interval, is much larger than the critical wavelength $2\pi/k_c$. This circumstance gives a possibility to apply the method of multiscale expansions (see, e.g., [27, 41, 42]).

We assume that the functions $U^{(n)}$ depend on the set of variables *considered as independent*:

$$x_k = \varepsilon^k x, \ t_k = \varepsilon^k t, \ k = 0, 1, 2, \dots \tag{1.34}$$

Actually, only one additional spatial scale x_1 will be used below; hence we will neglect the dependence of functions on x_k with $k > 1$. Formally, we substitute the following expansions into equation (1.30):

$$\frac{\partial}{\partial x} = \sum_{l=0}^{1} \varepsilon^l \frac{\partial}{\partial x_l}, \ \frac{\partial}{\partial t} = \sum_{l=0}^{\infty} \varepsilon^l \frac{\partial}{\partial t_l}. \tag{1.35}$$

The operator F and the bilinear form N are expanded in powers of ε, i.e.,

$$F\left(\frac{\partial}{\partial x}, \frac{\partial}{\partial y}, R\right) = \sum_{l,m=0}^{\infty} \left(\varepsilon \frac{\partial}{\partial x_1}\right)^l \left(\sum_{n=1}^{\infty} \varepsilon^n R^{(n)}\right)^m F^{(l,m)}\left(\frac{\partial}{\partial x_0}, \frac{\partial}{\partial y}\right). \tag{1.36}$$

Later on, we use the definitions

$$F_p^{(l,m)} = F^{(l,m)}\left(ipk_c, \frac{\partial}{\partial y}\right), \tag{1.37}$$

$$N_{p,q}^{(0)}\left(f_1(y),f_2(y)\right) = \sum_i G_i\left(y,ipk_c,\frac{\partial}{\partial y}\right)f_1 \cdot H_i\left(y,iqk_c,\frac{\partial}{\partial y}\right)f_2. \quad (1.38)$$

At the first order in ε, we obtain a homogeneous boundary value problem

$$\left(S\frac{\partial}{\partial t_0} - F^{(0,0)}\right)U^{(1)} = 0 \quad \text{in } D; \quad (1.39)$$

$$U^{(1)} = 0 \quad \text{on } \Gamma, \quad (1.40)$$

which coincides with the linearized problem, given by (1.8) and (1.9), for the disturbance of the base state at $R = R_c$. In the case of a nondegenerate oscillatory short-wavelength instability, we construct the solution

$$U^{(1)} = a^{(0)}u_1^{(1)}(y)\exp\left[i(k_cx_0 - \omega_ct_0)\right] + \left[a^{(0)}\right]^*u_{-1}^{(1)}(y)\exp\left[-i(k_cx_0 - \omega_ct_0)\right], \quad (1.41)$$

where the N-component vector $u_1^{(1)}(y)$ is the solution $\tilde{u}(y;k,R)$ of linearized problems (1.8) and (1.9) at $k = k_c$, $R = R_c$; $\omega_c = -\sigma_i(k_c)$ is the frequency of the critical disturbance; $u_{-1}^{(1)} = \left[u_1^{(1)}\right]^*$. The unknown function $a^{(0)} = a^{(0)}(x_1,t_1,t_2,\ldots)$ is an envelope function of the wave train formed by disturbances with wavenumbers from an interval of the width $O(\varepsilon)$ around k_c.

The solution at the second order in ε has the structure

$$U^{(2)} = \sum_{l=-2}^{2} v_l^{(2)}(y,x_1,t_1,\ldots)\exp[il(k_cx_0 - \omega_ct_0)], \quad (1.42)$$

where the functions $v_l^{(2)}$ satisfy the equations:

$$-\left(F_1^{(0,0)} + i\omega_cS\right)v_1^{(2)} = \frac{\partial a^{(0)}}{\partial x_1}F_1^{(1,0)}u_1^{(1)} + a^{(0)}\left(R^{(1)}F_1^{(0,1)} - S\frac{\partial}{\partial t_1}\right)u_1^{(1)}, \quad (1.43)$$

$$-F_0^{(0,0)}v_0^{(2)} = |a^{(0)}|^2\left[N_{1,-1}^{(0)}\left(u_1^{(1)},u_1^{(1)}\right) + N_{-1,1}^{(0)}\left(u_{-1}^{(1)},u_1^{(1)}\right)\right], \quad (1.44)$$

$$-\left(F_2^{(0,0)} + 2i\omega_cS\right)v_2^{(2)} = \left[a^{(0)}\right]^2 N_{1,1}^{(0)}\left(u_1^{(1)},u_1^{(1)}\right), \quad (1.45)$$

$$v_{-1}^{(2)} = v_1^{(2)*}, \quad v_{-2}^{(2)} = v_2^{(2)*}. \quad (1.46)$$

We do not write explicitly the boundary conditions which are always homogeneous.

The *solvability condition* for inhomogeneous linear equation (1.43) (see Appendix A) gives the first-order amplitude equation

$$\frac{\partial a^{(0)}}{\partial t_1} = R^{(1)}\sigma_0 a^{(0)} - \omega_1\frac{\partial a^{(0)}}{\partial x_1}, \quad (1.47)$$

where

$$\sigma_0 = -\frac{\left(u^c, F_1^{(0,1)} u_1^{(1)}\right)}{\left(u^c, S u_1^{(1)}\right)} = \left(\frac{\partial \sigma}{\partial R}\right)_c, \tag{1.48}$$

$$\omega_1 = \frac{\left(u^c, F_1^{(1,0)} u_1^{(1)}\right)}{\left(u^c, S u_1^{(1)}\right)} = -\left(\frac{\partial \sigma_i}{\partial k}\right)_c. \tag{1.49}$$

Here

$$(f,g) \equiv \sum_{n=1}^{N} \int_{y_-}^{y_+} dy f_n^* g_n, \tag{1.50}$$

u^c is the solution of the linear problem *adjoint* to the problem for $u_1^{(1)}$. The proof of relations (1.48) and (1.49) can be found in [30].

The saturation of the growth of $a^{(0)}$ is possible only if $R^{(1)} = 0$; in that case

$$a^{(0)} = a^{(0)}(x_1 - \omega_1 t_1, t_2, \ldots). \tag{1.51}$$

We eliminate the dependence of $a^{(0)}$ on t_1 using the reference frame moving along x_1 with the group velocity of waves ω_1. The solutions to equations (1.43)–(1.45) can be written in the form:

$$v_1^{(2)} = -i\frac{\partial a^{(0)}}{\partial x_1}\left(\frac{\partial \tilde{u}(y,z;k)}{\partial k}\right)_c + a^{(1)} u_1^{(1)}(y,z), \tag{1.52}$$

$$v_0^{(2)} = |a^{(0)}|^2 u_0^{(2)}(y,z), v_2^{(2)} = \left[a^{(0)}\right]^2 u_2^{(2)}(y,z), \tag{1.53}$$

where $a^{(1)}$ is a new function of slow variables.

Finally, at the third order, we obtain the equation which has solutions of the form

$$U^{(3)} = \sum_{l=-3}^{3} v_l^{(3)}(y,z,x_1,t_2,\ldots) \exp\left[il(k_c x_0 - \omega_c t_0)\right]. \tag{1.54}$$

The solvability condition to the problem for $v_1^{(3)}$ leads to the CGLE governing the evolution of the amplitude function $a^{(0)}$:

$$\frac{\partial a^{(0)}}{\partial t_2} = R^{(2)} \sigma_0 a^{(0)} + \sigma_2 \frac{\partial^2 a^{(0)}}{\partial x_1^2} - \kappa |a^{(0)}|^2 a^{(0)}, \tag{1.55}$$

where [30]

$$\sigma_2 = -\frac{1}{2}\left(\frac{\partial^2 \sigma}{\partial k^2}\right)_c, \tag{1.56}$$

$$\kappa = \left(u^c, S u_1^{(1)}\right)^{-1} \left[u^c, N_{0,1}^{(0)}\left(u_0^{(2)}, u_1^{(1)}\right) + N_{1,0}^{(0)}\left(u_1^{(1)}, u_0^{(2)}\right)\right. \\ \left. + N_{2,-1}^{(0)}\left(u_2^{(2)}, u_{-1}^{(1)}\right) + N_{-1,2}^{(0)}\left(u_{-1}^{(1)}, u_2^{(2)}\right)\right]. \tag{1.57}$$

The coefficient κ describing the nonlinear interaction of disturbances is called *Landau constant*.

The coefficient $R^{(2)}$ in (1.55) depends on the definition of ε. In the case of a *supercritical instability*, where the bifurcating solutions exist in the region $R > R_c$, it is convenient to define $\varepsilon^2 = R - R_c$ and then $R^{(2)} = 1$ and $R^{(n)} = 0$ for any other n. Other ways of definition are possible; for instance, one can define $\varepsilon = \delta k$, etc.

The parameter ε can be eliminated using the definition

$$a = \varepsilon a^{(0)} + \varepsilon^2 a^{(1)} + \dots, \tag{1.58}$$

which allows rewriting the system of equations (1.47) and (1.55) in a compact form

$$\frac{\partial a}{\partial t} = (R - R_c)\sigma_0 a + \sigma_1 \frac{\partial a}{\partial x} + \sigma_2 \frac{\partial^2 a}{\partial x^2} - \kappa |a|^2 a \tag{1.59}$$

or

$$\frac{\partial a}{\partial t} = \left[\hat{\sigma}\left(k_c - i\frac{\partial}{\partial x}, R \right) - \sigma(k_c, R_c) \right] a - \kappa |a|^2 a, \tag{1.60}$$

where $\hat{\sigma}$ is the operator obtained from expansions (1.17) and (1.18) by substituting $-i\partial/\partial x$ instead of $k - k_c$.

The construction of a single equation which describes processes with different characteristic time scales is called "restitution procedure." The strong and weak sides of that approach will be discussed in more detail for another example in Section 4.1.3.

Note that *mechanical equilibrium* instability problems, in a contradistinction to *flow* instability problems, are usually characterized by a reflection symmetry, in addition to the translational symmetry. In the language of the group theory, the problem of bifurcations of spatially periodic solutions is characterized by $O(2)$ symmetry, rather than $SO(2)$ symmetry [39]. The modulated traveling waves moving in opposite directions are described by two complex Ginzburg–Landau equations with a nonlocal coupling between them. For details, see [43–45].

1.4 Classification of Longwave Amplitude Equations

1.4.1 Amplitude Equations for Type III Instabilities

Let us consider now a long-wavelength instability corresponding to the dispersion relation, determined by (1.17) and (1.18), with $k_c = 0$. Because of property (1.15), there are two main possibilities: 1) nondegenerate stationary instability (the growth rate $\sigma(0, R)$ and the eigenfunction $\tilde{u}(0, R)$ are real) and 2) twofold degenerate oscillatory instability (there are two unstable modes $\tilde{u}(0, R)$ and $\tilde{u}^*(0, R)$ with growth rates $\sigma(0, R)$ and $\sigma^*(0, R)$, respectively).

In the former case, the real solution of boundary value problems (1.39), (1.40) can be written in the form

$$U^{(1)} = a^{(0)} u_1^{(1)}(y),$$ (1.61)

where the amplitude function $a^{(0)}$ and the eigenfunction $u_1^{(1)} = \tilde{u}(0, R_c)$ are real; in the latter case, we have

$$U^{(1)} = a^{(0)} u_1^{(1)}(y) \exp(-i\omega_c t_0) + a^{(0)*} u_{-1}^{(1)}(y) \exp(i\omega_c t_0),$$ (1.62)

where $\omega_c = -\sigma_i(0), u_1^{(1)} = \tilde{u}(0, R_c), u_{-1}^{(1)} = \tilde{u}^*(0, R_c)$.

The evolution equations for the amplitude function can be derived similarly to the case of a short-wavelength instability. For an oscillatory instability, the direct derivation [25] leads exactly to CGLE (1.55). In the case of stationary instability, the amplitude function and the eigenfunction of the linear problem are real. In that case, the quadratic term is not forbidden generally, and the nonlinear quadratic term cannot be balanced with the linear term unless

$$a^{(0)} = 0,$$ (1.63)

hence $U^{(1)} = 0$. Taking

$$U^{(2)} = a^{(1)} u_1^{(1)}(y),$$

one obtains the *Fisher–Kolmogorov–Petrovsky–Piskunov equation* (FKPPE) [46, 47]

$$\frac{\partial a^{(1)}}{\partial t_1} = R^{(2)} \sigma_0 a^{(1)} + \sigma_2 \frac{\partial^2 a^{(1)}}{\partial x_1^2} - \bar{\kappa} \left[a^{(1)} \right]^2,$$ (1.64)

where the coefficients σ_0, σ_2 and $\bar{\kappa}$ are real constants.

It should be emphasized, however, that in hydrodynamic stability problems for viscous fluid (including the Boussinesq convection), the coefficient $\bar{\kappa}$ vanishes because of specific features of nonlinear terms. In that case, $a^{(0)} \neq 0$, and the amplitude equation can be written in the form:

$$\frac{\partial a^{(0)}}{\partial t_2} = R^{(2)} \sigma_0 a^{(0)} + \sigma_2 \frac{\partial^2 a^{(0)}}{\partial x_1^2} + \alpha a^{(0)} \frac{\partial a^{(0)}}{\partial x_1} - \kappa \left[a^{(0)} \right]^3$$ (1.65)

(α, κ are constants). In the presence of a reflection symmetry $x \to -x$, $\alpha = 0$, and one obtains the *Allen–Cahn equation* (ACE)

$$\frac{\partial a^{(0)}}{\partial t_2} = R^{(2)} \sigma_0 a^{(0)} + \sigma_2 \frac{\partial^2 a^{(0)}}{\partial x_1^2} - \kappa \left[a^{(0)} \right]^3,$$ (1.66)

which is a real version of GLE (1.55). Using definition (1.58), one can write equations (1.63) and (1.64) in the form

$$\frac{\partial a}{\partial t} = (R - R_c) \sigma_0 a + \sigma_2 \frac{\partial^2 a}{\partial x^2} - \bar{\kappa} a^2.$$ (1.67)

Similarly, equations (1.65) and (1.66) can be reduced to the form

$$\frac{\partial a}{\partial t} = (R - R_c)\sigma_0 a + \sigma_2 \frac{\partial^2 a}{\partial x^2} + \alpha a \frac{\partial a}{\partial x} - \kappa a^3 \qquad (1.68)$$

and

$$\frac{\partial a}{\partial t} = (R - R_c)\sigma_0 a + \sigma_2 \frac{\partial^2 a}{\partial x^2} - \kappa a^3, \qquad (1.69)$$

respectively.

In some problems, nonlinearities of higher orders should be taken into account.

1.4.2 Amplitude Equations for Type II Instabilities

Let us consider now the case of a long-wavelength instability characterized by dispersion relation (1.25) and (1.26) for infinitesimal disturbances.

We postpone the derivation of amplitude equations to next chapters where some particular problems are considered. In the present section, we discuss only the possible structure of these equations.

We expect that the amplitude equation has the form of the conservation law,

$$\frac{\partial a}{\partial t} = -\frac{\partial q}{\partial x} \qquad (1.70)$$

where $q(a, \partial a/\partial x, \ldots)$ is the flux of the conserved quantity. The expression for $q = q_l + q_n$ contains linear and nonlinear terms.

The amplitude equations of that kind are rather diverse. Let us mention several typical examples.

1.4.2.1 Longwave Instabilities of the Type II_o

The linear development of an oscillatory type II instability (II_o) is governed by expressions (1.25), (1.26) for the linear growth rate,

$$\sigma_r(k) = \sigma_2(R - R_c)k^2 - \sigma_4 k^4 + \ldots, \quad \sigma_i = \sigma_1 k - \sigma_3 k^3, \qquad (1.71)$$

where

$$\sigma_1 = \left(\frac{\partial \sigma_i}{\partial k}\right)_c + \left(\frac{\partial^2 \sigma_i}{\partial k \partial R}\right)_c (R - R_c), \quad (1.72)$$

$$\sigma_2 = \frac{1}{2}\left(\frac{\partial^3 \sigma_r}{\partial k^2 \partial R}\right)_c, \quad \sigma_3 = -\frac{1}{6}\left(\frac{\partial^3 \sigma_i}{\partial k^3}\right)_c, \quad \sigma_4 = -\frac{1}{24}\left(\frac{\partial^4 \sigma_r}{\partial k^4}\right)_c, \ldots \quad (1.73)$$

Hence, the linearized amplitude equation $\partial a/\partial t = -\partial q_l/\partial x$ looks as

$$\frac{\partial a}{\partial t} = \sigma_1 \frac{\partial a}{\partial x} + \sigma_3 \frac{\partial^3 a}{\partial x^3} - \sigma_2(R - R_c)\frac{\partial^2 a}{\partial x^2} - \sigma_4 \frac{\partial^4 a}{\partial x^4} + \dots \qquad (1.74)$$

Taking $R - R_c = R_2\varepsilon^2$ and eliminating the group velocity term by transformation $X = \varepsilon(x + \sigma_1 t)$, we obtain:

$$\frac{\partial a}{\partial t} = \varepsilon^3 \sigma_3 \frac{\partial^3 a}{\partial X^3} - \varepsilon^4 \left(\sigma_2 R_2 \frac{\partial^2 a}{\partial X^2} + \sigma_4 \frac{\partial^4 a}{\partial X^4}\right) + \dots \qquad (1.75)$$

The terms of the lowest order in $q_n(a)$ are:

$$q_n(a) = -\kappa_1 a^2 - \varepsilon \kappa_2 \frac{\partial}{\partial X}(a^2) + \dots \qquad (1.76)$$

A balance between the dispersion term, $\varepsilon^3 \sigma_3 \partial^3 a/\partial X^3$, and the strongest nonlinear term in $-\partial q_n/\partial X$, which is $\varepsilon \kappa_1 \partial(a^2)/\partial X$, is reached if $a = \varepsilon^2 a_0$, $a_0 = O(1)$. Introducing $T = \varepsilon^3 t$, we obtain the *dissipation-modified Korteweg–de Vries equation*,

$$\frac{\partial a_0}{\partial T} = \sigma_3 \frac{\partial^3 a_0}{\partial X^3} + \kappa_1 \frac{\partial(a_0^2)}{\partial X} + \varepsilon \left[-\sigma_2 R_2 \frac{\partial^2 a_0}{\partial X^2} - \sigma_4 \frac{\partial^4 a_0}{\partial X^4}3 + \kappa_2 \frac{\partial^2(a_0^2)}{\partial X^2}\right]. \qquad (1.77)$$

Actually, this equation has been derived in many physical problems, for instance, for oscillatory Marangoni convection (Section 3.3), instability of a fluid film on a slightly inclined plane (Section 5.1.2), and Eckhaus instability of periodic waves (Section 7.2.1).

If the dispersion is small ($\sigma_3 = \varepsilon \bar{\sigma}_3$ $\bar{\sigma}_3 = O(1)$) because of some reasons (for instance, this feature is characteristic for instability of waves in a film on an inclined plane; see Section 5.1.5), it is balanced by the nonlinear term as $a = \varepsilon^3 a_0$, $a_0 = O(1)$; with $T = \varepsilon^4 t$, we obtain the *Kawahara equation* (KE) [48–50]

$$\frac{\partial a_0}{\partial T} = \sigma_2 R_2 \frac{\partial^2 a}{\partial X^2} + \bar{\sigma}_3 \frac{\partial^3 a}{\partial X^3} + \sigma_4 \frac{\partial^4 a}{\partial x^4} + \kappa_1 \frac{\partial(a^2)}{\partial X} \qquad (1.78)$$

(all the terms in this equation are of the same order).

1.4.2.2 Longwave Instabilities of the Type II_m

If the dispersion vanishes ($\sigma_3 = 0$), the KE is reduced to the famous *Kuramoto–Sivashinsky equation* (KSE)

$$\frac{\partial a_0}{\partial T} = \sigma_2 R_2 \frac{\partial^2 a_0}{\partial X^2} + \sigma_4 \frac{\partial^4 a_0}{\partial X^4} + \kappa_1 \frac{\partial(a^2)}{\partial X} \qquad (1.79)$$

typical for instability of oscillations in reaction–diffusion systems [25, 51] and instabilities of fronts [37]. Note that this amplitude equation does not include any characteristics of the wave frequency (except for definition of X); hence it is valid also in the case of a monotonic type II instability, II_m.

In the context of front instability, the KSE appears in the form

$$\frac{\partial h}{\partial t} = \sigma_2(R - R_c)\frac{\partial^2 h}{\partial x^2} + \sigma_4\frac{\partial^4 h}{\partial x^4} + \kappa_1\left(\frac{\partial h}{\partial x}\right)^2, \tag{1.80}$$

where h is the deflection of the front, $a = \partial h/\partial x$ (see Section 6.1.5.1). Equation (1.80) describes also the nonlinear development of the Benjamin–Feir instability of homogeneous oscillations (Section 7.2.1.3).

An equation of the same type can be obtained if the instability region is narrow because of large σ_4: $R - R_c = O(1), \sigma_4 = O(\varepsilon^{-2})$, rather than because of small $R - R_c$. This case is characteristic for a film flow instability in a fluid with large surface tension [52] (Section 5.1.4).

Some additional symmetry properties of the problem can forbid the nonlinearity $\kappa_1\partial(a^2)/\partial x$. For instance, if there is no preferred spatial direction, the amplitude equation should be invariant under the transformation $x \rightarrow -x$. In this case one obtains the *Sivashinsky equation* (SE)

$$\frac{\partial a}{\partial t} = \sigma_2(R - R_c)\frac{\partial^2 a}{\partial x^2} + \sigma_4\frac{\partial^4 a}{\partial x^4} + \kappa_2\frac{\partial^2(a^2)}{\partial x^2} \tag{1.81}$$

(see [53]). If in addition the symmetry $a \rightarrow -a$ takes place, one gets the *Cahn–Hilliard equation* (CHE)

$$\frac{\partial a}{\partial t} = \sigma_2(R - R_c)\frac{\partial^2 a}{\partial x^2} + \sigma_4\frac{\partial^4 a}{\partial x^4} + \kappa_3\frac{\partial^2(a^3)}{\partial x^3}, \tag{1.82}$$

where κ_3 is a constant (see Sections 3.1.1, 3.2.2, 5.2).

Let us finish the list of long-wavelength amplitude equations mentioning the case where the nonlinear terms cannot depend on the amplitude function itself but depend only on its derivatives. One example, already mentioned above, is the instability of a front, which is insensitive to a translation, $h \rightarrow h+$constant; hence the amplitude equation cannot include h itself. The same feature is characteristic also for the instability of a convective parallel flow between heat-insulating boundaries. The term of the lowest order is

$$q_n = \mu\left[\left(\frac{\partial a}{\partial x}\right)^2\right] \tag{1.83}$$

(a is proportional to the temperature averaged across the layer). In the case of the instability of a conductive state between thermally insulated boundaries, because of the symmetry of the term $\partial q_n/\partial x$ with respect to reflection $x \rightarrow -x$,

$$q_n = \mu_1 \left[\left(\frac{\partial a}{\partial x} \right)^3 \right] + \mu_2 \frac{\partial}{\partial x} \left[\left(\frac{\partial a}{\partial x} \right)^2 \right] \tag{1.84}$$

(see Section 3.2.1).

In the case of a nonanalytical dependence of the growth rate on the wavenumber, (1.29), the appropriate amplitude equation contains a nonlocal term [37] known as the *Hilbert transform*:

$$\frac{\partial h(x,t)}{\partial t} = \sigma_1 \int_{-\infty}^{\infty} |k| e^{ik(x-z)} h(z,t) \frac{dkdz}{2\pi} + \sigma_2 \frac{\partial^2 h}{\partial x^2} + \kappa_1 \left(\frac{\partial h}{\partial x} \right)^2 . \tag{1.85}$$

Front dynamics governed by (1.85) has been studied in [54].

We do not discuss here the case (1.28) where we have no small parameter (see [34]).

In this section, for the sake of simplicity, we have focused on two-dimensional problems and obtained $1 + 1$-dimensional amplitude equations. Exactly the same types of equations can be obtained for motion in cylindrical regions (see the next section). In planar layers, one can introduce amplitude functions depending on two spatial coordinates, and we have to generalize the amplitude equations discussed in this section to the case of $2 + 1$ dimensions. Such a generalization, however, is not unique. First of all, it is important whether the original problem is rotationally symmetric or not. Also, in many cases some new long-scale variables should be taken into account (pressure, poloidal stream function, etc.). In that case, one obtains parabolic evolution equations coupled with some elliptic equations. Hence we do not discuss these generalizations here and postpone them to following sections, where specific physical problems are considered.

References

1. A.A. Abrikosov, Sov. Phys. JETP **5**, 1174 (1957)
2. M. Seul, R. Wolfe, Phys. Rev. Lett. **68**, 2460 (1992)
3. F.S. Bates, M.F. Schultz, A.K. Khandpur, S. Foster, J.H. Rosedale, K. Almdahl, K. Mortensen, Faraday Discuss. **98**, 7 (1994)
4. G. Springholtz, V. Holy, M. Pinczolits, G. Bauer, Science **282**, 734 (1998)
5. H. Masuda, K. Fukuda, Science **268**, 1466 (1995)
6. A. Huczko, Appl. Phys. A **70**, 365 (2000)
7. Y. Chen, A. Pepin, Electrophoresis **22**, 187 (2001)
8. B. Bertotti, P. Farinella, D. Vokrouhlický, *Physics of the Solar System* (Springer Science + Business Media, Dordrecht, 2003)
9. J.D. Murray, *Mathematical Biology* (Springer, Berlin, 1989)
10. T. Ackemann, T. Lange, Appl. Phys. B **72**, 21 (2001)
11. Q. Ouyang, H.L. Swinney, Nature (London) **352**, 61 (1991)
12. A. Joets, R. Ribotta, J. Phys. (Paris) **47**, 595 (1986)
13. W.S. Edwards, S. Fauve, J. Fluid Mech. **278**, 123 (1994)
14. P. Kolodner, C.M. Surko, H. Williams, Physica D **37**, 319 (1989)
15. C.H. Blohm, H.C. Kuhlmann, J. Fluid Mech. **450**, 67 (2002)

16. P. Le Gal, A. Pocheau, V. Croquette, Phys. Rev. Lett. **54**, 2501 (1985)
17. A.M. Rucklidge, Proc. Roy. Soc. Lond. A **453**, 107 (1997)
18. M.C. Cross, P.C. Hohenberg, Rev. Mod. Phys. **65**, 851 (1993)
19. L.M. Pismen, *Patterns and Interfaces in Dissipative Dynamics* (Springer, Berlin/Heidelberg, 2006)
20. R. Hoyle, *Pattern Formation, An Introduction to Methods* (Cambridge University Press, Cambridge, 2006)
21. M. Cross, H. Greenside, *Pattern Formation and Dynamics in Nonequilibrium Systems* (Cambridge University Press, Cambridge, 2009)
22. A.C. Newell, J.A. Whitehead, J. Fluid Mech. **38**, 279 (1969)
23. L.A. Segel, J. Fluid Mech. **38**, 203(1969)
24. K. Stewartson, J.T. Stuart, J. Fluid Mech. **48**, 529 (1971)
25. Y. Kuramoto, T. Tsuzuki, Prog. Theor. Phys. **55**, 356 (1976)
26. I.S. Aranson, L. Kramer, Rev. Mod. Phys. **74**, 99 (2002)
27. H. Haken, *Synergetics: Introduction and Advanced Topics* (Springer, Berlin/Heidelberg, 2004)
28. Y. Kuramoto, *Chemical Oscillations, Waves, and Turbulence* (Springer, New York, 1984)
29. D.D. Joseph, *Stability of Fluid Motions*, vol. I, II (Springer, New York, 1976)
30. R.K. Dodd, J.C. Eilbeck, J.D. Gibbon, H.C. Morris, *Solitons and Nonlinear Wave Equations* (Academic, London, 1982)
31. P. Manneville, *Dissipative Structure and Weak Turbulence* (Academic, Boston, 1990)
32. V.P. Starr, *Physics of Negative Viscosity Phenomena* (McGraw-Hill, New York, 1968)
33. G. Sivashinsky, V. Yakhot, Phys. Fluids **28**, 1040 (1985)
34. U. Frisch, Z.S. She, P.L. Sulem, Physica D **28**, 382 (1987)
35. L.D. Landau, E.M. Lifshitz, *Fluid Mechanics* (Pergamon Press, Oxford, 1987)
36. Y. Zeldovich, G. Barenblatt, V. Librovich, G. Makhviladze, *Mathematical Theory of Combustion and Explosions* (Plenum Press, New York, 1985)
37. G.I. Sivashinsky, Acta Astronaut. **4**, 1177 (1977)
38. S.V. Shklyaev, Fluid Dyn. **36**, 682 (2001)
39. J.D. Crawford, E. Knobloch, Physica D **31**, 1 (1988)
40. S. Fauve, Pattern forming instabilities, in *Hydrodynamics and Nonlinear Instabilities*, ed. by C. Godrèche, P. Manneville (Cambridge University Press, 1998), pp. 387–491
41. A.H. Nayfeh, *Perturbation Methods* (Wiley, New York, 1973)
42. J. Kevorkian, J.D. Cole, *Multiple Scale and Singular Perturbation Methods* (Springer, New York, 1994)
43. E. Knobloch, J. De Luca, Nonlinearity **3**, 975 (1990)
44. B.J. Matkowsky, V.A. Volpert, Physica D **179**, 183 (2003)
45. V.A. Volpert, A.A. Nepomnyashchy, L.G. Stanton, A.A. Golovin, SIAM J. Appl. Dyn. Syst. **7**, 265 (2008)
46. R.A. Fisher, Ann. Eugenics **7**, 355 (1937)
47. A. Kolmogorov, I. Petrovskii, N. Piskunov, Moscow University Bull. Math. **1**, 1 (1937)
48. G.M. Homsy, Lect. Appl. Math. **15**, 191 (1974)
49. J. Topper, T. Kawahara, J. Phys. Soc. Jpn. **44**, 663 (1978)
50. T. Kawahara, Phys. Rev. Lett. **51**, 381 (1983)
51. T. Yamada, Y. Kuramoto, Prog. Theor. Phys. **56**, 681 (1976)
52. A.A. Nepomnyashchy, Fluid Dyn. **9**, 354 (1974)
53. G.I. Sivashinsky, Physica D **8**, 243 (1983)
54. O. Thual, U. Frisch, M. Hénon, J. Phys. **46**, 1485 (1985)

Chapter 2
Convection in Cylindrical Cavities

The thermal convection in a viscous fluid heated from below, which provides a classical example of the pattern formation in a nonequilibrium system, is a subject of several monographs [1–6]. In contrast to the cited books, we focus on the specific features of *large-scale* convective motions produced by *long-wavelength* instabilities of the conductive state. There are two main physical origins of convection instabilities: the buoyancy force due to an inhomogeneous density distribution in a gravity field (*buoyancy* or *Rayleigh convection*) and the thermocapillary effect caused by the dependence of the surface tension on temperature (*surface tension driven* or *Marangoni convection*).

We start with the consideration of buoyancy convection in a cylindrical cavity, which gives an opportunity to discuss in details the phenomena of coarsening and emergence of one-dimensional (1D) patterns. Indeed, the amplitude functions introduced within this chapter vary along the axial coordinate only. Therefore, a cylindrical cavity provides a nice starting point for the analysis which will be followed by the generalization of the amplitude equations to the case of two-dimensional (2D) patterns in the next chapter.

2.1 Free Convection in a Horizontal Cylinder

Let us consider the long-wavelength convection in a Boussinesq fluid with the mean density ρ_{*0}, temperature volume expansion coefficient β_*, kinematic viscosity coefficient ν_*, and thermal diffusivity κ_* filling an infinite horizontal cylinder (see Figure 2.1) with the characteristic transverse scale d_*. We assume that the temperature distribution at the rigid boundary of the cylinder Γ is fixed: $T_{*\Gamma} = -A_* z_*$. In this case, an equilibrium (conductive) state is possible that is characterized by the stationary temperature field $T_{*0}(x_*, y_*, z_*) = -A_* z_*$ and the pressure field $p_{*0}(x_*, y_*, z_*) = \rho_{*0} g_*(z_* + \beta_* A_* z_*^2/2) + const$. Hereafter we analyze the dynamics of perturbations of the temperature T_* and pressure p_* from T_{*0} and p_{*0},

© Springer Science+Business Media, LLC 2017   19
S. Shklyaev, A. Nepomnyashchy, *Longwave Instabilities and Patterns in Fluids*,
Advances in Mathematical Fluid Mechanics,
https://doi.org/10.1007/978-1-4939-7590-7_2

respectively. We choose d_*, d_*^2/κ_*, κ_*/d_*, $\rho_{*0}v_*\kappa_*/d_*^2$, and A_*d_* as scales of the length, time, velocity, pressure, and temperature. The dimensionless boundary value problem governing the buoyancy convection reads:

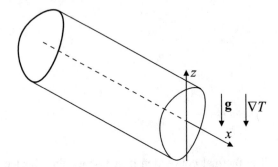

Fig. 2.1 Problem geometry: horizontal cylinder heated from below

$$\frac{1}{P}\left[\frac{\partial \mathbf{v}}{\partial t}+(\mathbf{v}\cdot\nabla)\mathbf{v}\right]=-\nabla p+\Delta\mathbf{v}+RT\mathbf{e}_z, \tag{2.1}$$

$$\nabla\cdot\mathbf{v}=0, \tag{2.2}$$

$$\frac{\partial T}{\partial t}+\mathbf{v}\cdot\nabla T-\mathbf{v}\cdot\mathbf{e}_z=\Delta T; \tag{2.3}$$

$$\mathbf{v}=0,\ T=0 \text{ at } \Gamma. \tag{2.4}$$

Here $R=g_*\beta_*A_*d_*^4/v_*\kappa_*$ is the Rayleigh number, $P=v_*/\kappa_*$ is the Prandtl number, and $\mathbf{e}_z$ is the unit vector along the z-axis. It is obvious that p can be determined up to an arbitrary additive function $p_0(t)$, because only ∇p appears in (2.1) rather than p itself. We search for solutions $(\mathbf{v},T,\nabla p)$ bounded as $x\rightarrow\pm\infty$ and assume that the mean flux along the cylinder vanishes:

$$\langle v_x(y,z)\rangle=0, \tag{2.5}$$

where

$$\langle f(y,z)\rangle\equiv\int_S dydz f(y,z) \tag{2.6}$$

(S is the cross-section of the cylinder). Later on, we use the definition $U=(\mathbf{v},p,T)$.

The stability of the conductive state $\mathbf{v}=0$, $p=0$, $T=0$ can be studied by means of a linearization of system (2.1)–(2.5). Because of the invariance of the problem with respect to translations of x and t, we can introduce normal disturbances (see Section 1.2):

$$U(\mathbf{r},t)=u_n(y,z;k,R)\exp[ikx+\sigma_n(k,R)t]. \tag{2.7}$$

According to general features of convective instability in a cavity heated from below [3], the eigenvalues $\sigma_n(k,R)$ are real, if $R>0$. For any normal mode, the equation

$\sigma_n(k, R) = 0$ defines a neutral curve $R = R_n(k)$ separating regions of decaying and growing disturbances.

It is known [3] that a competition of two instability modes takes place for the convection in a horizontal cylinder. One of them is a long-wavelength mode (the minimum of the neutral stability curve $R = R_c$ is situated at $k = 0$; see Section 1.2), whereas another one is short-wavelength (it has a minimum a at $k = k_c \neq 0$). In the present section, we restrict ourselves to consideration of flow generated by a long-wavelength instability mode (the nonlinear interaction between large-scale and short-scale motions is considered in Section 7.3). Here and below, we omit the subscript n, which enumerates the instability modes.

Let us try to guess the form of the amplitude equation which governs the convective motions in the horizontal cylinder near the threshold.

The instability under consideration is a type III instability (see Section 1.4), which is characterized by the following Taylor expansion of the growth rate σ near the point $(0, R_c)$:

$$\sigma = \left(\frac{\partial \sigma}{\partial R} \right)_c (R - R_c) + \frac{1}{2} \left(\frac{\partial^2 \sigma}{\partial k^2} \right)_c k^2 + \dots, \quad \left(\frac{\partial \sigma}{\partial R} \right)_c > 0, \quad \left(\frac{\partial^2 \sigma}{\partial k^2} \right)_c < 0. \quad (2.8)$$

The disturbance with the wavenumber $k = 0$ corresponds to a homogeneous rotation of the fluid (clockwise or counterclockwise, depending on the sign of the disturbance) in the whole cylinder. Such a disturbance grows with the maximum growth rate (otherwise, there is no long-wavelength instability). The perturbations with $k \neq 0$ produce a spatially periodic change of the rotation direction along the cylinder.

If $R - R_c = \varepsilon^2 \ll 1$, the disturbances with wavenumbers

$$|k| < k_m = \left[-2 \frac{(\partial \sigma / \partial R)_c}{(\partial^2 \sigma / \partial k^2)_c} \right]^{1/2} \varepsilon = O(\varepsilon) \quad (2.9)$$

grow, and the growth rates

$$\sigma(k) < \sigma(0) = \left(\frac{\partial \sigma}{\partial R} \right)_c \varepsilon^2 = O\left(\varepsilon^2 \right). \quad (2.10)$$

Thus, for the description of convective motion, it is natural to introduce rescaled variables $t_2 = \varepsilon^2 t$, $x_1 = \varepsilon x$.

According to the discussion in Section 1.4, the amplitude function $a^{(0)}(x_1, t_2)$ should be real and satisfy equation (1.65). However, it should be taken into account that the system of equations (2.1)–(2.4) is invariant with respect to the simultaneous inversion of the coordinate x and the x-component of the velocity v_x. Since for the linear normal mode corresponding to $k = 0$, $v_x = 0$, this transformation is equivalent to $x \to -x$, $a^{(0)} \to a^{(0)}$. Owing to this symmetry property, the coefficient α should vanish, and we obtain the Allen–Cahn equation (1.66).

The direct derivation of the amplitude equation is presented in Section 2.1.1 that can be omitted by the reader who is not interested in technical details.

2.1.1 Derivation of the Allen–Cahn Equation

It is known [3] that the amplitude of the convective motion near the stability threshold is proportional to $(R - R_c)^{1/2}$; therefore, one can introduce the following rescaled perturbation fields:

$$T = \varepsilon\Theta, \; v_y\mathbf{e}_y + v_z\mathbf{e}_z = \varepsilon\mathbf{w}, \; p = \varepsilon\Pi, \; v_x = \varepsilon^2 u \qquad (2.11)$$

($\mathbf{e}_y$ is the unit vectors along the y-axis). Under these rescalings, equations (2.1)–(2.5) are transformed to the following form:

$$\frac{\varepsilon}{P}\left[\varepsilon\frac{\partial\mathbf{w}}{\partial t_2} + (\mathbf{w}\cdot\nabla_\perp)\mathbf{w} + \varepsilon^2 u\frac{\partial\mathbf{w}}{\partial x_1}\right] = -\nabla_\perp\Pi + \nabla_\perp^2\mathbf{w} + \varepsilon^2\frac{\partial^2\mathbf{w}}{\partial x_1^2}$$
$$+(R_c + \varepsilon^2)\Theta\mathbf{e}_z, \qquad (2.12)$$

$$\frac{\varepsilon}{P}\left(\varepsilon\frac{\partial u}{\partial t_2} + \mathbf{w}\cdot\nabla_\perp u + \varepsilon^2 u\frac{\partial u}{\partial x_1}\right) = -\frac{\partial\Pi}{\partial x_1} + \nabla_\perp^2 u + \varepsilon^2\frac{\partial^2 u}{\partial x_1^2}, \qquad (2.13)$$

$$\nabla_\perp\cdot\mathbf{w} + \varepsilon^2\frac{\partial u}{\partial x_1} = 0, \qquad (2.14)$$

$$\varepsilon^2\frac{\partial\Theta}{\partial t_2} + \varepsilon\mathbf{w}\cdot\nabla_\perp\Theta + \varepsilon^3 u\frac{\partial\Theta}{\partial x_1} - w_z = \nabla_\perp^2\Theta + \varepsilon^2\frac{\partial^2\Theta}{\partial x_1^2}; \qquad (2.15)$$

$$\mathbf{w} = 0, \; u = 0, \; \Theta = 0 \text{ at } \Gamma; \qquad (2.16)$$

$$\langle u\rangle = 0. \qquad (2.17)$$

Here $\nabla_\perp$ is the two-dimensional gradient operator:

$$\nabla_\perp = \mathbf{e}_y\frac{\partial}{\partial y} + \mathbf{e}_z\frac{\partial}{\partial z}. \qquad (2.18)$$

The functions $\mathbf{w}$, u, Θ, $\nabla_\perp\Pi$, $\partial\Pi/\partial x_1$ are assumed to be bounded.

According to the scheme described in Section 1.3, we introduce an expansion

$$\mathbf{w} = \mathbf{w}_0 + \varepsilon\mathbf{w}_1 + \varepsilon^2\mathbf{w}_2 + \dots \qquad (2.19)$$

and similar expansions for other fields of perturbations. We substitute these expansions into equations (2.12)–(2.17) and separate the terms of different order in ε.

At the zeroth order, we obtain a linear homogeneous boundary problem determining the critical Rayleigh number R_c:

$$-\nabla_\perp\Pi_0 + \nabla_\perp^2\mathbf{w}_0 + R_c\Theta_0\mathbf{e}_z = 0, \; \nabla_\perp\cdot\mathbf{w}_0 = 0, \qquad (2.20)$$

$$\nabla_\perp^2\Theta_0 + w_{0z} = 0; \qquad (2.21)$$

$$\mathbf{w}_0 = 0, \Theta_0 = 0 \text{ at } \Gamma. \qquad (2.22)$$

The solution $U_0 = (\mathbf{w}_0, \Pi_0, \Theta_0)$ to this problem can be written in the form

$$\mathbf{w}_0 = a_0(x_1, t_2)\tilde{\mathbf{w}}_0(y, z), \tag{2.23}$$

$$\Theta_0 = a_0(x_1, t_2)\tilde{\Theta}_0(y, z), \tag{2.24}$$

$$\Pi_0 = a_0(x_1, t_2)\hat{\Pi}_0(y, z) + b_0(x_1, t_2), \tag{2.25}$$

where $a_0(x_1, t_2)$, $b_0(x_1, t_2)$ are arbitrary functions at this stage. The function $\hat{\Pi}_0(y, z)$ is not defined uniquely; it can be fixed by an additional condition $\langle \hat{\Pi}_0 \rangle = 0$.

The boundary value problem for u_0 is separated from the rest of equations:

$$\nabla_\perp^2 u_0 - \frac{\partial \Pi_0}{\partial x_1} = 0; \tag{2.26}$$

$$u_0 = 0 \text{ at } \Gamma; \tag{2.27}$$

$$\langle u_0 \rangle = 0. \tag{2.28}$$

Solving that problem allows eliminating the amplitude function b_0. The solution to equation (2.26) can be written in the form

$$u_0 = \frac{\partial a_0(x_1, t_2)}{\partial x_1}\hat{u}_0(y, z) + \frac{\partial b_0(x_1, t_2)}{\partial x_1}\bar{u}_0(y, z), \tag{2.29}$$

where $\hat{u}_0$ and $\bar{u}_0$ are solutions of the following boundary value problems:

$$\nabla_\perp^2 \hat{u}_0 = \hat{\Pi}_0, \ \nabla_\perp^2 \bar{u}_0 = 1; \ \hat{u}_0 = 0, \ \bar{u}_0 = 0 \text{ at } \Gamma. \tag{2.30}$$

The condition (2.28) provides:

$$\frac{\partial a_0(x_1, t_2)}{\partial x_1}\langle \hat{u}_0 \rangle + \frac{\partial b_0(x_1, t_2)}{\partial x_1}\langle \bar{u}_0 \rangle = 0; \tag{2.31}$$

thus,

$$b_0 = -a_0 \frac{\langle \hat{u}_0 \rangle}{\langle \bar{u}_0 \rangle} + c(t_2) \tag{2.32}$$

(the function $c(t_2)$ gives no contribution into the pressure gradient, and it will be dropped). Finally, we obtain:

$$\Pi_0 = a_0(x_1, t_2)\tilde{\Pi}_0(y, z), \ u_0 = \frac{\partial a_0(x_1, t_2)}{\partial x_1}\tilde{u}_0(y, z), \tag{2.33}$$

where

$$\tilde{\Pi}_0 = \hat{\Pi}_0 - \frac{\langle \hat{u}_0 \rangle}{\langle \bar{u}_0 \rangle}, \ \tilde{u}_0 = \hat{u}_0 - \bar{u}_0 \frac{\langle \hat{u}_0 \rangle}{\langle \bar{u}_0 \rangle}. \tag{2.34}$$

At the first order in ε, we obtain a linear nonhomogeneous boundary value problem

$$-\nabla_\perp \Pi_1 + \nabla_\perp^2 \mathbf{w}_1 + R_c \Theta_1 \mathbf{e}_z = \frac{1}{P}\mathbf{w}_0 \cdot \nabla_\perp \mathbf{w}_0, \nabla_\perp \cdot \mathbf{w}_1 = 0, \tag{2.35}$$

$$\nabla_\perp^2 \Theta_1 + w_{1z} = \mathbf{w}_0 \cdot \nabla_\perp \Theta_0; \qquad (2.36)$$

$$\mathbf{w}_1 = 0, \; \Theta_1 = 0 \text{ at } \Gamma. \qquad (2.37)$$

The problem (2.35)–(2.37) is solvable, if its right-hand side is orthogonal to the solution of the linear problem adjoint to (2.20)–(2.22). If we define the inner product of functions $U_1(y,z) = (\mathbf{w}_1, \Pi_1, \Theta_1)$ and $U_2(y,z) = (\mathbf{w}_2, \Pi_2, \Theta_2)$ as

$$(U_1, U_2) = \langle \mathbf{w}_1 \cdot \mathbf{w}_2 + \Pi_1 \Pi_2 + R_c \Theta_1 \Theta_2 \rangle, \qquad (2.38)$$

we find out that the linear problem (2.20)–(2.22) is self-adjoint. Hence, the solvability condition has the form (see Appendix A):

$$\left\langle \frac{1}{2P} \tilde{\mathbf{w}}_0 \cdot \nabla_\perp |\tilde{\mathbf{w}}_0|^2 + R_c \tilde{\Theta}_0 \tilde{\mathbf{w}}_0 \cdot \nabla_\perp \tilde{\Theta}_0 \right\rangle = 0 \qquad (2.39)$$

and it is satisfied *identically*.

This is the result of specific features of nonlinearities in equations (2.12), (2.13), and (2.15) characteristic for the Boussinesq convection and leading to a *pitchfork* bifurcation of solutions (see Appendix B). In the case of a generic quadratic nonlinearity, we would come to a contradiction at his point. Actually, in the generic case, a *transcritical* bifurcation takes place with the amplitude of the bifurcating solution proportional to $R - R_c$, rather than $(R - R_c)^2$.

The boundary value problem

$$\nabla_\perp^2 u_1 - \frac{\partial \Pi_1}{\partial x_1} = \frac{1}{P} \mathbf{w}_0 \cdot \nabla_\perp u_0; \; u_1 = 0 \text{ at } \Gamma; \; \langle u_1 \rangle = 0 \qquad (2.40)$$

determines u_1 and fully determines the function Π_1. The structure of the solution is:

$$\mathbf{w}_1 = a_0^2(x_1, t_2) \tilde{\mathbf{w}}_1(y, z) + a_1(x_1, t_2) \tilde{\mathbf{w}}_0(y, z), \qquad (2.41)$$

$$\Theta_1 = a_0^2(x_1, t_2) \tilde{\Theta}_1(y, z) + a_1(x_1, t_2) \tilde{\Theta}_0(y, z), \qquad (2.42)$$

$$\Pi_1 = a_0^2(x_1, t_2) \tilde{\Pi}_1(y, z) + a_1(x_1, t_2) \tilde{\Pi}_0(y, z), \qquad (2.43)$$

$$u_1 = a_0(x_1, t_2) \frac{\partial a_0(x_1, t_2)}{\partial x_1} \tilde{u}_1(y, z) + \frac{\partial a_1(x_1, t_2)}{\partial x_1} \tilde{u}_0(y, z), \qquad (2.44)$$

where $\mathbf{w}_1, \Theta_1, \Pi_1$, and u_1 are solutions of corresponding nonhomogeneous linear problems and a_1 is a new amplitude function.

Finally, the problem at the second order in ε

$$-\nabla_\perp \Pi_2 + \nabla_\perp^2 \mathbf{w}_2 + R_c \Theta_2 \mathbf{e}_z = \frac{1}{P} \left[\mathbf{w}_0 \cdot \nabla_\perp \mathbf{w}_1 + \mathbf{w}_1 \cdot \nabla_\perp \mathbf{w}_0 + \frac{\partial \mathbf{w}_0}{\partial t_2} \right]$$

$$- \frac{\partial^2 \mathbf{w}_0}{\partial x_1^2} - \Theta_0 \mathbf{e}_z, \qquad (2.45)$$

$$\nabla_\perp \cdot \mathbf{w}_2 = -\frac{\partial u_0}{\partial x_1}, \qquad (2.46)$$

$$\nabla_\perp^2 \Theta_2 + w_{2z} = \mathbf{w}_0 \cdot \nabla_\perp \Theta_1 \mathbf{w}_1 \cdot \nabla_\perp \Theta_0 + \frac{\partial \Theta_0}{\partial t_2} - \frac{\partial^2 \Theta_0}{\partial x_1^2}; \quad (2.47)$$

$$\mathbf{w}_2 = 0, \; \Theta_2 = 0 \text{ at } \Gamma; \quad (2.48)$$

is solvable if

$$\alpha_1 \frac{\partial a}{\partial t_2} = \alpha_2 a + \alpha_3 \frac{\partial^2 a}{\partial x_1^2} + \alpha_4 a^3, \quad (2.49)$$

where

$$\alpha_1 = \left\langle \frac{1}{P} |\tilde{\mathbf{w}}_0|^2 + R_c \tilde{\Theta}_0^2 \right\rangle, \quad (2.50)$$

$$\alpha_2 = \left\langle \tilde{\Theta}_0 \tilde{w}_{0z} \right\rangle = -\left\langle \tilde{\Theta}_0 \nabla_\perp^2 \tilde{\Theta}_0 \right\rangle, \quad (2.51)$$

$$\alpha_3 = \left\langle |\tilde{\mathbf{w}}_0|^2 + R_c \tilde{\Theta}_0^2 + \tilde{u}_0 \tilde{\Pi}_0 \right\rangle, \quad (2.52)$$

$$\alpha_4 = -\left\langle \frac{1}{P} \tilde{\mathbf{w}}_0 \cdot (\tilde{\mathbf{w}}_0 \cdot \nabla_\perp \tilde{\mathbf{w}}_1 + \tilde{\mathbf{w}}_1 \cdot \nabla_\perp \tilde{\mathbf{w}}_0) \right\rangle$$
$$+ R_c \left\langle \tilde{\Theta}_0 (\tilde{\mathbf{w}}_0 \cdot \nabla_\perp \tilde{\Theta}_1 + \tilde{\mathbf{w}}_1 \cdot \nabla_\perp \tilde{\Theta}_0) \right\rangle. \quad (2.53)$$

It is obvious that $\alpha_1 > 0$ and $\alpha_2 > 0$. In the case of a longwave instability,

$$\frac{\alpha_3}{\alpha_1} = -\frac{1}{2} \left(\frac{\partial^2 \sigma(k,R)}{\partial k^2} \right)_c > 0. \quad (2.54)$$

It was proved [3] that the onset of convection is supercritical, hence $\alpha_4 < 0$.
By means of the scale transformation

$$T = \frac{\alpha_2}{\alpha_1} t_2, \; X = \sqrt{\frac{\alpha_2}{\alpha_3}} x_1, \; A = \sqrt{-\frac{\alpha_4}{\alpha_2}} a \quad (2.55)$$

one can rewrite the amplitude equation as follows:

$$\frac{\partial A}{\partial T} = \frac{\partial^2 A}{\partial X^2} + A - A^3, \quad (2.56)$$

thus excluding all the coefficients. Equation (2.56) is a real version of the CGLE, which is called the Allen–Cahn equation (ACE); see (1.66).

As mentioned above, solutions of the original problem have to be finite; that leads to the following condition for A:

$$|A| < \infty \quad (2.57)$$

as $X \to \pm\infty$.

Equation (2.56) is the simplest amplitude equation for a single real one-component amplitude function, which is relevant to the pattern formation phenomenon. More exactly, it describes the phenomenon of *coarsening* [7], as it will be shown below. However, a number of important ideas of the pattern formation theory can be introduced on the ground of that equation.

2.1.2 Basic Properties of the Allen–Cahn Equation

2.1.2.1 Lyapunov Functional

Let us define the functional

$$F = \int_{-\infty}^{\infty} L(X)dX, \qquad (2.58)$$

where

$$L(X) = \frac{1}{2}\left(\frac{\partial A}{\partial X}\right)^2 + \frac{1}{4}(A^2 - 1)^2 \geq 0. \qquad (2.59)$$

Impose a variation of function $A(X)$, $A(X) + \delta A(X)$, such that $\delta A(X) \to 0$ as $X \to \pm\infty$. The linear part of the functional variation is

$$\delta F = \int_{-\infty}^{\infty} \frac{\partial A}{\partial X}\frac{\partial \delta A}{\partial X}dX + \int_{-\infty}^{\infty} (A^3 - A)\delta A dX. \qquad (2.60)$$

Integrating the first term by parts, we find that

$$\delta F = \int_{-\infty}^{\infty} L_A(X,T)\delta A dX, \qquad (2.61)$$

where

$$L_A(X,T) = -\frac{\partial^2}{\partial X^2} + A^3 - A. \qquad (2.62)$$

The function $L_A(X,T)$ is called the variational derivative of the functional F with respect to function $A(X,T)$, and it is denoted as

$$L_A(X,T) = \frac{\delta F}{\delta A}.$$

Thus, equation (2.56) can be written as

$$\frac{\partial A}{\partial T} = -\frac{\delta F}{\delta A}, \qquad (2.63)$$

i.e., it has a *potential* structure. The functional $F[A(X)]$ is the *Lyapunov functional* (also called "energy"). The Lyapunov functional decreases monotonically with time until a stationary state is reached:

$$\frac{dF}{dT} = -\int dX \left(\frac{\partial A}{\partial T}\right)^2 \leq 0. \qquad (2.64)$$

If $F < \infty$ at $T = 0$, the asymptotic state of the system is a stationary state with finite F.

2.1.2.2 Stationary Patterns

Let us consider stationary solutions $A = A(X)$ of (2.56) which satisfy the ordinary differential equation

$$\frac{d^2A}{dX^2} + A - A^3 = 0; \quad |A| < \infty \text{ at } X \to \pm\infty. \tag{2.65}$$

The functional $F[A(X)]$ defined by (2.58) plays the role of the action of a fictitious "particle" with the coordinate A moving in "time" X in the potential $V(A) = -\frac{1}{4}(A^2 - 1)^2$; the function $L(X)$ determined by (2.59) is the Lagrange function of that particle. The orbits of the dynamical system (2.65) in the phase space $(A, dA/dX)$ are shown schematically in Figure 2.2.

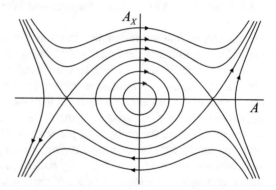

Fig. 2.2 Phase portrait for (2.65) in the plane (A, A_X)

It is obvious that system (2.65) has three fixed points corresponding to uniform fluid motions. The solution $A(X) = 0$ describes the *quiescent state*, and the solutions $A(X) = \pm 1$ describe a *uniform rotation* of the fluid (counterclockwise or clockwise). The latter solutions correspond to the absolute minima of the Lyapunov functional $F = 0$. For the solution $A = 0$, the integral (2.58) diverges; however, we can calculate the integral (2.58) over a finite interval $-l \leq X \leq l$ and characterize that stationary state by its *mean Lyapunov functional density*

$$\bar{L} = \lim_{l \to \infty} \frac{1}{2l} \int_{-l}^{l} L(X) dX \tag{2.66}$$

which is equal to $1/4$ for the quiescent state. The separatrices

$$A_\pm(X) = \pm \tanh \frac{X - \xi}{\sqrt{2}}, \quad \xi = const \tag{2.67}$$

describe *domain walls* separating semi-infinite domains with different directions of rotation. We shall call solutions A_+ and A_-, respectively, a *kink* and an *antikink*. For these solutions, the Lyapunov functional $F_0 = 2\sqrt{2}/3$ is finite.

Other finite orbits are closed and describe *spatially periodic patterns* which can be considered as periodic arrays of domains with alternating signs of A. They are expressed through the elliptic Jacobi function $sn(x)$:

$$A(X) = r \, sn \, \alpha(X - \xi); \quad \xi = const; \tag{2.68}$$

$$0 < r < 1, \quad \alpha^2 = 1 - r^2/2. \tag{2.69}$$

The modulus of the elliptic Jacobi function in equation (2.68) is $k_J = r/\sqrt{2 - r^2}$; the spatial half-period of the pattern is $l = K(k_J)/\alpha$, where $K(k_J)$ is the complete elliptic integral of the second kind; l is the length of each domain with a constant sign of A. All periodic patterns have a finite mean density of the Lyapunov functional.

2.1.2.3 Stability of Stationary Patterns and Wavenumber Selection

We have found that the system has an infinite number of steady states corresponding to local extrema of the functional F. However, not all of them can be actually realized in experiments. The problem of the selection of solutions that are actually obtained as a result of evolution starting from certain realistic initial conditions is called the *pattern selection* problem. In the case under consideration, the competing patterns can be characterized by their spatial period $2l$ or the wavenumber π/l; hence, in this case, the problem of pattern selection is reduced to a *wavenumber selection*. In the case of an infinite region, the only relevant selection factor is the *stability* property. A feasible stationary pattern should be stable; in other words, it should correspond to a local *minimum* of the functional F:

$$\delta^2 F \geq 0. \tag{2.70}$$

Local stability of a stationary pattern is determined by the spectrum of the eigenvalue problem for infinitesimal perturbations in the form $\tilde{A}(X)e^{\sigma T}$ superimposed on the stationary solution $A(X)$. The spectral problem obtained by linearization of (2.56) is

$$\sigma \tilde{A} = \frac{\partial^2 \tilde{A}}{\partial X^2} + (1 - 3A^2)\tilde{A}, \quad |\tilde{A}(\pm\infty)| < \infty. \tag{2.71}$$

The inequality (2.70) is equivalent to the requirement that all eigenvalues are nonpositive: $\sigma \leq 0$.

It is obvious that the solution $A = 0$ is unstable: $\sigma(\tilde{k}) = 1 - \tilde{k}^2$ for $\tilde{A} = e^{i k X}$, and the solutions $A = \pm 1$ are stable; $\sigma(\tilde{k}) = -2 - \tilde{k}^2$ for the same $\tilde{A}$. Stability of other stationary patterns can be investigated by means of a known property of eigenfunctions of the one-dimensional Schrödinger equation: the wave function of the ground state is nondegenerate and has no zeros [8]. Problem (2.71) always has a solution

$$\tilde{A} = \frac{\partial A}{\partial X}, \quad \sigma = 0 \tag{2.72}$$

corresponding to a homogeneous translation. In the case of kink solutions (2.67), function (2.72) has no zeros, and it is indeed the "ground state" eigenfunction. Hence, there are no solutions with $\sigma > 0$, and kink solutions are stable. In the case of spatially periodic solutions (2.68), the function (2.72) changes its sign. This means that there is another eigenfunction of (2.71) without zeros that has $\sigma > 0$. Thus, all the periodic solutions are unstable.

We find that the stability criterion selects nonperiodic stationary solutions (in some sense, the selected wavenumber is equal to zero).

2.1.3 Pattern Coarsening as Defects Dynamics

It follows from the above results that the final state should be homogeneous or contain a single defect, i.e., a domain wall. At the same time, it is clear that the decay of the quiescent state $A = 0$ (which has a diverging Lyapunov functional in an infinite region), due to spontaneous growth of spatially disordered perturbations, can produce an infinite number of domains with alternating signs of A. Such a state can be characterized by a certain *density of defects* decreasing with time or by a *mean domain length* which grows with time. Early stages of pattern evolution depend on the details of the initial state, and they are not amenable to an analytical description. Such a decription is possible, however, at late stages of the evolution, when the mean distance between domain walls becomes large compared to unity. In the latter case, it is convenient to apply the approach first used in [9] for description of soliton interaction. To the problem under consideration, that approach was applied by Kawasaki and Ohta [10].

We consider a nonstationary state of the system as a set of mutually interacting domain walls (2.67) centered at $\xi_i = \xi_i(T)$, $(i = 1, 2, \ldots)$ and slowly moving because of their interaction. Let us derive the equations of motion for centers of domain walls for a particular case of a pair of interacting domain walls (see Figure 2.3).

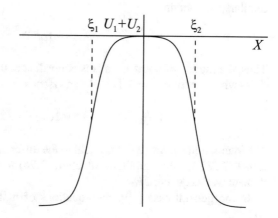

Fig. 2.3 Two interacting domain walls, the sum of U_1 and U_2 given by (2.75) and (2.76)

Near the center of the left domain wall, the solution of (2.56) can be presented in the form

$$A(X,\tau) = 1 + U(X,\tau), \tag{2.73}$$

$$U(X,\tau) = U_1(X,\tau) + U_2(X,\tau) + \sum_{n=1}^{\infty} U^{(n)}(X,\tau), \tag{2.74}$$

where

$$U_1(X,\tau) = \tanh\frac{X - \xi_1(\tau)}{\sqrt{2}} - 1, \tag{2.75}$$

$$U_2(X,\tau) = -\tanh\frac{X - \xi_2(\tau)}{\sqrt{2}} - 1 \approx -2\exp\sqrt{2}(X - \xi_2(\tau)), \tag{2.76}$$

and $U^{(n)} = O(\delta^n)$ are nth order interactive correction terms; $\delta \ll 1$ is a small parameter connected with the overlap of the kinks' tails ; ξ_i are functions of the "slow" time variable $\tau = \delta T$. For $U^{(1)}$ we obtain the equation

$$\frac{\partial^2 U^{(1)}}{\partial X^2} - \left(2 + 6U_1 + 3U_1^2\right) U^{(1)} = \frac{\partial U_1}{\partial \tau} + 6U_1 U_2 + 3U_1^2 U_2. \tag{2.77}$$

The solution $U^{(1)}$ is bounded if the right-hand side of equation (2.77) is orthogonal to the solution $\partial U_1/\partial X$ of the homogeneous (self-adjoint) problem; thus we obtain:

$$\frac{d\xi_1}{d\tau}\int_{-\infty}^{\infty} dX\left(\frac{\partial U_1}{\partial X}\right)^2 = 6\int_{-\infty}^{\infty} dX U_1 \frac{\partial U_1}{\partial X} U_2 + 3\int_{-\infty}^{\infty} dX U_1^2 \frac{\partial U_1}{\partial X} U_2. \tag{2.78}$$

Substituting the expressions (2.75) and (2.76), we find:

$$\frac{2\sqrt{2}}{3}\frac{d\xi_1}{d\tau} = 16e^{-\sqrt{2}(\xi_2 - \xi_1)}. \tag{2.79}$$

Similarly, we obtain

$$\frac{2\sqrt{2}}{3}\frac{d\xi_2}{dt} = -16e^{-\sqrt{2}(\xi_2 - \xi_1)}. \tag{2.80}$$

Thus, the motion of domain walls is equivalent to the motion of two attracting particles with mobilities $b = 3/(2\sqrt{2})$ in a viscous medium. Their interaction energy is

$$W(\xi_2 - \xi_1) = -8\sqrt{2}\exp\left[-\sqrt{2}(\xi_2 - \xi_1)\right]. \tag{2.81}$$

The interaction energy (2.81) is equal to the difference between the Lyapunov functional (2.58) for the amplitude function (2.74) and the "energy" of two isolated domain walls equal to $2F_0$.

In the general case [10], the equations of motion for the centers of domain walls are

$$\frac{2\sqrt{2}}{3}\frac{d\xi_i}{d\tau} = -\frac{\partial U}{\partial \xi_i}, \tag{2.82}$$

where

$$U = \sum_i W(\xi_i - \xi_{i-1}), \tag{2.83}$$

$$W(\xi_i - \xi_{i-1}) = -8\sqrt{2}\exp\left[-\sqrt{2}(\xi_i - \xi_{i-1})\right]. \tag{2.84}$$

Note that $U = F - NF_0$, where N is the number of domain walls.

The system (2.82) has a family of stationary solutions

$$\xi_i = a + jl, \quad j = 0, \pm 1, \pm 2, \ldots \tag{2.85}$$

corresponding to periodic patterns (2.68) with the spatial period $2l$. Because of the attractive interaction, all these solutions are unstable with respect to an infinitesimal perturbation of the type

$$\tilde{\xi}_i = (-1)^i \xi e^{\sigma \tau} \tag{2.86}$$

corresponding to an oncoming motion of two sublattices of "kinks" and "antikinks." Substituting ansatz (2.86) into equations (2.82) linearized near the stationary solution (2.85) yields

$$\sigma = 96 \exp(-l\sqrt{2}) > 0. \tag{2.87}$$

The same result can be obtained when solving the Schrödinger equation (2.71) directly by means of the perturbation theory, but checking the stability of the "mechanical system" (2.82) is certainly much simpler.

Because of the exponential interaction law, only the attraction to the nearest neighbor is important for each "particle." Thus we consider the interaction of a pair of domain walls governed by the equations (2.79) and (2.80). Setting $l = \xi_2 - \xi_1$, we find that the distance between domain walls is governed by the equation

$$\frac{2\sqrt{2}}{3}\frac{dl}{d\tau} = -32e^{-l\sqrt{2}}. \tag{2.88}$$

If $l(0) = l_0 \gg 1$, the solution is

$$l = l_0 + \frac{1}{\sqrt{2}}\ln\left(1 - 48e^{-\sqrt{2}l_0}\tau\right). \tag{2.89}$$

The distance between two domain walls becomes $O(1)$ at

$$\tau \sim t_0 = \frac{1}{48}e^{\sqrt{2}l_0} + O(1). \tag{2.90}$$

Finally, the domain walls reach the distance of order $O(1)$ and annihilate. Equation (2.88) does not describe that last stage of the domain wall evolution, but it is clear that the duration of that stage is of $O(1)$; thus it is much less than t_0.

Thus, (2.90) gives an estimate of the time necessary for annihilation of domain walls originally separated by the distance l_0. Vice versa, for any $\tau \gg 1$, only the domain walls with the original separation greater than $\sim \ln(48\tau)/\sqrt{2}$ can survive. That means that *coarsening* takes place. In a large but finite system with the length l_0, the formula (2.90) gives the evolution time necessary for reaching a final steady state.

Thus, late stages of evolution of the system toward a stable stationary state, when the system contains domain walls on large distances from each other, can be described as an exponentially slow motion of domain walls and their annihilation.

2.1.4 Stationary Patterns in the Presence of Inhomogeneities

Since the interaction between domain walls at large distance is exponentially weak, even a weak external factor can substantially influence the process of coarsening. As an example, we consider the action of a long-wavelength spatially periodic modulation of the linear growth coefficient (local Rayleigh number) that can be caused, for example, by inhomogeneities in the temperature distribution on the circumference of a horizontal cylinder. Modified equation (2.56) can be written then as

$$\frac{\partial A}{\partial T} = \frac{\partial^2 A}{\partial X^2} + [1 + qf(X)]A - A^3, \qquad (2.91)$$

where $q > 0$ is a parameter proportional to the modulation amplitude and $f(X)$ is a periodic function with the period l_0.

Stationary configurations for $A(X)$ satisfy

$$\frac{d^2 A}{dX^2} + [1 + qf(X)]A - A^3 = 0. \qquad (2.92)$$

This is equivalent to the equation of a nonlinear oscillator with a periodic parametric forcing (the coordinate X plays the role of time). The Lagrange function of this oscillator coincides with the density of the Lyapunov functional

$$L = \frac{1}{2}\left(\frac{\partial A}{\partial X}\right)^2 + \frac{1}{4}(A^2 - 1)^2 + \frac{1}{2}qf(X)(A^2 - 1). \qquad (2.93)$$

Equation (2.91) defines a reversible mapping of the plane upon itself:

$$u_{n+1} = F(u_n) \qquad (2.94)$$

where u_n is a vector with the components $A(X_n)$, $\partial A(X_n)/\partial X$, and $X_n = X_0 + nl_0$ (X_0 is a real constant, n is an integer).

The problem (2.94) is generally non-integrable and may possess the following types of orbits (see, e.g., [11]):

1. *Periodic orbits*: $u_{n+N} = u_n$ for a certain $N \geq 1$. One can divide these orbits into two classes: elliptic and hyperbolic orbits. This type of orbits corresponds to spatially periodic *commensurate patterns* [12].
2. *Homoclinic* and *heteroclinic* orbits which are double asymptotic to periodic ones. These orbits describe patterns approaching the commensurate ones as $X \to \pm\infty$, but containing some defects at finite X.
3. *Quasiperiodic orbits* corresponding to incommensurate structures that arise because of the action of the external l_0-periodic perturbation on a $2l$-periodic pattern in the case of incommensurate periods. If the conditions of the KAM theorem are satisfied [11], the orbits of this type cover a certain continuous closed curve that is topologically equivalent to the circle and is mapped on itself by the mapping (2.94) (the winding number is irrational). Besides, there exist quasiperiodic orbits covering a finite number of disjoined closed curves. Beyond a certain critical value of q, each of these curves loses its smoothness [13].
4. *Chaotic* orbits.

Of this variety of solutions, we have to choose *stable* solutions satisfying the condition $\delta^2 F \geq 0$.

We shall restrict ourselves to patterns that can be represented as a set of distant domain walls centered at $X = \xi_i$. Similarly to the previous subsection, we can derive equations valid for small $\delta = O(e^{-l_0\sqrt{2}})$ that describe the motion of domain walls, taking into account both the energy of their interaction and the energy corresponding to the action of inhomogeneities on the domain walls:

$$U = \sum_i V(\xi_i) + \sum_i W(\xi_i - \xi_{i-1}), \tag{2.95}$$

where $W(\xi_i - \xi_{i-1})$ is given by equation (2.84), and

$$V(\xi_i) = \frac{q}{2} \int_{-\infty}^{\infty} dX f(X) \cosh^{-2} \frac{\xi_i - X}{\sqrt{2}}. \tag{2.96}$$

Thus, the problem of finding stationary solutions of (2.91) is equivalent to finding equilibrium configurations of a particle chain interacting according to the law (2.84) and put under the action of the external potential (2.96). This model resembles the well-known Frenkel–Kontorova model (see, e.g., [13]). The most essential difference is the *exponential decay* of the interaction force at large distances that gives rise to a "rupture" of chains.

In the domain wall approximation, the abovementioned classes of orbits have a simple interpretation. $2l_0$-periodic solutions of the problem (2.92) with a sign-conserving function $A(X)$ correspond to states without domain walls. Other commensurate patterns are formed by periodic sequences of domain walls. Let us consider a particular type of commensurate patterns corresponding to equidistant chains of domain walls separated by a distance $l = ml_0$ (where m is an integer) situated in minima of the potential (2.96). It is easy to find the growth rate of the mode (2.86) for such a pattern, taking into account the interaction between the domain wall and the inhomogeneity. For the sake of simplicity, we assume

$$f(X) = -\sum_n \delta(X - nl_0),\tag{2.97}$$

which corresponds to a periodic sequence of short "inactive" segments (with a low Rayleigh number). Then

$$V(\xi_i) = -\frac{q}{2}\sum_n \cosh^{-2}\frac{\xi_i - nl_0}{\sqrt{2}}.\tag{2.98}$$

Hence, the inhomogeneities attract domain walls of both signs. We find

$$\sigma = 96e^{-l\sqrt{2}} - \frac{3}{4\sqrt{2}}q.\tag{2.99}$$

Thus, a periodic pattern with equidistant domain walls is stable with respect to small perturbations if

$$q > q_\infty = 128\sqrt{2}e^{-l\sqrt{2}} \approx 180e^{-l\sqrt{2}}.\tag{2.100}$$

Stable periodic configurations do not necessarily contain equidistant chains of domain walls. Generally, such configurations are characterized by a certain set of integers $(m_1, m_2, \ldots, m_N)$ defining the distances $l_0 m_j$, $j = 1, \ldots, N$ between potential minima holding domain walls, different for the given pattern. Because of the exponential dependence of the interaction force on the distance, it is sufficient to consider the attraction between pairs of *nearest* particles. It is clear that the threshold of instability is determined by a configuration of two nearest neighbors held by the minima of the potential at a distance $l \approx l_0 \min(m_n)$. Because of their attraction, both particles are displaced with respect to those minima by some distances $\pm X$. It is easy to show that all other particles of the given configuration, which are situated at distances larger than l from other particles, are shifted from the minima of the potential by exponentially small distances only. The energy of this two-particle configuration is

$$U_2 = -q\cosh^{-2}\frac{X}{\sqrt{2}} - 8\sqrt{2}e^{-l\sqrt{2}}e^{2\sqrt{2}X}.\tag{2.101}$$

Using this expression we find that the considered configuration has two equilibrium states (one of them is stable), if

$$q > q_2 = 16\sqrt{2}\frac{1+t}{t(1-t)^3}e^{-l\sqrt{2}} \approx 187\sqrt{2}e^{-l\sqrt{2}} \approx 264e^{-l\sqrt{2}},\tag{2.102}$$

where $t = (\sqrt{7}-2)/3$.

If m_n equals to the minimum value for some $k > 2$ neighboring numbers $n, n+1, \ldots, n+k-1$, the existence and stability conditions for such a configuration should be modified. The corresponding stability threshold values q_k are between q_∞ and q_2. For instance, the configuration containing three particles situated in vicinities of potential minima at distances l has the energy

$$U_3 = -\frac{q}{2} \sum_{i=1}^{3} \cosh^{-2} \frac{X_i}{\sqrt{2}} - 8\sqrt{2} e^{-l\sqrt{2}} (e^{-(X_2-X_1)\sqrt{2}} + e^{-(X_3-X_2)\sqrt{2}}), \qquad (2.103)$$

where X_i is the displacement of the ith particle with respect to the corresponding potential minimum. The equilibrium state $X_1 = -X_3, X_2 = 0$ exists at $q > q'_3 = 108\sqrt{2} e^{-l\sqrt{2}} \approx 152 e^{-l\sqrt{2}}$. However, this state is unstable in a certain region $q'_3 < q < q_3$ with respect to a displacement of even- and odd-numbered particles in opposite directions (similarly to a pattern containing an infinite number of equidistant particles). The stability threshold can be found as a solution of an algebraic equation of the fifth degree and is approximately equal to

$$q_3 \approx 197 e^{-l\sqrt{2}}. \qquad (2.104)$$

In any case, a periodic configuration of domain walls situated near the potential minima is stable at $q > q_2(l)$ if distances between domain walls are not less than l.

Homoclinic and heteroclinic orbits describe irregular sequences of domain walls situated between semi-infinite regions where domain walls are either absent or form spatially periodic patterns. Chaotic orbits describe irregular sequences of domain walls, and quasiperiodic orbits correspond to sequences with quasiperiodic ordering. It is obvious that due to the exponential decay of interaction between domain walls, the existence and stability of any configuration are determined by the minimum distance between neighboring particles, exactly like in the case of periodic patterns. Thus, at $q > q_2(l)$, not only periodic but *any* configurations with the minimum interparticle distance larger than l is stable.

Note that quasiperiodic orbits on smooth KAM tori are unstable, because the neutral disturbance $\tilde{A}$ corresponding to an infinitesimal change of the parameter characterizing the position of the orbit on the torus changes its sign, and hence it does not correspond to the largest eigenvalue of the linearized problem. However, a quasiperiodic orbit can become stable when the torus smoothness is broken [13].

In a similar way, we can consider another typical case, described by the inhomogeneous equation

$$\frac{\partial A}{\partial T} = \frac{\partial^2 A}{\partial X^2} + A - A^3 + qf(X), \qquad (2.105)$$

where the function $f(X)$ is again periodic with the period l_0. For the problem of convection in a horizontal cylinder, this corresponds to the case where the boundary temperature modulation depends on the transversal horizontal coordinate rather than on the vertical coordinate. In that case the local Rayleigh number is constant, and spatially periodic lateral heating generates a periodic convection flow even in the subcritical region.

The density of the Lyapunov functional is

$$L = \frac{1}{2} \left(\frac{\partial A}{\partial X} \right)^2 + \frac{1}{4}(A^2 - 1)^2 - qf(X)A. \qquad (2.106)$$

The equation of this type was investigated by Coullet et al. [14]. That was the first study bringing attention to a possibility of generating spatial chaos by pinning domain walls on inhomogeneities.

Let $f(X) = dF(X)/dX$, where $F(X)$ is a non-monotonic bounded function. That means that the external perturbation $f(X)$ is not sign-conserving, and hence it does not impose a preferred sign of the amplitude (or a preferred direction of the fluid rotation). The force exerted on the domain wall by the inhomogeneity can be calculated from the potential

$$V(\xi) = \pm \frac{q}{\sqrt{2}} \int_{-\infty}^{\infty} dX F(X) \cosh^{-2} \frac{X - \xi}{\sqrt{2}}. \tag{2.107}$$

It is obvious that a kink prefers to stay in the vicinity of a minimum of the function $F(X)$, while an antikink prefers maxima of $F(X)$.

For $q \sim \exp(-\sqrt{2}l)$ there is no essential qualitative difference between the potentials (2.96) and (2.107) from the point of view of the general picture of existence and stability of different types of configurations. The situation is different at $q \sim l^{-1}$. For (2.91), a state with domain walls always has larger energy than a state with a constant sign of A. It should be noted, however, that at large q a domain wall situated in a region of the potential minimum, where $1 + qf(X)$ is negative, has an exponentially small energy. In the case of (2.105), at q larger than a certain critical value q_0, configurations with the sign of $A(X)$ coinciding with the sign of $f(X)$ become preferable. Such configurations contain domain walls situated in the points where $f(X)$ changes its sign. For instance, let $f(X) < 0$ in the interval $l_1 - l_0 < X < 0$, and $f(X) > 0$ in the interval $0 < X < l_1$, and assume $\int_{l_1 - l_0}^{l_1} dX f(X) = 0$. Neglecting the interaction between domain walls, we find that the difference $\Delta \bar{L}$ between the mean density of the Lyapunov functional of a "multidomain" state with coinciding signs of $A(X)$ and $f(X)$, and that of a "unidomain" state with a constant sign of $A(X)$, is

$$\Delta \bar{L} = \frac{1}{l_0} \left(2F_0 - 2q \int_0^{l_1} dX f(X) \right), \tag{2.108}$$

where $F_0 = 2\sqrt{2}/3$ is the energy of the domain wall. Hence, the value

$$q_0 = \frac{F_0}{\int_0^{l_1} dX f(X)} \tag{2.109}$$

corresponds to the point of a phase transition from a defect-free state to a state with the maximum number of domain walls. At $q > q_0$, islands of the unidomain state enhancing the energy of the system play the role of defects. Finally, at some $q > q_1$, pairs of solutions corresponding to unidomain states of both signs disappear. After that, there are no stable states except a multidomain state with a period determined by the external perturbation.

2.1.5 Higher-Order Corrections

The amplitude equation (2.56) describes the phenomena with the characteristic time scale $O(\varepsilon^{-2})$. Now we discuss what happens on the time scale $O(\varepsilon^{-3})$. To that end, we find out, without straightforward derivations, what kinds of nonlinear terms are allowed in the next order in ε. First, we have to note that the problem (2.12)–(2.17) is symmetric with respect to the transformation

$$x' = -x, \ u' = -u \tag{2.110}$$

that corresponds, according to the definition of the amplitude function, to the transformation

$$X' = -X \tag{2.111}$$

in the amplitude equation. The symmetry property

$$y' = -y, \ z' = -z, \ w' = -w, \ \Theta' = -\Theta, \tag{2.112}$$

corresponding to the transformation

$$A' = -A \tag{2.113}$$

in the amplitude equation exists only if *the shape of the boundary Γ has the corresponding symmetry*. The lowest-order amplitude equation (2.56) is symmetric with respect to both abovementioned transformations by chance, because of the specific feature (2.39) of the nonlinear terms. In the next orders in ε, the amplitude equation should be symmetric with respect to the inversion of the coordinate X, but generally not symmetric with respect to the transformation (2.113). Using the representation for the amplitude function like (1.58), we can write an amplitude equation containing terms of two different orders in ε:

$$\frac{\partial A}{\partial T} = \frac{\partial^2 A}{\partial X^2} + A - A^3 + \varepsilon \left[\beta_1 A^4 + \beta_2 A \frac{\partial^2 A}{\partial X^2} + \beta_3 \left(\frac{\partial A}{\partial X} \right)^2 \right]. \tag{2.114}$$

The coefficients β_i are nonzero, if the shape of the cross section of the cylinder and hence the eigenfunctions of the linear problem are not symmetric to the transformation (2.112). It is worth noting that (2.114) is not potential and cannot be written in the form (2.63). Because of the absence of symmetry (2.113), the domains with $A = \pm 1 + O(\varepsilon)$ are not equivalent, and the solitary domain walls $A = \pm \tanh \frac{X-\xi}{\sqrt{2}} + \varepsilon A_1$ move with velocities $O(\varepsilon)$ that can be calculated from the solvability condition of the equation for A_1:

$$\frac{2\sqrt{2}}{3} \frac{d\xi}{dt} = \pm \varepsilon \beta, \ \beta = -\frac{2}{15} \beta_1 + \frac{4}{15} \beta_2 - \frac{8}{15} \beta_3. \tag{2.115}$$

In the absence of inhomogeneities, the "disadvantageous" domains are forced out completely, and the motion with $A = const$ is established in the whole region. In the presence of inhomogeneities, the domain walls are pinned by sufficiently strong inhomogeneities (in the case (2.97), as $q > 2\varepsilon\beta$).

2.1.6 Convection in a Non-Boussinesq Fluid

In Section 2.1.1, we have obtained ACE (2.56) rather than most generic FKPPE (1.64). The ACE has a symmetry $A \to -A$; the clockwise and counterclockwise motions of the fluid are fully equivalent. What is the origin of that symmetry?

First of all, system of equations (2.12)–(2.17) is invariant under the reflection $y \to -y$, $w_y \to -w_y$. In other words, for any flow, the motion which is observed in a vertical mirror is also allowed by the equations. Therefore, if the shape of the cylinder cross section is symmetric with respect to the reflection transformation $y \to -y$, the symmetry $A \to -A$ is prescribed by that symmetry, and the appearance of a term A^2 in the amplitude equation is not possible.

However, in the derivation presented above, we did not use the symmetry of the cylinder. The disappearance of the quadratic term in the amplitude equation is due to identity (2.39), which is caused by a specific feature of the nonlinearities in equations (2.12), (2.13), and (2.15). If the system has any other nonlinearities, the quadratic term in the amplitude equation is possible.

As an example, consider the convection in a non-Boussinesq fluid with the viscosity linearly depending on the temperature, $v(T) = v_0 + v_1 T$. That leads to an additional term

$$\bar{v}\frac{\partial}{\partial x_j}\left[T\left(\frac{\partial v_j}{\partial x_i} + \frac{\partial v_i}{\partial x_j}\right)\right] \quad (i, j = 1, 2, 3), \tag{2.116}$$

in the right-hand side of the equation for v_i; here $\bar{v} = v_1 A_* d_* / v_0$.

The Case of a Weak Non-Boussinesq Effect

If $\bar{v}$ is small, $\bar{v} = \varepsilon \bar{v}_1$, then an additional nonlinear term

$$\bar{v}_1 \frac{\partial}{\partial x_j}\left[\Theta_0\left(\frac{\partial w_{0j}}{\partial x_i} + \frac{\partial w_{0i}}{\partial x_j}\right)\right] \quad (i, j = 1, 2)$$

appears in the right-hand side of the equation for the transverse velocity in the second order in ε. Due to the modification of (2.45), the amplitude equation (2.49) includes an additional term $\bar{\beta}a^2$, where

$$\bar{\beta} = -\bar{v}_1 \left\langle \frac{\partial \tilde{w}_{0i}}{\partial x_j}\left(\frac{\partial \tilde{w}_{0j}}{\partial x_i} + \frac{\partial \tilde{w}_{0i}}{\partial x_j}\right)\tilde{\Theta}_0 \right\rangle \quad (i, j = 1, 2). \tag{2.117}$$

By rescaling, we get a modified ACE

$$\frac{\partial A}{\partial T} = \frac{\partial^2 A}{\partial X^2} + A + \beta A^2 - A^3, \tag{2.118}$$

where $\beta = \bar{\beta}\sqrt{-\alpha_4/\alpha_2^3}$.

The amplitudes corresponding to clockwise and counterclockwise rotations,

$$A_\pm = \frac{\beta}{2} \pm \sqrt{1 + \frac{\beta^2}{4}}$$

are not equal any further.

Equation (2.118) has Lyapunov functional density

$$L(A) = \frac{1}{2}\left(\frac{\partial A}{\partial X}\right)^2 + \frac{1}{4}(A^2 - 1)^2 - \frac{\beta}{3}A^3. \tag{2.119}$$

Depending on the sign of β, either A_+ or A_- provides the minimum value of the Lyapunov functional. The domain wall ("kink") between semi-infinite domains with different directions of rotation is not motionless, as in the case of the ACE (2.56), but moves with a definite velocity $V = V(\beta)$. The shape of the domain wall $A = A(z)$, $z = x - VT$, is governed by the equation

$$\frac{d^2 A}{dz^2} + V\frac{dA}{dz} + A + \beta A^2 - A^3 = 0. \tag{2.120}$$

Let us choose

$$A(-\infty) = A_+, \ A(+\infty) = A_-. \tag{2.121}$$

Note that (2.120) is formally equivalent to the equation of motion for a particle in the potential

$$U(A) = -\frac{1}{4}(A^2 - 1)^2 + \frac{\beta}{3}A^3 \tag{2.122}$$

(see Figure 2.4) with the friction coefficient V. Note that $L = (1/2)(\partial A/\partial X)^2 - U(A)$; therefore, the motion with the lower value of the Lyapunov functional density corresponds to a higher value of U.

The heteroclinic solution for problems (2.120) and (2.121) corresponds to the trajectory which starts on one peak of the potential, $A = A_+$, at "time" $z \to -\infty$ and reaches the other peak of the potential, $A = A_-$, at $z \to \infty$. Obviously, in the case $U(A_+) > U(A_-)$, the "friction coefficient" V has to be positive, i.e., the domain with $A = A_+$, which corresponds to the lower value of the Lyapunov functional density L, ousts the domain with $A = A_-$, which corresponds to a higher value of the Lyapunov functional density. Integrating (2.120) over z from $-\infty$ to $+\infty$, one finds that

$$V = \frac{U(A_+) - U(A_-)}{\int_{-\infty}^{\infty} [A'(z)]^2 \, dz}. \tag{2.123}$$

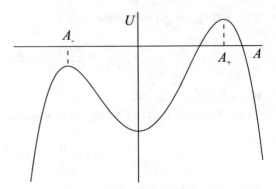

Fig. 2.4 Potential $U(A)$ given by (2.122)

In the case $|\beta| \ll 1$, the domain wall velocity can be found analytically. One can construct the asymptotic expansion for the domain wall solution in the form

$$A = A_0 + \beta A_1 + \dots, \quad V = \beta V_1 + \dots \tag{2.124}$$

and substitute (2.124) into (2.120). At the leading order, we find

$$A_0'' + A_0 - A_0^3 = 0$$

and hence

$$A_0 = \tanh \frac{z}{\sqrt{2}}.$$

At the next order, we obtain equation

$$A_1'' + A_1(1 - 3A_0^2) = -V_1 A_0' - \beta_1 A_0^2;$$

its solvability condition (orthogonality to A_0') yields

$$V_1 = -\beta_1 \frac{\int_{-\infty}^{\infty} A_0^2 A_0' dz}{\int_{-\infty}^{\infty} \left[A_0'\right]^2 dz} = \frac{\beta_1}{\sqrt{2}}.$$

Finally, the solution $A = A_+$ is established in the whole finite region.

2.1.6.1 The Case of a Strong Non-Boussinesq Effect

If $\bar{v}$ is not small, scaling (2.11) has to be reconsidered. Because of the stronger nonlinearity, the correct scaling is

$$T = \varepsilon^2 \Theta, \quad v_y \mathbf{e}_y + v_z \mathbf{e}_z = \varepsilon^2 \mathbf{w}, \quad p = \varepsilon^2 \Pi, \quad v_x = \varepsilon^3 u. \tag{2.125}$$

The solution of the zeroth order problem (2.23)–(2.25) is not changed, but at the second order in ε, (2.116) is the only nonlinear term, and the solvability condition is just

$$\alpha_1 \frac{\partial a}{\partial t_2} = \alpha_2 a + \alpha_3 \frac{\partial^2 a}{\partial x_1^2} + \bar{\beta} a^2,$$

where $\bar{\beta}$ is given by (2.117).

Thus, the most generic amplitude equation, FKPPE (1.64), is obtained. By the obvious rescaling, it can be transformed to the standard form,

$$\frac{\partial A}{\partial T} = \frac{\partial^2 A}{\partial X^2} + A + \beta A^2. \tag{2.126}$$

Equation (2.126) has been derived formerly in the context of biological applications [15, 16], where $A \geq 0$ and $\beta < 0$ by definition. In that case the problem has a global solution. However, in our case there is no reason to discard solutions with sign$A = $ signβ, which blow up during a finite time.

Thus, we have got an example of a generic problem which cannot be fully described by the asymptotic methods. Actually, the initial problem may have a global solution, but it is not close to the quiescent state even near the threshold. Another example of that kind is the Sivashinsky equation (1.81), which is discussed in Section 3.2.2.

In conclusion, the onset of the free convection in a horizontal cylinder considered above is a typical example of a type III monotonic instability. The generic equation for that kind of instability in the presence of the $x \to -x$ symmetry is the Fisher–Kolmogorov–Petrovskii–Piskunov equation (FKPPE) which is plagued with the finite-time blowup. However, in the case of a Boussinesq fluid, due to specific features of nonlinearities, the generic transcritical bifurcation is replaced by a pitchfork bifurcation, and, correspondingly, the FKPPE is replaced by the Allen–Cahn equation. The latter equation, which serves as a paradigmatic example of coarsening and inhomogeneity-induced pattern formation, has to be modified if the $x \to -x$ symmetry equation is violated, e.g., because of a flow along the cylinder. That case is considered in the next session.

2.2 Mixed Convection in a Horizontal Cylinder

We assume now that there is a longitudinal pressure gradient $dp_0/dx = Re$ (for the non-dimensionalization, we use the same units as in Section 2.1) generating a parallel longitudinal flow $\mathbf{v} = ReU_0(y,z)\mathbf{e}_x$ along the horizontal cylinder. In the absence of convection, the velocity profile can be calculated from the equation

$$\nabla_{\perp}^2 U_0 = 1; \tag{2.127}$$
$$U_0 = 0 \text{ at } \Gamma. \tag{2.128}$$

It is obvious that the function U_0 coincides with the function $\bar{u}_0$ defined by problem (2.30). This parallel flow influences neither the equilibrium temperature field nor the pressure distribution across the cylinder.

In the presence of the longitudinal flow, the evolution of finite-amplitude disturbances is governed by the following system of equations instead of (2.1)–(2.4):

$$\frac{1}{P}\left[\frac{\partial \mathbf{v}}{\partial t} + ReU_0 \frac{\partial \mathbf{v}}{\partial x} + Re(\mathbf{v} \cdot \nabla U_0)\mathbf{e}_x + (\mathbf{v} \cdot \nabla)\mathbf{v}\right] = -\nabla p + \Delta \mathbf{v} + RT\mathbf{e}_z, \qquad (2.129)$$

$$\nabla \cdot \mathbf{v} = 0, \qquad (2.130)$$

$$\frac{\partial T}{\partial t} + ReU_0 \frac{\partial T}{\partial x} + \mathbf{v} \cdot \nabla T - \mathbf{v} \cdot \mathbf{e}_z = \Delta T; \qquad (2.131)$$

$$\mathbf{v} = 0, \; T = 0 \text{ at } \Gamma. \qquad (2.132)$$

For the sake of simplicity, we will fix the longitudinal flux of the fluid:

$$\langle v_x(y,z)\rangle = 0 \qquad (2.133)$$

(an alternative possibility is to fix the mean pressure gradient).

It is obvious that because of the advection by the basic flow, a spatially periodic long-wavelength disturbance (with a wavenumber $k \neq 0$) propagates with some phase velocity, thus generating a frequency of order k. That is why the description of long-wavelength convection motions characterized by the spatial variable $x_1 = \varepsilon x$ involves now two time scales $t_1 = \varepsilon t$ and $t_2 = \varepsilon^2 t$. However, the dependence of the solution on t_1 can be eliminated, if the reference frame moving with the group velocity of waves is used.

What is the amplitude equation, which governs the evolution of nonlinear disturbances in that case? Because the longitudinal flow obviously violates the symmetry $x \to -x$, the amplitude equation should have the form (1.65) (see Chapter 1).

In the next subsection, we present a direct derivation of the amplitude equation. That subsection can be omitted by the reader not interested in technical details.

2.2.1 Derivation of the Amplitude Equation

Because of the term $Re\mathbf{v} \cdot \nabla U_0 \mathbf{e}_x$ in (2.129), the disturbances of the longitudinal component of the velocity and the longitudinal pressure gradient are of the same order as the disturbance of $\mathbf{v}_\perp$. Hence, instead of (2.11), we assume:

$$T = \varepsilon\Theta, \; v_y\mathbf{e}_y + v_z\mathbf{e}_z = \varepsilon\mathbf{w}, \; v_x = \varepsilon u. \qquad (2.134)$$

We split p from the very beginning into two parts:

$$p = G(x_1, t_1, t_2) + \varepsilon\Pi(x_1, y, z, t_1, t_2), \qquad (2.135)$$

where $G(x_1,t_1,t_2)$ does not depend on y and z (only $\partial G/\partial x_1$ but not G itself is bounded as $x_1 \to \pm\infty$), and $\langle \Pi(x_1,y,z,t_1,t_2) \rangle = 0$.

Equations (2.12)–(2.17) are replaced by the following system of equations:

$$\frac{\varepsilon}{P}\left[\frac{\partial \mathbf{w}}{\partial t_1} + \varepsilon\frac{\partial \mathbf{w}}{\partial t_2} + ReU_0\frac{\partial \mathbf{w}}{\partial x_1} + (\mathbf{w}\cdot\nabla_\perp)\mathbf{w} + \varepsilon u\frac{\partial \mathbf{w}}{\partial x_1}\right]$$
$$= -\nabla_\perp\Pi + \nabla_\perp^2\mathbf{w} + \varepsilon^2\frac{\partial^2\mathbf{w}}{\partial x_1^2} + (R_c + \varepsilon^2)\Theta\mathbf{e}_z, \quad (2.136)$$

$$\frac{\varepsilon}{P}\left(\frac{\partial u}{\partial t_1} + \varepsilon\frac{\partial u}{\partial t_2} + ReU_0\frac{\partial u}{\partial x_1} + Re\mathbf{w}_0\cdot\nabla_\perp U_0 + \mathbf{w}\cdot\nabla_\perp u + \varepsilon u\frac{\partial u}{\partial x_1}\right)$$
$$= -\frac{\partial G}{\partial x_1} - \varepsilon\frac{\partial\Pi}{\partial x_1} + \nabla_\perp^2 u + \varepsilon^2\frac{\partial^2 u}{\partial x_1^2}, \quad (2.137)$$

$$\nabla_\perp\cdot\mathbf{w} + \varepsilon^2\frac{\partial u}{\partial x_1} = 0, \quad (2.138)$$

$$\varepsilon\frac{\partial\Theta}{\partial t_1} + \varepsilon^2\frac{\partial\Theta}{\partial t_2} + Re\varepsilon U_0\frac{\partial\Theta}{\partial x_1} + \varepsilon\mathbf{w}\cdot\nabla_\perp\Theta + \varepsilon^2 u\frac{\partial\Theta}{\partial x_1} - w_z$$
$$= \nabla_\perp^2\Theta + \varepsilon^2\frac{\partial^2\Theta}{\partial x_1^2}; \quad (2.139)$$

$$\mathbf{w} = 0,\; u = 0,\; \Theta = 0 \text{ at } \Gamma; \quad (2.140)$$

$$\langle u \rangle = 0. \quad (2.141)$$

As usual, we introduce expansions $f = f_0 + \varepsilon f_1 + \varepsilon^2 f_2 + \dots$ for all the variables.

At the zeroth order, the system of equations can be split into two parts. The equations for w_0, Π_0, and Θ_0 form an eigenvalue boundary problem that is absolutely identical to (2.20)–(2.22), and its solution can be written in the form:

$$\mathbf{w}_0 = a_0(x_1,t_1,t_2)\tilde{\mathbf{w}}_0(y,z),\quad \Theta_0 = a_0(x_1,t_1,t_2)\tilde{\Theta}_0(y,z), \quad (2.142)$$
$$\Pi_0 = a_0(x_1,t_1,t_2)\tilde{\Pi}_0(y,z). \quad (2.143)$$

The separate boundary problem

$$\nabla_\perp^2 u_0 - \frac{\partial G_0}{\partial x_1} = \frac{Re}{P}\mathbf{w}_0\cdot\nabla_\perp U_0;\; u_0 = 0 \text{ at } \Gamma \quad (2.144)$$

determines the solution

$$u_0 = Re\, a_0(x_1,t_1,t_2)\hat{u}_0(y,z) + \frac{\partial G_0}{\partial x_1}\bar{u}_0(y,z), \quad (2.145)$$

where

$$\nabla_\perp^2\hat{u}_0 = \frac{1}{P}\mathbf{w}_0\cdot\nabla_\perp U_0,\; \nabla_\perp^2\bar{u}_0 = 1;\; \hat{u}_0 = 0,\; \bar{u}_0 = 0 \text{ at } \Gamma. \quad (2.146)$$

The longitudinal pressure gradient $\partial G_0/\partial x_1$ is coupled to a_0 because of zero-flux condition (2.141) at the zeroth order; this, in turn, allows us to finalize the calculation of u_0:

$$\frac{\partial G_0}{\partial x_1} = -Re a_0(x_1,t_1,t_2)\frac{\langle \hat{u}_0 \rangle}{\langle \bar{u}_0 \rangle}, \tag{2.147}$$

$$u_0 = a_0(x_1,t_1,t_2)Re\tilde{u}_0(y,z), \quad \tilde{u}_0 = \hat{u}_0 - \bar{u}_0\frac{\langle \hat{u}_0 \rangle}{\langle \bar{u}_0 \rangle}. \tag{2.148}$$

At the first order in ε, we obtain a linear nonhomogeneous boundary value problem for $\mathbf{w}_1, \Theta_1, \Pi_1$:

$$-\nabla_\perp \Pi_1 + \nabla_\perp^2 \mathbf{w}_1 + R_c \Theta_1 \mathbf{e}_z = \frac{1}{P}\left[\frac{\partial \mathbf{w}_0}{\partial t_1} + Re U_0 \frac{\partial \mathbf{w}_0}{\partial x_1} + (\mathbf{w}_0 \cdot \nabla_\perp)\mathbf{w}_0\right], \tag{2.149}$$

$$\nabla_\perp \cdot \mathbf{w}_1 + \frac{\partial u_0}{\partial x_1} = 0, \tag{2.150}$$

$$\nabla_\perp^2 \Theta_1 + w_{1z} = \frac{\Theta_0}{\partial t_1} + Re U_0 \frac{\partial \Theta_0}{\partial x_1} + \mathbf{w}_0 \cdot \nabla_\perp \Theta_0; \tag{2.151}$$

$$\mathbf{w}_1 = 0, \ \Theta_1 = 0 \text{ at } \Gamma. \tag{2.152}$$

The solvability condition to this problem is (see Section 2.1):

$$\alpha_1 \frac{\partial a_0}{\partial t_1} + \alpha_5 \frac{\partial a_0}{\partial x_1} = 0, \tag{2.153}$$

where α_1 is described by formula (2.50),

$$\alpha_5 = \frac{Re}{P}\langle |\tilde{\mathbf{w}}_0|^2 U_0 \rangle - \frac{Re}{P}\langle \tilde{\Pi}_0 \tilde{u}_0 \rangle + Re R_c \langle \tilde{\Theta}_0^2 U_0 \rangle. \tag{2.154}$$

To eliminate the dependence of the amplitude function on t_1, we perform the transformation to the reference frame moving with the group velocity of waves

$$Re\, c = \frac{\alpha_5}{\alpha_1}, \tag{2.155}$$

introducing the new variable

$$\tilde{x}_1 = x_1 - Re\, c\, t_1. \tag{2.156}$$

In this reference frame, the structure of the solution is:

$$\mathbf{w}_1 = a_0^2(\tilde{x}_1,t_2)\tilde{\mathbf{w}}_1(y,z) + Re\frac{\partial a_0}{\partial \tilde{x}_1}(\tilde{x}_1,t_2)\hat{\mathbf{w}}_1(y,z) + a_1(x_1,t_2)\tilde{\mathbf{w}}_0(y,z), \tag{2.157}$$

$$\Theta_1 = a_0^2(\tilde{x}_1,t_2)\tilde{\Theta}_1(y,z) + Re\frac{\partial a_0}{\partial \tilde{x}_1}(\tilde{x}_1,t_2)\hat{\Theta}_1(y,z) + a_1(x_1,t_2)\tilde{\Theta}_0(y,z), \tag{2.158}$$

$$\Pi_1 = a_0^2(\tilde{x}_1,t_2)\tilde{\Pi}_1(y,z) + Re\frac{\partial a_0}{\partial \tilde{x}_1}(\tilde{x}_1,t_2)\hat{\Pi}_1(y,z) + a_1(x_1,t_2)\tilde{\Pi}_0(y,z). \tag{2.159}$$

The boundary value problem

$$\nabla_\perp^2 u_1 - \frac{\partial G_1}{\partial x_1} = \frac{1}{P}\left(\frac{\partial u_0}{\partial t_1} + \mathbf{w}_0 \cdot \nabla_\perp u_0\right) +$$
$$\frac{Re}{P}\left(U_0 \frac{\partial u_0}{\partial x_1} + \mathbf{w}_1 \cdot \nabla_\perp U_0\right) + \frac{\partial \Pi_0}{\partial x_1}; \quad (2.160)$$

$$u_1 = 0 \text{ at } \Gamma; \quad (2.161)$$

$$\langle u_1 \rangle = 0, \quad (2.162)$$

determines the solution u_1 with the following structure:

$$u_1 = Re a_0^2(\tilde{x}_1, t_2)\tilde{u}_1(y,z) + \frac{\partial a_0}{\partial \tilde{x}_1}(\tilde{x}_1, t_2)(Re^2 \hat{u}_1(y,z) + \bar{u}_1(y,z))$$
$$+ Re a_1(x_1, t_2)\tilde{u}_0(y,z). \quad (2.163)$$

and a similar solution for $\partial G_1/\partial x_1$.

Finally, at the second order, we obtain the following problem for $\mathbf{w}_2$, Θ_2, and Π_2:

$$-\nabla_\perp \Pi_2 + \nabla_\perp^2 \mathbf{w}_2 + R_c \Theta_2 \mathbf{e}_z = \frac{Re}{P}(U_0 - c)\frac{\partial \mathbf{w}_1}{\partial \tilde{x}_1}$$
$$+ \frac{1}{P}\left[(\mathbf{w}_0 \cdot \nabla_\perp)\mathbf{w}_1 + \mathbf{w}_1 \cdot \nabla_\perp \mathbf{w}_0 + \frac{\partial \mathbf{w}_0}{\partial t_2} + u_0 \frac{\partial \mathbf{w}_0}{\partial \tilde{x}_1}\right] - \frac{\partial^2 \mathbf{w}_0}{\partial \tilde{x}_1^2} - \Theta_0 \mathbf{e}_z, \quad (2.164)$$

$$\nabla_\perp \cdot \mathbf{w}_2 = -\frac{\partial u_1}{\partial \tilde{x}_1}, \quad (2.165)$$

$$\nabla_\perp^2 \Theta_2 + w_{2z} = Re(U_0 - c)\frac{\partial \Theta_1}{\partial \tilde{x}_1} + \mathbf{w}_0 \cdot \nabla_\perp \Theta_1 + \mathbf{w}_1 \cdot \nabla_\perp \Theta_0$$
$$+ \frac{\partial \Theta_0}{\partial t_2} + u_0 \frac{\partial \Theta_0}{\partial \tilde{x}_1} - \frac{\partial^2 \Theta_0}{\partial x_1^2}; \quad (2.166)$$

$$\mathbf{w}_2 = 0, \ \Theta_2 = 0 \text{ at } \Gamma. \quad (2.167)$$

The system (2.164)–(2.167) is solvable if

$$\alpha_1 \frac{\partial a_0}{\partial t_2} = \alpha_2 a_0 + \alpha_3 \frac{\partial^2 a_0}{\partial \tilde{x}_1^2} + \alpha_4 a_0^3 + \alpha_6 \frac{\partial(a_0^2)}{\partial \tilde{x}_1}, \quad (2.168)$$

where α_1, α_2, and α_4 are defined by formulas (2.50), (2.51), and (2.53), respectively, and

$$\alpha_3 = \left\langle |\tilde{\mathbf{w}}_0|^2 + R_c \tilde{\Theta}_0^2 + \bar{u}_1 \tilde{\Pi}_0 \right\rangle \quad (2.169)$$
$$+ Re^2 \left\langle \hat{u}_1 \tilde{\Pi}_0 - (U_0 - c)\left(\frac{1}{P}\tilde{\mathbf{w}}_0 \cdot \hat{\mathbf{w}}_1 + R_c \tilde{\Theta}_0 \hat{\Theta}_1\right)\right\rangle, \quad (2.170)$$
$$\alpha_6 = -Re\left\langle (U_0 - c)\left(\frac{1}{P}\tilde{\mathbf{w}}_0 \cdot \tilde{\mathbf{w}}_1 + R_c \tilde{\Theta}_0 \tilde{\Theta}_1\right)\right\rangle$$

$$-Re\left\langle \frac{1}{P}[(\hat{\mathbf{w}}_1 \cdot \nabla_\perp)\tilde{\mathbf{w}}_0 + (\tilde{\mathbf{w}}_0 \cdot \nabla_\perp)\hat{\mathbf{w}}_1] + R_c\left(\hat{\mathbf{w}}_1 \cdot \nabla_\perp \tilde{\Theta}_0 + \tilde{\mathbf{w}}_0 \cdot \nabla_\perp \hat{\Theta}_1\right)\right\rangle$$

$$-\frac{1}{2}\left\langle \frac{1}{P}\tilde{u}_0|\tilde{\mathbf{w}}_0|^2 + R_c\tilde{u}_0\tilde{\Theta}_0^2 - 2\tilde{u}_1\tilde{\Pi}_0\right\rangle. \quad (2.171)$$

By means of a scale transformation, we obtain the amplitude equation:

$$\frac{\partial A}{\partial \tau} = \frac{\partial^2 A}{\partial X^2} + A - A^3 + 2\alpha A \frac{\partial A}{\partial X}, \quad (2.172)$$

$$\alpha = \frac{2\alpha_6\alpha_2}{(-\alpha_3\alpha_4)^{1/2}}, \quad (2.173)$$

that can be named *Cessi–Young equation* (CYE) (see [17]).

2.2.2 Stationary Patterns

The stationary solutions $A(X)$ of the equation (2.172) satisfy the system of ordinary differential equations

$$\frac{dB}{dX} = A(A^2 - 1 - 2\alpha B), \quad \frac{dA}{dX} = B. \quad (2.174)$$

The orbits of the dynamical system (2.174) governed by the equation

$$\frac{dB}{dA} = \frac{A(A^2 - 1 - 2\alpha B)}{B} \quad (2.175)$$

are shown schematically in Figure 2.5. The structure of the set of bounded stationary solutions is quite similar to one of the ACE, but the symmetry with respect to transformations (2.111) and (2.113) is lost. However, there exists a symmetry with respect to the transformation

$$X' = -X, \ A' = -A \quad (2.176)$$

There are three fixed points that describe the parallel flow $A(X) = 0$ and spatially homogeneous spiral flows $A(X) = \pm 1$. Also, there is an infinite number of spatially periodic stationary solutions with periods $l = 2\pi/q$, $0 < q < 1$. It is clear from the symmetry properties that these solutions are odd functions of $X - \xi$ for definite values of the constant ξ; hence,

$$\int_0^{2\pi/k} A(X)dX = 0. \quad (2.177)$$

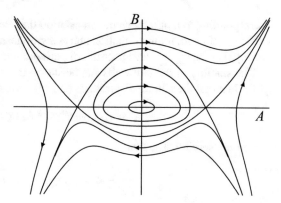

Fig. 2.5 Phase portrait for (2.174)

Finally, there are two kinds of stationary domain walls ("kink" and "antikink")

$$A_\pm(X) = \tanh \beta_\pm (X - \xi), \quad \xi = const, \tag{2.178}$$

$$\beta_\pm = \frac{1}{2}(\alpha \pm \sqrt{\alpha^2 + 2}). \tag{2.179}$$

In order to investigate the stability of a stationary solution $A(X)$, we have to consider the eigenvalue problem for disturbances $\tilde{A}(X)\exp\sigma\tau$ evolving on the background of $A(X)$:

$$\sigma\tilde{A} = \frac{d^2\tilde{A}}{dX^2} + 2\alpha\frac{d}{dX}(A\tilde{A}) + (1 - 3A^2)\tilde{A}; \tag{2.180}$$

$$|\tilde{A}| < \infty \text{ as } X \to \infty. \tag{2.181}$$

The solution $A = 0$ is unstable: $\sigma(\tilde{k}) = 1 - \tilde{k}^2$ for $\tilde{A} = \exp(i\tilde{k}X)$; the solutions $A = \pm 1$ are stable: $\sigma(\tilde{k}) = -2 - \tilde{k}^2 \pm 2i\alpha$ for $\tilde{A} = \exp(i\tilde{k}X)$. In the case of a spatially periodic solution $A(X)$, we transform the equation (2.180) to the form of a 1D Schrödinger equation by means of the transformation

$$\tilde{A}(X) = C(X)\exp\left(-\alpha\int_\xi^X A(X')dX'\right) \tag{2.182}$$

(because of property (2.177), $\int_\xi^X A(X')dX'$ is a bounded periodic function):

$$\sigma C = \frac{d^2 C}{dX^2} + \left(1 - 3A^2 + \alpha\frac{dA}{dX} - \alpha^2 A^2\right)C. \tag{2.183}$$

Now we can use the same argument as in Section 2.1. Problem (2.183) has always a neutral solution

$$C = \frac{dA}{dX}\exp\left(\alpha\int_\xi^X A(X')dX'\right), \quad \sigma = 0, \tag{2.184}$$

corresponding to a homogeneous translation that changes its sign. This implies the existence of another eigenfunction that is sign-conserving and has $\sigma > 0$. It is noteworthy that the eigenvalue spectrum of periodic stationary solutions is real.

The stability of domain walls can be shown in a similar way as the consequence of the fact that the neutral disturbance (2.184) corresponding to the solution (2.178)

$$C = \beta_\pm \cosh^{\alpha/\beta_\pm - 2} \beta_\pm (X - \xi), \tag{2.185}$$

is sign-conserving. The quantity

$$\frac{\alpha}{\beta_\pm} - 2 = -\alpha^2 - 2 \pm \sqrt{\alpha^2 + 2} \tag{2.186}$$

is negative for any α; hence, function (2.185) is bounded.

The bounded function $\tilde{A}(X)$ generates a non-bounded function C, if $\tilde{A}$ tends to zero slower than $\exp(-\alpha X)$, as $X \to \infty$. Such solutions slip out from our analysis. However, one can find from (2.183) that if $\tilde{A} \sim \exp(-\kappa X)$ as $X \to \infty$, then

$$\sigma = \kappa^2 - 2\alpha\kappa - 2 = (\kappa - \alpha)^2 - \alpha^2 - 2. \tag{2.187}$$

Hence, for $0 < \kappa < \alpha$, $\sigma < 0$, and such disturbances cannot generate an instability.

Thus, as in the case $\alpha = 0$, homogeneous states and domain walls are the only stable stationary patterns.

2.2.3 Kink Dynamics

The interaction between domain walls can be considered similarly to Section 2.1. Let $\alpha > 0$ (otherwise, we perform an appropriate change of variables). In this case, $\beta_+ > |\beta_-|$, and we have "narrow" kinks and "wide" antikinks. As in Section 2.1, we consider a kink–antikink pair that consists of a kink situated near the point $\xi = \xi_1(\tau)$ and an antikink near the point $\xi = \xi_2(\tau)$. Near the left kink, instead of (2.77), we obtain the following equation:

$$\frac{\partial^2 B^{(1)}}{\partial X^2} - \left(2 + 6B_1 + 3B_1^2\right) B^{(1)} + 2\alpha \frac{\partial}{\partial X} \left((1 + B_1) B^{(1)}\right)$$

$$= -\frac{\partial B_1}{\partial \tau} - 6B_1 B_2 - 3B_1^2 B_2 + 2\alpha \frac{\partial}{\partial X} (B_1 B_2), \tag{2.188}$$

$$B_1 = \tanh \beta_+ (X - \xi_1(\tau)) - 1, \tag{2.189}$$

$$B_2 \approx -2\exp 2|\beta_-| (X - \xi_2(\tau)). \tag{2.190}$$

Unlike (2.77), the operator in the left-hand side of (2.188) is not self-adjoint. However, the solution $B_c(X)$ of the adjoint problem

$$\frac{\partial^2 B_c}{\partial X^2} - \left(2 + 6B_1 + 3B_1^2\right) B_c - 2\alpha (1 + B_1) \frac{\partial B_c}{\partial X} = 0 \tag{2.191}$$

can be found explicitly:

$$B_c = \cosh^{-\beta_+^{-2}} \beta_+ (X - \xi_1(\tau)). \qquad (2.192)$$

The solvability condition to equation (2.188)

$$\frac{d\xi_1}{d\tau} \int_{-\infty}^{\infty} dX B_c \frac{\partial B_1}{\partial X} = \int_{-\infty}^{\infty} dX B_c \left(6B_1 B_2 + 3B_1^2 B_2 - 2\alpha \frac{\partial}{\partial X}(B_1 B_2) \right) \qquad (2.193)$$

gives rise to the following equation of motion:

$$\frac{d\xi_1}{d\tau} = F(\alpha)e^{-2|\beta_-|(\xi_2 - \xi_1)}, \qquad (2.194)$$

$$F(\alpha) = \frac{\Gamma\left(\frac{1}{\beta_+^2}\right)\Gamma\left(\frac{1}{2\beta_+^2} + \frac{\alpha}{\beta_+}\right)}{\beta_+^3 \Gamma^2\left(1 + \frac{1}{2\beta_+^2}\right)} \left(5 + 2\alpha^2 + 2|\beta_-|^2\right). \qquad (2.195)$$

In a similar way we find the equation of motion for the right domain walls:

$$\frac{d\xi_2}{d\tau} = -F(-\alpha)e^{-2\beta_+(\xi_2 - \xi_1)}, \qquad (2.196)$$

where $F(-\alpha)$ is obtained from $F(\alpha)$ by the transformation $\alpha \to -\alpha$, $\beta_+ \to |\beta_-|$, $|\beta_-| \to \beta_+$.

It should be emphasized that the interaction of domain walls is attractive but asymmetric. In the case where distances between kinks are large, the attraction of a narrow kink by a wide antikink is the main phenomenon. The motion of a wide antikink due to the attraction to a narrow kink is much slower.

By means of (2.194) and (2.196), we find once again that stationary periodic solutions (chains of kinks and antikinks) are always unstable with respect to countermoving kink and antikink sublattices. The growth rate of the unstable disturbance of the periodic solution with a period $2l$ that can be obtained formally from (2.194) and (2.196) is given by

$$\sigma = 4|\beta_-|F(\alpha)\exp(-2|\beta_-|l) + 4\beta_+ F(-\alpha)\exp(-2\beta_+ l). \qquad (2.197)$$

In fact, the latter term is exponentially small and should be dropped, if $\alpha = O(1)$:

$$\sigma \approx 4|\beta_-|F(\alpha)\exp(-2|\beta_-|l). \qquad (2.198)$$

As in the case $\alpha = 0$, the evolution of the system can be considered as a slow motion of domain walls and their eventual annihilation. In a finite region, a unidomain state is established during the finite but exponentially long time. This prediction coincides with results of numerical simulations [17].

2.2.4 Higher-Order Corrections

The symmetry (2.176) is not characteristic for the original problem (2.129)–(2.133); hence, in the next order, we have to add all the possible terms:

$$\frac{\partial A}{\partial \tau} = \frac{\partial^2 A}{\partial X^2} + A - A^3 + 2\alpha A \frac{\partial A}{\partial X}$$

$$+\varepsilon \left[\delta_1 \frac{\partial A}{\partial X} + \delta_2 \frac{\partial^3 A}{\partial X^3} + \delta_3 A \frac{\partial^2 A}{\partial X^2} + \delta_4 \left(\frac{\partial A}{\partial X} \right)^2 + \delta_5 A^2 \frac{\partial A}{\partial X} + \delta_6 A^4 \right]. \quad (2.199)$$

Because of the terms $O(\varepsilon)$ in (2.199), the stationary solutions do not exist anymore. Single kink and antikink move with different velocities. In the case of a pair, one can dispose a kink and an antikink in such a way (on distances $\sim |\ln(\varepsilon)|$) that the difference of velocities is exactly compensated by their interaction. In this case, one gets a kink–antikink bound state, which is a limit solution for a family of traveling periodic waves, as $k \to 0$. However, because of the attractive interaction of kinks and antikinks, all these structures are unstable. Finally, one kind of domains spreads and another one is forced out, and a homogeneous state is set eventually in the whole region.

In the section above, we have considered the general case, where two kinds of nonlinearities are present: a cubic nonlinearity that leads to the instability saturation and a non-potential quadratic nonlinearity violating the $x \to -x$ symmetry. In the next section, we investigate the case where the cubic saturation term is absent.

2.3 Convection in a Vertical Cylinder

The CYE considered in Section 2.2 is the general amplitude equation for a long-wavelength convection instability in the absence of a conservation law, and we can expect that this equation is valid for different convection problems, where the symmetry $X \to -X$ is broken (e.g., convection in an inclined cylinder), except for problems where the instability mode is not unique. Recall that in Boussinesq convection the quadratic term proportional to A^2 is absent, and the coefficient α_4 in the cubic term proportional to A^3 cannot be positive, because the subcritical instability of the quiescent state is forbidden according to the Sorokin's theorem (see [3, 18]).

However, there is a class of problems where the coefficient α_4 is equal to zero identically, because the homogeneous disturbance does not generate nonlinear terms. The typical example of such a problem is convection in a vertical cylinder [19].

We consider a vertical cylinder (Figure 2.6) with the fixed dimensional temperature distribution $T_{*\Gamma} = -A_* z_*$ on the rigid boundary. The equilibrium temperature and pressure fields and the equations governing the convection are identical to those presented in Section 2.1, see (2.1)–(2.4), but the condition of zero mean flux is now

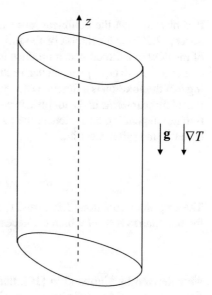

Fig. 2.6 Problem geometry: vertical cylinder heated from below

as follows:

$$\langle v_z(x,y)\rangle = 0, \tag{2.200}$$

where $\langle f(x,y)\rangle \equiv \int_S dxdy f(x,y)$.

The system is homogeneous in z-direction, and instead of (2.7) we have the following structure of normal disturbances:

$$U(\mathbf{r},t) = u_n(x,y;k,R)\exp[ikz + \sigma_n(k,R)t]. \tag{2.201}$$

There exists a long-wavelength instability of the equilibrium state [3], and the critical disturbance with $k = 0$ is a *parallel flow*:

$$\mathbf{v} = (0,0,\tilde{w}(x,y))e^{\sigma t}, T = \tilde{T}(x,y)e^{\sigma t}, \tag{2.202}$$

that is determined by the system of equations

$$\sigma\tilde{w} = \nabla_\perp^2 \tilde{w} + R\tilde{T}, \tag{2.203}$$
$$\sigma\tilde{T} - \tilde{w} = \nabla_\perp^2 \tilde{T}; \tag{2.204}$$
$$\tilde{w} = 0, \ \tilde{T} = 0 \text{ at } \Gamma; \tag{2.205}$$
$$\langle\tilde{w}\rangle = 0, \tag{2.206}$$

where

$$\nabla_\perp^2 = \frac{\partial^2}{\partial x^2} + \frac{\partial^2}{\partial y^2}. \tag{2.207}$$

It is obvious that the nonlinear terms are identically zero for such a parallel flow; hence, (2.202) is a solution of the full nonlinear problem. Only an inhomogeneity of the flow in z-direction can produce nonlinear terms.

For $R - R_c = O(\varepsilon^2)$ we can derive the amplitude equation in a usual way, assuming that the solution is a function of x, y, $z_1 = \varepsilon z$, and $t_2 = \varepsilon^2 t$. The only difference is that nonlinear terms in the amplitude equation should vanish for the amplitude function not depending on z. Hence, the cubic term does not appear, and in the generic case we find (after a scaling):

$$\frac{\partial A}{\partial \tau} = \frac{\partial^2 A}{\partial Z^2} + A + \frac{\partial (A^2)}{\partial Z}. \tag{2.208}$$

The amplitude equation (2.208) and the boundary condition for the amplitude A on the boundaries $Z = \pm l$ which are necessary in the case of a finite-length cylinder:

$$Z = \pm l : A = 0, \tag{2.209}$$

were derived by Normand in [19], that is why we name (2.208) as the *Normand equation* (NE). In an infinite region, (2.208) is physically meaningless, because it possesses an unphysical solution $A(\tau) = A(0) \exp \tau$ describing an unlimited growth of the intensity of a parallel flow.

We omit the derivation of (2.208) and (2.209), proceeding to the analysis of their properties.

2.3.1 Lyapunov Functional

Unlike (2.172), equation (2.208) is *potential*. In order to demonstrate this fact, we perform the *Hopf–Cole transformation*

$$A = \frac{\partial}{\partial Z}(\ln \Psi) \tag{2.210}$$

that transforms (2.208) into the *nonlinear reaction–diffusion equation*

$$\frac{\partial \Psi}{\partial \tau} = \frac{\partial^2 \Psi}{\partial Z^2} + \Psi \ln \Psi + C\Psi, \tag{2.211}$$

where C is an arbitrary constant (its arbitrariness corresponds to the transformation $\Psi' = C'\Psi$, $C' = const$ that does not change A). We set $C = 1/2$ below. Boundary conditions (2.209) give:

$$Z = \pm l : \frac{\partial \Psi}{\partial Z} = 0. \tag{2.212}$$

Equation (2.211) can be written in the form

$$\frac{\partial \Psi}{\partial \tau} = -\frac{\delta F}{\delta \Psi}, \tag{2.213}$$

$$F = \int L(Z)dZ, \tag{2.214}$$

$$L(Z) = \frac{1}{2}\left(\frac{\partial \Psi}{\partial Z}\right)^2 - \frac{1}{2}\Psi^2 \ln \Psi. \tag{2.215}$$

Hence,

$$\frac{dF}{d\tau} \leq 0; \tag{2.216}$$

in any finite region, the system tends to a certain stationary solution, as $\tau \to \infty$.

2.3.2 Stationary Solutions

Integration of the equation

$$\frac{d^2\Psi}{dZ^2} + \Psi \ln \Psi + \frac{1}{2}\Psi = 0 \tag{2.217}$$

results in the equation

$$\frac{1}{2}\left(\frac{d\Psi}{dZ}\right)^2 + \frac{1}{2}\Psi^2 \ln \Psi = E. \tag{2.218}$$

The larger is E, the smaller is the mean value of the Lyapunov functional F. Depending on E, (2.218) determines a constant solution $\Psi = 1/\sqrt{e}$ corresponding to the quiescent state ($E = E_{min} = -1/4e$), periodic solutions with different spatial periods ($E_{min} < E < 0$; the larger is the period, the smaller is the mean value of the Lyapunov functional), and a homoclinic solution

$$\Psi = \exp\left[-\frac{1}{4}(Z - \zeta)^2\right], \quad \zeta = const \tag{2.219}$$

corresponding to a non-bounded stationary solution of (2.208)

$$A = -\frac{1}{2}(Z - \zeta). \tag{2.220}$$

In the original variables $A(Z)$ and $B(Z) = dA/dZ$, the stationary solutions correspond to orbits described by the equation

$$\frac{dB}{dA} = -\frac{A(1 + 2B)}{B}. \tag{2.221}$$

Integrating (2.221), we find

$$A^2 - \frac{1}{2}\ln(1+2B) + B = C, \ C = const. \tag{2.222}$$

The orbits in the plane (A, B) are shown in Figure 2.7.

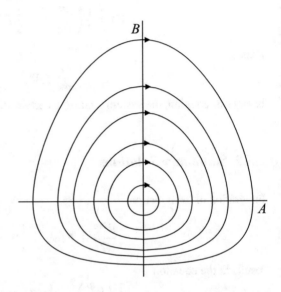

Fig. 2.7 Phase portrait for
(2.222)

It should be noted that (2.208) resembles the Burgers equation, with the only additional linear term in the right-hand side. The large-period stationary solutions look like chains of Burgers shocks separated by fragments described by a linear function $A = -(Z - \zeta)/2$.

Boundary condition (2.212) is satisfied for the constant solutions and for periodic solutions with the periods l_n, where

$$n\frac{l_n}{2} = 2l, \ n = 1, 2, \ldots \tag{2.223}$$

The nontrivial stationary solutions exist if $l > \pi/2$. Since the orbit can start at $Z = -l$ either with the maximum or with the minimum value of Ψ, there are two solutions for every n. For $n = 1$, the function Ψ changes monotonically in the region $-l < Z < l$; hence, $A(Z)$ is sign-conserving. Otherwise, $A(Z)$ has $n - 1$ zeros in the region $-l < Z < l$. For even values of n, $A(Z)$ is an odd function of Z.

2.3.3 Stability of Stationary Solutions

For disturbances $\tilde{\Psi} \exp \sigma \tau$ on the background Ψ, we find

$$\frac{d^2\tilde{\Psi}}{dZ^2} + \tilde{\Psi}\left(\ln\Psi + \frac{3}{2}\right) = \sigma\tilde{\Psi}; \qquad (2.224)$$

$$\frac{d\tilde{\Psi}}{dZ} = 0 \text{ at } Z = \pm l. \qquad (2.225)$$

For the solution $\Psi = 1/\sqrt{e}$, which corresponds to a quiescent state, one obtains:

$$\tilde{\Psi} = \cos k_m(Z+l), \ k_m = \frac{m\pi}{2l}, \ m = 1, 2, \ldots; \ \sigma_m = 1 - k_m^2; \qquad (2.226)$$

hence, the solution is unstable if $l > \pi/2$, i.e., when periodic solutions of problem (2.211) and (2.212) appear. For periodic solutions, we can use the same argumentation as in the preceding sections. Equation (2.224) has always the solution $\tilde{\Psi} = d\Psi/dZ, \sigma = 0$. If $\tilde{\Psi}$ is not sign-conserving, the solution Ψ is unstable; otherwise it is stable. Taking into account the relation (2.210), one can see that only two sign-conserving (i.e., unidomain) solutions for $A(Z)$ corresponding to $n = 1$ in (2.223) are stable, and all the remaining solutions are unstable.

For $n \geq 3$, the unstable stationary solution contains "shocks" inside the region $-l < Z < l$. The instability of such a "shock" looks like its exponential self-acceleration [20]. Indeed, one can impose a disturbance $\tilde{A}$ localized near the shock, such that $\langle\tilde{A}\rangle = \int_{-l}^{l} \tilde{A}dZ$ is nonzero, and $d\tilde{A}/dZ$ is negligible near the ends of this region. Obviously, such a disturbance displaces the position of the shock. Integrating (2.208) over the region, we end up with

$$\frac{d\langle\tilde{A}\rangle}{d\tau} = \langle\tilde{A}\rangle. \qquad (2.227)$$

Therefore, the displacement of the shock moves with an acceleration. Finally, all the shocks existing in the initial state stick together and reach the ends of the region. After the elimination of internal shocks, the solution looks like a slowly evolving superposition of $n = 1$ and $n = 2$ modes and eventually tends to the $n = 1$ state.

We do not discuss here the case where because of the symmetry of eigenfunctions the term $\partial(A^2)/\partial Z$ vanishes and higher nonlinearities become relevant referring the reader to [17, 19].

In conclusion, convection in a cylindrical cavity gives a plethora of scenarios of nonlinear development for a monotonic type III instability. However, in a spatially homogeneous case, all these scenarios are finished by pattern coarsening and establishing either a flow homogeneous with respect to the longitudinal coordinate or a unicellular convection flow. Thus, no pattern formation in the usual sense is predicted.

References

1. S. Chandrasekhar, *Hydrodynamic and Hydromagnetic Instability* (Oxford University Press, Oxford, 1961)
2. D.D. Joseph, *Stability of Fluid Motions*, vols. I, II (Springer, New York, 1976)
3. G.Z. Gershuni, E.M. Zhukhovitsky, *Convective Stability of Incompressible Fluid* (Keter, Jerusalem, 1976)
4. P. Colinet, J.-C. Legros, M.G. Velarde, *Nonlinear Dynamics of Surface-Tension-Driven Instabilities* (Wiley, New York, 2001)
5. A.A. Nepomnyashchy, M.G. Velarde, P. Colinet, *Interfacial Phenomena and Convection* (Chapman and Hall/CRC Press, Boca Raton, 2002)
6. R.V. Birikh, V.A. Briskman, M.G. Velarde, J.-C. Legros, *Liquid Interfacial Systems. Oscillations and Instability* (Marcel Dekker, New York, 2003)
7. A.A. Nepomnyashchy, C.R. Phys. **16**, 267 (2015)
8. L.D. Landau, E.M. Lifshitz, *Quantum Mechanics* (Pergamon Press, Oxford, 1965)
9. K.A. Gorschkov, L.A. Ostrovsky, E.N. Pelinovsky, Proc. IEEE **62**, 1511 (1974)
10. K. Kawasaki, T. Ohta, Physica A **116**, 573 (1982)
11. A.J. Lichtenberg, M.A. Lieberman, *Regular and Chaotic Dynamics* (Springer, New York, 2010)
12. P. Coullet, Phys. Rev. Lett. **56**, 724 (1986)
13. M. Peyrard, S. Aubry, J. Phys. C **16**, 1593 (1983)
14. P. Coullet, C. Elphick, D. Repaux, Phys. Rev. Lett. **58**, 431 (1987)
15. R.A. Fisher, Ann. Eugenics **7**, 355 (1937)
16. A. Kolmogorov, I. Petrovskii, N. Piskunov, Moscow University Bull. Math. **1**, 1 (1937)
17. P. Cessi, W.R. Young, J. Fluid Mech. **237**, 57 (1992)
18. L.D. Landau, E.M. Lifshitz, *Fluid Mechanics* (Pergamon Press, Oxford, 1987)
19. C. Normand, J. Fluid Mech. **143**, 223 (1984)
20. A.B. Mikishev, G.I. Sivashinsky, Phys. Lett. A **175**, 409 (1993)

Chapter 3
Convection in Liquid Layers

In the previous chapter, we considered convection in cavities with a fixed boundary temperature distribution, which gives an example of instability "without a conservation law" (type III): a spatially homogeneous disturbance with the wavenumber $k = 0$ has a nonzero growth rate. In the present chapter, we present examples of systems "with a conservation law" (type II instabilities). In Section 3.1 we consider the Rayleigh–Bénard convection in a layer with poorly conducting solid boundaries, where the conservation of the mean temperature in the layer is weakly violated by the heat transfer on the boundaries. The solution of that classical problem was done far ago in works of Busse and Riahi [1], Chapman and Proctor [2, 3], Gertsberg and Sivashinsky [4], and Pismen [5] (see also [6]). Section 3.2 is devoted to the analysis of the longwave pattern formation caused by monotonic surface tension-driven instabilities. Oscillations and waves produced by different kinds of oscillatory Marangoni instabilities are considered in Section 3.3.

3.1 Rayleigh–Bénard Convection in a Layer with Poorly Conducting Boundaries

In the previous chapter, we dealt with instabilities "without conservation law" (type III). In the present section, we start considering instabilities that belong to type II briefly overviewed in Section 1.4.2. Our basic physical system is a horizontal liquid layer heated from below. In a contradistinction to the convection in cylindrical cavities, where the geometry prescribes one-dimensional patterns, the flows in a horizontal liquid layer can form different kinds of periodic patterns (rolls, squares, hexagons) or even more complex "superlattices" or "quasicrystals." In the case of shortwave instabilities, where the patterns near the threshold are characterized by a definite wavenumber $k_c \neq 0$, the principles of pattern selection are well-known (see, e.g., [7] and Appendix C). Below we consider the pattern formation in the case of

© Springer Science+Business Media, LLC 2017
S. Shklyaev, A. Nepomnyashchy, *Longwave Instabilities and Patterns in Fluids*,
Advances in Mathematical Fluid Mechanics,
https://doi.org/10.1007/978-1-4939-7590-7_3

the longwave instabilities, where the linear stability theory does not provide a clear criterion for the wavenumber selection.

Let us consider a liquid layer sandwiched between two solid plates fixed on the distance d_*. The layer is heated from below so that in the absence of motion, the stationary temperature field is $T_{*0}(z_*) = -A_* z_*$. Using the same choice of scales as in Section 2.1.1, we obtain the following nondimensional system of equations which coincides with that presented in the previous section (equations (2.1)–(2.3)),

$$\frac{1}{P}\left[\frac{\partial \mathbf{v}}{\partial t} + (\mathbf{v} \cdot \nabla)\mathbf{v}\right] = -\nabla p + \Delta \mathbf{v} + RT\mathbf{e}_z, \tag{3.1}$$

$$\nabla \cdot \mathbf{v} = 0, \tag{3.2}$$

$$\frac{\partial T}{\partial t} + \mathbf{v} \cdot T - \mathbf{v} \cdot \mathbf{e}_z = \Delta T. \tag{3.3}$$

Recall that T is the deviation of the temperature from the linear profile $T_{*0}(z_*)$. Similarly, the no-slip condition on the rigid boundaries gives

$$\mathbf{v} = 0 \text{ at } z = 0, 1. \tag{3.4}$$

Here we will take a more general boundary condition for temperature than that used in the previous section taking into account the finite heat conductivity of solid plates adjacent to the liquid layer. Strictly speaking, we have to solve the heat equations inside the solid plates and match the temperature fields in the solid and in the liquid using the condition of the continuity of the temperature

$$T|_{z=0+} = T|_{z=0-}, \; T|_{z=1-} = T|_{z=1+} \tag{3.5}$$

and that of the heat flux continuity,

$$\frac{\partial T}{\partial z}\bigg|_{z=0+} = \tilde{\lambda}\frac{\partial T}{\partial z}\bigg|_{z=0-}, \; \frac{\partial T}{\partial z}\bigg|_{z=1-} = \tilde{\lambda}\frac{\partial T}{\partial z}\bigg|_{z=1+}, \tag{3.6}$$

where $\tilde{\lambda} = \lambda_s/\lambda_l$ is the ratio of heat conductivities of the solid and the liquid. However, in the case of slowly growing longwave disturbances, it is possible to formulate a *one-sided* model imposing the boundary conditions of the third kind on the boundaries of the liquid layer:

$$\frac{\partial T}{\partial z} = BiT, \; z = 0; \; \frac{\partial T}{\partial z} = -BiT, \; z = 1. \tag{3.7}$$

The value of the Biot number Bi can be calculated by solving the heat equations inside the solid place. For slowly evolving longwave temperature disturbances,

$$Bi = \frac{\lambda_s d}{\lambda_l h_s}, \tag{3.8}$$

where h_s is the thickness of the plate (see, e.g., [8]).

3.1.1 The Limit of Zero Biot Number: Cahn–Hilliard Equation

3.1.1.1 Derivation of the Cahn–Hilliard Equation

The problem under consideration has no "luxury" of the one-dimensional geometry characteristic for the convection in a cylinder. It has a *rotational symmetry* in the horizontal plane; therefore beyond the instability threshold, disturbances with different directions of the wavevectors grow eventually forming ordered or disordered three-dimensional patterns. Still, we will start with the derivation of a one-dimensional longwave amplitude equation assuming that the flow is two-dimensional, i.e., functions $\mathbf{v} = (u, 0, w)$, T, and p do not depend on y. That may correspond to the development of convection in a rectangular cavity which is long in x-direction but relatively short in y-direction. In that case, we can define the stream function Ψ by the relations

$$u = \frac{\partial \Psi}{\partial z}, \quad w = -\frac{\partial \Psi}{\partial x} \tag{3.9}$$

and rewrite the system of equations (3.1)–(3.3) as

$$\frac{1}{P}\left(\frac{\partial \Delta \Psi}{\partial t} + \frac{\partial \Psi}{\partial z}\frac{\partial \Delta \Psi}{\partial x} - \frac{\partial \Psi}{\partial x}\frac{\partial \Delta \Psi}{\partial z}\right) = \Delta^2 \Psi - R\frac{\partial T}{\partial x}, \tag{3.10}$$

$$\frac{\partial T}{\partial t} + \frac{\partial \Psi}{\partial z}\frac{\partial T}{\partial x} - \frac{\partial \Psi}{\partial x}\frac{\partial T}{\partial z} = \Delta T. \tag{3.11}$$

The boundary conditions for the stream function Ψ on the rigid boundaries of the layer are:

$$\Psi = \frac{\partial \Psi}{\partial z} = 0, \, z = 0, 1. \tag{3.12}$$

For temperature, conditions (3.7) are retained.

In the present section, we consider the limit $B \to 0$. Note that these conditions mean the negligibly small ratio of the system parameters (3.8) rather than a true heat insulation of the liquid layer: the heat flux across the layer, $\lambda_s A_* \neq 0$. The boundary conditions (3.7) become

$$\frac{\partial T}{\partial z} = 0, \, z = 0, 1. \tag{3.13}$$

According to the linear stability theory for the mechanical equilibrium state [3], the convection is developed for $R > R_0 = 720$ due to a long-wavelength instability. Because of the boundary condition (3.13), the integral of T over the layer does not change with time; thus there is a *conservation law* which rules out the growth of a spatially homogeneous temperature disturbance. Therefore, relations (1.22) and (1.25) are characteristic for the unstable mode; relation (1.26) is irrelevant, because the growth rate is real.

Near the instability threshold, $R - R_0 \ll 1$, the disturbances with the wavenumbers $0 < k < k_m(R)$, $k_m \sim (R - R_0)^{1/2}$ have positive values of the growth rate. It is natural to use k_m for rescaling the variable x. We define $x_1 = k_m x$, which leads to a formal replacement

$$\frac{\partial}{\partial x} = k_m \frac{\partial}{\partial x_1} \tag{3.14}$$

in the governing equations. Then

$$R = R_0 + R_2 k_m^2 + O(k_m^4). \tag{3.15}$$

The growth rate $\sigma \sim k_m^4$; therefore the appropriate rescaling of the time variable is $t_4 = k_m^4 t$; hence

$$\frac{\partial}{\partial t} = k_m^4 \frac{\partial}{\partial t_4}. \tag{3.16}$$

We construct the solution in the form of series,

$$\Psi = \sum_{n=0}^{\infty} k_m^n \Psi_n, \ T = \sum_{n=0}^{\infty} k_m^n T_n. \tag{3.17}$$

Let us substitute expansions (3.17) into (3.11)–(3.13), and equate the terms of the same order in k_m. We demand the boundness of solutions at $x_1 \to \pm\infty$.

From the zeroth order equations, we find

$$\Psi_0 = 0, \ T_0 = T_0(x_1, t_4). \tag{3.18}$$

At the first order in k_m, we obtain:

$$\Psi_1 = \frac{R_0}{24} \frac{\partial T_0}{\partial x_1} (z^4 - 2z^3 + z^2), \ T_1 = T_1(x_1, t_4). \tag{3.19}$$

At the second order, we find:

$$R_0 = 720, \ \Psi_2 = 30 \frac{\partial T_1}{\partial x_1} (z^4 - 2z^3 + z^2), \tag{3.20}$$

$$T_2 = \frac{\partial^2 T_0}{\partial x_1^2} \left(z^6 - 3z^5 + \frac{5}{2} z^4 - \frac{1}{2} z^2 \right) + \left(\frac{\partial T_0}{\partial x_1} \right)^2 (6z^5 - 15z^4 + 10z^3) + \bar{T}_2(x_1, t_4). \tag{3.21}$$

Equations at the third order allow to determine Ψ_3 and T_3:

$$\Psi_3 = \frac{\partial^3 T_0}{\partial x_1^3} \left(\frac{1}{7} z^{10} - \frac{5}{7} z^9 + \frac{15}{14} z^8 - 3z^6 + 6z^5 - 5z^4 + \frac{10}{7} z^3 + \frac{1}{14} z^2 \right) +$$

$$\frac{\partial T_0}{\partial x_1} \frac{\partial^2 T_0}{\partial x_1^2} \left(\frac{20}{7} z^9 - \frac{90}{7} z^8 + \frac{120}{7} z^7 - \frac{200}{7} z^3 + \frac{150}{7} z^2 \right) +$$

$$\frac{1}{P} \frac{\partial T_0}{\partial x_1} \frac{\partial^2 T_0}{\partial x_1^2} \left(\frac{50}{7} z^9 - \frac{225}{7} z^8 + 60z^7 - 60z^6 + 30z^5 - \frac{50}{7} z^3 + \frac{15}{7} z^2 \right) +$$

$$30\frac{\partial \bar{T}_2}{\partial x_1}(z^4 - 2z^3 + z^2) + \frac{R_2}{24}\frac{\partial T_0}{\partial x_1}(z^4 - 2z^3 + z^2), \tag{3.22}$$

$$T_3 = \frac{\partial^2 T_1}{\partial x_1^2}\left(z^6 - 3z^5 + \frac{5}{2}z^4 - \frac{1}{2}z^2\right) + \frac{\partial T_0}{\partial x_1}\frac{\partial T_1}{\partial x_1}(12z^5 - 30z^4 + 20z^3) + \bar{T}_3(x_1, t_4). \tag{3.23}$$

Finally, the solvability condition for the problem at the fourth order gives a closed equation for $T(x_1, t_4)$ [6]:

$$\frac{\partial T_0}{\partial t_4} + \frac{17}{462}\frac{\partial^4 T_0}{\partial x_1^4} + \frac{R_2}{720}\frac{\partial^2 T_0}{\partial x_1^2} - \frac{10}{7}\frac{\partial}{\partial x_1}\left[\left(\frac{\partial T_0}{\partial x_1}\right)^3\right] = 0. \tag{3.24}$$

Taking into account that k_m is by definition the wavenumber of a neutral infinitesimal disturbance and, hence, the linearized problem has to have a solution $T_0 = \exp(ix_1)$, we find that

$$R_2 = \frac{2040}{77}.$$

In order to convert the obtained amplitude equation to a standard form, we differentiate it with respect to x_1 and define

$$\tau = \frac{462}{17}t_4, \quad \frac{\partial T_0}{\partial x_1} = A\left(\frac{17}{660}\right)^{1/2}. \tag{3.25}$$

We find:

$$\frac{\partial A}{\partial \tau} + \frac{\partial^2}{\partial X^2}\left(\frac{\partial^2 A}{\partial X^2} + A - A^3\right) = 0. \tag{3.26}$$

The quantity A is proportional to the horizontal component of the heat flux in the layer, $-\partial T_0/\partial x_1$. If there is no mean horizontal heat flux, i.e., T_0 is bounded as $x \to \pm\infty$, the averaged value of A has to be equal to 0.

Equation (3.26) is called the *Cahn–Hilliard equation* [9] (see (1.82)).

3.1.1.2 Basic Properties of the Cahn–Hilliard Equation

Conservation Law

Equation (3.26) is similar to the Allen–Cahn equation (2.56) discussed in Section 2, but there are some important differences. First, equation (3.26) can be written in the conservative form,

$$\frac{\partial A}{\partial \tau} + \frac{\partial J}{\partial X} = 0, \tag{3.27}$$

where

$$J = \frac{\partial}{\partial X}\left(\frac{\partial^2 A}{\partial X^2} + A - A^3\right). \tag{3.28}$$

If the layer is situated between distant boundaries $X = \pm l$, then

$$\frac{d}{dT}\int_{-l}^{l} A dX = -J \Big|_{X=-l}^{X=l}, \tag{3.29}$$

which is the manifestation of the conservation law (in an integral form). If J vanishes at the lateral boundaries of the layer,

$$\frac{d}{dT}\int_{-l}^{l} A dX = 0. \tag{3.30}$$

Lyapunov Functional

Equation (3.26) can be written also as

$$\frac{\partial A}{\partial \tau} = \frac{\partial^2}{\partial X^2}\frac{\delta F}{\delta A}, \tag{3.31}$$

where $F[A(X)]$ is the same Lyapunov functional

$$F(\tau) = \int L(X,\tau)dX, \ L = \frac{1}{2}\left(\frac{\partial A}{\partial X}\right)^2 - V(A), \ V(A) = -\frac{1}{4}(A^2 - 1)^2, \tag{3.32}$$

as in the case of the Allen–Cahn equation (cf. (2.58), (2.59)). The Lyapunov functional decreases monotonically with time, if J vanishes at the infinity:

$$\frac{dF}{d\tau} = \int dX \left[\frac{\partial A}{\partial X}\frac{\partial^2 A}{\partial X \partial \tau} - V'(A)\frac{\partial A}{\partial \tau}\right] = \int dX \frac{\partial A}{\partial \tau}\left[-\frac{\partial^2 A}{\partial X^2} - V'(A)\right] =$$

$$\int dX \left[\frac{\partial^2 A}{\partial X^2} + V'(A)\right]\frac{\partial^2}{\partial X^2}\left[\frac{\partial^2 A}{\partial X^2} + V'(A)\right] = -\int dX \left|\frac{\partial}{\partial X}\left[\frac{\partial^2 A}{\partial X^2} + V'(A)\right]\right|^2 \leq 0. \tag{3.33}$$

Thus, similar to the case of the Allen–Cahn equation (see Section 2.1.2.1), the system evolves toward a stationary state.

Stationary Solutions

Stationary solutions $A = A(X)$ of (3.26) satisfy the ordinary differential equation

$$\frac{d^2}{dX^2}\left(\frac{d^2 A}{dX^2} + A - A^3\right) = 0; |A| < \infty \text{ at } X \to \pm\infty, \tag{3.34}$$

hence

$$\frac{d^2 A}{dX^2} + A - A^3 = DX + C, \tag{3.35}$$

where D and C are constant. Because we are interested in bounded solutions of (3.34), we have to take $D = 0$. Thus, the set of stationary solutions of the one-dimensional Cahn–Hilliard equation is wider than that of the Allen–Cahn equation (cf. (2.65)), because constant C is generally nonzero. For $C \neq 0$, bounded solutions of equation (3.35) are either periodic with a nonzero averaged value of A or correspond to a solitary wave with $A(+\infty) = A(-\infty) \neq 0$. However, recall that the amplitude A is proportional to the horizontal gradient of the temperature; see (3.25). Therefore, in the absence of a mean horizontal heat flux, one has to fix $C = 0$. In that case, the stationary solutions of the Cahn–Hilliard equation are identical to those of the Allen–Cahn equation (see (2.67)–(2.69)). A description of stationary solutions of the Cahn–Hilliard equation in a finite interval can be found in [2, 10].

Stability of Stationary Solutions

Computation of the Growth Rate

Let us investigate the stability of periodic stationary solutions (2.68) that can be presented in the form

$$A(X) = \sqrt{\frac{2k_J^2}{1 + k_J^2}} \, \mathrm{sn} \frac{X - \xi}{\sqrt{1 + k_J^2}}, \tag{3.36}$$

where sn is the Jacobi elliptic function (elliptic sine), k_J is its modulus, and ξ is a constant which is taken equal to zero later on, without a loss of generality. The spatial period of the solution is

$$L(k_J) = 4K(k_J)\sqrt{1 + k_J^2} \tag{3.37}$$

where $K(k_J)$ is the complete elliptic integral of the first kind.

Linearizing (3.25), we obtain the following eigenvalue problem for infinitesimal disturbances $\tilde{A}(X)\exp(\sigma T)$ on the background of stationary solution (3.36):

$$\sigma\tilde{A} + \frac{d^4\tilde{A}}{dX^4} + \frac{d^2}{dX^2}[(1 - 3A^2)\tilde{A}] = 0, \ |\tilde{A}(\pm\infty)| < \infty. \tag{3.38}$$

Because A^2 is a periodic function with the period equal to $L/2$, bounded solutions of (3.38) have the form of the Floquet–Bloch function:

$$\tilde{A}(X) = f(X, q)\exp(iqX), \tag{3.39}$$

where $f(X, q)$ is a periodic function with the period $L/2$ and q is a real number (quasi-wavenumber) defined modulo $4\pi/L$ (one can choose $|q| \leq 2\pi/L$). Function f satisfies the following equation:

$$\sigma f + \left(\frac{d}{dX} + iq\right)^4 f + \left(\frac{d}{dX} + iq\right)^2 [(1 - 3A^2)f] = 0, \qquad (3.40)$$

$$f(X + L/2) = f(X). \qquad (3.41)$$

Actually, the conclusion on the instability of all the stationary solutions can be reached by the consideration of the longwave modulations of the stationary pattern with $q \ll 1$ [6]. Let us present $f(X)$ and σ in the form of a series in q,

$$f = \sum_{n=0}^{\infty} f_n q^n, \ \sigma = \sum_{n=0}^{\infty} \sigma_n q^n.$$

At the zeroth order in q, we obtain the following problem,

$$\sigma_0 f_0 + \frac{d^4 f_0}{dX^4} + \frac{d^2}{dX^2}[(1 - 3A^2)f_0] = 0,$$

$$f_0(X + L/2) = f(X_0).$$

The solution with $\sigma_0 = 0$,

$$f_0 = 1 - k_j^2 + 2k_j^2 \mathrm{cn}^2 \frac{X}{\sqrt{1 + k_j^2}}, \qquad (3.42)$$

which corresponds to a homogeneous translation of the stationary pattern, is of major interest. Here cn is the Jacobi elliptic function (elliptic cosine).

At the first order, we find:

$$\frac{d^4 f_1}{dX^4} + \frac{d^2}{dX^2}[(1 - 3A^2)f_1] = \sigma_1 f_0 - 4i\frac{d^3 f_0}{dX^3} - 2i\frac{d}{dX}[(1 - 3A^2)f_0], \qquad (3.43)$$

$$f_1(X + L/2) = f_1(X).$$

Integrating both parts of (3.43) with respect to X over the period and taking into account the periodicity of functions $A^2(X)$ and $f_0(X)$, we find:

$$\sigma_1 \int_0^{L/2} f_0 dX = 0.$$

Because

$$\int_0^{L/2} f_0 dX \neq 0,$$

we obtain $\sigma_1 = 0$.

Finally, at the second order in q, we get:

$$\frac{d^4 f_2}{dX^4} + \frac{d^2}{dX^2}[(1 - 3A^2)f_2] = \sigma_2 f_0 - 4i\frac{d^3 f_1}{dX^3}$$

$$-2i\frac{d}{dX}[(1-3A^2)f_1]+6\frac{d^2f_0}{dX^2}+(1-3A^2)f_0,$$

$$f_2(X+L/2)=f_2(X).$$

The integration over the period determines the growth rate:

$$\sigma_2=\frac{\int_0^{L/2}(1-3A^2)f_0dX}{\int_0^{L/2}f_0dX}. \tag{3.44}$$

Substituting expressions (3.36) and (3.42) into (3.44), we find:

$$\sigma_2=\frac{(1-k_J^2)^2}{(1+k_J^2)[2E(k_J)/K(k_J)-1+k_J^2]}, \tag{3.45}$$

where $K(k_J)$ and $E(k_J)$ are the complete elliptic integrals of the first and second kind, correspondingly. Because $\sigma_2>0$ in the whole interval $0<k_J<1$, all the periodic stationary solutions (3.36) are unstable. This conclusion is compatible with the general properties of the eigenvalue spectrum for the Cahn–Hilliard equation [11] (see the next subsection).

The kink and antikink solutions (2.67)) are the only stable stationary solutions in the case of an infinite region. They describe domain walls between regions characterized by different direction of the horizontal component of the temperature gradient and, according to (3.19), between long-scale cells of the convective flow.

Langer's Theorem

Though we have managed to investigate the stability of the periodic solutions (3.36) by a direct analytical computation of the growth rate, it is worth describing the general method of stability investigation developed by Langer [11] that can be applied in more difficult cases.

Let us consider a problem

$$\frac{\partial Z}{\partial \tau}+\Delta[\Delta Z-f(Z)]=0 \tag{3.46}$$

(the dimension of the space is not important). It is obvious that (3.26) is a particular case of (3.46). Equation (3.46) is considered in a finite region with some lateral boundary conditions. Let us define the inner product (Φ,Ψ) for any functions satisfying those boundary conditions in such a way that Δ is a self-adjoint operator:

$$(\Phi,\Delta\Psi)=(\Delta\Phi,\Psi). \tag{3.47}$$

For instance, we can consider periodic boundary conditions with arbitrary large period and take

$$(\Phi,\Psi)\equiv\int\Phi^*\Psi d\mathbf{r}. \tag{3.48}$$

Let us check the stability of a stationary solution $Z_0(\mathbf{r})$ satisfying the equation

$$\Delta[\Delta Z_0 - f(Z_0)] = 0. \tag{3.49}$$

For normal disturbances $\tilde{Z} \exp(-\lambda_n \tau)$, we find

$$\Delta[\Delta \tilde{Z}_n - W(\mathbf{r})\tilde{Z}_n] = \lambda_n \tilde{Z}_n, \tag{3.50}$$

$$W(\mathbf{r}) = \frac{df}{dZ}\Big|_{Z=Z_0(\mathbf{r})}. \tag{3.51}$$

Later on, we assume $\lambda_1 \leq \lambda_2 \leq \ldots$. The solution is unstable if $Re\,\lambda_1 < 0$.

Let us consider an auxiliary Schrödinger equation

$$[-\Delta + W(\mathbf{r})\Psi_n] = \Lambda_n \Psi_n. \tag{3.52}$$

There exists the following fundamental result obtained by Langer [11]: if the problem (3.52) has negative eigenvalues, then the problem (3.50) has negative eigenvalues, too.

Applying Langer's theorem to stability of stationary solutions of equation (3.26), we obtain the stability problem identical to the stability problem for stationary solutions of the Allen–Cahn equation (see Section 2.1.2) and find that all stationary solutions except kinks are unstable.

3.1.1.3 Kink Dynamics

Let us consider the Cahn–Hilliard equation (3.26) in an extended but finite region $-l < X < l$. Assume that the region includes several large-scale convective cells divided by distant domain walls (kinks and antikinks).

Though the stationary kink and antikink solutions of the Allen–Cahn equation and those of the Cahn–Hilliard equation are identical, the temporal dynamics of multikink systems is different. The origin of that difference is the conservation law (3.30) which is valid only for the Cahn–Hilliard equation. In the case of the Allen–Cahn equation, the kink dynamics is based on the effective attraction between a kink and an antikink, which is governed by equations (2.79) and (2.80). That attraction leading to a decrease of the distance between kinks governed by equation (2.88) is impossible in the case of the Cahn–Hilliard equation, because it would violate the conservation law (3.30). Instead, the kinks can move *in pairs* (see Figure 3.1). Thus, we cannot expect that the motion of Cahn–Hilliard kinks is governed by *separate* equations of motion for each kink in the spirit of (2.82). The motions of different kinks are strongly correlated.

As an example, consider a system of kinks with coordinates ξ_{2m-1}, $m = 1, \ldots, n$ and antikinks with coordinates ξ_{2m}, $j = 1, \ldots, n-1$ placed in the region $-l < x < l$, $-l < \xi_1 < \xi_2 < \ldots \xi_{2n-1} < l$. Denote $\xi_0 \equiv -l$ and $\xi_{2n} \equiv l$. If the distances between kinks are large, the value of $A(x)$ is approximately -1 on the intervals $\xi_{2m} < X < \xi_{2m+1}$, $m = 0, \ldots, n-1$ and 1 on the intervals $\xi_{2m+1} < X < \xi_{2m+2}$, $j = 0, \ldots, n-1$. Thus,

$$\int_0^L A(X)dX \approx -(\xi_1 - \xi_0) + (\xi_2 - \xi_1) + \ldots + (\xi_{2n} - \xi_{2n-1}) = 2\sum_{j=1}^{2n-1}(-1)^j\xi_j.$$

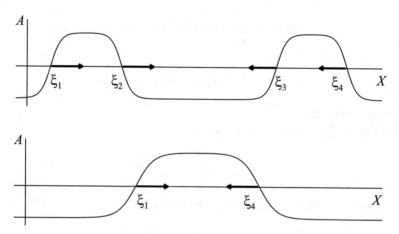

Fig. 3.1 Dynamics of two kink–antikink pairs governed by the Cahn–Hilliard equation.

A similar relation is obtained in that case of an even number of kinks and antikinks and in the case where the coordinates of antikinks have odd subscripts, while the coordinates of antikinks have even subscripts. The conservation law (3.30) prescribes a global relation that couples the velocities of kinks and antikinks

$$\sum_{j=1}^{N}(-1)^j\frac{d\xi_j}{d\tau} = 0, \tag{3.53}$$

where N is the total number of domain walls.

Let us describe briefly how the equations of motions of domain walls are obtained in the case of the Cahn–Hilliard equation. A detailed derivation can be found in [12]. For the sake of simplicity, let us impose the boundary conditions

$$A_X = A_{XXX} = 0$$

at the distant lateral boundaries $X = \pm l$; hence $J|_{X=l} = J|_{X=-l} = 0$, and the conservation law (3.30) is valid. Inverting the operator $\partial^2/\partial X^2$, we rewrite (3.26) as

$$\frac{1}{2}\int_{-l}^{l} dY|X - Y|A_\tau(Y,\tau) + A_{XX} + A - A^3 + C_1(\tau)X + C(\tau) = 0, \tag{3.54}$$

where we have used an integral expression for the nonlocal operator $(\partial/\partial X)^{-2}$. Differentiating (3.54) with respect to X,

$$\frac{1}{2}\int_{-l}^{l} dY \operatorname{sign}(X-Y)A_\tau(Y,\tau)+J(X,\tau)+C_1(\tau)=0, \tag{3.55}$$

and using the boundary conditions $J(\pm l,\tau)=0$, we find:

$$X=l, \quad \frac{1}{2}\int_{-l}^{l} dY A_\tau(Y,\tau)+C_1(\tau)=0$$

and

$$X=-l, \quad -\frac{1}{2}\int_{-l}^{l} dY A_\tau(Y,\tau)+C_1(\tau)=0,$$

therefore $C_1(\tau)=0$.

Let us consider now the system of kinks,

$$A\approx A_j=\tanh\left[\frac{X-\xi_j(\tau)}{\sqrt{2}}\right]$$

for X around ξ_j with an odd value of j, and antikinks,

$$A\approx A_j=-\tanh\left[\frac{X-\xi_j(\tau)}{\sqrt{2}}\right]$$

for even values of j. In the whole region, we can present the leading-order solution as

$$A(X,\tau)\approx A_i(X,\tau)+\sum_{j>i}[A_j(X,\tau)-(-1)^{j-1}]+\sum_{j<i}[A_j(X,\tau)-(-1)^j], \tag{3.56}$$

where i is arbitrary and the leading-order expressions for A_j are

$$A_j(X,\tau)=(-1)^{j-1}\tanh\frac{X-\xi_j(\tau)}{\sqrt{2}}. \tag{3.57}$$

Therefore,

$$A_\tau=-\sum_{j=1}^{N} A_{j,X}\frac{d\xi_j(\tau)}{d\tau}. \tag{3.58}$$

Using the same arguments as in Section 2.1.3, we can write an equation for the correction to A_i and impose its solvability condition which is the orthogonality of the right-hand side to the homogeneous solution $A_{i,X}$.

The difference between equation (3.54) (with $C_1=0$) and Allen–Cahn equation (2.56) is twofold: the former equation contains a nonlocal operator $(\partial/\partial X)^{-2}$ acting on A_τ and an additional term with $C(t)$. Therefore, the abovementioned orthogonality condition, instead of (2.82), gives the following coupled system of equations of motion for kinks:

$$-\sum_j \int_{-l}^{l} dX A_{i,X} \left(\frac{\partial}{\partial X}\right)^{-2} A_{j,X} \frac{d\xi_j}{d\tau} = -\frac{\partial U}{\partial \xi_i} - C(\tau) \int_{-l}^{l} A_{i,X} dX, \qquad (3.59)$$

where U is determined by formula (2.85). In the limit $l \to \infty$, we obtain the following system of equations:

$$-\sum_{j=1}^{N} b_{ij} \frac{d\xi_j}{d\tau} = 16 \left(e^{-(x_{i+1}-x_i)\sqrt{2}} - e^{-(x_i-x_{i-1})\sqrt{2}}\right) + 2(-1)^i C(\tau), \quad i = 1, \ldots, N,$$

$$(3.60)$$

where

$$b_{ij} = \frac{1}{4}(-1)^{i-j} \int_{-\infty}^{\infty} dX \int_{-\infty}^{\infty} dY \cosh^{-2}\frac{X-\xi_i}{\sqrt{2}} |X-Y| \cosh^{-2}\frac{Y-\xi_i}{\sqrt{2}} \approx 2(-1)^{i-j}|\xi_i - \xi_j|.$$

$$(3.61)$$

System (3.60) has to be solved together with equation (3.53).

As an example, consider a pair of kink and antikink, which is governed by the following system of equations:

$$2(\xi_2 - \xi_1)\frac{d\xi_1}{d\tau} = 16e^{-(\xi_2-\xi_1)\sqrt{2}} - 2C(\tau),$$

$$2(\xi_2 - \xi_1)\frac{d\xi_2}{d\tau} = -16e^{-(\xi_2-\xi_1)\sqrt{2}} + 2C(\tau),$$

$$-\frac{d\xi_1}{d\tau} + \frac{d\xi_2}{d\tau} = 0.$$

The solution of that system is

$$C = 8e^{-(\xi_2-\xi_1)}, \quad \frac{d\xi_1}{d\tau} = \frac{d\xi_2}{d\tau} = 0.$$

Thus, in a contradistinction to the case of the Allen–Cahn equation, the kinks do not move.

In the case of two kink–antikink pairs, the mutual interaction of kinks leads to a decrease of the distance between the pairs (see Figure 3.1) and annihilation of colliding kink and antikink (not described by the asymptotic theory presented above).

More examples can be found in [12].

3.1.2 Low Biot Number

In the case of small Biot number Bi, the linear stability theory shows that the critical wavenumber k_c is finite but small, $k_c = O(B^{1/4})$. Thus, the natural small parameter of the problem characterizing the spatial scale of the pattern is $Bi^{1/4}$. However, in order to keep the notations similar to those obtained in the case of the zero Biot number, we shall use a formal small parameter ε which is proportional to $Bi^{1/4}$ but not necessarily equal to it,

$$Bi = b\varepsilon^4,\tag{3.62}$$

where b is a certain positive number. That small parameter will be used in the way similar to parameter k_m, i.e., we define $x_1 = \varepsilon x$, $t_4 = \varepsilon^4 t$, and $R = R_0 + R_2\varepsilon^2$ and construct the solution in the form

$$\Psi = \sum_{n=0}^{\infty} \varepsilon^n \Psi_n, \quad T = \sum_{n=0}^{\infty} \varepsilon^n T_n$$

but using (3.7) rather than (3.13). The latter modification of the boundary condition creates an additional term in the amplitude equation [4]:

$$\frac{\partial T_0}{\partial t_4} + \frac{17}{462}\frac{\partial^4 T_0}{\partial x_1^4} + \frac{R_2}{720}\frac{\partial^2 T_0}{\partial x_1^2} - \frac{10}{7}\frac{\partial}{\partial x_1}\left[\left(\frac{\partial T_0}{\partial x_1}\right)^3\right] + 2bT_0 = 0.\tag{3.63}$$

If we define $\varepsilon = (R/R_0 - 1)^{1/2}$, then $R_2 = R_0 = 720$. By mean of the rescaling of variables, we can reduce equation (3.63) to the standard form,

$$\frac{\partial A}{\partial \tau} + \frac{\partial}{\partial X}\left[\frac{\partial^3 A}{\partial X^3} + \frac{\partial A}{\partial X} - \left(\frac{\partial A}{\partial X}\right)^3\right] + \alpha A = 0,\tag{3.64}$$

where α is a positive coefficient proportional to b. The obtained *Gertsberg–Sivash-insky equation* has a variational structure:

$$\frac{\partial A}{\partial \tau} = -\frac{\delta F}{\delta A},\tag{3.65}$$

where

$$F[A(X)] = \int\left(\frac{1}{2}A_{XX}^2 - \frac{1}{2}A_X^2 + \frac{1}{4}A_X^4 + \frac{\alpha}{2}A^2\right)dX.\tag{3.66}$$

Therefore, the temporal evolution of the system leads to a certain stationary pattern.

Linearizing (3.66), we find that the growth rate σ of the disturbance with the wavenumber K is determined by the relation

$$\sigma(K) = -\alpha + K^2 - K^4.\tag{3.67}$$

Thus, the most unstable mode has the wavenumber $K_0 = 1/\sqrt{2}$, and the instability takes place as $\alpha < \alpha_0 = 1/4$. Recall that formerly we applied the transformation $x_1 = \varepsilon x$, $\varepsilon \ll 1$; therefore $K = O(1)$ corresponds actually to longwave disturbances. In a contradistinction to the case of the Cahn–Hilliard equation, there is no unlimited coarsening, i.e., unbounded growth of the spatial scale. The numerical simulation reveals the formation of a stationary cellular structure [4].

For studying pattern formation, it may be convenient to carry out an additional rescaling $X = \bar{X}\sqrt{2}$, $\tau = 4\bar{\tau}$, which makes the critical wavenumber equal to 1 and defines $\delta^2 = 1 - 4\alpha$. Dropping the bars, we rewrite equation (3.64) in the form

$$\frac{\partial A}{\partial \tau} = \delta^2 A - \left(1 + \frac{\partial^2}{\partial X^2}\right)^2 A + \frac{\partial}{\partial X}\left[\left(\frac{\partial A}{\partial X}\right)^3\right], \tag{3.68}$$

which resembles the Swift–Hohenberg equation [13] but has another nonlinear term.

Let us mention also the modification of the amplitude equation in the case of a weak temperature dependence of nondimensional viscosity,

$$v = \left[1 + \varepsilon\mu\left(\frac{1}{2} - z + T\right)\right].$$

The term $\Delta^2 \Psi$ in (3.10) is replaced by

$$(v\Psi_{xx})_{xx} + 2(v\Psi_{xz})_{xz} + (v\Psi_{zz})_{zz},$$

which leads to an additional term $-\mu\partial/\partial x_1(T_0\partial T_0/\partial x_1)$ in the left-hand side of equation (3.63). After rescaling, the amplitude equation becomes

$$\frac{\partial A}{\partial \tau} + \frac{\partial}{\partial X}\left[\frac{\partial^3 A}{\partial X^3} + \frac{\partial A}{\partial X} - \gamma A\frac{\partial A}{\partial X} - \left(\frac{\partial A}{\partial X}\right)^3\right] + \alpha A = 0, \tag{3.69}$$

where γ is proportional to μ. The temporal evolution of an initial disturbance leads to formation of stationary cells which have no symmetry $A \to -A$.

3.1.3 Two-Dimensional Patterns

3.1.3.1 Amplitude Equations

Let us consider now the full three-dimensional problem, where the temperature field T_0 is a function of two slowly changing coordinates, $x_1 = \varepsilon x$ and $y_1 = \varepsilon y$, where $\varepsilon = O(Bi^{1/4})$. Because of the rotational symmetry of the problem, it seems to be natural to replace equation (3.63) with its rotationally invariant modification,

$$\frac{\partial T_0}{\partial t_4} + \frac{17}{462}\Delta_\perp^2 T_0 + \frac{R_2}{720}\Delta_\perp T_0 - \frac{30}{7}|\nabla_\perp T_0|^2\Delta_\perp T_0 + 2bT_0 = 0, \tag{3.70}$$

where

$$\nabla_\perp = \left(\frac{\partial}{\partial x_1}, \frac{\partial}{\partial y_1}\right), \quad \Delta_\perp = \nabla_\perp^2.$$

Actually, such an equation can be derived but only in the limit of $P \to \infty$ [4]. In the case of a finite Prandtl number, the dynamics of the system cannot be described by a single partial differential equation for the temperature. The reason is the nonlocal dependence of the velocity field on the leading-order temperature field [5].

Indeed, the relation $\nabla \cdot \mathbf{v} = 0$ allows to present the velocity field as $\mathbf{v} = \nabla \times \Phi$, where Φ is a vector potential of the velocity field. In the two-dimensional case considered formerly, $\Phi = -\Psi e_y$, and Ψ is "slaved" to T_0 according to formulas (3.18)–(3.20) and (3.22). In the three-dimensional case, vector Φ has all three components. Because vector Φ is defined up to a gradient of a scalar function, it can be presented using only two scalar functions,

$$\Phi = \psi e_z + \nabla \times (\chi e_z), \tag{3.71}$$

where ψ and χ are toroidal potential and poloidal potential, respectively. Note that in the two-dimensional case

$$\psi = 0,$$

and the scalar stream function

$$\Psi = \frac{\partial \chi}{\partial x}.$$

The relation between the velocity components and potentials is given by the formulas

$$v_x = \frac{\partial \psi}{\partial y} + \frac{\partial^2 \chi}{\partial x \partial z}, \ v_y = -\frac{\partial \psi}{\partial x} + \frac{\partial^2 \chi}{\partial y \partial z}, \ v_z = -\left(\frac{\partial^2 \chi}{\partial x^2} + \frac{\partial^2 \chi}{\partial y^2}\right). \tag{3.72}$$

Therefore, the boundary conditions for the potentials on the solid boundaries, in the absence of a net flow through the layer, can be chosen as

$$\psi = 0, \ \chi = \frac{\partial \chi}{\partial z} = 0 \text{ at } z = 0, 1. \tag{3.73}$$

Taking $(\mathbf{curl})_z$ and $(\mathbf{curl}\ \mathbf{curl})_z$ of both sides of the Navier–Stokes equation (3.1) and introducing $x_1 = \varepsilon x$, $y_1 = \varepsilon y$, one obtains [5]:

$$\nabla_\perp^2 \left(\frac{\partial^4 \chi}{\partial z^4} - RT\right) = -2\varepsilon^2 \nabla_\perp^4 \frac{\partial^2 \chi}{\partial z^2} + O(\varepsilon^4), \tag{3.74}$$

$$\nabla_\perp^2 \frac{\partial^2 \chi}{\partial z^2} = \varepsilon^2 \left(-\nabla_\perp^4 \psi + P^{-1}\nabla_\perp \nabla_\perp^2 \chi \wedge \nabla_\perp \frac{\partial^2 \chi}{\partial z^2}\right) + O(\varepsilon^4), \tag{3.75}$$

where

$$\nabla_\perp f \wedge \nabla_\perp g \equiv f_y g_x - f_x g_y.$$

Equation for temperature is written as

$$\varepsilon^4 \frac{\partial T}{\partial t_4} + \varepsilon^2 \nabla_\perp T \cdot \nabla_\perp \frac{\partial \chi}{\partial z} - \varepsilon^2 T \frac{\partial \chi}{\partial z} + \varepsilon^2 \nabla_\perp T \wedge \nabla_\perp \psi = \frac{\partial^2 T}{\partial z^2} + \nabla_\perp^2 T - \nabla_\perp^2 \chi. \tag{3.76}$$

One can see that the disturbance of T directly determines the disturbance of χ similarly to (3.20) and (3.22) (below $R_0 = 720$ is substituted):

$$\chi_0 = 30 T_0 (z^4 - 2z^3 + z^2), \qquad (3.77)$$

$$\chi_2 = \nabla_\perp^2 T_0 \left(\frac{1}{7} z^{10} - \frac{5}{7} z^9 + \frac{15}{14} z^8 - 3z^6 + 6z^5 - 5z^4 + \frac{10}{7} z^3 + \frac{1}{14} z^2 \right) +$$

$$(\nabla_\perp T_0)^2 \left(\frac{10}{7} z^9 - \frac{45}{7} z^8 + \frac{60}{7} z^7 - \frac{100}{7} z^3 + \frac{75}{7} z^2 \right) +$$

$$30 T_2 (z^4 - 2z^3 + z^2) + \frac{R_2}{24} T_0 (z^4 - 2z^3 + z^2). \qquad (3.78)$$

The leading-order term for the toroidal potential is obtained from the equation

$$\nabla_\perp^2 \frac{\partial^2 \psi_1}{\partial z^2} = P^{-1} \nabla_\perp \nabla_\perp^2 \chi_0 \wedge \nabla_\perp \frac{\partial^2 \chi_0}{\partial z^2}$$

with boundary conditions

$$\psi_1 = 0 \text{ at } z = 0, 1.$$

We find that

$$\psi_1 = \Psi_1(x_1, y_1, t_4) \frac{900}{P} \left(\frac{3}{14} z^8 - \frac{6}{7} z^7 + \frac{19}{15} z^6 - \frac{4}{5} z^5 + \frac{1}{6} z^4 + \frac{1}{105} z \right), \qquad (3.79)$$

where $\Psi_1(x_1, y_1, t_4)$ satisfies the elliptic equation

$$\nabla_\perp^2 \Psi_1 = \nabla_\perp \nabla_\perp^2 T_0 \wedge \nabla_\perp T_0. \qquad (3.80)$$

The solvability condition for the problem at the fourth order in ε, in addition to the terms given by (3.70), contains the term

$$\frac{15}{7P} \nabla_\perp T_0 \wedge \nabla_\perp \Psi_1.$$

A simple rescaling of variables brings the system of governing equations to the form [5]:

$$\frac{\partial A}{\partial \tau} = \delta^2 A - (1 + \nabla_\perp^2 A)^2 A + \frac{1}{3} \nabla_\perp \cdot (|\nabla_\perp A|^2 \nabla_\perp A) - \frac{1}{2P} \nabla_\perp A \wedge \nabla_\perp B, \qquad (3.81)$$

$$\nabla_\perp^2 B = \nabla_\perp \nabla_\perp^2 A \wedge \nabla_\perp A. \qquad (3.82)$$

The amplitude function A describes the temperature distribution in the layer, while the amplitude function B characterizes a large-scale toroidal flow induced by pattern distortions.

3.1.3.2 Pattern Selection

Equations (3.81) and (3.82) are valid for arbitrary $\delta = O(1)$, and they can be simulated numerically. In the present section, we consider only the vicinity of the critical parameter value, $\delta^2 \ll 1$, where $|A| \ll 1$, $|B| \ll 1$.

Linearizing (3.81), we find that the growth rate of the disturbance with the wavevector $\mathbf{K} = (K_x, K_y)$ is determined by the relation

$$\sigma(K_x, K_y) = \delta^2 - (1 - K^2)^2, \tag{3.83}$$

hence it is positive inside the ring $K_-^2 < K^2 < K_+^2$, where $K_{\pm}^2 = 1 \pm \delta$. Note that the width of that ring is small with respect to its radius; therefore one can apply the methods developed formerly for studying patterns generated by shortwave instabilities (see Appendix C).

In the case $0 < \delta \ll 1$, it is natural to expect that the nontrivial solution A of equations (3.81) and (3.82) is proportional to a square root of the governing parameter δ^2, i.e., $A = O(\delta)$, while $B = O(\delta^2)$. Also, because the maximum value of the linear growth rate $\sigma_m = \delta^2$, we can expect that at least at the linear stage of the growth, the characteristic time scale of growing disturbances is $T = \delta^2 \tau$. Let us construct bounded solutions of equations (3.81) and (3.82) in the infinite region $\mathbf{X} \in R^2$ in the form:

$$A(\mathbf{X}, T) = \delta A_1(\mathbf{X}, T) + \delta^2 A_2(\mathbf{X}, T) + \ldots, \quad B(\mathbf{X}, T) = \delta^2 B_1(\mathbf{X}, T) + \delta^3 B_2(\mathbf{X}, T) + \ldots \tag{3.84}$$

We substitute (3.84) into (3.81) and (3.82) and take into account that $\partial/\partial \tau = \delta^2 \partial/\partial T$.

At the leading order, $O(\delta)$, we find:

$$-(1 + \nabla^2)^2 A_1 = 0. \tag{3.85}$$

A bounded solution can be presented as a sum (integral) of a finite (infinite) number of plane waves with wavevectors $\mathbf{K}_n$, $|\mathbf{K}_n| = 1$. Below, we shall assume that the number of plane waves is finite. Because the order parameter A is real, we get

$$A_1(\mathbf{r}, T) = \sum_{n=1}^{N} \left[a_n(T) e^{i\mathbf{K}_n \cdot \mathbf{r}} + a_n^*(T) e^{-i\mathbf{K}_n \cdot \mathbf{r}} \right], \tag{3.86}$$

where $*$ means complex conjugate. The reader can find the general analysis for arbitrary N elsewhere [7]. Here we consider the interaction of two orthogonal roll systems taking $N = 2$ in (3.86) and assuming $\mathbf{K}_1 \perp \mathbf{K}_2$ (see Figure 3.2):

$$A_1 = \left(a_1 e^{i\mathbf{K}_1 \cdot \mathbf{r}} + a_1^* e^{-i\mathbf{K}_1 \cdot \mathbf{vr}} \right) + \left(a_2 e^{i\mathbf{K}_2 \cdot \mathbf{r}} + a_2^* e^{-i\mathbf{K}_2 \cdot \mathbf{r}} \right), \tag{3.87}$$

$$K_1^2 = K_2^2 = 1, \quad \mathbf{K}_1 \cdot \mathbf{K}_2 = 0.$$

Obviously,

$$\nabla_{\perp}^2 A_1 = -A_1,$$

therefore,

$$\nabla_{\perp}^2 B_2 = 0;$$

hence $B_2 = 0$ (recall that B_2 is defined up to a constant and there is no net flux along the layer).

Substituting

$$\nabla_\perp A_1 = i\mathbf{K}_1 \left(a_1 e^{i\mathbf{K}_1 \cdot \mathbf{r}} - a_1^* e^{-i\mathbf{K}_1 \cdot \mathbf{r}}\right) + i\mathbf{K}_2 \left(a_2 e^{i\mathbf{K}_2 \cdot \mathbf{r}} - a_2^* e^{-i\mathbf{K}_2 \cdot \mathbf{r}}\right),$$

$$\begin{aligned}
|\nabla_\perp A_1|^2 &= - \left(a_1^2 e^{2i\mathbf{K}_1 \cdot \mathbf{r}} - 2|a_1|^2 + a_1^{*2} e^{-2i\mathbf{K}_1 \cdot \mathbf{r}}\right) \\
&\quad - \left(a_2^2 e^{2i\mathbf{K}_2 \cdot \mathbf{r}} - 2|a_2|^2 + a_2^{*2} e^{-2i\mathbf{K}_2 \cdot \mathbf{r}}\right)
\end{aligned}$$

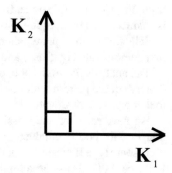

Fig. 3.2 Basic wavevectors of square patterns.

into the right-hand side of the equation for A_3, we find the following solvability conditions:

$$\frac{da_1}{dT} = (1 - |a_1|^2 - \frac{2}{3}|a_2|^2)a_1; \quad \frac{da_2}{dT} = (1 - |a_2|^2 - \frac{2}{3}|a_1|^2)a_2, \tag{3.88}$$

or

$$\frac{dR_1}{dT} = (1 - R_1^2 - \frac{2}{3}R_2^2)R_1; \quad \frac{dR_2}{dT} = (1 - R_2^2 - \frac{2}{3}R_1^2)R_2; \tag{3.89}$$

$$\frac{d\theta_1}{dT} = 0; \quad \frac{d\theta_2}{dT} = 0,$$

where we present the complex amplitudes in the form $A_n = R_n \exp(i\theta_n)$, $R_n \geq 0$, $n = 1, 2$.

System (3.89), which has four stationary solutions:

(i) $R_1 = R_2 = 0$ (quiescent state);
(ii) $R_1 = 1$, $R_2 = 0$ (rolls);
(iii) $R_1 = 0$, $R_2 = 1$ (rolls);
(iv) $R_1 = R_2 = \sqrt{3/5}$ (squares),

is a particular example of the system

$$\frac{dI_n}{dT} = \frac{1}{2}(1 - M_{nn}I_n - \sum_{m \neq n} M_{nm}I_m)I_n, \; n = 1, \ldots, N, \tag{3.90}$$

where $I_n = R_n^2$ and M_{nm} are elements of a symmetric matrix satisfying the conditions

$$M_{11} = M_{22} = \ldots = M_{nn} > 0.$$

System (3.90) belongs to the class of problems of *the competition of species* [14], which is considered in more detail in Appendix C. The stability of stationary solutions is determined by the value of $\kappa = M_{12}/M_{11}$. If $\kappa > 1$ ("competition"), only roll patterns (ii) and (iii) are stable [15]. The relation $|\kappa| < 1$ corresponds to the case of a *symbiosis of species*: solutions (ii) and (iii) are saddle points, while the solution (iv) is a stable node. In our case $\kappa = 2/3$, hence, the stable stationary state is characterized by a "symbiosis" of two roll systems, i.e., it is a square pattern.

Selection of the square pattern in a layer with poorly conducting boundaries was first demonstrated by Busse and Riahi [1] and Proctor [3].

The analysis presented above leaves open the question on the stability of the stationary square pattern with respect to disturbances beyond ansatz (3.87). We return to that question in Chapter 7.

We have revisited the classical problem of the Rayleigh–Bénard convection in a layer with poorly conducting boundaries, which is a paradigmatic example of a monotonic type II instability in the presence of a (weakly violated) conservation law. In the case of a Boussinesq liquid, only cubic nonlinearities appear in the amplitude equation; in a weakly non-Boussinesq case, a quadratic term is also present. In the next section, we consider flows in layers with a free surface or an interface between two fluids. New physical effects and new instability mechanisms will be included.

3.2 Bénard–Marangoni Convection

It is well-known that the buoyancy is not the sole origin of convective instability. If the upper boundary of the layer is open to a gas, the *thermocapillary effect*, i.e., the dependence of the surface tension γ on the temperature T_*, becomes a significant effect. Typically, the surface tension decreases when the temperature grows (the *normal* thermocapillary effect: $\gamma_T = -d\gamma/dT_* > 0$).

Assume that d_* is the layer thickness and θ is the temperature difference across the layer. On the first stage, let us disregard the buoyancy effect (its influence on the surface tension-driven convection will be discussed in Section 3.2.4). The flow in the layer is governed by equations

$$\frac{\partial \mathbf{v}_*}{\partial t_*} + (\mathbf{v}_* \cdot \nabla_*)\mathbf{v}_* = -\frac{1}{\rho}\nabla_* p_* + \nu \Delta_* \mathbf{v}_*, \tag{3.91}$$

$$\frac{\partial T_*}{\partial t_*} + (\mathbf{v}_* \cdot \nabla_*)T_* = \chi \Delta_* T_*, \tag{3.92}$$

$$\nabla_* \cdot \mathbf{v}_* = 0 \tag{3.93}$$

with boundary conditions on the "free surface" $z_* = h_*(x_*, y_*)$,

$$(p_* - \rho g h_*) - \gamma \left(\frac{1}{R_{*1}} + \frac{1}{R_{*2}} \right) = \sigma'_{*ik} n_i n_k, \tag{3.94}$$

$$\sigma'_{*ik} \tau_i^{(l)} n_k - \frac{\partial \gamma}{\partial x_{*i}} \tau_i^{(l)} = 0, \; l = 1, 2, \tag{3.95}$$

$$\frac{\partial h_*}{\partial t_*} + v_{*x} \frac{\partial h_*}{\partial x_*} + v_{*y} \frac{\partial h_*}{\partial y_*} = v_{*z}. \tag{3.96}$$

For temperature and heat flux on the upper boundary, we use an empirical condition

$$\kappa \frac{\partial T_*}{\partial x_{*i}} n_i = -K(T_* - T_g), \tag{3.97}$$

where K is the heat exchange coefficient and T_g is a characteristic temperature of the ambient gas.

The nondimensional system of equations is (3.1)–(3.3) with $R = 0$. The boundary conditions on the free surface $z = h$ (when the influence of a gas is disregarded) are written in the nondimensional form as follows [16]:

$$p - Gah - Ca(1 - \delta_\gamma T) \left(\frac{1}{R_1} + \frac{1}{R_2} \right) = \sigma'_{ik} n_i n_k, \tag{3.98}$$

$$\sigma'_{ik} \tau_i^{(l)} n_k + M \frac{\partial T}{\partial x_i} \tau_i^{(l)} = 0, \; l = 1, 2, \tag{3.99}$$

$$\frac{\partial h}{\partial t} + v_x \frac{\partial h}{\partial x} + v_y \frac{\partial h}{\partial y} = v_z, \tag{3.100}$$

$$\frac{\partial T}{\partial x_i} n_i = -Bi(T - \bar{T}_g), \tag{3.101}$$

where

$$\sigma'_{ik} = \left(\frac{\partial v_i}{\partial x_k} + \frac{\partial v_k}{\partial x_i} \right)$$

is the viscous stress tensor, $\mathbf{n}$ is the normal vector directed into the gas phase, $\tau^{(l)}$ ($l = 1, 2$) are the tangent vectors to the surface, R_1 and R_2 are the curvature radii of the surface, $Ga = g d_*^3 / v\chi$ is the *modified Galileo number*, $Ca = \gamma_0 d_* / \eta \chi$ is the *modified inverse capillary number*, $M = \gamma_T \theta d_* / \rho v \chi$ is the *Marangoni number*, $Bi = K d_* / \kappa$ is the *Biot number*, $\delta_\gamma = \gamma_T \theta / \gamma_0$, and $\bar{T}_g = T_g / \theta$. We use the expression "modified" in order to distinguish between the parameters defined above and the standard definitions of the Galileo number, $G = g d_*^3 / v^2$, and inverse capillary number, $C = \gamma_0 d / \eta v$. The ratio of the Rayleigh number to the Marangoni number can be written as $R/M = (d_*/d_c)^2$, where $d_c^2 = \gamma_T / g \beta \rho$. In "thick" layers, $d \gg d_c$, the buoyancy mechanism of instability prevails, and the thermocapillary effect can be neglected. In "thin" layers, $d_* \ll d_c$, the thermocapillary effect plays the dominant role, and the buoyancy is not important.

It is necessary to emphasize that the boundary conditions formulated above, which contain the modified Galileo number, are *incompatible* with the dynamic equation (3.1) written in the Boussinesq approximation, if the Rayleigh number $R \neq 0$. The Boussinesq approximation is based on the assumption of small relative deviations of density: $\delta_\beta = \beta\theta = R/Ga \ll 1$ (see [3]). If the latter condition is violated, the consideration of non-Boussinesq corrections in the equation of motion and in the continuity equation is mandatory [17]. Later on, we either assume that $Ga \gg 1$ and disregard the surface deformation or assume that $Ga = O(1)$ and disregard R.

3.2.1 Poorly Conducting Boundaries

3.2.1.1 Amplitude Equations

Similar to the case of the buoyancy convection, poorly conducting boundaries are a reason for a longwave type of the convective instability. A longwave expansion similar to those described above can be applied. In the limit $Ga \to \infty$, the deformation of the free surface can be neglected. Assuming that the Biot number on the solid bottom interface is small, $Bi = be^4$, and that one free surface vanishes, one obtains the following one-dimensional equation for the leading-order temperature distribution (*cf.* (3.63)) [18]:

$$\frac{\partial T_0}{\partial t_4} + \frac{1}{15}\frac{\partial^4 T_0}{\partial x_1^4} + \frac{\partial}{\partial x_1}\left\{\left[1 - \frac{48}{35}\left(\frac{\partial T_0}{\partial x_1}\right)^2 + \frac{13}{10}\frac{\partial^2 T_0}{\partial x_1^2}\right]\frac{\partial T_0}{\partial x_1}\right\} + bT_0 = 0. \quad (3.102)$$

The crucial difference between the present equation and the Gertsberg–Sivashinsky equation is the presence of a *quadratic nonlinear term*. Actually, that term is generic, and its absence in the case of the Boussinesq buoyancy convection is a specific, non-generic feature of the latter problem.

The two-dimensional generalization of equation (3.102) is nontrivial even in the limit $P \to \infty$, because the one-dimensional quadratic nonlinear term can be generalized in two different ways, $\nabla_\perp \cdot (\nabla_\perp^2 T_0 \nabla_\perp T_0)$ and $\nabla_\perp^2 |\nabla_\perp T_0|^2$ [19]. As explained in the previous section, in the case of finite P, a toroidal velocity potential, Ψ_0, is produced. Also, if the modified Galileo number Ga and modified inverse capillary number Ca are finite, the temperature inhomogeneity creates a surface deformation H_0, which generates a number of nonlinear terms [20].

Finally, one obtains the following system of equations:

$$\frac{\partial T_0}{\partial t_4} + \nabla_\perp T_0 \wedge \nabla_\perp \Psi_1 + \Delta_\perp T_0 - \Delta_\perp H_0 + \frac{1}{15}\Delta_\perp^2 T_0 - \frac{48}{35}\nabla_\perp \cdot (|\nabla_\perp T_0|^2 \nabla_\perp T_0) +$$

$$\left(\frac{1}{10} + \frac{1}{5P}\right)\nabla_\perp \cdot (\Delta_\perp T_0 \nabla_\perp T_0) + \left(\frac{3}{5} + \frac{1}{10P}\right)\Delta_\perp T_0 |\nabla_\perp T_0|^2 +$$

$$2\nabla_\perp \cdot (H\nabla_\perp T_0) + bT_0 = 0, \quad (3.103)$$

$$\Delta_\perp \Psi_1 = \frac{312}{35P}\nabla_\perp\Delta_\perp T_0 \wedge \nabla_\perp T_0 - 24\nabla_\perp H_0 \wedge \nabla_\perp T_0,$$

$$Ga\Delta_\perp H_0 - Ca\Delta_\perp^2 H_0 = -72\Delta_\perp T_0.$$

It is convenient to rescale the variables and present system (3.103) in the following form [20]:

$$\frac{\partial A}{\partial \tau} + \nabla_\perp A \wedge \nabla_\perp B + 2\Delta_\perp A - 2\Delta_\perp H + \Delta_\perp^2 A - \nabla_\perp \cdot (|\nabla_\perp A|^2 \nabla_\perp A)+$$

$$\lambda \nabla_\perp \cdot (\Delta_\perp A \nabla_\perp A) + \mu \Delta_\perp |\nabla_\perp A|^2 + v\nabla_\perp \cdot (H\nabla_\perp A) + \tilde{b}A = 0, \qquad (3.104)$$

$$\Delta_\perp B = p^{-1}\nabla_\perp \Delta_\perp A \wedge \nabla_\perp A - q\nabla_\perp H \wedge \nabla_\perp A, \qquad (3.105)$$

$$\Delta_\perp(gH - c\Delta_\perp H) = -\Delta_\perp A, \qquad (3.106)$$

where

$$\lambda = \frac{\sqrt{7}}{8}\left(1 + \frac{2}{P}\right), \; \mu = \frac{3\sqrt{7}}{4}\left(1 + \frac{1}{6P}\right), \; v = \frac{2\sqrt{7}}{6}, \qquad (3.107)$$

$$p = \frac{2P}{13}, \; q = \frac{7}{3}, \; g = \frac{Ga}{72}, \; c = \frac{5Ca}{48}, \; \tilde{b} = \frac{4}{15}b.$$

It should be noted that the "one-sided" model based on boundary conditions (3.98)–(3.101) ignores completely the influence of the gas phase on the instability of the liquid. While the viscous stresses created by the gas can usually be neglected, the penetration of the temperature disturbances into the gas phase may be relevant. Corresponding corrections to equations (3.104)–(3.106) can be found in [20].

3.2.1.2 Pattern Selection

The pattern selection for a Rayleigh–Bénard convection in a layer with poorly conducting boundaries was considered in Section 3.1.3.2. It was found that in the case of Boussinesq convection, a square pattern with basic wavevectors $K_1 \perp K_2$ is selected.

As mentioned above, the crucial difference between the Rayleigh–Bénard problem and the Bénard–Marangoni problem is the presence of quadratic terms in the amplitude equations (3.104) which violate the symmetry $\{A \rightarrow -A, B \rightarrow B, H \rightarrow -H\}$. For patterns with wavevectors satisfying the resonant condition $K_1 + K_2 + K_3 = 0$ (the basic example is the hexagonal pattern; see Figure 3.3), because of the relations

$$(-K_2) + (-K_3) = K_1, \; (-K_3) + (-K_1) = K_2, \; (-K_1) + (-K_2) = K_3,$$

quadratic nonlinear terms in the equations generate nonzero quadratic terms in the solvability condition already at the second order of the small-amplitude expansion.

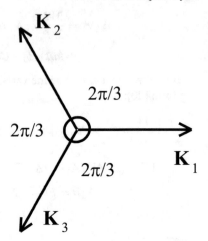

Fig. 3.3 Basic wavevectors of hexagonal patterns

A special role of the hexagonal patterns was first established in the PhD thesis of Busse [21] (see also [22]) who studied pattern formation by a shortwave Rayleigh–Bénard instability in a weakly non-Boussinesq fluid. Because of the presence of a quadratic term in the bifurcation equation for the hexagonal pattern, the pitchfork bifurcation is replaced by a transcritical bifurcation. One of the types of hexagons (either "up-hexagons" with the fluid moving upward in the middle of a hexagonal cell or "down-hexagons" with the opposite direction of the fluid motion) exists in the region of the linear stability of the motionless (conductive) state. In the limit of large g and c, the direction of the subcritical motion is determined by the sign of the parameter $\lambda - \mu$ [19]. Using expressions (3.107), one can find that up-hexagons exist in the subcritical parameter region, when $P > P_c = 0.2$, and down-hexagons exist there, when $P < P_c$.

The small-amplitude solution bifurcating into the stability region of the motionless state is unstable. Strictly speaking, the analysis of the competition of different kinds of patterns and finding a stable selected pattern by means of the perturbation theory is possible only if the quadratic term contains a small parameter, so that the contribution of that term is shifted to the third order (*cf.* (3.69)). In the case of the Marangoni instability, the quadratic term in the bifurcation equation is small only near the point $P = P_c$, where the transition from the hexagonal patterns and other kinds of patterns (square patterns or roll patterns) takes place when the pattern amplitudes are small. Generally, the quadratic terms are not small; therefore there may exist finite-amplitude subcritical solutions that cannot be found by the perturbation theory methods. Nevertheless, an approximate approach using a formal expansion of the solution in powers of an artificial small parameter and combination of quadratic and cubic terms of the solvability conditions obtained in different orders may give reasonable predictions. The rules of the pattern selection are summarized in Appendix C.2.3.

For Marangoni convection in a layer with poorly conducting boundaries, the corresponding analysis has been carried out in [20]. For a water–air system, the competition between up-hexagons and squares has been predicted in the case of large values of g and c (weakly deformable interface). With the decrease of g and c (significant deformations of the interface), new patterns become stable, namely, rolls, down-hexagons, and even a quasiperiodic dodecagonal structure (the stability criteria for the latter structure are described in [23]; see also Section C.3.3). For even smaller values of g and c, squares become subcritical; therefore the small-amplitude approach does not work anymore.

A direct numerical simulation of equation (3.104) for infinite p, g, and c (i.e., $B = 0$, $H = 0$) performed by Shtilman and Sivashinsky [24] and Pontes et al. [25] showed that hexagonal patterns appear near the instability threshold. The direction of motion (flow up or down in the center of the hexagons) depends on the sign of the parameter $\lambda - \mu$, in accordance with the theory of Knobloch [19]. Near the point $\lambda = \mu$, competition between hexagonal and square patterns is observed.

Numerical simulation of the full system (3.104)–(3.106) has been carried out in [26] for large and moderate values of g and c. The value of p was always large. In the case of large Galileo number and capillary number, when the mean flow described by the variable B is suppressed, regular hexagonal patterns divided by domain walls (chains of penta–hepta defects) are observed. Far from the threshold, a transition to square patterns takes place. For moderate g and c, the mean flow causes a strong distortion of hexagonal patterns and formation of skewed hexagons close to the instability threshold. Far from the instability threshold, one observes a chaotic evolution of irregular cellular patterns.

3.2.2 Deformational Instability: Weakly Nonlinear Approach

If the Galileo number is not large, another type of longwave Marangoni instability plays the principal role. While the longwave nature of the instability at small Biot number is caused by the (weakly broken) conservation law for the averaged temperature, the longwave instability at moderate values of the Galileo number is connected with the conservation of the liquid volume. The active variable is now the local layer thickness $h(x,y,t)$. There exist an infinite number of stationary base solutions corresponding to a motionless liquid layer of arbitrary thickness d_*. A longwave inhomogeneity of the layer thickness evolves in time slowly (due to the action of the gravity and surface tension). For the growth rate, one finds $\lim_{k \to 0} \sigma(k) = 0$, and this circumstance is favorable for the development of a longwave instability. That instability can be explained qualitatively in the following way.

Let us consider a layer heated from below with a local depression of the fluid level. The depressed area of the surface is closer to the heated lower boundary, and hence it is warmer. The corresponding thermocapillary tangential stress generates a motion away from the depression that strengthens this depression. In the case of longwave surface deformations, the only factor that can control such an instability is gravity (because both capillarity and dissipative effects are negligible in the longwave limit). Because of this, the longwave limit of the critical Marangoni number

$M_c = M(0)$ is proportional to the modified Galileo number Ga. The deformational instability becomes important for very thin layers and under microgravity conditions.

First, let us discuss briefly the description of the deformational instability in the framework of the "one-sided" approach. That means that the flow in the gas layer is ignored, while the heat transfer is modeled by a constant heat transfer coefficient. For more details, see [16, 27, 28].

Assume that the deviation of the nondimensional local thickness from the main thickness of the layer, $\tilde{h} = h - 1$, is small, because the Marangoni number M is close to the critical value M_c. In that case, the nonlinear evolution of the interfacial deformation is governed at the leading order by the *Funada equation* [29]:

$$\frac{\partial \tilde{h}}{\partial t} = -\Delta_\perp \left[a\Delta_\perp \tilde{h} + b(M - M_c)\tilde{h} - c\tilde{h}^2 \right]. \tag{3.108}$$

where $a > 0, b > 0, c$ are some constant real coefficients. That equation, which is also known as the *Sivashinsky equation* [30], can be presented in the form

$$\frac{\partial \tilde{h}}{\partial t} = \Delta_\perp \frac{\delta F}{\delta \tilde{h}}, \tag{3.109}$$

with the functional

$$F\{\tilde{h}\} = \int \left(\frac{a}{2}(\nabla_\perp \tilde{h})^2 - \frac{b}{2}(M - M_c)\tilde{h}^2 + \frac{c}{3}\tilde{h}^3 \right) dxdy, \tag{3.110}$$

which contains a cubic nonlinearity, and hence it is not bounded from below. Therefore, a blowup in finite time is possible [31]; in that case the assumption of small $\tilde{h}$ is violated. The physical reason of that circumstance is the fact that the longwave instability is nonsaturable: a blowup corresponds to the formation of dry spots [32–37]. The absence of saturation is unavoidable in the framework of the one-layer approach. Indeed, let us take a layer with mean thickness corresponding to the critical value of the Marangoni number and consider a longwave modulation of its local thickness $h(x,y)$. The local *critical* Marangoni number is proportional to the Galileo number and hence to $h^3(x,y)$. The *actual* local Marangoni number is proportional to $h(x,y)$. Thus, instability is enhanced in the regions where the thickness decreases.

3.2.2.1 Two-Layer Problem Formulation

In the present section, we present a detailed description of the deformational Marangoni instability using the two-layer approach. We consider a system of two layers of different immiscible fluids that fill a gap between two rigid horizontal plates $z_* = -d_2$ and $z_* = d_1$. The fluids are separated by a deformable interface, which is described by the equation $z_* = h_*(x_*, y_*, t_*)$. The upper fluid (marked by the subscript 1) occupies the region $h_* < z_* < d_1$, and the lower fluid (subscript 2) occupies the region $-d_2 < z_* < h_*$. The origin is chosen in such a way that $\langle h_* \rangle = 0$

(here averaging is performed with respect to spatial coordinates). The system is heated or cooled from below. We assume that

$$T_* = 0 \text{ as } z_* = d_1; \ T_* = s\Theta \text{ as } z = -d_2, \tag{3.111}$$

where Θ is a characteristic temperature difference, $s = \pm 1$. The m-th fluid ($m = 1, 2$) has density ρ_m, kinematic viscosity ν_m, dynamic viscosity $\eta_m = \rho_m \nu_m$, heat diffusivity κ_m, and heat conductivity λ_m. All these parameters are assumed to be constant. The gravity acceleration is equal to g. The surface tension coefficients γ on the interface is assumed to be a linear function of the temperature: $\gamma = \gamma_0 - \gamma_T (T_* - \bar{T}_{*0})$, where $\bar{T}_{*0}$ is the temperature of the interface in the quiescent state.

In the present section, we choose $d_* = d_1 + d_2$, d_*^2/ν_1, ν_1/d_*, $\rho_1 \nu_1^2/d_*^2$ and Θ as the scales of length, time, velocity, pressure, and temperature. Let us introduce nondimensional parameters:

$$\rho = \frac{\rho_1}{\rho_2}, \ \nu = \frac{\nu_1}{\nu_2}, \ \eta = \frac{\eta_1}{\eta_2} = \rho\nu, \ \kappa = \frac{\kappa_1}{\kappa_2}, \ \lambda = \frac{\lambda_1}{\lambda_2}, \tag{3.112}$$

$$G = \frac{g d_*^3 \rho_1 (\rho_2 - \rho_1)}{\eta_1^2} \text{ (Galileo number)}, \tag{3.113}$$

$$M = \frac{\gamma_T \Theta d_*}{\eta_1 \nu_1} \text{ (modified Marangoni number)}, \tag{3.114}$$

$$C = \frac{\gamma a}{\eta_1 \nu_1} \text{ (inverse capillary number)}, \tag{3.115}$$

$$P = \frac{\nu_1}{\kappa_1} \text{ (Prandtl number)}, \ H = \frac{d_2}{d_*}. \tag{3.116}$$

The nondimensional equations governing the flow and heat transfer in both fluids are (see, e.g., [38]):

$$\frac{\partial \mathbf{v}_1}{\partial t} + (\mathbf{v}_1 \cdot \nabla)\mathbf{v}_1 = -\nabla p_1 + \Delta \mathbf{v}_1, \ \nabla \cdot \mathbf{v}_1 = 0, \tag{3.117}$$

$$\frac{\partial T_1}{\partial t} + \mathbf{v}_1 \cdot \nabla T_1 = \frac{1}{P} \Delta T_1, \tag{3.118}$$

$$\frac{\partial \mathbf{v}_2}{\partial t} + (\mathbf{v}_2 \cdot \nabla)\mathbf{v}_2 = -\rho \nabla p_2 + \frac{1}{\nu} \Delta \mathbf{v}_2, \ \nabla \cdot \mathbf{v}_2 = 0, \tag{3.119}$$

$$\frac{\partial T_2}{\partial t} + \mathbf{v}_2 \cdot \nabla T_2 = \frac{1}{\kappa P} \Delta T_1, \tag{3.120}$$

where p_m, $m = 1, 2$ are differences between the total pressure and the hydrostatic pressure $-\rho_m g z$. The boundary conditions are:

$$z = 1 - H : \mathbf{v}_1 = 0, T_1 = 0. \tag{3.121}$$

$$z = -H : \mathbf{v}_2 = 0, T_2 = s. \tag{3.122}$$

$$z = h :$$

$$p_1 - p_2 + \frac{1}{R}(Ca - MT) + Gah = S_{ik}n_i n_k; \tag{3.123}$$

$$S_{ik}\tau_i^{(l)}n_k - M\tau_i^{(l)}\frac{\partial T_1}{\partial x_i} = 0, \ l = 1,2; \tag{3.124}$$

$$\mathbf{v}_1 = \mathbf{v}_2; \tag{3.125}$$

$$\frac{\partial h}{\partial t} + v_{1x}\frac{\partial h}{\partial x} + v_{1y}\frac{\partial h}{\partial y} = v_{1z}; \tag{3.126}$$

$$T_1 = T_2; \tag{3.127}$$

$$\left(\lambda\frac{\partial T_1}{\partial x_i} - \frac{\partial T_2}{\partial x_i}\right)n_i = 0, \tag{3.128}$$

where $1/R$ is the local curvature of the interface, $\mathbf{n}$ is normal vector, and $\tau^{(l)}$ ($l = 1, 2$) are two tangent vectors:

$$S_{ik} = \left(\frac{\partial v_{1i}}{\partial x_k} + \frac{\partial v_{1k}}{\partial x_i}\right) - \eta^{-1}\left(\frac{\partial v_{2i}}{\partial x_k} + \frac{\partial v_{2k}}{\partial x_i}\right). \tag{3.129}$$

The boundary problem (3.117)–(3.128) always has a solution corresponding to the mechanical equilibrium:

$$h = 0; \ \mathbf{v}_m^e = 0, \ p_m^e = 0, \ m = 1,2; \tag{3.130}$$

$$T_1^e = s\frac{1 - H - z}{1 + (\lambda - 1)H}, \ T_2^e = s\frac{1 - H - \lambda z}{1 + (\lambda - 1)H}, \tag{3.131}$$

which becomes unstable, when the Marangoni number M is sufficiently large.

3.2.2.2 Linear Stability Theory

The linear stability theory of the equilibrium state in a two-layer system with respect to stationary disturbances was developed by Smith [39]. Similar to the one-layer problem, there are two main instability modes: a short-wavelength instability (critical wavenumber $k_c \neq 0$) and a long-wavelength instability (critical wavenumber $k_c = 0$).

For the *short-wavelength* instability, the deformation of the boundary is strongly suppressed, if G and C are large (see Section 2.3.2). Stationary hexagonal convection patterns are observed near the threshold, in accordance with the general nonlinear theory [40, 41].

In the long-wavelength region, the deformation of the boundary is a prevailing factor. This *deformational* type of Marangoni instability is the subject of the present section.

Because of the *conservation law* (the mean value of $h(x,y)$ is conserved in the case of incompressible fluids), the dependence of the growth rate on the wavenumber in the long-wavelength region is described by the formula:

$$\sigma = \frac{1}{2}\frac{\partial^3 \sigma}{\partial k^2 \partial M}k^2(M - M_0) + \frac{1}{24}\frac{\partial^4 \sigma}{\partial k^4}k^4 + \ldots, \tag{3.132}$$

(see (1.25)). The neutral stability curve calculated by Smith has the following long-wavelength asymptotics [38]:

$$sM(k) = sM_0(1 + k^2 N + O(k^4)), \tag{3.133}$$

where

$$sM_0 = \frac{2G}{3\kappa}\frac{H(1 - H)[1 + (\eta - 1)H][1 + (\lambda - 1)H]^2}{(1 - H)^2 - \eta H^2}. \tag{3.134}$$

The expression (3.134) shows that in the case $H < 1/\sqrt{\eta}$, the long-wavelength instability takes place by heating from below ($s = 1$), and in the case $H > 1/\sqrt{\eta}$, it takes place by heating from above ($s = -1$). Inside an interval of a definite sign s, depending on parameters η and λ, the function $sM_0(H)$ defined by equation (3.134) can be either monotonic or non-monotonic. In the case of non-monotonicity, $sM_0(H)$ has a local maximum at a certain value $H = H_-$ and a local minimum at a certain value $H = H_+$, $H_- < H_+$. The coefficient N can be positive or negative. In the latter case, the neutral curve has a maximum in the point $k = 0$, and there is no long-wavelength instability. Later on, we shall assume that $N > 0$ and discuss the case $s = 1$.

3.2.2.3 The Structure of the Amplitude Equation

In order to guess the anticipated form of a one-dimensional amplitude equation governing the problem, we have to take into account two circumstances: a) there is no dispersion, because the instability is stationary, and b) the amplitude equation should be invariant to the transformation $x \to -x$. Using the arguments presented in Section 1, we find that the problem should be governed by the *Sivashinsky equation* (1.81), which can be rewritten in the form

$$\frac{\partial h}{\partial t} = \sigma_2(M - M_0)\frac{\partial^2 h}{\partial x^2} + \sigma_4\frac{\partial^4 h}{\partial x^4} + \kappa_2\frac{\partial^2(h^2)}{\partial x^2}, \tag{3.135}$$

where coefficients σ_2 and σ_4 correspond to the data of the linear stability theory.

Moreover, the coefficient κ_2 also can be determined from the results of the linear theory. Let us consider a small constant deviation h of the interface, corresponding to a change of the parameter H: $H \to H + h$. According to formula (3.134), this change leads to a shift of the critical Marangoni number:

$$M_0(H + h) = M_0(H) + K_1 h + O(h^2), \quad K_1 = \frac{dM_0}{dH}. \tag{3.136}$$

At the same time, equation (3.135) predicts that the instability threshold for the stationary solution $h = const$ is equal to $M_0 + (2\kappa_2/\sigma_2)h$. Hence,

$$\kappa_2 = -\frac{\sigma_2 K_1(H)}{2}. \tag{3.137}$$

Near the extrema of the function $M_0(H)$, the coefficients K_1 and κ_2 vanish, and we have to consider the correction of the next order:

$$M_0(H+h) = M_0(H) + \frac{1}{2}K_2 h^2 + O(h^3), \quad K_2 = \frac{d^2 M_0}{dH^2}. \tag{3.138}$$

In this case, the appropriate amplitude equation is the *Cahn–Hilliard equation* (see (1.82)):

$$\frac{\partial h}{\partial t} = \sigma_2(M - M_0)\frac{\partial^2 h}{\partial x^2} + \sigma_4 \frac{\partial^4 h}{\partial x^4} + \kappa_3 \frac{\partial^2(h^3)}{\partial x^2}. \tag{3.139}$$

Similarly, comparing the stability threshold (3.138) with the prediction made using the equation (3.139), we find

$$\kappa_3 = -\frac{\sigma_2 K_2(H)}{6}. \tag{3.140}$$

Because of the rotational isotropy of the problem, we can expect that the three-dimensional convective motions will be described by two-dimensional amplitude equations that we can obtain from (3.135) and (3.139) replacing $\partial^2/\partial x^2$ by $\Delta_\perp = \partial^2/\partial x^2 + \partial^2/\partial y^2$.

The formal derivation of amplitude equations describing the nonlinear evolution of disturbances generated by the deformational Marangoni instability mode is presented in the next subsection.

3.2.2.4 Derivation of Amplitude Equations

Let us consider disturbances with characteristic wavenumbers $O(\varepsilon)$ and $\varepsilon \ll 1$ and introduce "slow" spatial variables $x_1 = \varepsilon x$, $y_1 = \varepsilon y$. In order to have a possibility to consider both supercritical and subcritical regions, we introduce parameter M_2 that can have an arbitrary sign: $M = M_0 + \varepsilon^2 M_2$. For the time scale, we use the estimate given by linear theory: $t_4 = \varepsilon^4 t$. The estimates obtained from the balance of linear and nonlinear term (which is assumed) prescribe the following transformation of variables:

$$h = \varepsilon^2 \bar{h}, \quad T_1 = s\frac{1 - H - z}{1 + (\lambda - 1)H} + \varepsilon^2 \Theta_1, \quad T_2 = s\frac{1 - H - \lambda z)}{1 + (\lambda - 1)H} + \varepsilon^2 \Theta_2, \tag{3.141}$$

$$p_m = \varepsilon^2 P_m, \quad u_m = \varepsilon^3 U_m, \quad v_m = \varepsilon^3 V_m, \quad w_m = \varepsilon^4 W_m (m = 1, 2) \tag{3.142}$$

(u, v, w are x-, y-, and z-components of velocity).

Substituting the expansions (3.141) and (3.142) together with relations

$$M = M_0 + \varepsilon^2 M_2, \tag{3.143}$$

$$\frac{\partial}{\partial x} = \varepsilon \frac{\partial}{\partial x_1}, \frac{\partial}{\partial y} = \varepsilon \frac{\partial}{\partial y_1}, \frac{\partial}{\partial t} = \varepsilon^4 \frac{\partial}{\partial t_4} + \ldots \tag{3.144}$$

into the boundary problem (3.117)–(3.128), we obtain a modified boundary problem containing the parameter ε^2. Let us expand the functions $\bar{h}$, Θ_m, P_m, U_m V_m W_m in powers of ε^2:

$$\bar{h} = h^{(0)} + \varepsilon^2 h^{(2)} + \ldots \tag{3.145}$$

etc., and substitute (3.145) into the modified boundary problem. Now we will collect the terms of the same order in ε^2.

At the lowest order, we obtain the linear eigenvalue problem that gives expression (3.134) for sM_0 and the following solution:

$$\Theta_1^{(0)} = h^{(0)} \frac{s(\lambda - 1)(z - 1 + H)}{[1 + (\lambda - 1)H]^2}, \quad P_1^{(0)} = h^{(0)} \frac{G\eta H^2}{(1 - H)^2 - \eta H^2}, \tag{3.146}$$

$$U_1^{(0)} = \frac{\partial h^{(0)}}{\partial x_1} \cdot \frac{G\eta H^2}{(1 - H)^2 - \eta H^2} \left[\frac{1}{2}(z - 1 + H)^2 + \frac{1}{3}(1 - H)(z - 1 + H) \right], \tag{3.147}$$

$$V_1^{(0)} = \frac{\partial h^{(0)}}{\partial y_1} \cdot \frac{G\eta H^2}{(1 - H)^2 - \eta H^2} \left[\frac{1}{2}(z - 1 + H)^2 + \frac{1}{3}(1 - H)(z - 1 + H) \right], \tag{3.148}$$

$$W_1^{(0)} = \Delta_1 h^{(0)} \cdot \frac{G\eta H^2}{(1 - H)^2 - \eta H^2} \left[-\frac{1}{6}(z - 1 + H)^3 + \frac{1}{6}(1 - H)(z - 1 + H)^2 \right], \tag{3.149}$$

$$\Theta_2^{(0)} = h^{(0)} \frac{s\lambda(\lambda - 1)(z + H)}{[1 + (\lambda - 1)H]^2}, \quad P_2^{(0)} = h^{(0)} \frac{G(1 - H)^2}{(1 - H)^2 - \eta H^2}, \tag{3.150}$$

$$U_2^{(0)} = \frac{\partial h^{(0)}}{\partial x_1} \cdot \frac{G\eta(1 - H)^2}{(1 - H)^2 - \eta H^2} \left[\frac{1}{2}(z + H)^2 - \frac{1}{3}H(z + H) \right], \tag{3.151}$$

$$V_2^{(0)} = \frac{\partial h^{(0)}}{\partial y_1} \cdot \frac{G\eta(1 - H)^2}{(1 - H)^2 - \eta H^2} \left[\frac{1}{2}(z + H)^2 - \frac{1}{3}H(z + H) \right], \tag{3.152}$$

$$W_2^{(0)} = \Delta_1 h^{(0)} \cdot \frac{G\eta(1 - H)^2}{(1 - H)^2 - \eta H^2} \left[-\frac{1}{6}(z + H)^3 + \frac{1}{6}H(z + H)^2 \right], \tag{3.153}$$

where $\Delta_1 = \partial^2/\partial x_1^2 + \partial^2/\partial y_1^2$.

The solvability condition for the equation at the next order in ε^2 leads to an evolution equation for the function $h^{(0)}$ that can be written in the form

$$\frac{\partial h^{(0)}}{\partial t_2} = \sigma_2 M_2 \Delta_1 h^{(0)} + \sigma_4 \Delta_1^2 h^{(0)} + \kappa_2 \Delta_1 (h^{(0)2}), \tag{3.154}$$

where

$$\sigma_2 = -s\lambda\eta\,\frac{H^2(1-H)^2[(1-H)^2-\eta H^2]}{2[1+(\lambda-1)H]^2\{[(1-H)^2-\eta H^2]^2+4\eta H(1-H)\}},\qquad(3.155)$$

$$\sigma_4 = \sigma_2 M_0 N < 0 \ (\text{if } N > 0), \ \kappa_2 = -\frac{1}{2}\sigma_2 K_1.\qquad(3.156)$$

By means of the transformation

$$\tau = -\frac{\sigma_2^2}{\sigma_4}t_2,\ X = \left(\frac{\sigma_2}{\sigma_4}\right)^{1/2}x_1,\ Y = \left(\frac{\sigma_2}{\sigma_4}\right)^{1/2}y_1,\ Z = \left(\frac{\kappa_2}{\sigma_2}\right)h^{(0)},\qquad(3.157)$$

equation (3.154) can be transformed to the *Sivashinsky equation*:

$$\frac{\partial Z}{\partial\tau}+\Delta_\perp^2 Z+M_2\Delta_\perp Z+\Delta_\perp(Z^2)=0,\qquad(3.158)$$

where

$$\Delta_\perp = \frac{\partial^2}{\partial X^2}+\frac{\partial^2}{\partial Y^2}.\qquad(3.159)$$

Since variable Z is proportional to the interface deflection from the mean position, we have

$$\langle Z\rangle = 0.\qquad(3.160)$$

Equation (3.158) was derived in the one-dimensional form for the one-layer problem by Pukhnachev [42] and by Funada [29].

Let us discuss now the case where $K_1 = dM_0/dH = 0$, which corresponds to maxima and minima of function $M_0(H)$ $(H = H_\pm)$. The appropriate scaling of variables is now:

$$h = \varepsilon\bar{h},\ T_1 = s\,\frac{1-H-z}{1+(\lambda-1)H}+\varepsilon\Theta_1,\ T_2 = s\,\frac{1-H-\lambda z)}{1+(\lambda-1)H}+\varepsilon\Theta_2,\qquad(3.161)$$

$$p_m = \varepsilon P_m,\ u_m = \varepsilon^2 U_m,\ v_m = \varepsilon^2 V_m,\ w_m = \varepsilon^3 W_m\,(m=1,\,2),\qquad(3.162)$$

and the expansions are performed in powers of ε:

$$\bar{h} = h^{(0)}+\varepsilon h^{(1)}+\varepsilon^2 h^{(2)}+\dots\qquad(3.163)$$

At the zeroth order, we obtain the results identical to (3.134) and (3.146)–(3.153). In the first order in ε, we find $K_1 = 0$ as a solvability condition, and in the second order in ε, we obtain the evolution equation for $h^{(0)}$:

$$\frac{\partial h^{(0)}}{\partial t_2} = \sigma_2 M_2\Delta_1 h^{(0)}+\sigma_4\Delta_1^2 h^{(0)}+\kappa_3\Delta_1(h^{(0)3}),\qquad(3.164)$$

where σ_2 and σ_4 are the same as in equation (3.154) and

$$\kappa_3 = -\frac{\sigma_2 K_2}{6}, \; K_2 = \frac{d^2 M_0}{dH^2}. \tag{3.165}$$

In the point of the minimum of $M_0(H)$, $H = H_+$, we have $K_2 > 0$, $\kappa_3 < 0$; in the point of the maximum of $M_0(H)$, $H = H_-$, we have $K_2 < 0$, $\kappa_3 > 0$.

By rescaling of independent variables and with

$$Z = \left(\frac{|\kappa_3|}{\sigma_2} \right)^{1/2} h^{(0)}, \tag{3.166}$$

we obtain the *Cahn–Hilliard equation*

$$\frac{\partial Z}{\partial \tau} + \Delta_\perp^2 Z + M_2 \Delta_\perp Z - \mathrm{sgn}(\kappa_3) \Delta_\perp (Z^3) = 0. \tag{3.167}$$

As in the previous case,

$$\langle Z \rangle = 0. \tag{3.168}$$

Finally, let us consider the region $H = H_\pm + O(\varepsilon)$, where K_1 is not equal to zero but small: $K_1 = K_2(H - H_\pm) = O(\varepsilon)$. In this region we obtain instead of (3.164):

$$\frac{\partial h^{(0)}}{\partial t_2} = \sigma_2 M_2 \Delta_1 h^{(0)} + \sigma_4 \Delta_1^2 h^{(0)} + 3[(H - H_\pm)/\varepsilon] \kappa_3 \Delta_1 (h^{(0)2}) + \kappa_3 \Delta_1 (h^{(0)3}),$$
$$\tag{3.169}$$
$$\langle h^{(0)} \rangle = 0. \tag{3.170}$$

Let us define

$$\varepsilon \bar{h}^{(0)} = \varepsilon h^{(0)} + (H - H_\pm), \varepsilon^2 \bar{M}_2 = \varepsilon^2 M^2 + \frac{1}{2}(H - H_\pm)^2 K_2. \tag{3.171}$$

It is obvious that $\varepsilon \bar{h}^{(0)}$ and $\varepsilon^2 \bar{M}_2$ correspond to deviations of actual local values of H and M from $H_\pm$, $M_0(H_\pm)$. Now equation (3.169) becomes similar to (3.164):

$$\frac{\partial \bar{h}^{(0)}}{\partial t_2} = \sigma_2 \bar{M}_2 \Delta_1 \bar{h}^{(0)} + \sigma_4 \Delta_1^2 \bar{h}^{(0)} + \kappa_3 \Delta_1 (\bar{h}^{(0)3}), \tag{3.172}$$

but instead of (3.170) we have

$$\langle \bar{h}^{(0)} \rangle = (H - H_\pm)\varepsilon^{-1}. \tag{3.173}$$

By rescaling of variables, we obtain:

$$\frac{\partial Z}{\partial \tau} + \Delta_\perp^2 Z + \bar{M}_2 \Delta_\perp Z - \mathrm{sgn}(\kappa_3) \Delta_\perp (Z^3) = 0, \tag{3.174}$$

$$\langle Z \rangle = \bar{Z}, \tag{3.175}$$

where

$$Z = \left(\frac{|\kappa_3|}{\sigma_2} \right)^{1/2} \bar{h}^{(0)}, \tag{3.176}$$

$$\bar{Z} = \left(\frac{|\kappa_3|}{\sigma_2} \right)^{1/2} (H - H_\pm)\varepsilon^{-1}. \tag{3.177}$$

3.2.2.5 Lyapunov Functional

Both Sivashinsky equation (3.158) and Cahn–Hilliard equation (3.167) have a *Lyapunov functional*

$$F(Z) = \int L(Z) dx dy, \tag{3.178}$$

but the connection between the amplitude equation and this functional has now the form

$$\frac{\partial Z}{\partial \tau} = \Delta_\perp \frac{\delta F}{\delta Z} \tag{3.179}$$

rather than

$$\frac{\partial Z}{\partial \tau} = -\frac{\delta F}{\delta Z} \tag{3.180}$$

(see Section 2.2.1.2). The Lyapunov functional is

$$F(Z) = \int dx dy \left[\frac{1}{2} \nabla_\perp^2 Z - \frac{1}{2} M_2 Z^2 - \frac{1}{3} Z^3 \right] \tag{3.181}$$

for the Sivashinsky equation and

$$F(Z) = \int dx dy \left[\frac{1}{2} \nabla_\perp^2 Z - \frac{1}{2} M_2 Z^2 + \frac{1}{4} \operatorname{sgn} \kappa_3 Z^4 \right] \tag{3.182}$$

for the Cahn–Hilliard equation. We find that

$$\frac{dF}{d\tau} = -\int dx dy \left(\nabla_\perp \frac{\delta F}{\delta Z} \right)^2 \le 0. \tag{3.183}$$

It should be noted that functionals (3.181) and (3.182) in the case $\operatorname{sgn} \kappa_3 < 0$ ($H = H_-$) *are not bounded from below*. Therefore, the problem (3.179) has generally no global solution, and the time evolution leads to a singularity in a finite time. Only in the case (3.182), $\operatorname{sgn} \kappa_3 > 0$ ($H = H_+$) we can expect that the nonlinear evolution will lead to a certain bounded stationary state.

The physical reason of such a situation is quite clear. Let us take a regular point $H \ne H_\pm$ where $K_1(H) = dM_0/dH \ne 0$ and consider a long-wavelength deformation h of the interface. Because the volume of the fluid is conserved, the ratios of layer thicknesses are larger than H in some regions and smaller than H in other regions.

Since $K_1(H) = dM_0/dH \neq 0$, in some of those regions, the local threshold value of Marangoni number becomes smaller than M_0. In those regions, we can expect a nonsaturated growth of the disturbance and a subcritical finite-amplitude instability. If $K_1(H) = 0$, $K_2(H) < 0$, the finite-amplitude instability arises for $h > 0$ as well as for $h < 0$. Only if $K_1(H) = 0$ and $K_2(H) > 0$ (hence $H = H_+$), we can expect a saturation of the instability in the supercritical region.

As the matter of fact, the singularity means that our *weakly nonlinear* amplitude equations based on the estimates (3.141) and (3.142) cannot describe the final stage of the nonlinear evolution of disturbances, because those estimates are violated in finite time. The equations governing a *strongly nonlinear* $(h = O(1))$ long-wavelength interface deformation in a two-layer system are known [43]. The description of the strongly nonlinear approach to studying the development of the deformational instability for one-layer and two-layer fluid systems is presented in Sections 3.2.3 and 3.3.3.2.

3.2.2.6 Stationary Solutions of Sivashinsky Equation

Though we do not expect the existence of stable regimes of convection in the case governed by equation (3.158), it is instructive to calculate stationary solutions of this problem. Let us consider bounded one-dimensional stationary solutions that satisfy the equation

$$\frac{d^2}{dX^2}\left(\frac{d^2Z}{dX^2} + M_2Z + Z^2\right) = 0; \tag{3.184}$$

$$\langle Z \rangle = 0. \tag{3.185}$$

Problems (3.184) and (3.185) have a class of periodic solutions:

$$Z(X) = -\frac{M_2}{2} - \frac{2-q^2}{3q^2}A + \frac{A}{q^2}\mathrm{dn}^2\sqrt{\frac{A}{6}}\frac{X-\xi}{q}, \tag{3.186}$$

where

$$A = \max_x H - \min_x H = \frac{3q^2M_2}{2[q^2 - 2 + 3E(q)/K(q)]} > 0, \tag{3.187}$$

dn is an elliptical Jacobi function with the modulus q, $E(q)$ and $K(q)$ are complete elliptical integrals, and ξ is an arbitrary constant. The spatial period of the function $H(X)$ is

$$L = \frac{2\pi}{k}, \ k = \frac{\pi\sqrt{A/6}}{qK(q)}. \tag{3.188}$$

Let us consider first the supercritical region $M_2 > 0$ (i.e., $M > M_0$). In this region the condition $A > 0$ leads to the relation $q^2 - 2 + 3E(q)/K(q) > 0$ that is satisfied at $0 < q < q_0$, $q_0 \approx 0.98038$. The dependence of the amplitude A on the wavenumber k is given parametrically by formulas (3.186) and (3.187). Let us emphasize that solutions exist in the *stability region* $k > K_0 = M_2^{1/2}$ of the quiescent state.

If $M_2 = 0$ $(M = M_0)$, the solution (3.186) satisfies the condition (3.185) for the only value $q = q_0$ irrespective of the value of A. The dependence of the amplitude on the wavenumber is self-similar and has the form

$$A = 6 \left(\frac{q_0 K(q_0)}{\pi} \right)^2 k^2 \tag{3.189}$$

for all k.

In the subcritical region $M_2 < 0$ $(M < M_0)$, we have $q_0 < q < 1$. The solution (3.186) exists for any wavenumber. The minimal value of A is reached as $k \to 0$ $(q \to 1)$ for the solitary wave solution:

$$Z = \frac{3}{2} |M_2| \cosh^{-2} \frac{|M_2|^{1/2}(X - \xi)}{2}. \tag{3.190}$$

Let us note that q is close to unity for any wavenumbers; hence the form of the stationary solution strongly differs from a sinusoidal one and resembles a chain of solitary waves.

It is quite obvious that all the solutions described in the previous subsection are unstable, because they bifurcate from the neutral curve $M_2 = k^2$ into the region of larger k and smaller M_2, where the quiescent state is stable (see, e.g., [44]). These solutions can be considered as saddle points in the functional space. The stable manifold of each saddle point separates the regions of growing and decaying finite-amplitude disturbances imposed on the quiescent state. Roughly speaking, $A(k)$ determines such a "threshold" amplitude that the disturbance with the period $2\pi/k$ decays if its amplitude is smaller than $A(k)$ and grows otherwise (certainly, actually the shape of the disturbance is important).

The same result can be found using Langer's theory (see Section 3.1.1.2). One obtains the following auxiliary Schrödinger problem:

$$\left[-\frac{d^2}{dX^2} - (M_2 Z(X) + Z^2(X)) \right] \Psi = \Lambda \Psi. \tag{3.191}$$

Now one can use the standard arguments: there exists solution $\Psi = dZ/dX$, $\Lambda = 0$, that is not sign-preserving. Hence, there exists a sign-preserving eigenfunction with $\Lambda < 0$. Therefore, the stationary solution of the original problem is unstable.

3.2.2.7 Stationary Solutions of Cahn–Hilliard Equation and Their Stability

We have analyzed the dynamics of the Cahn–Hilliard equation (3.167) in Section 3.1.1. Here we just note that the kink solutions of the Cahn–Hilliard equation

$$Z = \pm (M_2)^{1/2} \tanh \left[\left(\frac{M_2}{2} \right)^{1/2} (X - \xi) \right], \tag{3.192}$$

describe a steady steplike surface deformation, supported by a convective roll centered at $X = \xi$. Taking into account (3.166), we find that at large distances from that "step," the ratios of thicknesses of layers are equal to $H_+ \pm [6(M - M_c(H_+))/K_2]^{1/2}$ and are *outside the instability region*

$$H_1 = H_+ - [2(M - M_c)/K_2]^{1/2} < H < H_2 = H_+ + [2(M - M_c)/K_2]^{1/2}. \quad (3.193)$$

Thus, the system avoids the unstable interval of ratios (3.193) by means of generation of a domain wall, separating regions with stable ratios of thicknesses.

The results obtained in the regular case $H \neq H_+$ and in the special case $H = H_+$ are quite different. That is why it is interesting to consider the intermediate case $H - H_+ = O(\varepsilon)$ governed by problems (3.174) and (3.175). We remind that Z is proportional to the local deflection of the ratio of thicknesses from H_+ rather than from the actual mean value H, that is why $\langle Z \rangle$ is nonzero. Similarly, $\bar{M}_2$ is proportional to $M - M_0(H_+)$ rather than $M - M_0(H)$.

For stationary solutions of the problem,

$$\frac{d^2}{dX^2}\left(\frac{d^2Z}{dX^2} + \bar{M}_2 Z - Z^3\right) = 0, \quad (3.194)$$

$$\langle Z \rangle = \bar{H} \quad (3.195)$$

we find:

$$\frac{d^2Z}{dX^2} + \bar{M}_2 Z - Z^3 = C_1 X + C_2, \quad (3.196)$$

where C_1 and C_2 are constants. Because we are interested only in bounded solutions, $C_1 = 0$. Hence, we obtain the problem

$$\frac{d^2Z}{dX^2} + \frac{\partial U}{\partial Z} = 0, \quad (3.197)$$

where the potential $U(Z)$ is

$$U(Z) = \frac{1}{2}\bar{M}_2 Z^2 - \frac{1}{4}Z^4 - C_2 Z. \quad (3.198)$$

The bounded solutions with $< Z > \neq 0$ exist if

$$\bar{M}_2 > 0, \ 0 < |C_2| < 2(\bar{M}_2/3)^{3/2}. \quad (3.199)$$

It is obvious that for fixed $\bar{M}_2$ and C_2, the equation has a set of periodic solutions, such that the integration constant

$$E = \frac{1}{2}\left(\frac{dZ}{dX}\right)^2 + U(Z) \quad (3.200)$$

is situated in the interval $E_- < E < E_+$ and a solitary wave solution for $E = E_+$. The explicit expressions for the solutions in terms of elliptic functions were found in [10].

Let us note however that for all solutions with $\langle Z \rangle \neq 0$, the function dZ/dX is *not sign-conserving*. Hence, Langer's theory predicts an instability of all the stationary solutions *in the infinite region*.

However, it is necessary to take into account that under real conditions the layers have always a finite horizontal size. Let us consider, for the sake of simplicity, the "reflecting" boundary conditions

$$X = \pm l : \quad \frac{\partial Z}{\partial X} = \frac{\partial^3 Z}{\partial X^3} = 0 \tag{3.201}$$

corresponding to zero contact angle and impermeable boundaries (for large l, the precise form of boundary conditions is actually not very important). In that case we can use the same solutions but their spatial periods are quantized. It is obvious that there are exactly two monotonic solutions $Z(X)$ with sign-conserving derivative dZ/dX corresponding to a fragment with the length $2l$ taken from the spatially periodic solution with the period $4l$. For large l, those solutions can be approximated up to exponentially small terms by kink solutions

$$Z = \pm(\bar{M}_2)^{1/2} \tanh \left[\left(\frac{\bar{M}_2}{2} \right)^{1/2} (X - \xi_\pm) \right], \tag{3.202}$$

where the constants $\xi_\pm$ are determined by the condition (3.195):

$$\xi_\pm = \mp l\bar{Z}/(\bar{M}_2)^{1/2}. \tag{3.203}$$

The solutions exist in the whole region

$$H_+ - [6(M - M_c(H_+))/K_2]^{1/2} < H < H_+ + [6(M - M_c(H_+))/K_2]^{1/2} \tag{3.204}$$

that is larger than the linear instability interval (3.193); hence, for any $H \neq H_+$, there is a subcritical instability of the quiescent state. In full analogy with the gas–liquid separation near the critical point (i.e., the original application of the Cahn–Hilliard equation), the solutions describe the separation of the whole region into two subregions, one of a "low-level phase" and another one of a "high-level phase." The thicknesses $H_+ \pm [6(M - M_c(H_+))/K_2]^{1/2}$ are independent on H; only the *volumes* of both phases depend on H.

Hence, in the region of small $H - H_+$, the Marangoni long-wavelength instability is subcritical, but it does not destroy the liquid layers and produces a stationary relief. For finite $H - H_+$ the method of the small parameter is incapable of solving the problem.

3.2.3 Deformational Instability: Strongly Nonlinear Approach

Another approach, first applied for the treatment of the deformational Marangoni instability by Davis [33, 34], is valid for finite $M - M_c$ and h. It is based on the assumption that the surface tension parameter Ca is large. In order to explain the essence of that approach, we start with the consideration of an isothermic film motion.

3.2.3.1 Lubrication Approximation

In the case of *film flows,* when the fluid system is thin in a certain ("transverse") direction and extended in other ("longitudinal") directions, the nonlinear models governing three-dimensional flows with a deformable interface can be drastically simplified. A film flow is strongly affected by interfacial phenomena and *enslaved* to the interface deformation.

To explain the idea of the longwave asymptotic approach, let us consider an isothermic flow in a thin film governed by equations (3.91) and (3.93) with boundary conditions (3.94)–(3.96) on the free surface

$$z_* = h_*(x_*, y_*, t_*), \tag{3.205}$$

and with no-slip condition $\mathbf{v}_* = 0$ on the rigid surface $z_* = 0$. Assume that the characteristic spatial scales in the directions x_* and y_* are much larger than that in the direction z_*, i.e., the solution depends on the scaled horizontal coordinates $\bar{x}_* = \varepsilon x_*$ and $\bar{y}_* = \varepsilon y_*$, $\varepsilon \ll 1$. Also, it is assumed that the solution depends on the scaled time variable $\bar{t}_* = \varepsilon^2 t_*$.

At the leading order, the evolution of the system is governed by the following system of equations and boundary conditions:

$$-\bar{\nabla}_{*\perp} P_* + \eta \frac{\partial^2 \mathbf{V}_*}{\partial z_*^2} = 0, \quad \frac{\partial P_*}{\partial z_*} = 0, \tag{3.206}$$

$$\bar{\nabla}_{*\perp} \cdot \mathbf{V}_* + \frac{\partial W_*}{\partial z_*} = 0; \tag{3.207}$$

$$z = h: P_* - \rho g h_* = 0, \quad \frac{\partial \mathbf{V}_*}{\partial z_*} = 0, \tag{3.208}$$

$$\frac{\partial h_*}{\partial \bar{t}_*} + \mathbf{V}_* \cdot \bar{\nabla}_{*\perp} h_* = W_*; \tag{3.209}$$

$$z = 0: \mathbf{V}_* = 0, \quad W_* = 0, \tag{3.210}$$

where $\mathbf{V}_*$, W_*, and P_* are the leading-order terms of the expansions

$$p_* = P_* + \ldots, \quad v_{*x}\mathbf{e}_x + v_{*y}\mathbf{e}_y = \varepsilon \mathbf{V}_* + \ldots, \quad v_{*z} = \varepsilon^2 W_* + \ldots, \quad \bar{\nabla}_{*\perp} = \mathbf{e}_x \frac{\partial}{\partial \bar{x}_*} + \mathbf{e}_y \frac{\partial}{\partial \bar{y}_*}.$$

Here $\mathbf{e}_x$ and $\mathbf{e}_y$ are unit vectors of the horizontal axes x_* and y_*.

The horizontal velocity components of a flow in a thin film form locally a Poiseuille flow

$$\mathbf{V}_* = \frac{1}{\eta}\bar{\nabla}_{*\perp}P_*\left(\frac{z_*^2}{2} - z_*h_*(\bar{x}_*,\bar{y}_*,\bar{t}_*)\right) \tag{3.211}$$

generated by the longitudinal pressure gradient $\bar{\nabla}_{*\perp}P_*$ (the pressure does not depend on the transverse coordinate z_* at the leading order). Solving the continuity equation (3.207) with respect to W and using the corresponding boundary condition, we obtain:

$$W(\bar{x}_*,\bar{y}_*,h_*,\bar{t}_*) = -\int_0^{h_*}\bar{\nabla}_{*\perp}\cdot\mathbf{V}_*dz_*. \tag{3.212}$$

By means of (3.212), we rewrite the kinematic condition (3.209) in the following form:

$$\frac{\partial h_*}{\partial \bar{t}_*} + \bar{\nabla}_{*\perp}\cdot\bar{\mathbf{q}}_* = 0, \tag{3.213}$$

where

$$\bar{\mathbf{q}}_* = \int_0^{h_*}\mathbf{V}_*dz_* = -\frac{1}{3\eta}h_*^3\bar{\nabla}P. \tag{3.214}$$

Substituting $P_* = \rho g h_*$, we obtain a closed evolution equation for $h(\bar{x}_*,\bar{y}_*,\bar{t}_*)$.

Returning to the initial variables, we obtain the following evolution equation:

$$\frac{\partial h_*}{\partial t_*} + \nabla_{*\perp}\cdot\mathbf{q}_* = 0, \tag{3.215}$$

$$\mathbf{q}_* = -\frac{1}{3\eta}h_*^3\nabla_{*\perp}p_*, \quad p_* = \rho g h_*. \tag{3.216}$$

The approach described above can be applied for the derivation of thin-film evolution equations under the action of different physical factors. In the presence of a capillary pressure,

$$p_* = \rho g h_* - \gamma_0 \nabla^2_{*\perp}h_*, \tag{3.217}$$

where γ_0 is the surface tension coefficient. Let us emphasize that the surface tension is assumed to be strong; therefore the term with the surface tension, otherwise small, is included.

If a tangential stress τ is applied at the interface, then it creates an additional Couette flow, so that

$$\mathbf{q}_* = -\frac{1}{3\eta}h_*^3\nabla_{*\perp}p_* + \frac{1}{2\eta}h_*^2\tau. \tag{3.218}$$

Various applications of this longwave approximation described above (which is called *lubrication approximation*) to different physical problems are described in [27].

3.2.3.2 Intermolecular Forces

In order to correctly describe the rupture of the film caused by an insaturable deformational instability, it is necessary to extend our model. In the case of very thin (but
still macroscopic) films, when the film thickness is less than about 100 nm, a new
physical phenomenon has to be incorporated. It is necessary to take into account
the *long-range intermolecular forces* acting between molecules of the liquid and
substrate [45, 46]. It is essential that these forces act on distances *largely* relative to
interatomic distances. Hence, despite their microscopic (quantum) origin [47–49],
they can be incorporated into a macroscopic theory. It can be shown that when the
pair potential between molecules is $U_*(r_*) \sim 1/r_*^n$, the effective energy of interaction between two planar surfaces at distance h_* is $f(h_*) \sim 1/h_*^{n-4}$. In the framework
of the continuum approach, the intermolecular forces manifest themselves as "surface forces" or *"disjoining pressure"* $\Pi_*(h_*) = df(h_*)/dh_*$ (see, e.g., [50, 51]),
which can be considered as a certain external normal stress imposed on the free
surface; that is, (3.217) is replaced by

$$p_* = \rho g h_* - \gamma_0 \nabla_{1*}^2 h_* + \Pi_*(h_*). \tag{3.219}$$

The sign of the disjoining pressure can be either positive or negative.

If the film is formed by an apolar fluid, the only relevant kind of long-range
intermolecular interaction is the van der Waals interaction with $U_*(r_*) \sim 1/r_*^6$. In
that case, the disjoining pressure can be taken as

$$\Pi_*(h_*) = A/6\pi h_*^3, \tag{3.220}$$

where A is the *Hamaker constant* [45, 52, 53]. Note that the Hamaker constant can
be positive or negative. The law (3.220) is supported by many experimental data
obtained at $h < 30$ nm (see [46]). For larger distances between molecules, another
law of intermolecular interaction is expected, due to the effect of electromagnetic
retardation, $U_*(r_*) \sim 1/r_*^7$. That leads to the prediction

$$\Pi_*(h_*) = \frac{B}{10\pi h_*^4}, \tag{3.221}$$

which is supported by experiments done in the interval 40 nm$< h <$80 nm [46].

In the case of an electrolyte solution, the disjoining pressure includes an electrostatic component caused by a double electric layer on the liquid–gas surface characterized by the Debye screening length $1/\kappa$. In the simplest case, when the surface
charge of the surface equals zero, the electrostatic contribution to the disjoining
pressure in the region $\kappa h_* \ll 1$ is given by the Langmuir formula

$$\Pi_{*e}(h_*) = (\pi \varepsilon_0/8h_*^2)(kT_*/ez)^2, \tag{3.222}$$

where ε_0 is the dielectric constant, k is the Boltzmann constant, T_* is the temperature, e is the electron charge, and z is the ion valency. For $\kappa h_* \sim 1$, an approximate
formula

$$\Pi_{*e}(h) = Ce^{-\kappa h_*} \tag{3.223}$$

can be used.

Another possible component of the disjoining pressure, caused by a structural modification of polar liquids in thin films, has the form

$$\Pi_{*s}(h) = K\exp(-h_*/\lambda_*), \tag{3.224}$$

where, in the case of water, the force magnitude K is of order 10^7 N/m^2 and the length λ_* varies from 2.3 to 3.3 nm (for more details, see [54]).

At very short (atomic) distances, the *steric repulsion* becomes relevant, which is due to overlapping electron shells, and varies as $U_* \sim 1/r_*^{12}$, which yields $f(h_*) \sim 1/h_*^8$, $\Pi_*(h_*) \sim 1/h_*^9$.

In the literature, one can find some combinations of the expressions presented above. A more detailed discussion of the disjoining pressure in the context of the thin-film dynamics can be found in the book [55].

Disjoining pressure dominates at small film thicknesses, and therefore it governs the stability and wettability of ultrathin films. Typically, the gravity term can be neglected for an ultrathin film.

Another example of the situation in which the disjoining pressure has to be taken into account is the motion near the *contact line*. In the latter case, one has to take into account the dependence of the disjoining pressure not only on the film thickness h but also on the film slope h_x [56, 57].

The latter problem demands one more modification of the standard model of a viscous fluid: the no-slip condition $\mathbf{v} = 0$ is violated near the contact line [27, 58]. Indeed, the standard no-slip condition leads to singularities [59, 60]. Note that a real molecular slippage on the solid surface is justified by molecular dynamic simulations [61–64]. Violation of the slip condition changes the velocity profile of the film flow and hence the relation (3.218) between $\mathbf{q}_*$, $\nabla_* p_*$, and τ_*. For instance, if the slippage condition for the tangential velocity component $u_* = v_{*x}$ is

$$u_* - \beta \frac{\partial u_*}{\partial z_*} = 0,$$

then

$$\mathbf{q}_* = -\frac{1}{\eta}\left(\frac{h_*^3}{3} + \beta h_*^2\right)\nabla p_* + \frac{1}{\eta}\left(\frac{h_*^2}{2} + \beta h_*\right)\tau_* \tag{3.225}$$

(see [27]).

3.2.3.3 Nonisothermal Films

Let us return now to the deformational Marangoni instability in a liquid with a strong surface tension.

If a thermocapillary effect is taken into account, then

$$\tau_* = \nabla_* \gamma = -\gamma_T \nabla_* T_*^I, \tag{3.226}$$

where T_*^I is the interfacial temperature. Its dependence on the local thickness h_* is obtained by solving the heat transfer equation with appropriate boundary condition in the longwave limit. For instance, for a film with a fixed temperature T_0 on the bottom and the condition (3.97) on the free surface, one obtains the relation

$$T_*^I(h_*) = T_g + \frac{\kappa(T_0 - T_g)}{\kappa + Kh_*}. \tag{3.227}$$

If the surface tension is strong ($Ca \gg 1$), the deformation instability appears only in the region of longwaves even for finite values of $M - M_c$. Using nondimensional parameters defined above, one obtains the following closed nonlinear equation, which describes the dynamics of a heated ultrathin film on a horizontal solid substrate [27]:

$$\frac{\partial h}{\partial t} + \nabla_\perp \cdot \left\{ \frac{h^3}{3} [Ca\Delta_\perp \nabla_\perp h - Ga\nabla_\perp h - \nabla_\perp \Pi(h)] + \frac{MBih^2}{2(1+Bih)^2} \nabla_\perp h \right\} = 0, \tag{3.228}$$

where $\Pi(h)$ is a (dimensionless) disjoining pressure. The term containing Ca appears in the leading-order amplitude equation, if $Ca = O(\varepsilon^{-2})$. Note that without that term the problem would be ill-posed.

It is important that (3.228) can be presented as a *generalized Cahn–Hilliard equation*

$$\frac{\partial h}{\partial t} + \nabla_\perp \cdot \left\{ Q(h)\nabla \left[Ca\Delta_\perp^2 h - \frac{df(h)}{dh} \right] \right\} = 0, \tag{3.229}$$

where

$$Q(h) = \frac{h^3}{3}, \tag{3.230}$$

$$\frac{df(h)}{dh} = Gah - \frac{3}{2}MBi \left[\ln\left(\frac{h}{1+Bih} \right) + \frac{1}{1+Bih} \right] - \Pi(h). \tag{3.231}$$

Hence, the evolution equation of a film on a horizontal substrate can be written in the *variational form*

$$\frac{\partial h}{\partial t} = \nabla_\perp \left[Q(h)\nabla_\perp \frac{\delta F[h]}{\delta h} \right], \tag{3.232}$$

where

$$F[h] = \int d\mathbf{r} \left[\frac{W_0}{2}(\nabla_\perp h)^2 + f(h) \right] \tag{3.233}$$

is the *Lyapunov functional* of the problem. Because

$$\frac{dF}{dt} = \int d\mathbf{r} \frac{\partial h}{\partial t} \frac{\delta F[h]}{\delta h} = -\int Q(h) \left[\nabla_\perp \frac{\delta F[h]}{\delta h} \right]^2 d\mathbf{r} \le 0,$$

the time evolution of the system leads to a quiescent state corresponding to a minimum of the functional $F[h]$, while persistent oscillations and chaos are impossible.

3.2.4 The Role of the Buoyancy Effect

In the examples considered above, the surface tension inhomogeneity was the mechanism of the instability. However, the instability of the mechanical equilibrium state in a heated liquid layer can be caused also by the buoyancy. Recall that the buoyancy convection is often described in the framework of the *Boussinesq approximation* (see [65]): the liquid is incompressible, the density is constant, and a buoyancy force proportional to a local temperature disturbance acts on the liquid. The basic criterion of the applicability of that approximation is $\delta_\beta = \beta_T A_* d_* \ll 1$, where A_* is the temperature gradient, d_* is the thickness of the layer, and β_T is the heat expansion coefficient.

Assume that a liquid layer with a free surface is under the action of gravity and the condition $\delta_\beta \ll 1$ is satisfied. What is the interplay between the buoyancy and Marangoni instability mechanisms?

The answer is different for deformational and nondeformational modes.

For nondeformational modes the buoyancy effect and the thermocapillary effect may be equally relevant. By heating from below, both instability mechanisms act jointly and create a monotonic instability. For instance, for the longwave instability in a layer between poorly conducting boundaries, the instability threshold is described by the relation of Nield [66],

$$R/R_c + M/M_c = 1. \tag{3.234}$$

Here $R_c = 720$ is the critical value of R for a pure buoyancy convection, and $M_c = 48$ is the threshold value of M for a thermocapillary convection.

When considering the surface deformation, we have to take into account that $R = \delta_\beta Ga$. Hence, if $Ga = O(1)$, which is necessary for the observation of a deformational Marangoni mode, then $R = O(\delta_\beta) \ll 1$; thus the buoyancy effect can be disregarded. If $R = O(1)$, which is necessary for the appearance of a buoyancy instability, then it means $Ga = O(\delta_\beta^{-1}) \gg 1$; hence the deformation of the surface, which is proportional to Ga^{-1}, is a small non-Boussinesq effect [65]. If the surface deformation is taken into account, other non-Boussinesq corrections should be also incorporated (e.g., the correction to the continuity equation violating the incompressibility of the liquid). Otherwise, spurious instabilities can be obtained (see [17]).

However, there is an exception. If there is a two-liquid system with close liquid densities, the deformation can be significant even in the region of the validity of the Boussinesq approximation [67, 68]. Both monotonic and oscillatory instability modes exist. In the limit of a zero Biot number, similar to the case considered in Section 3.1.1, a coupled system of strongly nonlinear equations for longwave surface deformations and temperature disturbances has been derived [69]. However, it turns out that the obtained system has no stable solution, because the shortwave instability, which leads to the development of oscillatory cellular convection, is stronger [69].

In this section we have demonstrated that the Marangoni instability can be of type II due to two different reasons: (i) approximate conservation of the mean temperature in the case of poorly conducting boundaries and (ii) conservation of liquid volume by surface deformation. Correspondingly, there are two modes of longwave instabilities, (i) a mode not related to the surface deformations and (ii) a deformational instability mode. While the former type of instability creates diverse patterns, the latter type of instability often produces a "blowup" which is the manifestation of the film rupture. The analysis presented above is still incomplete: due to the non-self-adjointness of the linear instability problem, an oscillatory instability is possible. Its development is the subject of the next section.

3.3 Marangoni Oscillations and Waves

Now we start a systematic investigation of patterns created by *oscillatory instabilities*. It will be continued in subsequent sections of this book. Generally, the most typical mechanisms that produce oscillatory instabilities of a motionless state are (i) energy supply (e.g., by mode mixing) to preexisting decaying oscillatory modes and (ii) negative feedback with delay. Oscillatory instabilities are extremely widespread in nature and engineering. Besides phenomena in liquids including Marangoni waves [70], convection in binary mixtures [71], and electroconvection [72], oscillatory patterns are observed in reaction–diffusion systems [73], cardiac tissue [74], granular matter [75], and optical systems [76].

In the present section, we consider several kinds of oscillatory instabilities which take place in a heated liquid layer with a free surface.

3.3.1 Transverse Oscillatory Instability by Heating from Above

3.3.1.1 Linear Stability Theory

As the first example of oscillatory instability caused by the destabilization of pre-existing oscillatory modes, we consider the spontaneous generation of waves by heating *from above*, which has been discovered by Levchenko and Chernyakov in 1981 [77] (see also [78]).

Infinitely Deep Layer

In order to include the case of an infinitely deep layer into consideration, in the definition of the nondimensional parameters presented above, we replace the layer thickness d_* by the capillary length $l_c = (\gamma/\rho g)^{1/2}$ (in that case, $Ga = Ca = gl_c^3/\nu\kappa$).

Two kinds of wavy motion are possible in a deep layer heated from above. The first kind is the *longitudinal*, or dilational, wave, first discovered in the pioneering works by Lucassen [79] and Lucassen–Reynders and Lucassen [80]. In order to understand the physical nature of that kind of waves, let us consider a fluid layer on a cooled rigid plate. When a liquid element rises to the free surface, it creates a cold spot. The surface tension gradient acts toward this spot, pushing the element to the bulk. Thus, some oscillations are generated. It can be shown that the dependence of the nondimensional frequency ω on the nondimensional wavenumber k in the limit of large M and k is determined by the relation

$$\omega^2 = \omega_{long}^2 = \frac{MP}{\sqrt{P+1}}k^2 \qquad (3.235)$$

(see [77]).

Another type of wave that is possible in a system with a deformable surface is the *transverse*, or capillary-gravity, wave. It is caused by the joint action of gravity and the surface tension. In the limit of small viscosity, the corresponding dispersion relation is

$$\omega^2 = \omega_{tr}^2 = GaP(k+k^3). \qquad (3.236)$$

In the high-frequency limit, when the decay of waves is neglected, the full dispersion relation of the system is just

$$(\omega^2 - \omega_{long}^2)(\omega^2 - \omega_{tr}^2) = 0. \qquad (3.237)$$

However, the next-order corrections to the dispersion relation (3.237) describe a certain *mixing* between two kinds of waves which is especially strong in the resonant case, where the frequencies of both waves are close. If the next-order corrections are taken into account, any solution $\omega(k)$ has a certain *imaginary part* that corresponds to decay or growth of waves in time. It turns out that an instability is developed with the growth of M and the neutral curve is

$$M = 2^{5/2}(GaP^2)^{3/4}\frac{(1+k^2)^{3/4}}{k^{1/4}}. \qquad (3.238)$$

It has a minimum at

$$k = k_c = \sqrt{5}/5. \qquad (3.239)$$

Layer of the Finite Depth

In the case of a liquid layer of finite thickness, the existence of transverse Marangoni waves by heating from the gas side was justified by Takashima [81]. A generalization of the theory in the case of a two-layer system with a deformable interface has been done in [82].

The instability described above becomes *longwave*, if the layer has a *stress-free* lower boundary. That is a simplified model of a flow in an upper layer of a two-layer liquid system, when the viscosity of the lower layer is small with respect to that of the upper layer. In that case, the minimum of the neutral stability curve is situated at $k = 0$, which provides the possibility to apply longwave expansions [83–86].

The problem under consideration belongs to the class of problems in which the growth of spatially homogeneous disturbances is forbidden by a *conservation law*. Indeed, the homogeneous change of the layer's thickness is impossible because of the conservation of the fluid volume. In this case, there exists the *Goldstone mode* described by formulas (1.22)–(1.26) (in those formulas, R should be replaced by M). That mode generates an instability when $\partial^2 \sigma_r / \partial k^2(0,M)$ becomes positive, which happens for $M > M_c = 12$. Near the threshold point $k = 0$, $M = M_c$, the growth rate can be expanded into a Taylor series (see (1.25) and (1.26)):

$$\sigma_r(k,M) = \sigma_{21}k^2(M - M_c) + \sigma_{40}k^4 + \cdots, \qquad (3.240)$$

$$\sigma_i(k,M) = \sigma_{10}k + \sigma_{11}k(M - M_c) + \sigma_{30}k^3 + \cdots, \qquad (3.241)$$

where

$$\sigma_{21} = \frac{P}{6}, \quad \sigma_{40} = -\frac{2}{105}(17GaP^2 + 204P^2 + 134P + 22),$$

$$\sigma_{10} = \sqrt{Ga + 12}, \quad \sigma_{11} = \frac{\sqrt{P}}{2\sqrt{GaP + 12}},$$

$$\sigma_{30} = \frac{Ca - GaP(8P/5 + 1/3) - (96P/5 + 56/5)}{2\sqrt{(Ga + 12)}}$$

(we use the definitions of parameters Ga and Ca given in Section 3.2). The instability interval is $0 < k < k_m$, where

$$k_m = \sqrt{\frac{\sigma_{21}(M - M_c)}{-\sigma_{40}}}. \qquad (3.242)$$

3.3.1.2 Nonlinear Waves

As indicated in Section 1.4.2, in the simplest case of a one-dimensional wave, the longwave expansions give rise to an evolution equation, which can be written in the form (1.70),

$$\frac{\partial h}{\partial t} = -\left(\frac{\partial q_l}{\partial x} + \frac{\partial q_n}{\partial x}\right). \tag{3.243}$$

Here, the "order parameter" $h(x,t)$ is the surface deformation,

$$-\frac{\partial q_l}{\partial x} = [\sigma_{10} + \sigma_{11}(M - M_c)]\frac{\partial h}{\partial x} - \sigma_{21}(M - M_c)\frac{\partial^2 h}{\partial x^2} - \sigma_{30}\frac{\partial^3 h}{\partial x^3} + \sigma_{40}\frac{\partial^4 h}{\partial x^4}, \tag{3.244}$$

and

$$q_n(h) = \delta_1 h^2 + \delta_2 \frac{\partial}{\partial x}(h^2) + \cdots, \tag{3.245}$$

where

$$\delta_1 = -\frac{3(Ga + 8)\sqrt{P}}{4\sqrt{Ga + 12}}, \quad \delta_2 = -2P.$$

Equation (3.243) is called the *dissipation-modified Korteweg–de Vries* (dmKdV) equation (see Section 1.4.2.1). By means of a scaling transformation of variables, in a moving reference frame, it can be reduced to the standard form (see [87])

$$H_T + H_{XXX} + 3(H^2)_X + \delta[H_{XX} + H_{XXXX} + D(H^2)_{XX}] = 0, \tag{3.246}$$

where a subscript means a partial derivative with respect to the corresponding variable,

$$\delta = \frac{\sqrt{-\sigma_{21}\sigma_{40}(M - M_c)}}{|\sigma_{30}|}, \quad D = -\frac{3\sigma_{30}\delta_2}{\sigma_{40}\delta_1}. \tag{3.247}$$

When δ vanishes, the ideal KdV equation is recovered [88], which has an infinite number of conservation laws. Recall that this equation has a family of traveling wave solutions

$$H(X,T) = H(\xi), \quad \xi = X - cT, \tag{3.248}$$

with arbitrary spatial period $L = 2\pi/q$:

$$H(\xi + 2\pi/q) = H(\xi). \tag{3.249}$$

If we take into account that

$$\langle H \rangle = \int_0^{2\pi/q} H(\xi)d\xi = 0 \tag{3.250}$$

(the mean surface deformation is equal to zero by definition), the corresponding solutions (cnoidal waves) are determined by the formulas [87, 89]

$$u(\xi) = \frac{2q^2 K^2}{\pi^2}\left[\mathrm{dn}^2\left(\frac{(\xi - \xi_0)qK}{\pi}\right) - \frac{E}{K}\right], \tag{3.251}$$

$$c = \frac{4q^2 K^2}{\pi^2}\left(2 - s^2 - \frac{3E}{K}\right), \tag{3.252}$$

where dn is the Jacobi's delta amplitude function with modulus s and $E = E(s)$ and $K = K(s)$ are complete elliptic integrals. The limit $s \to 1$ corresponds to solitary waves. Thus, in the case $\delta = 0$, the wavelength $L = 2\pi/q$ and the wave amplitude

$$A = H_{max} - H_{min} = \frac{2q^2 K^2 s^2}{\pi^2} \quad (3.253)$$

are independent parameters that can take arbitrary values.

When $\delta \neq 0$, the governing equation (3.246) contains three additional terms that describe, respectively, longwave instability of the type II_o, shortwave dissipation, and nonlinear dissipation. The conservation laws characteristic of the pure KdV equation are violated (except that of the conservation of fluid volume). For instance, when $H(X,T)$ is spatially periodic with period $L = 2\pi/q$, the time evolution of the squared deflection of the interface ("momentum") is governed by the equation

$$\frac{d}{dT} \int_0^L H^2 dX = \delta \left(\int_0^L H_X^2 dX - \int_0^L H_{XX}^2 dX + 2D \int_0^L HH_X^2 dX \right). \quad (3.254)$$

The right-hand side of equation (3.254) vanishes only for some definite values of the wave amplitude $A(q)$.

In the limit of small δ, the stationary values of the wave amplitude are calculated analytically (see [87, 89, 90]). There are three different types of behavior of $A(q)$, depending on D:

1. If $D \leq 5/4$, the function $A(q)$ is uniquely defined in the whole region $0 < q < 1$. However, the spatially periodic solutions with stationary value of the amplitude $A(q)$ are stable only inside a certain subinterval $q_- \leq q \leq q_+$.

2. If $5/4 < D < 2$, there are two solutions for $A(q)$ in a certain region $q_{min}(D) < q < 1$ (the lower branch and the upper branch): one solution for $q > 1$ and no solutions for $0 < q < q_{min}(D)$. Only the solutions on the lower branch in the interval $q_{min}(D) < q < 1$ are stable with respect to strictly periodic disturbances with the same spatial period $2\pi/q$. However, even these solutions are unstable with respect to disturbances violating the periodicity of the solution.

3. If $D \geq 2$, there is a unique solution for $q > 1$, which is unstable.

We shall postpone a detailed stability analysis for periodic solutions of the dissipation-modified Korteweg–de Vries equation to Section 5.1, where that equation is obtained in another physical context.

For finite values of δ, the traveling wave solutions of equation (3.246) were studied analytically in [91, 92]. Some numerical simulations of equation (3.246) were carried out in [93, 94].

In the nearly one-dimensional case, when some transverse modulations of one-dimensional waves are taken into account, the problem is governed by the *dissipation-modified Kadomtsev–Petviashvili* equation [85]. By means of a scaling transformation, this equation is reduced to the following form [95]

$$\{H_T + H_{XXX} + 3(H^2)_X + \delta[H_{XX} + H_{XXXX} + D(H^2)_{XX}]\} - 3sH_{YY} = 0, \quad (3.255)$$

where δ and D are determined by formulas (3.247),

$$s = \text{sign}\left(\frac{\lambda_{30}}{\lambda_{10}}\right). \tag{3.256}$$

In the framework of equation (3.256), it can be shown [95] that any one-dimensional waves are unstable with respect to transverse modulations.

In the general case, the waves generated by the instability can propagate in arbitrary directions simultaneously. The full system of longwave equations that describes such waves is rather complicated [86]. That system was used for the investigation of collisions of one-dimensional solitary waves at arbitrary incident angles, formerly observed in experiments [96, 97].

Note that the problem considered above is unrealistic in two aspects. First, the friction of the fluid at the bottom is completely neglected. The friction at the solid lower boundary suppresses instability in the limit $k \to 0$, so that the critical wave number k_c is nonzero. Also, phenomena in the gas phase adjacent to liquid are ignored.

The development of the nonlinear theory in the general case is a formidable task. However, the problem can be essentially simplified in the quite realistic limit $Ga \gg 1$, because in this case, the critical wavenumber k_c is small, and the longwave approach can be applied. Also, one can take into account the fact that the kinematic viscosity and thermal diffusivity of a gas are typically much larger than corresponding parameters of a fluid. Velarde et al. [98] have considered the limit where the characteristic wavenumber satisfied $k \sim \varepsilon \ll 1$, but $\nu \sim \kappa \sim Ga^{1/2}\varepsilon \gg 1$ (actually, the relation $\varepsilon \sim Ga^{-1/10}$ was selected). In this limit, another kind of dissipation-modified KdV equation was obtained, which after rescaling of variables can be written as

$$H_T + H_{XXX} + 3(H^2)_X + \int_{-\infty}^{+\infty} Q(X - X')H(X')dX' = 0. \tag{3.257}$$

The Fourier transform $\hat{Q}(k)$ of the kernel $Q(X - X')$ is a complicated function of parameters that we do not reproduce here.

3.3.2 Deformable Free Surface and Low Biot Number

Consider a liquid layer heated from below, and assume that the modified Galileo number $Ga = O(1)$, while the modified inverse capillary number Ca is large, $Ca = \varepsilon^{-2}\tilde{C}$. We expect that the deformational Marangoni instability is concentrated in the longwave region and apply longwave expansions based on the scaling

$$X = \varepsilon x, \ \tau = \varepsilon^2 t,$$

without any assumptions concerning the value of the Marangoni number, i.e., derive a strongly nonlinear equation for the surface deformation $h(\mathbf{X}, \tau)$. Assume also that the layer is heated by a fixed heat flux on the bottom and the Biot number at the free surface is small, $Bi = \varepsilon^2 \beta$. Thus, there is one more Goldstone mode corresponding the finite-amplitude temperature disturbance $F(\mathbf{X}, \tau)$. Applying the long-wave approach [99], one obtains the following system of nondimensional amplitude equations [100, 101]:

$$\frac{\partial h}{\partial \tau} = \nabla \cdot \left(\frac{h^3}{3} \nabla R + \frac{Mh^2}{2} \nabla f \right), \tag{3.258}$$

$$h \frac{\partial F}{\partial \tau} = -\beta f + \nabla \cdot \left(\frac{h^4}{8} \nabla R + \frac{Mh^3}{6} \nabla f + h \nabla F \right) \tag{3.259}$$

$$+ \left(\frac{h^3}{3} \nabla R + \frac{Mh^2}{2} \nabla f \right) \cdot \nabla f - \frac{1}{2} (\nabla h)^2,$$

where

$$R = Gah - \tilde{C} \Delta h$$

is the pressure disturbance and

$$-f = h - F$$

is a perturbation of the surface tension.

The linear stability analysis for a disturbance with the wavenumber $k = \varepsilon K$ reveals a monotonic instability mode with the neutral stability curve

$$M_m(K) = \frac{48(\beta + K^2)\tilde{G}}{K^2(72 + \tilde{G})}, \quad \tilde{G} = Ga + \tilde{C}K^2, \tag{3.260}$$

which gives a smooth transition between the limits known for the nondeformational mode ($M_m \to 48$ as $K \to \infty$) and the deformational mode ($M_m \to 2Ga/3$ as $\beta = 0$, $K \to 0$) [102]. Besides, an oscillatory mode exists with the neutral stability curve

$$M_o(K) = 3 + \frac{3\beta}{K^2} + \tilde{G} \tag{3.261}$$

and frequency

$$\Omega(K) = \frac{K^2}{12} \sqrt{(72 + \tilde{G})(M_m(K) - M_o(K))}. \tag{3.262}$$

The comparison of expressions (3.261) and (3.262) shows that the oscillatory instability takes place for sufficiently large β (that is why it was not found in the previous works where the scaling $Bi = O(\varepsilon^4)$ was postulated) and not too large G. The detailed analysis of the competition between monotonic and oscillatory linear modes can be found in [103].

A weakly nonlinear analysis, which determines the pattern selection, and the direct numerical simulation of equations (3.258) and (3.259) have been carried out in [101].

The derivation of amplitude equations and the analysis of instability have been done also using a more precise, two-layer, formulation of the problem, where the Biot number is not postulated but calculated using the parameters of the gas phase [104].

3.3.3 Systems with Two Deformable Interfaces

Another example of longwave oscillatory patterns is the Marangoni wave in a layered system with two deformable boundaries.

3.3.3.1 Three-Layer System Between Two Solid Plates

As an example of the application of a weakly nonlinear approach, let us consider a three-layer system between two solid plates. The linear stability theory [105] predicts the existence of two stationary instability boundaries $M = M_1$ and $M = M_2$ and in some cases also the oscillatory instability boundary $M = M_0$. In the latter case, the problem is governed by a system of *two coupled equations* for deformations of both interfaces h_1 and h_2. For instance, in the case when the relative thickness of the bottom layer is small in comparison with the thicknesses of the top layer and the middle layer [106], the corresponding coupled equations for the rescaled interfacial deformations H_1 and H_2 of both interfaces are:

$$\frac{\partial H_2}{\partial \tau} + \Delta(\Delta H_2 + \mu H_2 + \gamma H_2^2 - H_2^3 + H_1) = 0, \qquad (3.263)$$

$$\frac{\partial H_1}{\partial \tau} - \Delta H_2 = 0, \qquad (3.264)$$

where

$$\Delta = \frac{\partial}{\partial X^2} + \frac{\partial}{\partial Y^2}.$$

Direct numerical two-dimensional simulations of system (3.263)–(3.264) [107] show that the most typical wavy patterns are traveling rolls and traveling squares. Near the boundary between the regions of the abovementioned patterns, alternating rolls are observed. This pattern is a nonlinear superposition of two systems of standing waves with orthogonal wave vectors. The temporal phase shift between standing waves of different spatial orientations is equal to $T/4$, where T is the full period of oscillations. Thus, one observes some kind of roll patterns that change their orientation with the time interval $T/4$. Note that the alternating roll pattern is one of the generic wave patterns that appear in rotationally invariant systems due to a primary oscillatory instability of the spatially homogeneous state [108–110] (see Appendix D). Also, some spatially chaotic patterns were found. In the latter flow regime, the deformation of the upper interface displays irregular "spots" of a nearly flat interface that split and merge in a chaotic manner.

3.3.3.2 Two-Layer System with Two Deformable Boundaries

Marangoni Convection

Formulation of the Problem

Let us consider now a system of two superposed layers of immiscible liquids with different physical properties in the framework of the strongly nonlinear approach. The bottom layer (layer 1) rests on a solid substrate of temperature T_s; the top layer (layer 2) is in contact with the adjacent gas phase of temperature T_g. The deformable interfaces are described by equations $z_* = h_{*1}(x,y,t)$ (liquid–liquid interface) and $z_* = h_{*2}(x,y,t)$ (liquid–gas interface); see Figure 3.4. The ith liquid has density ρ_i, kinematic viscosity v_i, dynamic viscosity $\eta_i = \rho_i v_i$, thermal diffusivity χ_i, and heat conductivity κ_i. The surface tension coefficients on the lower and upper interfaces, γ_1 and γ_2, are linear functions of temperature T: $d\gamma_1/dT = -\alpha_1$, $d\gamma_2/dT = -\alpha_2$.

At the first stage, let us formulate the problem *in the absence* of intermolecular forces and the gravitational force, when the Marangoni effect is the only source of convective flow. The complete system of nonlinear equations governing Marangoni convection is written in the following form:

$$\frac{\partial \mathbf{v}_{*i}}{\partial t_*} + (\mathbf{v}_{*i}\nabla_*)\mathbf{v}_{*i} = -\frac{1}{\rho_i}\nabla_* P_{*i} + v_i \Delta_* \mathbf{v}_{*i}, \tag{3.265}$$

$$\frac{\partial T_{*i}}{\partial t_*} + \mathbf{v}_{*i}\nabla_* T_{*i} = \chi_i \Delta_* T_{*i}, \tag{3.266}$$

$$\nabla \cdot \mathbf{v}_{*i} = 0, \; i = 1,2. \tag{3.267}$$

The boundary conditions on the rigid boundary are as follows:

$$\mathbf{v}_{*1} = 0, \; T_{*1} = T_s \; \text{at} \; z_* = 0. \tag{3.268}$$

On the deformable interface $z = h_{*1}$, the following boundary conditions are imposed: the balance of normal stresses,

$$P_{*2} - P_{*1} + 2\gamma_1 K_1 = \left[-\eta_{*1}\left(\frac{\partial v_{*1i}}{\partial x_{*k}} + \frac{\partial v_{*1k}}{\partial x_{*i}}\right) + \eta_2\left(\frac{\partial v_{*2i}}{\partial x_{*k}} + \frac{\partial v_{*2k}}{\partial x_{*i}}\right)\right]n_{1i}n_{1k}, \tag{3.269}$$

the continuity of the velocity field,

$$\mathbf{v}_{*1} = \mathbf{v}_{*2}, \tag{3.270}$$

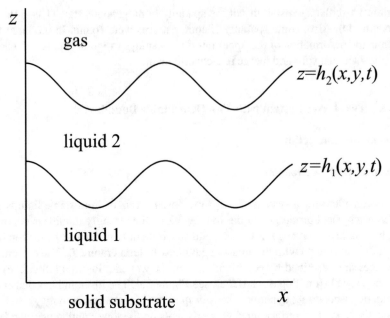

Fig. 3.4 Schematic of two-layer liquid system.

the kinematic equation for the interface motion,

$$\frac{\partial h_{*1}}{\partial t_*} + v_{*1x}\frac{\partial h_{*1}}{\partial x_*} + v_{*1y}\frac{\partial h_{*1}}{\partial y_*} = v_{*1z}, \tag{3.271}$$

the balance equations for tangential stresses which include the Marangoni stresses:

$$\left[-\eta_1\left(\frac{\partial v_{*1i}}{\partial x_{*k}} + \frac{\partial v_{*1k}}{\partial x_{*i}}\right) + \eta_2\left(\frac{\partial v_{*2i}}{\partial x_{*k}} + \frac{\partial v_{*2k}}{\partial x_{*i}}\right)\right]\tau_{1i}^{(l)}n_{1k} - \alpha_1\tau_{1i}^{(l)}\frac{\partial T_{*1}}{\partial x_{*i}} = 0, \ l = 1,2,$$

$$\tag{3.272}$$

the continuity of the temperature field,

$$T_{*1} = T_{*2}, \tag{3.273}$$

and the balance of normal heat fluxes,

$$\left(\kappa_1\frac{\partial T_{*1}}{\partial x_{*i}} - \kappa_2\frac{\partial T_{*2}}{\partial x_{*i}}\right)n_{1i} = 0. \tag{3.274}$$

Similar boundary conditions are imposed on the deformable interface $z_* = h_{*2}$:

$$-p_{*2} + 2\sigma_2 K_2 = -\eta_2\left(\frac{\partial v_{*2i}}{\partial x_{*k}} + \frac{\partial v_{*2k}}{\partial x_{*i}}\right)n_{2i}n_{2k}, \tag{3.275}$$

$$-\eta_2\left(\frac{\partial v_{*2i}}{\partial x_{*k}}+\frac{\partial v_{*2k}}{\partial x_{*i}}\right)\tau_{2i}^{(l)}n_{2k}-\alpha_2\tau_{2i}^{(l)}\frac{\partial T_{*2}}{\partial x_{*i}}=0,\ l=1,2,\tag{3.276}$$

$$\frac{\partial h_{*2}}{\partial t_*}+v_{*2x}\frac{\partial h_{*2}}{\partial x_*}+v_{*2y}\frac{\partial h_{*2}}{\partial y_*}=v_{*2z}.\tag{3.277}$$

For a heat flux on the liquid–gas interface, we use an empirical condition:

$$\kappa_2\frac{\partial T_{*2}}{\partial x_{*i}}n_{2i}=-q(T_{*2}-T_g),\tag{3.278}$$

where q is the heat exchange coefficient, which is assumed to be constant.

Derivation of Longwave Amplitude Equations

Let us assume that the solution of equations and boundary conditions (3.265)–(3.278) depends on the scaled horizontal coordinates $\mathbf{X}_{*\perp}=\varepsilon\mathbf{x}_{*\perp}$, while the appropriate scaled time variable is $\tau_*=\varepsilon^2 t_*$, $\varepsilon\ll 1$, and applies the long-wavelength expansions.

At the leading order, the evolution of the system is governed by the following equations and boundary conditions:

$$\frac{\partial^2\mathbf{u}_{*1\perp}}{\partial z_*^2}=0;\ \nabla_{*\perp}\cdot\mathbf{u}_{*1\perp}+\frac{\partial w_{*1}}{\partial z_*}=0;\ \frac{\partial^2 T_{*1}}{\partial z_*^2}=0;\ 0<z_*<h_{*1};\tag{3.279}$$

$$\frac{\partial^2\mathbf{u}_{*2\perp}}{\partial z_*^2}=0;\ \nabla_{*\perp}\cdot\mathbf{u}_{*2\perp}+\frac{\partial w_{*2}}{\partial z_*}=0;\ \frac{\partial^2 T_{*2}}{\partial z_*^2}=0;\ h_{*1}<z_*<h_{*2};\tag{3.280}$$

$$z_*=0:\ \mathbf{u}_{*1}=0;\ w_{*1}=0;\ T_{*1}=T_s;\tag{3.281}$$

$$z_*=h_{*1}:\ \mathbf{u}_{*1}=\mathbf{u}_{*2};\ w_{*1}=w_{*2};\tag{3.282}$$

$$\eta_2\frac{\partial\mathbf{u}_{*2}}{\partial z_*}-\eta_1\frac{\partial\mathbf{u}_{*1}}{\partial z_*}-\alpha_1\left(\nabla_{*\perp}T_{*1}+\nabla_{*\perp}h_{*1}\frac{\partial T_{*1}}{\partial z_*}\right)=0;\tag{3.283}$$

$$\frac{\partial h_{*1}}{\partial\tau}+\mathbf{u}_{*1}\cdot\nabla_{*\perp}h_{*1}=w_{*1};\tag{3.284}$$

$$T_{*1}=T_{*2};\ \kappa_1\frac{\partial T_{*1}}{\partial z_*}=\kappa_2\frac{\partial^2 T_{*2}}{\partial z_*};\tag{3.285}$$

$$z_*=h_{*2}:\ -\eta_2\frac{\partial\mathbf{u}_{*2}}{\partial z_*}-\alpha_2\left(\nabla_{*\perp}T_{*2}+\nabla_{*\perp}h_{*2}\frac{\partial T_{*2}}{\partial z_*}\right)=0;\tag{3.286}$$

$$\frac{\partial h_{*2}}{\partial\tau_*}+\mathbf{u}_{*2}\cdot\nabla_{*\perp}h_{*2}=w_{*2};\tag{3.287}$$

$$\kappa_2\frac{\partial T_{*2}}{\partial z_*}=-q(T_*-T_g),\tag{3.288}$$

where $\mathbf{u}_{*j}$ and w_j, $j = 1, 2$, are the leading-order terms in the expansions in powers of ε:

$$(v_{*xj}, v_{*yj}) = \varepsilon \mathbf{u}_{*j} + \ldots, \quad v_{*zj} = \varepsilon^2 w_{*j} + \ldots.$$

The leading-order term for pressure caused by the surface tensions is $O(\varepsilon^2)$; therefore the pressure does not contribute to the leading-order equations.

Solving the problem for the temperature fields, we obtain

$$T_{*1} = T_s - (T_s - T_g)Dq\kappa_2 z_*; \tag{3.289}$$

$$T_{*2} = T_s - (T_s - T_g)Dq[(\kappa_2 - \kappa_1)h_{*1} + \kappa_1 z_*], \tag{3.290}$$

where

$$D = [\kappa_1 \kappa_2 + q(\kappa_2 - \kappa_1)h_{*1} + q\kappa_1 h_{*2}]^{-1}. \tag{3.291}$$

The horizontal components of the flow velocities generated by the thermocapillary stresses are determined by the following formulas:

$$\mathbf{u}_{*1} = \frac{(T_s - T_g)\kappa_2}{\eta_1} \nabla_{*\perp} [D(\alpha_1 q h_1 - \alpha_2 \kappa_1)]z_*, \tag{3.292}$$

$$\mathbf{u}_{*2} = \frac{(T_s - T_g)\kappa_2}{\eta_2} \{-\alpha_2 \kappa_1 \nabla_{*\perp} D z_* +$$

$$\frac{h_{*1}}{\eta_1} \nabla_{*\perp} [D(\alpha_1 \eta_2 q h_{*1} - \alpha_2 (\eta_2 - \eta_1)\kappa_1)] \}. \tag{3.293}$$

Solving the continuity equations with respect to w_{*1} and w_{*2} with corresponding boundary conditions, we find that

$$w_{*1}(\mathbf{X}_{*\perp}, h_{*1}) = -\int_0^{h_{*1}} \nabla_{*\perp} \cdot \mathbf{u}_{*1} dz_*, \tag{3.294}$$

$$w_{*2}(\mathbf{X}_{*\perp}, h_{*2}) = -\int_0^{h_{*1}} \nabla_{*\perp} \cdot \mathbf{u}_{*1} dz_* + \int_{h_{*1}}^{h_{*2}} \nabla_{*\perp} \cdot \mathbf{u}_{*2} dz_*. \tag{3.295}$$

Using (3.294) and (3.295), we rewrite the kinematic conditions (3.284) and (3.287) in the following form:

$$h_{*1\tau_*} + \nabla_{*\perp} \cdot \int_0^{h_{*1}} \mathbf{u}_{*1} dz_* = 0; \tag{3.296}$$

$$h_{*2\tau_*} + \nabla_{*\perp} \cdot \left(\int_0^{h_{*1}} \mathbf{u}_{*1} dz_* + \int_{h_{*1}}^{h_{*2}} \mathbf{u}_2 dz_* \right) = 0, \tag{3.297}$$

where the subscript τ_* denotes the corresponding partial derivative. Substituting expressions for flow velocities obtained above into equations (3.296) and (3.297), we arrive at a closed system of equations that govern the evolution of a heated two-layer film under the action of the thermocapillary effect:

$$h_{*1\tau} + \nabla \cdot \mathbf{Q}_1^T = 0, \quad h_{*2\tau} + \nabla \cdot \mathbf{Q}_2^T = 0, \tag{3.298}$$

where

$$\mathbf{Q}_1^T = \frac{(T_s - T_g)\kappa_2}{2\eta_1} h_{*1}^2 \nabla[D(q\alpha_1 h_{*1} - \alpha_2 \kappa_1)], \qquad (3.299)$$

$$\mathbf{Q}_2^T = \frac{(T_s - T_g)\kappa_2}{2\eta_1 \eta_2} \{h_{*2}^2 \nabla[(-\alpha_2 \kappa_1 \eta_1)D] + \qquad (3.300)$$

$$(2h_{*2} - h_{*1})h_{*1}\nabla\{D[q\alpha_1 \eta_2 h_{*1} - \alpha_2 \kappa_1(\eta_2 - \eta_1)]\}\}.$$

It should be emphasized that equations (3.298) are *not sufficient* for the description of the Marangoni instability development. They are invariant with respect to the scaling transformation $\mathbf{X}_* \to c\mathbf{X}_*$, $\tau_* \to c^2\tau_*$, where c is an arbitrary constant. Therefore, in the framework of equations (3.298), the instability appears simultaneously for any wavenumbers. The anticipated dependence of the linear growth rate σ on the wavenumber k is $\sigma(k) \sim k^2$, i.e., the problem is ill-posed.

Contribution of Laplace Pressures, van der Waals Forces, and Gravity

As explained in Section 3.2.3, the strongly nonlinear longwave approach is justified only in the limit of *strong surface tension*, which suppresses shortwave deformations of the surface; therefore the instability takes place only in the region of longwaves. If σ_1, $\sigma_2 = O(\varepsilon^{-2})$, the disturbances of the pressure are $O(1)$ rather than $O(\varepsilon^2)$; thus they contribute to the evolution equations.

In the thin-film limit, the contribution of the van der Waals interactions should be added. For a two-layer film, one has to take into account that the total potential energy of the van der Waals interactions $f = f_1 + f_2 + f_{12}$ includes the energy f_1 of the substrate interaction with layer 2 across layer 1, the energy f_2 of the gas phase interaction with layer 1 across layer 2, and the energy f_{12} of the interaction between the gas phase and the substrate across two layers [45, 111]

$$f(h_{*1}, h_{*2}) = -\frac{A_{sg} - A_{s2} - A_{g1}}{12\pi h_{*2}^2} - \frac{A_{s2}}{12\pi h_{*1}^2} - \frac{A_{g1}}{12\pi(h_{*2} - h_{*1})^2}, \qquad (3.301)$$

where A_{sg}, A_{s2}, and A_{g1} are the Hamaker constants characterizing the interactions between the solid substrate and the gas across the two layers, between the solid substrate and liquid 2 across liquid 1, and between the gas phase and liquid 1 across liquid 2, respectively. The latter constants are determined by the dielectric permittivities of all the media as functions of the frequency [45, 112], and they depend mainly on the zero-frequency dielectric constants and high-frequency refractive indices of the media [45, 113].

The disjoining pressures $\Pi_1(h_{*1}, h_{*2})$ and $\Pi_2(h_{*1}, h_{*2})$ are computed as

$$\Pi_1(h_{*1}, h_{*2}) = \frac{\partial f}{\partial h_{*1}} = \frac{A_{s2}}{6\pi h_{*1}^3} - \frac{A_{g1}}{6\pi(h_{*2} - h_{*1})^3}, \qquad (3.302)$$

$$\Pi_2(h_{*1}, h_{*2}) = \frac{\partial f}{\partial h_{*2}} = \frac{A_{sg} - A_{s2} - A_{g1}}{6\pi h_{*2}^3} + \frac{A_{g1}}{6\pi(h_{*2} - h_{*1})^3}. \tag{3.303}$$

Including also the gravity, we obtain the following leading-order problem for finding pressures:

$$\frac{\partial p_{*m}}{\partial z_*} = -\rho_m g, \, m = 1, 2, \tag{3.304}$$

$$z = h_{*1} : \, p_{*1} - p_{*2} = -\gamma_1 \nabla_{*\perp}^2 h_{*1} + \Pi_1(h_{*1}, h_{*2}), \tag{3.305}$$

$$z = h_{*2} : \, p_{*2} = -\gamma_2 \nabla_{*\perp}^2 h_{*2} + \Pi_2(h_{*1}, h_{*2}). \tag{3.306}$$

That leads to the following dependences of the pressures p_{*1} and p_{*2} in each layer on the layers' thicknesses h_{*1} and h_{*2} [111]:

$$p_{*1} = -\gamma_1 \Delta_{*\perp} h_{*1} - \gamma_2 \Delta_{*\perp} h_{*2} + W_1(h_{*1}, h_{*2}), \tag{3.307}$$

$$p_{*2} = -\gamma_2 \Delta_{*\perp} h_{*2} + W_2(h_{*1}, h_{*2}), \tag{3.308}$$

where

$$W_1(h_{*1}, h_{*2}) = \frac{A_{sg} - A_{s2} - A_{g1}}{6\pi h_{*2}^3} + \frac{A_{s2}}{6\pi h_{*1}^3} + \rho_1 g h_{*1} + \rho_2 g(h_{*2} - h_{*1}), \tag{3.309}$$

$$W_2(h_{*1}, h_{*2}) = \frac{A_{sg} - A_{s2} - A_{g1}}{6\pi h_{*2}^3} + \frac{A_{g1}}{6\pi(h_{*2} - h_{*1})^3} + \rho_2 g h_{*2}. \tag{3.310}$$

In the absence of the thermocapillary stresses, one obtains the following system of coupled evolution equations for h_1 and h_2 [111, 114, 115]:

$$h_{*1\tau_*} + \nabla_{*\perp} \cdot \mathbf{Q}_1 = 0, \, h_{*2\tau_*} + \nabla_{*\perp} \cdot \mathbf{Q}_2 = 0, \tag{3.311}$$

where

$$\mathbf{Q}_1 = F_{11} \nabla p_{*1} + F_{12} \nabla p_{*2}, \, \mathbf{Q}_2 = F_{21} \nabla p_{*1} + F_{22} \nabla p_{*2}. \tag{3.312}$$

The pressures p_1 and p_2 are determined by expressions (3.307) and (3.308), and the mobility functions are

$$F_{11} = -\frac{1}{3\eta_1} h_{*1}^3; \, F_{12} = -\frac{1}{2\eta_1} h_{*1}^2 (h_{*2} - h_{*1}); \, F_{21} = \frac{1}{6\eta_1} h_{*1}^3 - \frac{1}{2\eta_1} h_{*1}^2 h_{*2};$$

$$F_{22} = (h_{*2} - h_{*1}) \left[h_{*1}^2 \left(\frac{1}{2\eta_1} - \frac{1}{3\eta_2} \right) + h_{*1} h_{*2} \left(-\frac{1}{\eta_1} + \frac{2}{3\eta_2} \right) - \frac{1}{3\eta_2} h_{*2}^2 \right].$$

The simultaneous action of Marangoni forces, capillarity, gravity, and van der Waals forces on the dynamics of a nonisothermic two-layer thin film is described by the combination of two flux terms in evolution equations:

$$h_{*1\tau_*} + \nabla_{*\perp} \cdot (\mathbf{Q}_1^T + \mathbf{Q}_1) = 0, \, h_{*2\tau_*} + \nabla_{*\perp} \cdot (\mathbf{Q}_2^T + \mathbf{Q}_2) = 0, \tag{3.313}$$

where $\mathbf{Q}_1^T$ and $\mathbf{Q}_2^T$ are determined by formulas (3.299) and (3.300), while $\mathbf{Q}_1$ and $\mathbf{Q}_2$ are determined by formulas (3.312).

Let us transform equations (3.313) to a nondimensional form. As the vertical length scale, we choose the equilibrium thickness of the lower layer, h_1^0. At the present stage, we do not fix the horizontal length, L^*. We choose

$$\tau^* = \frac{\eta_1 (L^*)^4}{\sigma_1^0 (h_1^0)^3} \tag{3.314}$$

as a time scale and

$$p^* = \frac{\sigma_1^0 h_1^0}{(L^*)^2} \tag{3.315}$$

as a pressure scale.

Equations (3.313) written in nondimensional form look as follows:

$$h_{1\tau} + \nabla_\perp \cdot \mathbf{q}_1 = 0, \quad h_{2\tau} + \nabla_\perp \cdot \mathbf{q}_2 = 0, \tag{3.316}$$

$$\mathbf{q}_1 = f_{11} \nabla_\perp p_1 + f_{12} \nabla_\perp p_2 + \mathbf{q}_1^T, \quad \mathbf{q}_2 = f_{21} \nabla_\perp p_1 + f_{22} \nabla_\perp p_2 + \mathbf{q}_2^T. \tag{3.317}$$

As usual, we assume that the dependence of interfacial tensions on the temperature is relatively weak and can be neglected in the boundary conditions for *normal stresses* (but not in those for tangential stresses where it is the source of a thermocapillary motion). The contributions of disjoining pressures and gravity are included:

$$p_1 = -\Delta_\perp h_1 - \gamma \Delta_\perp h_2 + w_1(h_1, h_2), \tag{3.318}$$

$$p_2 = -\gamma \Delta_\perp h_2 + w_2(h_1, h_2), \tag{3.319}$$

$$w_1 = \frac{\tilde{a}_0 - \tilde{a}_1 - \tilde{a}_2}{h_2^3} + \frac{\tilde{a}_1}{h_1^3} + g_1 h_1 + g_2(h_2 - h_1), \tag{3.320}$$

$$w_2 = \frac{\tilde{a}_0 - \tilde{a}_1 - \tilde{a}_2}{h_2^3} + \frac{\tilde{a}_2}{(h_2 - h_1)^3} + g_2 h_2. \tag{3.321}$$

The nondimensional expressions for the fluxes generated by the thermocapillary effect are

$$\mathbf{q}_1^T = \frac{M}{2} h_1^2 \nabla[d(Bih_1 - \alpha\kappa)], \tag{3.322}$$

$$\mathbf{q}_2^T = \frac{M}{2} \{-h_2^2 \nabla(d\eta\alpha\kappa) + (2h_2 - h_1)h_1 \nabla\{d[Bih_1 - \alpha\kappa(1 - \eta)]\}\}. \tag{3.323}$$

Here

$$M = \frac{\alpha_1 (T_s - T_g)}{\sigma_1^0} \left(\frac{L_*}{h_1^0}\right)^2 \tag{3.324}$$

is the modified Marangoni number,

$$Bi = \frac{q h_1^0}{\kappa_2} \tag{3.325}$$

is the Biot number,

$$g_1 = \frac{g(h_1^0)^3}{v_1^2}, \; g_2 = \frac{\rho_2}{\rho_1}g_1 \tag{3.326}$$

are the Galileo numbers, and

$$d = [\kappa + Bi(1 - \kappa)h_1 + Bi\kappa h_2]^{-1}, \tag{3.327}$$

$\eta = \eta_1/\eta_2, \; \kappa = \kappa_1/\kappa_2, \; \gamma = \gamma_2^0/\gamma_1^0, \; \alpha = \alpha_2/\alpha_1,$

$$\tilde{a}_0 = \text{sign}(A_{sg}) \left(\frac{L^*}{L_0^*}\right)^2, \; \tilde{a}_1 = \frac{A_{s2}}{|A_{sg}|}\left(\frac{L^*}{L_0^*}\right)^2, \; \tilde{a}_2 = \frac{A_{g1}}{|A_{sg}|}\left(\frac{L^*}{L_0^*}\right)^2, \tag{3.328}$$

where

$$L_0^* = (H_1^0)^2 \sqrt{6\pi\gamma_1^0/|A_{sg}|}. \tag{3.329}$$

The choice $L^* = L_0^*$ is convenient for the analysis of the instability induced by inter-molecular forces [111].

Stability

General Case

We consider a bilayer with plane interfaces located at $z = h_1^0$ and $z = h_2^0$. The corresponding *basic* solution of nondimensional equations (3.316) is

$$h_1 = 1, \; h_2 = h = 1 + a,$$

where $h = h_2^0/h_1^0$, $a = (h_2^0 - h_1^0)/h_1^0$.

In order to investigate the stability of the plane two-layer film, we substitute

$$h_1 = 1 + \tilde{h}_1, \; h_2 = 1 + a + \tilde{h}_2$$

into equations (3.316) and linearize them with respect to the variables $\tilde{h}_1, \tilde{h}_2$.

The solutions of the linear problem can be written as

$$\tilde{h}_j(X, Y, \tau) = \bar{h}_j e^{i\mathbf{k} \cdot \mathbf{R} + \lambda\tau}, \; j = 1, 2, \tag{3.330}$$

where $\mathbf{R} = (X, Y)$, $\mathbf{k} = (k_x, k_y)$ is the wave vector, λ is the growth rate, and $\bar{h}_j$, $j = 1, 2$, are constants. Substituting (3.330) into the linearized equations, we obtain the dispersion relation

$$\det(\mathbf{N} - \lambda\mathbf{I}) = 0, \tag{3.331}$$

which determines the eigenvalues $\lambda(\mathbf{k})$. Here $\mathbf{I}$ is the unit matrix, while the matrix $\mathbf{N}$ can be presented in the following form:

$$\mathbf{N} = \mathbf{B} + M\tilde{\mathbf{C}}.\tag{3.332}$$

The matrix $\mathbf{B}$ presents the contribution of the van der Waals forces, gravity, and surface tensions, and it has the following components:

$$B_{11} = -\frac{k^2}{3}(k^2 - 3\tilde{a}_1 + g_1 - g_2) - \frac{3k^2\tilde{a}_2}{2a^3};$$

$$B_{12} = -k^2\left(\frac{1}{3} + \frac{a}{2}\right)\left[\sigma k^2 - \frac{3(\tilde{a}_0 - \tilde{a}_1 - \tilde{a}_2)}{(a+1)^4} + g_2\right] + \frac{3k^2\tilde{a}_2}{2a^3};$$

$$B_{21} = -k^2\left(\frac{1}{3} + \frac{a}{2}\right)(k^2 - 3\tilde{a}_1 + g_1 - g_2) +$$

$$\left[\frac{1}{2} - \frac{\eta}{3} + (a+1)\left(-1 + \frac{2\eta}{3}\right) - \frac{\eta}{3}(a+1)^2\right]\frac{3k^2\tilde{a}_2}{a^3};$$

$$B_{22} = -k^2\left[\sigma k^2 - \frac{3(\tilde{a}_0 - \tilde{a}_1 - \tilde{a}_2)}{(a+1)^4} + g_2\right]\left[a(a+1) + \frac{1}{3}(1 + \eta a^3)\right]$$

$$+ k^2\frac{3\tilde{a}_2}{a^3}\left(\frac{1}{2} + a + \frac{\eta a^2}{3}\right).$$

One can see that in the framework of the linear stability theory, the action of gravity is equivalent to the replacement of the van der Waals coefficients $\tilde{a}_0$, $\tilde{a}_1$, and $\tilde{a}_2$ by renormalized coefficients $\hat{a}_0$, $\hat{a}_1$, and $\hat{a}_2$ determined by the following relations:

$$\hat{a}_1 = \tilde{a}_1 - (g_1 - g_2)/3;\ \hat{a}_2 = a_2,\tag{3.333}$$

$$\hat{a}_0 - \hat{a}_1 - \hat{a}_2 = \tilde{a}_0 - \tilde{a}_1 - \tilde{a}_2 - g_2(a+1)^4/3.\tag{3.334}$$

The second term on the right-hand side of (3.332) is caused by the thermocapillary effect. The corresponding matrix

$$\tilde{\mathbf{C}} = \frac{Bik^2\kappa}{2(\kappa + Bi + Bi\kappa a)^2}\mathbf{C},$$

where the elements of matrix $\mathbf{C}$ are as follows:

$$C_{11} = 1 + Bi(a+1) + \alpha(1 - \kappa),$$

$$C_{12} = \alpha\kappa - Bi,$$

$$C_{21} = \eta\alpha a^2(1 - \kappa) + (2a+1)[\alpha(1 - \kappa) + 1 + Bi(a+1)],$$

$$C_{22} = \eta\alpha\kappa a^2 + (2a+1)(\alpha\kappa - Bi).$$

Solving the quadratic equation (3.331), we obtain:

$$\lambda = \frac{1}{2}\left[\text{tr}(\mathbf{N}) \pm \sqrt{[\text{tr}(\mathbf{N})]^2 - 4\det(\mathbf{N})}\right],\tag{3.335}$$

where $\text{tr}(\mathbf{N})$ and $\det(\mathbf{N})$ are the trace and the determinant of the matrix $\mathbf{N}$. The basic state is stable if $\text{tr}(\mathbf{N}) < 0$ and $\det(\mathbf{N}) > 0$. In the absence of heating ($M = 0$), these conditions are satisfied, but they can be violated with the growth of $|M|$ for a definite way of heating (sign of M). The relations $\det(\mathbf{N}) = 0$, $\text{tr}(\mathbf{N}) < 0$ determine the monotonic instability boundary; the relations $\text{tr}(\mathbf{N}) = 0$, $\det(\mathbf{N}) > 0$ determine the oscillatory instability boundary.

Marangoni Instability in the Absence of Gravity

In the present subsection, we consider the case of sufficiently thick films for which the influence of the van der Waals interactions is weak in comparison with the Marangoni effect. Also, gravity is neglected.

It is reasonable to choose

$$L^* = \sqrt{\sigma_1^0 (h_1^0)^3 / \eta_1 \nu_1}. \tag{3.336}$$

Then the modified Marangoni number M, which is defined by (3.324), becomes

$$M = \frac{\alpha(T_s - T_g)h_1^0}{\eta_1 \nu_1}. \tag{3.337}$$

Note that

$$\left(\frac{L^*}{h_1^0}\right)^2 = \frac{\sigma_1^0 h_1^0}{\eta_1 \nu_1} \tag{3.338}$$

is rather large even for thin layers; therefore the scaling (3.336) is compatible with the longwave approach.

In the absence of heating ($M_\perp = 0$, $\mathbf{N} = \mathbf{B}$) and intermolecular interactions ($\tilde{a}_j = 0$, $j = 0, 1, 2$), the mechanical equilibrium state is stable, because the conditions $\text{tr}(\mathbf{B}) \leq 0$, $\det(\mathbf{B}) \geq 0$ are satisfied, but they can be violated with the growth of $|M_\perp|$ for a definite way of heating (sign of $M_\perp$).

First, let us consider the longwave limit ($k^2 \ll 1$) and neglect both the intermolecular interactions ($\tilde{a}_j = 0$) and the terms caused by the surface tensions, which are proportional to k^4 and hence are small with respect to Marangoni terms proportional to k^2. In that limit, the dispersion relation (3.331) becomes

$$\det\left[\frac{M_\perp Bi k^2 \kappa}{2(\kappa + Bi + Bi\kappa a)^2}\mathbf{C} - \lambda \mathbf{I}\right] = 0. \tag{3.339}$$

The eigenvalues can be presented in the form

$$\lambda(k) = \frac{M_\perp Bi k^2 \kappa}{2(\kappa + Bi + Bi\kappa a)^2}\Lambda, \tag{3.340}$$

where Λ satisfies the quadratic equation

$$\Lambda^2 - \text{tr}(\mathbf{C})\Lambda + \det(\mathbf{C}) = 0, \tag{3.341}$$

$$\text{tr}(\mathbf{C}) = 1 - Bia + \alpha(1 + \eta\kappa a^2 + 2\kappa a), \tag{3.342}$$

$$\det(\mathbf{C}) = \eta\alpha a^2(\kappa + Bi + Bi\kappa a). \tag{3.343}$$

Obviously, $\det(\mathbf{C})$ is always positive, while $\text{tr}(\mathbf{C})$ is positive when $Bi < Bi_c$ and negative when $Bi > Bi_c$; here

$$Bi_c = \frac{1 + \alpha(1 + \eta\kappa a^2 + 2\kappa a)}{a}. \tag{3.344}$$

It is clear from the expression

$$\Lambda = \frac{1}{2}\left[\text{tr}(\mathbf{C}) \pm \sqrt{(\text{tr}(\mathbf{C}))^2 - 4\det(\mathbf{C})}\right] \tag{3.345}$$

that the real part of Λ is positive for $Bi < Bi_c$ and negative for $Bi > Bi_c$. Therefore, the Marangoni instability exists for arbitrary small k either if heating is from below ($T_s > T_g$, $M_\perp > 0$) and $Bi < Bi_c$ or if heating is from above ($T_s < T_g$, $M_\perp < 0$) and $Bi > Bi_c$. Note that, in the absence of gravity, the critical Marangoni number is always $M_\perp = 0$. The instability is monotonic if $S = (\text{tr}(C))^2 - 4\det(C)$ is positive and oscillatory if S is negative. The expression for S can be presented in the following form:

$$S = a^2\{Bi^2 - 2[Bi_c + 2\eta\alpha(\kappa a + 1)] + Bi_c^2 - 4\eta\alpha\kappa\}. \tag{3.346}$$

One can see that S is negative (i.e., the instability is oscillatory) when $Bi_- < Bi < Bi_+$,

$$Bi_\pm = Bi_c + 2\eta\alpha(\kappa a + 1) \pm \sqrt{4\eta\alpha[Bi_c(\kappa a + 1) + \kappa + \eta\alpha(\kappa a + 1)^2]}, \tag{3.347}$$

and positive (i.e., the instability is monotonic) otherwise. Note that because the term

$$Bi_c^2 - 4\eta\alpha\kappa = [(1 - a\sqrt{\eta\alpha\kappa})^2 + \alpha(1 + 2\kappa a)][(1 + a\sqrt{\eta\alpha\kappa})^2 + \alpha(1 + 2\kappa a)]$$

in the expression for S is positive, both $Bi_\pm$ are positive. Thus, for any two-layer film, both monotonic and oscillatory instabilities are possible, depending on the Biot number.

Let us emphasize that the conclusions presented above are valid for instabilities in the limit $k \to 0$ but not for finite values of k. For instance, a monotonic instability by heating from below is possible in an interval $0 < k_- < k < k_+$ for the same value of Bi as an oscillatory instability by heating from above in an interval $0 < k < k_m$ (see formula (3.351) below). Recall that because of the rescaling of the horizontal coordinates done by the derivation of amplitude equations, $k = O(1)$ does not mean a shortwave instability: in the original variables, the wavenumber is $O(\varepsilon)$.

Marangoni Instability in the Presence of Gravity

Later on, we assume that $\rho_1 > \rho_2$, i.e., the density of the bottom liquid is higher than that of the top liquid, so that Rayleigh–Taylor instability is impossible. Therefore, $g_1 > g_2 > 0$.

Under the stabilizing action of gravity, the critical Marangoni numbers for both monotonic and oscillatory instabilities become nonzero. Indeed,

$$\mathrm{tr}(\mathbf{B}) = -\frac{k^2}{3}(k^2 + g_1 - g_2) - k^2(\sigma k^2 + g_2)\left[a(a+1) + \frac{1}{3}(1 + \eta a^3)\right] \leq 0, \quad (3.348)$$

$$\det(\mathbf{B}) = k^4(k^2 + g_1 - g_2)(\sigma k^2 + g_2)a^2\left(\frac{1}{12} + \frac{1}{9}\eta\right) \geq 0, \quad (3.349)$$

and therefore there is no instability when $M = 0$.

The *monotonic* instability boundary is determined by the relation $\det(\mathbf{N}) = \det(\mathbf{B} + M_{\perp m}\tilde{\mathbf{C}}) = 0$, which leads to the quadratic equation

$$\det(\tilde{\mathbf{C}})M_{\perp m}^2 + FM_{\perp m} + \det(\tilde{\mathbf{B}}) = 0, \quad (3.350)$$

where

$$F = B_{11}\tilde{C}_{22} + B_{22}\tilde{C}_{11} - B_{12}\tilde{C}_{21} - B_{21}\tilde{C}_{12}.$$

Note that $\det(\tilde{\mathbf{C}})$ is always positive (see (3.343)), while $\det(\mathbf{B})$ is positive in the absence of intermolecular forces (see (3.349)). Solving (3.350), we find that

$$M_{\perp m} = \frac{-F \pm \sqrt{F^2 - 4\det(\tilde{\mathbf{C}})\det(\mathbf{B})}}{2\det(\tilde{\mathbf{C}})}. \quad (3.351)$$

Expression (3.351) is $O(1)$ in the limit $k^2 \ll 1$. For any k^2, there are either two values of $M_{\perp m}$ (if $\det(\mathbf{B}) > 0$, both of them have the same sign) or there is no monotonic instability.

The *oscillatory* instability boundary is determined by the relation $\mathrm{tr}(\mathbf{N}) = \mathrm{tr}(\mathbf{B}) + M_{\perp o}\mathrm{tr}(\tilde{\mathbf{C}}) = 0$; hence

$$M_{\perp o} = -\frac{\mathrm{tr}(\mathbf{B})}{\mathrm{tr}(\tilde{\mathbf{C}})}, \quad (3.352)$$

under the condition that the expression

$$\omega^2 = \det(\mathbf{B} + M_{\perp o}\mathbf{C}),$$

which determines the squared frequency of oscillations, is positive. In the absence of van der Waals interactions,

$$M_{\perp o}(k) = \frac{2(\kappa + Bi + Bi\kappa a)^2}{Bi\kappa a(Bi_c - Bi)} \times$$

$$\left\{ \frac{1}{3}g_1 + \left[a(a+1) + \frac{1}{3}\eta a^3 \right] + k^2 \left[\frac{1}{3} + \sigma \left[a(a+1) + \frac{1}{3}(1+\eta a^3) \right] \right] \right\}, \quad (3.353)$$

where B_c is defined by expression (3.344). Note that oscillatory instability develops by heating from below ($M_\perp > 0$) if $Bi < Bi_c$ and by heating from above ($M_\perp < 0$) if $Bi > Bi_c$, similar to the case considered above.

Direct numerical simulations of nonlinear waves governed by systems (3.316) and (3.317) are described in [116].

We have considered a number of mechanisms leading to an oscillatory convection, among them the interaction of longitudinal and transverse waves, interaction of deformations and temperature disturbances, and interaction of deformations of different interfaces. In all the cases, we managed to obtained some systems of partial differential equations governing the pattern formation. However, that is not always possible. In the next section, we will consider the case where the construction of unified amplitude equations containing only *local* differentiation operators is impossible: the interaction of waves is essentially nonlocal, and another "philosophy" has to be applied. The physical problem most appropriate for the demonstration of those ideas is the convection in a binary solution in the presence of Soret effect.

References

1. F.H. Busse, N. Riahi, J. Fluid Mech. **96**, 243 (1980)
2. C.J. Chapman, M.R.E. Proctor, J. Fluid Mech. **101**, 759 (1980)
3. M.R.E. Proctor, J. Fluid Mech. **113**, 469 (1981)
4. V.L. Gertsberg, G.I. Sivashinsky, Prog. Theor. Phys. **66**, 1219 (1981)
5. L.M. Pismen, Phys. Lett. A **116**, 241 (1986)
6. A.A. Nepomnyashchy, *Fluid Dynamics*, Pt. 9. Proceeding of the Perm State Pedagogical Institute, vol. 152 (Perm State Pedagogical Institute, Perm, 1976, in Russian), p. 53
7. R. Hoyle, *Pattern Formation: An Introduction to Methods* (Cambridge University Press, Cambridge, 2006)
8. A. Podolny, A.A. Nepomnyashchy, A. Oron, Phys. Rev. E **76**, 026309 (2007)
9. J.W. Cahn, J.E. Hilliard, J. Chem. Phys. **28**, 258 (1958)
10. A. Novick-Cohen, L.A. Segel, Physica D **10**, 277 (1984)
11. J.S. Langer, Ann. Phys. **65**, 53 (1971)
12. K. Kawasaki, T. Ohta, Physica A **116**, 573 (1982)
13. J.W. Swift, P.C. Hohenberg, Phys. Rev. A **15**, 319 (1977)
14. J.D. Murray, *Mathematical Biology* (Springer, Berlin, 1989)
15. A. Schlüter, D. Lortz, F. Busse, J. Fluid Mech. **23**, 129 (1965)
16. A. Nepomnyashchy, I. Simanovskii, J.C. Legros, *Interfacial Convection in Multilayer Systems* (Springer, New York, 2012)
17. M.G. Velarde, A.A. Nepomnyashchy, M. Hennenberg, Adv. Appl. Mech. **37**, 167 (2001)
18. G.I. Sivashinsky, Physica D **4**, 227 (1982)
19. E. Knobloch, Physica D **41**, 450 (1990)
20. A.A. Golovin, A.A. Nepomnyashchy, L.M. Pismen, Physica D **81**, 117 (1995)
21. F.H. Busse, Das Stabilitätsverhalten der Zellularkonvektion bei endlicher Amplitude, Ph.D. thesis, Munich, 1962
22. F.H. Busse, J. Fluid Mech. **30**, 625 (1967)
23. B.A. Malomed, A.A. Nepomnyashchy, M.I. Tribelsky, Sov. Phys. JETP **69**, 388 (1989)
24. L. Shtilman, G. Sivashinsky, Physica D **52**, 477 (1991)

25. J. Pontes, C.I. Christov, M.G. Velarde, Int. J. Bifurcation Chaos **6**, 1883 (1996)
26. A.A. Golovin, A.A. Nepomnyashchy, L.M. Pismen, Int. J. Bifurcation Chaos **12**, 2487 (2002)
27. A. Oron, S.H. Davis, S.G. Bankoff, Rev. Mod. Phys. **69**, 931 (1997)
28. A.A. Nepomnyashchy, M.V. Velarde, P. Colinet, *Interfacial Phenomena and Convection* (Chapman and Hall/CRC, Boca Raton, 2002)
29. T. Funada, J. Phys. Soc. Jpn. **56**, 2031 (1987)
30. G.I. Sivashinsky, Physica D **8**, 243 (1983)
31. A.J. Bernoff, A.L. Bertozzi, Physica D **85**, 375 (1995)
32. S.J. VanHook, M.F. Schatz, W.D. McCormick, J.B. Swift, H.L. Swinney, Phys. Rev. Lett. **75**, 4397 (1995)
33. S.H. Davis, Rupture of thin liquid films, in *Waves on Fluid Interfaces*, ed. by R.E. Meyer (Academic, New York, 1983), p. 291
34. S.H. Davis, Ann. Rev. Fluid Mech. **19**, 403 (1987)
35. S. Krishnamoorthy, B. Ramaswamy, S.W. Joo, Phys. Fluids **7**, 2291 (1995)
36. W. Boos, A. Thess, Phys. Fluids **11**, 1484 (1999)
37. S. Shklyaev, A.V. Straube, A. Pikovsky, Phys. Rev. E **82**, 020601 (2010)
38. I.B. Simanovskii, A.A. Nepomnyashchy, *Convective Instabilities in Systems with Interface* (Gordon and Breach, London, 1993)
39. K.A. Smith, J. Fluid Mech. **24**, 401 (1966)
40. J.W. Scanlon, L.A. Segel, J. Fluid Mech. **30**, 149 (1967)
41. E.A. Kuznetsov, M.D. Spektor, J. Appl. Mech. Tech. Phys. **21**, 220 (1980)
42. V.V. Pukhnachev, The appearance of thermocapillary effect in a thin fluid layer, in *Hydrodynamics and Heat Mass Transfer of Fluid Flows with a Free Surface* (Institute of Thermophysics SB AS USSR, Novosibirsk, 1985, in Russian), p. 119
43. L.G. Badratinova, Continuous Media Dyn. (Novosibirsk) **69**, 3 (1985, in Russian)
44. D.D. Joseph, *Stability of Fluid Motions*, I and II (Springer, New York, 1976)
45. J.H. Israelachvili, *Intermolecular and Surface Forces* (Academic, New York, 1992)
46. B.V. Derjaguin, N.V. Churaev, V.M. Muller, *Surface Forces* (Consultants Bureau, New York, 1987)
47. E.M. Lifshitz, Sov. Phys. JETP **2**, 73 (1956)
48. I.E. Dzyaloshinskii, L.P. Pitaevskii, Sov. Phys. JETP **9**, 1282 (1959)
49. I.E. Dzyaloshinskii, E.M. Lifshitz, L.P. Pitaevskii, Sov. Phys. JETP **10**, 161 (1960)
50. M.B. Williams, S.H. Davis, J. Colloid Interface Sci. **90**, 220 (1982)
51. A. Sharma, E. Ruckenstein, J. Colloid Interface Sci. **113**, 456 (1986)
52. H.C. Hamaker, Physica **4**, 1058 (1937)
53. P.G. de Gennes, F. Brochard-Wyart, D. Quéré, *Capillary and Wetting Phenomena* (Springer, New York, 2004)
54. N.V. Churaev, Adv. Colloid Interface Sci. **103**, 197 (2003)
55. V.M. Starov, M.G. Velarde, C.J. Radke, *Wetting and Spreading Dynamics* (CRC Press, Boca Raton, 2007)
56. L.M. Hocking, Phys. Fluids **5**, 793 (1993)
57. Q.F. Wu, H. Wong, J. Fluid Mech. **506**, 157 (2004)
58. Y.D. Shikhmurzaev, J. Fluid Mech. **334**, 211 (1997)
59. E.B. Dussan, S.H. Davis, J. Fluid Mech. **65**, 71 (1974)
60. E.B. Dussan, Ann. Rev. Fluid Mech. **11**, 371 (1979)
61. J. Koplik, J.R. Banavar, J.F. Willemsen, Phys. Fluids A **1**, 781 (1989)
62. P.A. Thompson, M.O. Robbins, Phys. Rev. Lett. **63**, 766 (1989)
63. J. Barrat, L. Bocquet, Phys. Rev. Lett. **82**, 4671 (1999)
64. N.V. Priezjev, A.A. Darhuber, S.M. Troian, Phys. Rev. E. **71**, 041608 (2005)
65. P.G. Drazin, W.H. Reid, *Hydrodynamic Stability* (Cambridge University Press, Cambridge, 1982)
66. D.A. Nield, J. Fluid Mech. **71**, 441 (1975)
67. T.P. Lyubimova, Y.N. Parshakova, Fluid Dyn. **42**, 695 (2007)
68. T.P. Lyubimova, D.V. Lyubimov, Y.N. Parshakova, Eur. Phys. J. Spec. Top. **224**, 249 (2015)

69. T.P. Lyubimova, D.V. Lyubimov, Y. Parshakova, A. Ivantsov, Microgravity Sci. Technol. **23**, 143 (2011)
70. K. Schwarzenberger, T. Köllner, H. Linde, T. Boeck, S. Odenbach, K. Eckert, Adv. Colloid Interf. Sci. **206**, 344 (2014)
71. A. La Porta, C.M. Surko, Physica D **123**, 21 (1998)
72. M. Dennin, D.S. Cannell, G. Ahlers, Science **272**, 388 (1996)
73. Y. Kuramoto, *Chemical Oscillations, Waves, and Turbulence* (Springer, Berlin, 1984)
74. W.J. Rappel, F. Fenton, A. Karma, Phys. Rev. Lett. **83**, 456 (1999)
75. K. Roeller, S. Herminghaus, Phys. Rev. E **86**, 021301 (2012)
76. J. Lega, J.V. Moloney, A.C. Newell, Physica D **83**, 478 (1995)
77. E.B. Levchenko, E.B. Chernyakov, Sov. Phys. JETP **54**, 202 (1981)
78. P.L. Garcia-Ybarra, M.G. Velarde, Phys. Fluids **30**, 1649 (1987)
79. J. Lucassen, Trans. Faraday Soc. **64**, 2221, 2230 (1968)
80. E.H. Lucassen-Reynders, J. Lucassen, Adv. Colloid Interf. Sci. **2**, 347 (1969)
81. M. Takashima, J. Phys. Soc. Jpn. **50**, 2751 (1981)
82. A.Y. Rednikov, P. Colinet, M.G. Velarde, J.C. Legros, Phys. Rev. E **57**, 2872 (1998)
83. X.L. Chu, M.G. Velarde, Phys. Rev. A **43**, 1094 (1991)
84. A.N. Garazo, M.G. Velarde, Phys. Fluids **3**, 2295 (1991)
85. A.N. Garazo, M.G. Velarde, in *Proceedings of the VIII European Symposium on Materials and Fluid Sciences in Microgravity* (European Space Agency, Noordwijk, 1992)
86. A.A. Nepomnyashchy, M.G. Velarde, Phys. Fluids **6**, 187 (1994)
87. D.E. Bar, A.A. Nepomnyashchy, Physica D **86**, 90 (1995)
88. G.B. Whitham, *Linear and Nonlinear Waves* (Wiley, New York, 1974)
89. A.A. Nepomnyashchy, Proc. Perm State Univ. **362**, 114 (1976, in Russian)
90. T. Kawahara, S. Toh, Phys. Fluids **28**, 1636 (1985)
91. V.I. Nekorkin, M.G. Velarde, Int. J. Bifurcation Chaos **4**, 1135 (1994)
92. M.G. Velarde, V.I. Nekorkin, A.G. Maksimov, Int. J. Bifurcation Chaos **5**, 831 (1995)
93. C.I. Christov, M.G. Velarde, Physica D **86**, 323 (1995)
94. A. Oron, P. Rosenau, Phys. Rev. E **55**, R1267 (1997)
95. D.E. Bar, A.A. Nepomnyashchy, Physica D **132**, 411 (1999)
96. P.D. Weldman, H. Linde, M.G. Velarde, Phys. Fluids A **4**, 921 (1992)
97. H. Linde, X. Chu, M.G. Velarde, Phys. Fluids A **5**, 1068 (1993)
98. M.G. Velarde, A.Y. Rednikov, H. Linde, *Fluid Dynamics at Interfaces* (Cambridge University Press, Cambridge, 1999), p. 43
99. A. Oron, S.H. Davis, S.G. Bankoff, Rev. Mod. Phys. **69**, 931 (1997)
100. S. Shklyaev, M. Khenner, A.A. Alabuzhev, Phys. Rev. E **82**, R025302 (2010)
101. S. Shklyaev, M. Khenner, A.A. Alabuzhev, Phys. Rev. E **85**, 016328 (2012)
102. P.L. Garcia-Ybarra, J.L. Castillo, M.G. Velarde, Phys. Fluids **30**, 2655 (1987)
103. A.E. Samoilova, N.I. Lobov, Phys. Fluids **26**, 064101 (2014)
104. A.E. Samoilova, S. Shklyaev, Eur. Phys. J. Spec. Top. **224**, 241 (2015)
105. A.A. Nepomnyashchy, I.B. Simanovskii, Q. J. Mech. Appl. Math. **50**, 149 (1997)
106. I.L. Kliakhandler, A.A. Nepomnyashchy, I.B. Simanovskii, M.A. Zaks, Phys. Rev. E **58**, 5765 (1998)
107. A.A. Nepomnyashchy, I.B. Simanovskii, T. Boeck, A.A. Golovin, L.M. Braverman, A. Thess, Lect. Notes Phys. **628**, 21 (2003)
108. L.M. Pismen, Dyn. Stabil. Syst. **1**, 97 (1986)
109. J.W. Swift, Nonlinearity **1**, 333 (1988)
110. M. Silber, E. Knobloch, Nonlinearity **4**, 1063 (1991)
111. L.S. Fisher, A.A. Golovin, J. Colloid Interface Sci. **291**, 515 (2005)
112. E.M. Lifshitz, L.P. Pitaevskii, *Statistical Physics*, Part 2 (Pergamon, New York, 1980)
113. A. Pototsky, M. Bestehorn, D. Merkt, U. Thiele, J. Chem. Phys. **122**, 224711 (2005)
114. A. Pototsky, M. Bestehorn, D. Merkt, U. Thiele, Phys. Rev. E **70**, 025201 (2004)
115. D. Bandyopadhyay, R. Gulabani, A. Sharma, Ind. Eng. Chem. Res. **44**, 1259 (2005)
116. A. Nepomnyashchy, I. Simanovskii, J.C. Legros, *Interfacial Convection in Multilayer Systems*, 2nd edn. (Springer, New York, 2012)

Chapter 4
Convection in Binary Liquids: Amplitude Equations for Stationary and Oscillatory Patterns

While the Rayleigh–Bénard convection in a pure liquid serves as a paradigmatic example of stationary patterns generated by monotonic instability, convection in a binary liquid provides a basic example of an oscillatory instability generating wave patterns. Therefore within this chapter, we mainly focus on time-periodic patterns; the stationary convection is discussed only when it is qualitatively different from that in a pure liquid. The description of longwave oscillatory patterns is strongly different from both longwave stationary patterns and shortwave oscillatory patterns. Partial differential amplitude equations, which are local in space, cannot be derived in that case, and a different approach, based on the analysis of resonances, is needed.

4.1 Buoyancy Convection

In this section we consider longwave buoyancy convection in a binary liquid layer. This problem is characterized by the presence of two longwave modes: in addition to the monotonic mode, which is similar to that in a pure liquid (see Section 3.1), there appears the parameter range, where the longwave oscillatory mode is critical. We will see that the nonlinear development of that mode is quite different from those discussed for problems with deformable free surface or interface in Chapter 3; a local partial differential equation governing the evolution of the amplitude functions does not exist. Instead, the nonlinear evolution is governed by nonlocal equations— the solvability conditions for a certain linear nonhomogeneous problem.

In Section 4.1.1 we formulate the problem, derive the nonlocal amplitude equations, and discuss the weakly nonlinear analysis for patterns on three lattices: rhombic (in the Fourier domain), square, and hexagonal ones. The impact of non-Boussinesq effects, which produce additional quadratic nonlinear terms, is addressed in Section 4.1.2. The analysis presented in these sections follows and extends the early work by Pismen [1] on that subject. Finally, in Section 4.1.3, we discuss the restitution (reconstitution) procedure, which transforms the nonlocal am-

© Springer Science+Business Media, LLC 2017
S. Shklyaev, A. Nepomnyashchy, *Longwave Instabilities and Patterns in Fluids*,
Advances in Mathematical Fluid Mechanics,
https://doi.org/10.1007/978-1-4939-7590-7_4

plitude equations into a set of partial differential equations, but with different orders of the small parameter mixed. It is explained, in what sense that restitution has to be understood.

4.1.1 Boussinesq Convection

4.1.1.1 Formulation of the Problem

In this section the Rayleigh–Bénard convection is analyzed in a layer of binary mixture of thickness $2d_*$ heated from below.

The density of the binary mixture is assumed to be linear with respect to both temperature and solute concentration:

$$\rho_* = \rho_*^{(0)} \left[1 - \beta_T (T_* - T_*^{(0)}) + \beta_C (C_* - C_*^{(0)}) \right], \tag{4.1}$$

where $\rho_*^{(0)}$, $T_*^{(0)}$, and $C_*^{(0)}$ are the reference values of the density, temperature, and concentration. The solutal expansion coefficient β_C is positive (negative) if the solution density grows (decreases) with the increase in the solute concentration.

Similar to the case of a pure liquid, the longwave instability is developed when the heat conductivity of both solid walls is small in comparison with that of the liquid. We represent the liquid temperature as a superposition of the linear distribution produced by the uniform temperature gradient across the layer and the perturbation, $T_* = -Az_* + \Theta_*(\mathbf{r}_*, t_*)$. For the latter quantity, we make use of Newton's law of cooling for the temperature perturbations at both rigid walls with equal heat transfer coefficients q_*.

In the expression for the mass flux $\mathbf{J}_m$, we take into account the Soret effect:

$$\mathbf{J}_m = -\rho_*^{(0)} D(\nabla C_* + \alpha \nabla T_*), \tag{4.2}$$

where D and α are the mass diffusivity and the Soret coefficient of the binary mixture, respectively.

Extensive measurements of the Soret coefficient α for aqueous solutions of NaCl and KCl for various temperatures and solute concentrations have been performed by Gaeta et al. [2]. The important result of these experiments was the finding that α_* can be either positive or negative. Similar results were later obtained for the mixtures of water and ethanol [3]. We neglect the Dufour effect (a heat flux generated by a concentration gradient) that has been justified by numerous experiments [4–7]. At the solid boundaries, the normal component of $\mathbf{J}_m$ vanishes; therefore, the temperature gradient maintained over the layer creates the gradient of the solute concentration. Similar to the temperature field, the solute concentration is represented as a sum of two terms corresponding to the conductive temperature profile and the disturbance, $C_* = \alpha A z_* + \Sigma_*(\mathbf{r}_*, t_*)$.

The following equations and boundary conditions govern the buoyancy convection in the layer:

$$\nabla \cdot \mathbf{v}_* = 0, \qquad (4.3)$$

$$\rho_*^{(0)} \left[\partial_{t_*} \mathbf{v}_* + (\mathbf{v}_* \cdot \nabla) \mathbf{v}_* \right] = -\nabla p_* + \eta \nabla^2 \mathbf{v}_* - \rho_*^{(0)} \mathbf{g} (\beta_T \Theta_* - \beta_C \Sigma_*), \qquad (4.4)$$

$$\partial_{t_*} \Theta_* + \mathbf{v}_* \cdot \nabla \Theta_* - A w_* = \kappa \nabla^2 \Theta_*, \qquad (4.5)$$

$$\partial_{t_*} \Sigma_* + \mathbf{v}_* \cdot \nabla \Sigma_* + \alpha A w_* = D \nabla^2 (\Sigma_* + \alpha \Theta_*), \qquad (4.6)$$

$$\mathbf{v}_* = 0, \quad k_{th} \partial_{z_*} \Theta_* = \mp q \Theta_*, \quad \partial_{z_*} (\Sigma_* + \alpha_* \Theta_*) = 0 \text{ at } z_* = \pm d_*. \qquad (4.7)$$

The problem is nondimensionalized using d_*^2/κ, d_*, Ad_*, $Ad_* \beta_T/\beta_C$, κ/d_*, and $\eta \kappa / d_*^2$ as the scales for the time, length, temperature, solute concentration, the velocity, and the pressure, respectively. That yields:

$$\nabla \cdot \mathbf{v} = 0, \qquad (4.8)$$

$$P^{-1} \left[\mathbf{v}_t + (\mathbf{v} \cdot \nabla) \mathbf{v} \right] = -\nabla p + \nabla^2 \mathbf{v} + R(\Theta - \Sigma) \mathbf{e}_z, \qquad (4.9)$$

$$\Theta_t + \mathbf{v} \cdot \nabla \Theta - w = \nabla^2 \Theta, \qquad (4.10)$$

$$L^{-1} (\Sigma_t + \mathbf{v} \cdot \nabla \Sigma + \psi w) = \nabla^2 (\Sigma + \psi \Theta), \qquad (4.11)$$

$$\mathbf{v} = 0, \quad \Theta_z = \mp Bi \Theta, \quad (\Sigma + \psi \Theta)_z = 0 \text{ at } z = \pm 1. \qquad (4.12)$$

The problem is characterized by the following five dimensionless parameters:

$$R = \frac{g \beta_T A_* d^4}{\nu_* \kappa_*}, \quad Bi = \frac{q_* d}{k_{th}}, \quad \psi = \frac{\alpha_* \beta_C}{\beta_T}, \quad P = \frac{\nu_*}{\kappa_*}, \quad L = \frac{D_*}{\kappa_*}, \qquad (4.13)$$

representing the Rayleigh, Biot, separation, Prandtl, and Lewis numbers, respectively. The Biot number is assumed small which warrants the longwave instability. The Schmidt number $S = P/L$ is also used below.

The conductive state corresponds to motionless liquid, $\mathbf{v} = 0$, with zero deviations from the linear distributions of the temperature and concentration $\Theta = \Sigma = 0$.

4.1.1.2 Linear Stability Problem

To investigate the linear stability of the conductive state with respect to the longwave mode, we introduce small perturbations of the velocity, pressure, temperature, and solute concentration. Due to the arguments similar to those used in Chapter 3, we can restrict ourselves to 2D analysis: $\mathbf{v} = (u, 0, w)$ and the disturbances do not depend on y. For normal perturbations of any field f, the following representation is valid: $f(x, z, t) = \tilde{f}(z) \exp(\sigma t + ikx)$, where σ and k are the complex-valued growth rate and real wavenumber of perturbations, correspondingly. With this in view, the set of linearized equations is written in the following form:

$$ik\tilde{u} + \tilde{w}' = 0, \qquad (4.14)$$

$$\sigma P^{-1} \tilde{u} = -ik\tilde{p} + \Delta \tilde{u}, \qquad (4.15)$$

$$\sigma P^{-1}\tilde{w} = -\tilde{p}' + \Delta\tilde{w} + R(\tilde{\Theta} - \tilde{\Sigma}), \tag{4.16}$$

$$\sigma\tilde{\Theta} - \tilde{w} = \Delta\tilde{\Theta}, \tag{4.17}$$

$$L^{-1}\left(\sigma\tilde{\Sigma} + \psi\tilde{w}\right) = \Delta\left(\tilde{\Sigma} + \psi\tilde{\Theta}\right), \tag{4.18}$$

$$\tilde{u} = \tilde{w} = \tilde{\Theta}' \pm Bi\tilde{\Theta} = (\tilde{\Sigma} + \psi\tilde{\Theta})' = 0 \text{ at } z = \pm 1. \tag{4.19}$$

We discuss within this subsection only the longwave mode ($k \ll 1$) for small Bi. To that end, both the wavenumber and growth rate are rescaled as follows:

$$k = \varepsilon K, \ \sigma = \varepsilon^2\Lambda, \tag{4.20}$$

where $\varepsilon = Bi^{1/4} \ll 1$ (see Section 2.2.2). The velocity components are rescaled in the following way:

$$\tilde{w} = \varepsilon^2\tilde{W}, \ \tilde{u} = \varepsilon\tilde{U}, \tag{4.21}$$

whereas the rest of perturbations fields remain unscaled. Finally, we expand the perturbation fields, the rescaled growth rate Λ, and the Rayleigh number in powers of ε^2:

$$\tilde{f} = \tilde{f}_0 + \varepsilon^2\tilde{f}_2 + \ldots, \ \Lambda = \Lambda_0 + \varepsilon^2\Lambda_2 + \ldots, \ R = 45(r_0 + \varepsilon^2 r_2 + \ldots). \tag{4.22}$$

Substituting these series into (4.14)–(4.19) and collecting the terms of each order in ε (in fact, only even orders are nontrivial), one obtains at the zeroth order:

$$\tilde{W}_0' = -iK\tilde{U}_0, \ \tilde{P}_0' = 45r_0(\tilde{\Theta}_0 - \tilde{\Sigma}_0), \ \tilde{U}_0'' = iK\tilde{P}_0, \ \tilde{\Theta}_0'' = \tilde{\Sigma}_0'' = 0, \tag{4.23}$$

$$\tilde{U}_0 = \tilde{W}_0 = \tilde{\Theta}_0' = \tilde{\Sigma}_0' = 0 \text{ at } z = \pm 1. \tag{4.24}$$

The solution of this boundary value problem has the following form:

$$\tilde{\Theta}_0 = a_0, \ \tilde{\Sigma}_0 = b_0, \tag{4.25}$$

$$\tilde{P}_0 = 45r_0 h_0 z, \ \tilde{U}_0 = r_0 iKh_0 V_\parallel^{(R)}, \ \tilde{W}_0 = -r_0 K^2 h_0 V_\perp^{(R)}, \tag{4.26}$$

$$V_\parallel^{(R)}(z) = -\frac{15}{2}z(1 - z^2), \ V_\perp^{(R)}(z) = -\frac{15}{8}(1 - z^2)^2, \tag{4.27}$$

where the constant $-h_0 = b_0 - a_0$ is the dimensionless disturbance of the density.

The second-order problem reads:

$$W_2' = -iKU_2, \ \tilde{U}_2'' = iK\tilde{P}_2 + \left(K^2 + \Lambda_0 P^{-1}\right)\tilde{U}_0, \tag{4.28}$$

$$\tilde{P}_2' = \tilde{W}_0'' + 45\left[r_0(\tilde{\Theta}_2 - \tilde{\Sigma}_2) + r_2(\tilde{\Theta}_0 - \tilde{\Sigma}_0)\right], \tag{4.29}$$

$$\tilde{\Theta}_2'' = \left(K^2 + \Lambda_0\right)\tilde{\Theta}_0 - \tilde{W}_0, \tag{4.30}$$

$$\left(\tilde{\Sigma}_2 + \psi\tilde{\Theta}_2\right)'' = \psi K^2\tilde{\Theta}_0 + \left(K^2 + \Lambda_0 L^{-1}\right)\tilde{\Sigma}_0 + \psi L^{-1}\tilde{W}_0, \tag{4.31}$$

$$\tilde{U}_2 = \tilde{W}_2 = \tilde{\Theta}_2' = \tilde{\Sigma}_2' = 0 \text{ at } z = \pm 1. \tag{4.32}$$

Integration of (4.30) and (4.31) across the layer yields the solvability conditions for the corresponding boundary value problem:

$$\mathscr{L}_1(a_0, b_0) = \left(K^2 + \Lambda_0\right) a_0 - r_0 K^2 (a_0 - b_0) = 0, \qquad (4.33)$$

$$\mathscr{L}_2(a_0, b_0) = L\psi K^2 a_0 + \left(LK^2 + \Lambda_0\right) b_0 + r_0 \psi K^2 (a_0 - b_0) = 0. \qquad (4.34)$$

Nontrivial solutions of this linear homogeneous algebraic system are possible, if Λ solves the following quadratic equation

$$\Lambda_0^2 + \left[1 + L - r_0 (1 + \psi)\right] \Lambda_0 K^2 + (L - r_0 \psi_L) K^4 = 0, \quad \psi_L = L + \psi + \psi L. \quad (4.35)$$

This equation allows for both monotonic ($\Lambda_0 = 0$ at the stability boundary) and oscillatory ($\Lambda_0 = -i\Omega$ at the stability boundary) modes. For the monotonic mode, the stability threshold is given by

$$r_0^{(m)} = \frac{L}{\psi_L}. \qquad (4.36)$$

Following [8], this value can be recast from the standard $r_0 = 1$ by Nield [9]. Indeed, for the monotonic mode at $B = 0$ (or B being small to contribute to the corresponding order) at the stability boundary one can readily derive $\tilde{\Sigma} = -\psi L^{-1} \tilde{\Theta}$ comparing (4.17) and (4.18). Therefore, the combination $\tilde{\Theta} - \tilde{\Sigma}$, which appears in the product with R, can be rewritten as $\psi_L L^{-1} \tilde{\Theta}$. Thus, $R_{eff} = R\psi_L L^{-1} = 45 = 720/16$ according to [9], which provides (4.36).

For the oscillatory mode one obtains:

$$r_0^{(o)} = \frac{1 + L}{1 + \psi}, \qquad (4.37)$$

the frequency of neutral perturbations reads

$$\Omega = K^2 \tilde{\Omega}; \quad \tilde{\Omega}^2 = L \left(1 - \frac{r_0^{(o)}}{r_0^{(m)}}\right) = -\frac{\psi \left(1 + L + L^2\right) + L^2}{1 + \psi}. \qquad (4.38)$$

The oscillatory mode takes place if this expression is positive; hence,

$$-1 < \psi < \psi_c, \quad \psi_c = -\frac{L^2}{1 + L + L^2}. \qquad (4.39)$$

It is worth noting that the threshold values of the Rayleigh number for both modes do not depend on K, but the frequency does. Moreover, the waves are dispersive since $\Omega \sim K^2$.

Variation of $r_0^{(m,o)}$ with ψ for different values of the Lewis number is shown in Figure 4.1(a). The monotonic instability occurs at $\psi > \psi_c$ for heating from below and at $\psi < \psi_\infty^{(m)} = -L/(1 + L)$ for heating from above. (In fact, as we shall discuss in Section 4.2.1.2, for heating from below the situation is more complicated.)

The oscillatory mode is critical within the interval (4.39) for heating from below. At $\psi_\infty^{(m)} < \psi < \psi_c$, the neutral stability curve $r_0 = r_0^{(m)}$ corresponds to the stabilization of one of two monotonic modes, whereas the second one still has positive Λ_0. The variation of the rescaled growth rate Λ_0/K^2 with r in this interval of ψ is shown in Figure 4.1(b). If (4.35) is met, the solution of the homogeneous system

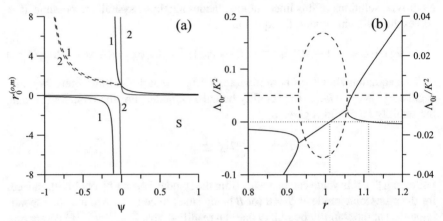

Fig. 4.1 Results on linear stability for the longwave buoyancy convection in a binary mixture. (a) Stability map in the plane $\psi - r_0^{(o,m)}$ for $L = 0.1$ (lines 1) and $L = 0.01$ (lines 2). The solid and dashed lines correspond to the monotonic (4.36) and oscillatory (4.37) mode, respectively. Domain of stability is marked "S." (b) Variation of the scaled growth rate Λ_0/K^2 with $r_0 = R/45$ at $L = 0.01$ and $\psi_\infty^{(m)} < \psi = -0.001 < \psi_c$: the solid and dashed lines show the real and imaginary parts of the growth rate, respectively; the dotted vertical lines correspond to $r_0^{(o)}$ (smaller value) and $r_0^{(m)}$ (larger value)

of algebraic equations (4.33) and (4.34) is determined up to an arbitrary factor. One can, for example, set $a_0 = 1$ and express b_0 from any of these two equations. Below, we do not prescribe any fixed value to a_0 and b_0.

Within the rest of this section, we discuss the oscillatory mode only. For the monotonic one, the analysis is quite similar to that for the Marangoni convection (Section 4.2.1); the corresponding comments are given, e.g., in Sections 4.2.1.2 and 4.2.1.3.

In order to calculate the correction to the growth rate Λ_2, one has to solve boundary value problems (4.28)–(4.32) and derive the solvability conditions at the fourth order. For the oscillatory mode, the solution to the second-order problem is

$$\tilde{\Theta}_2 = a_2 - r_0 K^2 h_0 \Theta_{l2}(z), \qquad (4.40)$$

$$\tilde{\Sigma}_2 = b_2 - r_0 \psi (1 + L^{-1}) K^2 h_0 \Theta_{l2}(z), \qquad (4.41)$$

$$\tilde{p}_2 = -K^2 P_{l2} h_0 + 45(r_0 h_2 + r_2 h_0)z, \qquad (4.42)$$

$$\tilde{U}_2 = -iK^3 \left(h_0 U_{l2}(z) + S^{-1} U_\Phi \tilde{\Phi} \right) + iK(r_2 h_0 + r_0 h_2) V_\parallel^{(R)}, \qquad (4.43)$$

$$\tilde{W}_2 = K^4 \left(h_0 W_{l2}(z) + S^{-1} W_\Phi \tilde{\Phi} \right) - (r_0 h_2 + r_2 h_0) K^2 V_\perp^{(R)}, \qquad (4.44)$$

$$T_{12}(z) = \frac{1}{16}\left(z^6 - 5z^4 + 7z^2 - \frac{31}{21}\right), \quad (4.45)$$

$$P_{12}(z) = \frac{45\psi_L r_0^2}{16L}\left(\frac{z^7}{7} - z^5 + \frac{7}{3}z^3 - \frac{31}{21}z\right) - r_0 V_{\parallel}^{(R)}, \quad (4.46)$$

$$U_{12}(z) = \frac{45\psi_L r_0^2}{16L}\left(\frac{z^9}{504} - \frac{z^7}{42} + \frac{7}{60}z^5 - \frac{31}{126}z^3 + \frac{127}{840}z\right) + 2U_\Phi, \quad (4.47)$$

$$W_{12}(z) = -\frac{45\psi_L r_0^2}{16L}\left(\frac{z^{10}}{5040} - \frac{z^8}{336} + \frac{7}{360}z^6 - \frac{31}{504}z^4 + \frac{127}{1680}z^2 - \frac{31}{1008}\right)$$
$$+ 2W_\Phi, \quad (4.48)$$

$$U_\Phi = -\frac{r_0}{8}(3z^5 - 10z^3 + 7z), \quad W_\Phi = \frac{r_0}{16}(z^6 - 5z^4 + 7z^2 - 3), \quad (4.49)$$

where $h_2 = a_2 - b_2$ and

$$\tilde{\Phi} = (1+\psi)a_0 - L^{-1}b_0. \quad (4.50)$$

At the fourth order, only boundary value problems for Θ_4 and Σ_4 are needed; they read

$$\tilde{\Theta}_4'' = \Lambda_2\tilde{\Theta}_0 + \left(K^2 + \Lambda_0\right)\tilde{\Theta}_2 - \tilde{W}_2, \quad (4.51)$$

$$\left(\tilde{\Sigma}_4 + \psi\tilde{\Theta}_4\right)'' = \psi K^2\tilde{\Theta}_2 + \Lambda_2 L^{-1}\tilde{\Sigma}_0 + \left(K^2 + \Lambda_0 L^{-1}\right)\tilde{\Sigma}_2 + \psi L^{-1}\tilde{W}_2, \quad (4.52)$$

$$\tilde{\Theta}_4' = \mp\tilde{\Theta}_0, \quad (\tilde{\Sigma}_4 + \psi\tilde{\Theta}_4)' = 0 \text{ at } z = \pm 1. \quad (4.53)$$

The solvability conditions for these two boundary value problems can be obtained via a usual routine—integrating the right-hand sides of the equations and accounting for the boundary conditions. This leads to the following set of algebraic equations for a_2 and b_2:

$$\mathcal{L}_1(a_2, b_2) = -(\Lambda_2 + 1)a_0 + \frac{2r_0 K^4}{231}A_\Phi + r_2 K^2 h_0, \quad (4.54)$$

$$\mathcal{L}_2(a_2, b_2) = -\Lambda_2 b_0 - \frac{2\psi r_0 K^4}{231}A_\Phi - \psi r_2 K^2 h_0, \quad (4.55)$$

$$A_\Phi = \left(5r_0\frac{\psi_L}{L} - 22\right)h_0 - \frac{11}{S}\tilde{\Phi}.$$

At r_0 given by (4.37) and $\Lambda_0 = -i\Omega$, the homogeneous part of (4.54) and (4.55) has a nontrivial solution. Thus, the right-hand side has to satisfy a certain solvability condition, which yields the following correction to the growth rate:

$$\Lambda_2 = -\frac{1}{2}\left(1 + i\frac{L - \psi r_0}{\tilde{\Omega}}\right) + \frac{r_2 K^2}{2r_0}\left(1 + L + i\frac{L - \tilde{\Omega}}{\tilde{\Omega}}\right) - \frac{1}{2}f(L, P, \psi)K^4, \quad (4.56)$$

$$f(L, P, \psi) = \frac{2}{231}\left(\left(5\tilde{\Omega}^2 L^{-1} + 17\right)\left(1 + L + \frac{i}{\tilde{\Omega}}(L - \tilde{\Omega}^2)\right) + \frac{f_P}{P}\right), \quad (4.57)$$

$$f_P = 11\left(r_0\psi_L + \frac{i}{\tilde{\Omega}}(1 + L)(L^2 + r_0\psi)\right). \quad (4.58)$$

The growth rate $\Lambda_2(K)$ with f_P-term omitted agrees with that given in [1] (see (83) there), up to unimportant complex conjugation. The only distinction is the sign inversion for the third (K-independent) term, which is caused by the typo in the cited paper. Contrary to the statement in Section VI of [1], the inertial term (f_P) has a nonzero real part, and therefore it changes the critical wavenumber.

Setting $\Lambda_{2r}(K) = 0$, one obtains the neutral stability curve $r_{2*}(K)$

$$r_{2*} = \frac{K^{-2} + f_r K^2}{1 + \psi}. \tag{4.59}$$

For finite P, f_r can be negative, which makes the longwave expansion invalid; however the Schmidt number S has to be unrealistically small, $S < 11/5$, in order to get negative f_r. We do not consider this situation any further dealing with $f_r > 0$ below. Therefore, $r_{2*}(K)$ has the only minimum

$$r_{2c} = \frac{2\sqrt{f_r}}{1 + \psi} \text{ at } K_c = f_r^{-1/4}. \tag{4.60}$$

4.1.1.3 Oscillatory Mode: Derivation of the Amplitude Equations

In order to study the nonlinear evolution of large-scale oscillatory regimes, we rescale the velocity and introduce the stretched coordinates and multiple time scales:

$$\mathbf{u} = \varepsilon \mathbf{U}, \ w = \varepsilon^2 W, \ (X,Y) = \varepsilon(x,y), \ \tau = \varepsilon^2 t, \ t_4 = \varepsilon^4 t, \ B = \varepsilon^4. \tag{4.61}$$

Also, we expand the Rayleigh number and fields of perturbations in powers of ε^2 as follows:

$$R = 45(r_0 + \varepsilon^2 r_2 + \ldots), \ f = f^{(0)} + \varepsilon^2 f^{(2)} + \ldots, \ f = (\Theta, \Sigma, \mathbf{U}, W, p). \tag{4.62}$$

These expansions being substituted into (4.8)–(4.12) yield at zero order of ε^2:

$$\mathbf{U}_{zz}^{(0)} = \nabla p^{(0)}, \ \nabla \cdot \mathbf{U}^{(0)} + W_z^{(0)} = 0, \tag{4.63}$$

$$p_z^{(0)} = 45r_0 \left(\Theta^{(0)} - \Sigma^{(0)} \right), \ \Theta_{zz}^{(0)} = 0, \ \Sigma_{zz}^{(0)} = 0, \tag{4.64}$$

$$U^{(0)} = W^{(0)} = \Theta_z^{(0)} = \Sigma_z^{(0)} = 0 \text{ at } z = \pm 1. \tag{4.65}$$

The solution of this problem reads

$$\Theta^{(0)} = F(X,Y,\tau,t_4), \ \Sigma^{(0)} = G(X,Y,\tau,t_4), \ p^{(0)} = 45r_0 hz, \tag{4.66}$$

$$U^{(0)} = r_0 V_\parallel^{(R)}(z) \nabla h, \ W^{(0)} = r_0 V_\perp^{(R)}(z) \nabla^2 h. \tag{4.67}$$

Here, similar to Section 4.1.1.2, $-h = G - F$ is the perturbation of the density and $V_{\parallel,\perp}^{(R)}$ are defined by (4.27); within the rest of this subsection, ∇ stands for the two-dimensional gradient with respect to (X,Y).

At the second order, we derive the following boundary value problem:

$$p_z^{(2)} = W_{zz}^{(0)} + 45\left[r_0(\Theta^{(2)} - \Sigma^{(2)}) + r_2(\Theta^{(0)} - \Sigma^{(0)})\right], \quad (4.68)$$

$$\mathbf{U}_{zz}^{(2)} - \nabla p^{(2)} = -\nabla^2\mathbf{U}^{(0)} + P^{-1}\left[\mathbf{U}_\tau^{(0)} + (\mathbf{U}^{(0)} \cdot \nabla)\mathbf{U}^{(0)} + W^{(0)}\mathbf{U}_z^{(0)}\right], \quad (4.69)$$

$$\nabla \cdot \mathbf{U}^{(2)} + W_z^{(2)} = 0, \quad (4.70)$$

$$\Theta_{zz}^{(2)} = -\nabla^2\Theta^{(0)} + \Theta_\tau^{(0)} + \mathbf{U}^{(0)} \cdot \nabla\Theta^{(0)} - W^{(0)}, \quad (4.71)$$

$$\Sigma_{zz}^{(2)} + \psi\Theta_{zz}^{(2)} = -\nabla^2\left(\Sigma^{(0)} + \psi\Theta^{(0)}\right)$$

$$+L^{-1}\left(\Sigma_\tau^{(0)} + \mathbf{U}^{(0)} \cdot \nabla\Sigma^{(0)} + \psi W^{(0)}\right); \quad (4.72)$$

$$\mathbf{U}^{(2)} = W^{(2)} = \Theta_z^{(2)} = \Sigma_z^{(2)} = 0 \text{ at } z = \pm 1. \quad (4.73)$$

Integrating (4.71) and (4.72) over the interval $-1 \le z \le 1$ and accounting for the corresponding boundary conditions (4.73), one arrives at the following solvability conditions:

$$\mathscr{L}_1(F,G) = F_\tau - (1 - r_0)\nabla^2 F - r_0\nabla^2 G = 0, \quad (4.74)$$

$$\mathscr{L}_2(F,G) = G_\tau - \psi(L + r_0)\nabla^2 F - (L - \psi r_0)\nabla^2 G = 0, \quad (4.75)$$

which are compatible with (4.33) and (4.34).

At r_0 larger than $r_0^{(o)}$ given by (4.37), the perturbations grow exponentially in the fast time τ; at $r_0 < r_0^{(o)}$ they decay. Near the stability threshold, according to (4.74) and (4.75), F and G oscillate in τ with the frequency Ω given by (4.38), where K is the perturbation wavenumber.

Therefore, below we set

$$r_0 = r_0^{(o)} = \frac{1 + L}{1 + \psi} \quad (4.76)$$

and represent the general solution to (4.74) and (4.75) as follows:

$$h = \sum_{\mathbf{k}} A_{\mathbf{k}}(t_4)e^{i(\mathbf{k} \cdot \mathbf{R} - \Omega\tau)} + c.c., \quad (4.77)$$

$$F = \frac{r_0}{1 - i\tilde{\Omega}}\sum_{\mathbf{k}} A_{\mathbf{k}}(t_4)e^{i(\mathbf{k} \cdot \mathbf{R} - \Omega\tau)} + c.c., \quad G = F - h, \quad (4.78)$$

where $A_{\mathbf{k}}(t_4)$ are the yet unknown Fourier harmonics for h, which evolve in the slow time t_4, and $\mathbf{R} = (X, Y)$ is the rescaled two-dimensional radius vector. To describe the evolution of these harmonics, we have to proceed to the fourth-order problem.

First, let us present the solution of the second-order problem:

$$\Theta^{(2)} = Q(X, Y, \tau, t_4) + (r_0 z - U_\Phi)\nabla h \cdot \nabla F + r_0 T_{l2}(z)\nabla^2 h, \quad (4.79)$$

$$\Sigma^{(2)} = R(X, Y, \tau, t_4) - (r_0 z - U_\Phi)\nabla h \cdot \nabla(\tilde{\Phi} - F)$$

$$-\psi r_0(1 + L^{-1})T_{l2}(z)\nabla^2 h, \quad (4.80)$$

$$p^{(2)} = 45(r_2 h + r_0(Q-R))z + \tilde{p}_2 + P_{l2}\nabla^2 h$$

$$+ \frac{45 r_0^2}{16} \nabla h \cdot \nabla \tilde{\Phi} \left(z^6 - 5z^4 + 15z^2 - \frac{55}{21} \right) - \frac{15 r_0^2}{7P}, \qquad (4.81)$$

$$\mathbf{U}^{(2)} = V_{\parallel}^{(R)} \nabla [r_2 h + r_0(Q-R)] + \nabla \nabla^2 (U_{l2} h + S^{-1} U_\Phi \tilde{\Phi})$$

$$+ \frac{45 r_0^2}{16} \nabla (\nabla h \cdot \nabla \tilde{\Phi}) \left(\frac{z^8}{56} - \frac{z^6}{6} + \frac{5}{4} z^4 - \frac{55}{42} z^2 + \frac{5}{24} \right)$$

$$+ \frac{225 r_0^2}{8P} \left[\left(\frac{z^8}{56} - \frac{z^6}{15} + \frac{z^4}{12} - \frac{4}{105} z^2 + \frac{1}{280} \right) \nabla [(\nabla h)^2] \right.$$

$$\left. - \left(\frac{3}{112} z^8 - \frac{7}{60} z^6 + \frac{5}{24} z^4 - \frac{19}{140} z^2 + \frac{29}{1680} \right) \nabla^2 h \nabla h \right]$$

$$- \frac{45 r_0^2}{14P} (z^2 - 1) \nabla \times (\mathbf{\Psi} \mathbf{e}_z), \qquad (4.82)$$

$$W^{(2)} = V_{\perp}^{(R)} \nabla^2 [r_2 h + r_0(Q-R)] + \nabla^4 (W_{l2} h + S^{-1} W_\Phi \tilde{\Phi})$$

$$- \frac{45 r_0^2}{16} \left(\frac{z^9}{504} - \frac{z^7}{42} + \frac{z^5}{4} - \frac{55}{126} z^3 + \frac{5}{24} z \right) \nabla^2 (\nabla h \cdot \nabla \tilde{\Phi})$$

$$- \frac{225 r_0^2}{4P} \left[\left(\frac{z^9}{504} - \frac{z^7}{105} + \frac{z^5}{60} - \frac{4}{315} z^3 + \frac{z}{280} \right) \nabla^2 [(\nabla h)^2] \right.$$

$$\left. - \left(\frac{z^9}{336} - \frac{z^7}{60} + \frac{z^5}{24} - \frac{19 z^3}{420} + \frac{29}{1680} z \right) \nabla \cdot (\nabla^2 h \nabla h) \right], \qquad (4.83)$$

where T_{l2}, P_{l2}, $U_{l2,\Phi}$, and $W_{l2,\Phi}$ are given by (4.45)–(4.49) and

$$\tilde{\Phi} = (1 + \psi) F - L^{-1} G \qquad (4.84)$$

in agreement with (4.50).

In (4.81), $\tilde{p}_2$ is a z-independent field (see [10]); its gradient is determined by the relation

$$\nabla \tilde{p}^{(2)} = -\frac{45 r_0^2}{7P} (\nabla \times (\tilde{\Psi} \mathbf{e}_z) + \nabla^2 h \nabla h).$$

Here $\tilde{\Psi}$ is proportional to the toroidal potential of the velocity; it satisfies the Poisson equation:

$$\nabla^2 \tilde{\Psi} = h_Y \nabla^2 h_X - h_X \nabla^2 h_Y. \qquad (4.85)$$

At the fourth order, we only have to demand the solvability conditions for the heat and mass balance equations:

$$\Theta_{zz}^{(4)} = -\nabla^2 \Theta^{(2)} + \dot{\Theta}^{(0)} + \Theta_\tau^{(2)}$$

$$+ \mathbf{U}^{(0)} \cdot \nabla \Theta^{(2)} + \mathbf{U}^{(2)} \cdot \nabla \Theta^{(0)} + W^{(0)} \Theta_z^{(2)} - W^{(2)}, \qquad (4.86)$$

$$\Sigma_{zz}^{(4)} + \psi \Theta_{zz}^{(4)} = -\nabla^2 \left(\Sigma^{(2)} + \psi \Theta^{(2)} \right) + L^{-1} \left(\Sigma_\tau^{(0)} + \dot{\Sigma}^{(2)} \right)$$

$$+L^{-1}\left(\mathbf{U}^{(0)}\cdot\nabla\Sigma^{(2)}+\mathbf{U}^{(2)}\cdot\nabla\Sigma^{(0)}+W^{(0)}\Sigma_z^{(2)}+\psi W^{(2)}\right); \tag{4.87}$$

$$\Theta_z^{(4)}=\mp\Theta^{(0)},\ \left(\Sigma^{(4)}+\psi\Theta^{(4)}\right)_z=0\ \text{at}\ z=\pm1, \tag{4.88}$$

where the dot denotes the time derivative with respect to t_4.

Integrating (4.86) and (4.87) over the interval $-1\le z\le1$ and taking into account the boundary conditions (4.88), we get:

$$\mathscr{L}_1(Q,R)=-\dot{F}-F-r_2\nabla^2h+\frac{2r_0}{231}\nabla^4\left((5r_0\psi_L L^{-1}-22)h+\frac{11}{S}\tilde{\Phi}\right)$$

$$+\frac{10r_0^2}{7}\nabla\cdot[(\nabla h\cdot\nabla F)\nabla h]-\frac{15r_0^2}{7P}J(F,\Psi), \tag{4.89}$$

$$\mathscr{L}_2(Q,R)=-\dot{G}+\psi r_2\nabla^2h-\frac{2\psi r_0}{231}\nabla^4\left((5r_0\psi_L L^{-1}-22)h+\frac{11}{S}\tilde{\Phi}\right)$$

$$-\frac{10r_0^2}{7}\nabla\cdot[(\nabla h\cdot\nabla(\tilde{\Phi}-F))\nabla h]-\frac{15r_0^2}{7P}J(G,\tilde{\Psi}), \tag{4.90}$$

where $J(f,g)=f_X g_Y-f_Y g_X$ is a Jacobian of the mapping from (X,Y) to (f,g); the fields $\tilde{\Phi}$ and $\tilde{\Psi}$ are defined by (4.84) and (4.85), respectively.

We have obtained a linear nonhomogeneous problem for Q and R with the homogeneous part having a nontrivial solution and the right-hand sides depending on F and G (or $h=F-G$). Therefore, one has to demand the solvability conditions for this problem, thus deriving the amplitude equations for F and G or keeping in mind their Fourier representation, (4.77) and (4.78), for the Fourier amplitudes $A_\mathbf{k}(t_4)$. In a contradistinction to the problems considered in the previous chapters, the nonlinear evolution of dispersive waves generated by an oscillatory instability in the presence of a conservation law is not governed by partial differential equations. It is described by *nonlocal* amplitude equations, which can be transformed to a set of ordinary differential equations for the Fourier amplitudes (Landau equations).

At the leading order, only the *resonant* groups of waves, which satisfy dispersion relation (4.38), contribute to the nonlinear interaction determined by the solvability conditions for (4.89) and (4.90). The nonresonant waves produce only the corresponding nonresonant additive terms in Q and R according to (4.89) and (4.90). According to dispersion relation (4.38), the frequency of the resulting wave has to be proportional to its squared wavenumber with the same coefficient $\tilde{\Omega}$. Therefore, the following resonant conditions

$$\mathbf{k}=\mathbf{k}_1+s_2\mathbf{k}_2+s_3\mathbf{k}_3, \tag{4.91}$$

$$k^2=k_1^2+s_2 k_2^2+s_3 k_3^2, \tag{4.92}$$

have to hold true in order to contribute to the nonlinear interactions. Here $s_{2,3}=\pm1$, for the waves bearing the minus sign, the complex conjugate of the amplitude appears in the nonlinear term.

4.1.1.4 Oscillatory Mode: Weakly Nonlinear Analysis

Generally speaking, problems (4.74), (4.75) and (4.89), (4.90) are appropriate for finite values of F and G or Fourier amplitudes $A_\mathbf{k} = O(1)$. However, following [1], we present here only a weakly nonlinear analysis. A discussion of finite amplitude regimes for a similar problem is given in Section 4.2.1.4.

In compliance with Section 4.1.1.2, the minimum of the neutral stability curve $r_2 = r_{2*}(K)$ is given by (4.60). Near that minimum, nonlinear terms are small, and hence the growth or decay of perturbations is determined by the relations

$$\dot{A}_\mathbf{k} = \gamma(K)A_\mathbf{k}, \ \gamma(K) = \Lambda_2(K), \tag{4.93}$$

where the growth rate Λ_2 is given by (4.56).

Slightly above this value, $r_2 - r_{2c} = \delta^2 \ll 1$, only perturbations with the critical wavenumber K_c grow. Therefore, solution (4.77) includes only the Fourier harmonics with the same wavenumber; the corresponding amplitudes are $O(\delta)$.

$$A_\mathbf{k} = \delta\tilde{A}_\mathbf{k} \text{ at } |\mathbf{k}| = K_c \tag{4.94}$$

and $A_\mathbf{k} = o(\delta)$ otherwise. In particular, that means that the terms related to $\tilde{\Psi}$ vanish, because $\nabla^2 h = -K_c^2 h$ and the right-hand side of (4.85) vanishes.

Rescaling of $A_\mathbf{k}$ has to be supplemented by the expansion of the growth rate and introduction of two time scales:

$$\gamma = \gamma_0 + \delta^2 \gamma_2, \ t_{40} = t_4, \ t_{42} = \delta^2 t_4, \tag{4.95}$$

where $\gamma_0 = \gamma(r_2 = r_{2c})$ is purely imaginary; $\gamma_2 = \partial\gamma/\partial r_2$; the time t_{40} corresponds to the oscillations with the frequency γ_{0i}, and t_{42} is the typical time of the nonlinear evolution.

Substituting these expansions into (4.89) and (4.90), we obtain at the leading order

$$\partial_{t_{40}}\tilde{A}_\mathbf{k} = \gamma_0\tilde{A}_\mathbf{k} \tag{4.96}$$

for each amplitude. This equation describes uncoupled oscillations of the amplitudes with the frequency γ_{0i}.

At the first-order problem (with respect to δ^2), the cubic nonlinear interactions are involved. It was shown by Pismen [1] that only two types of the nonlinear interactions, which satisfy (4.91) and (4.92), are possible for the waves of identical wavenumber:

(i) Interaction of waves with the wavevectors $\mathbf{k}$, $\mathbf{q}$, and $\mathbf{q}$, such that a term proportional to $\tilde{A}_\mathbf{k}|\tilde{A}_\mathbf{q}|^2$ gives a contribution to the evolution equation for $\tilde{A}_\mathbf{k}$ with the coefficient of nonlinear interaction N_θ, where θ is the angle between the interacting waves $(\cos\theta = \mathbf{k}\cdot\mathbf{q}/K_c^2)$; later on, we use the notation

$$\tilde{A}_\mathbf{k}|\tilde{A}_\mathbf{q}|^2 \rightarrow \tilde{A}_\mathbf{k}$$

for such interaction

(ii) Interaction of waves with the wavevectors $\mathbf{q}$, $-\mathbf{q}$, and $-\mathbf{k}$

$$\tilde{A}_{\mathbf{q}}\tilde{A}_{-\mathbf{q}}\tilde{A}^*_{-\mathbf{k}} \to \tilde{A}_{\mathbf{k}}$$

with the coefficient of nonlinear interaction K_θ (a similar notation is used). In both cases, $s_2 = -s_3 = 1$ in (4.91) and (4.92).

Performing the detailed calculation, one arrives at the following expressions for the coefficients of nonlinear interaction:

$$N_\theta = \beta + (2\alpha + \beta)\cos^2\theta, \; K_\theta = 2(\alpha + \beta\cos^2\theta), \qquad (4.97)$$

$$\alpha = \frac{5r_0^2 K_c^4}{7L}\left(1 + L + \frac{i}{\tilde{\Omega}}(L^2 + 1 - r_0)\right), \; \beta = -\frac{10r_0^2 K_c^4}{7\tilde{\Omega}}i. \qquad (4.98)$$

It is clear that $N_\theta = N_{\pi+\theta}$, and the same relation is correct for K_θ. The only exception is the self-interaction coefficient $N_0 = \frac{1}{2}N_\pi = \alpha + \beta$; the factor $\frac{1}{2}$ follows from the combinatorial arguments.

The detailed explanation of (4.97) is given in [11] (see Appendix B there). The analysis in the cited paper is performed for arbitrary system with the longwave oscillatory mode, which exhibits two following properties: (i) quadratic dispersion relation takes place and (ii) only gradients of the amplitude functions rather than functions themselves influence the nonlinear dynamics.

Summarizing, one arrives at the following set of Landau equations at the first order:

$$\partial_{t_{42}}\tilde{A}_{\mathbf{k}} = \left(\gamma_2 - \sum_{\mathbf{q}} N_{\theta_{\mathbf{kq}}}|\tilde{A}_{\mathbf{q}}|^2\right)\tilde{A}_{\mathbf{k}} - \frac{1}{2}\sum_{\mathbf{q}} K_{\theta_{\mathbf{kq}}}\tilde{A}_{\mathbf{q}}\tilde{A}_{-\mathbf{q}}\tilde{A}^*_{-\mathbf{k}}, \qquad (4.99)$$

where $\theta_{\mathbf{kq}}$ is the angle between two wavevectors. All possible wavevectors are taken into account in the sums. Therefore, each pair of counter-propagating waves enters in the sum twice; that is why one half appears in that term.

In what follows, we consider separately three important cases, patterns belonging to rhombic (in the Fourier space), square, and hexagonal lattices, whose wavevectors are given by (see Figure 4.2):

$$\mathbf{k}_1^{(R,S,H)} = K_c(1,0); \; \mathbf{k}_2^{(R)} = K_c(\cos\theta, \sin\theta); \; \mathbf{k}_2^{(S)} = K_c(0,1);$$

$$\mathbf{k}_{2,3}^{(H)} = \frac{1}{2}K_c(-1, \sqrt{3}). \qquad (4.100)$$

Dealing with the rhombic lattice, one can set $\theta < \pi/2$ without loss of generality. Square lattice is the limiting case $\theta = \pi/2$ of the rhombic lattice, but due to some circumstances, it has to be studied separately, see Sections 4.1.2 and 4.2.1.4.

In order to simplify the formulas, we rename the amplitudes of these "base" waves according the rule:

$$A_{1,2}^{(R,S,H)} = \tilde{A}_{\pm\mathbf{k}_1^{(R,S,H)}}; \; B_{1,2}^{(R,S,H)} = \tilde{A}_{\pm\mathbf{k}_2^{(R,S,H)}}; \; C_{1,2}^{(H)} = \tilde{A}_{\pm\mathbf{k}_3^{(H)}}; \qquad (4.101)$$

see Figure 4.2. For the sake of brevity, within the corresponding paragraphs, the superscripts (R), (S), and (H) for the complex amplitudes are omitted.

Rhombic Lattice

For the rhombic lattice, (4.99) is reduced to

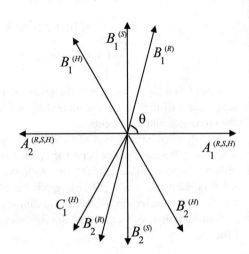

Fig. 4.2 The wavevectors and amplitudes of the base waves for rhombic (R), square (S), and hexagonal (H) lattices, (4.101). The wavevector $\mathbf{k}_3^{(H)}$ and the corresponding amplitude $C_2^{(H)}$ are not shown in order to prevent overburdening the figure

$$\dot{A}_1 = \left[\gamma_2 - N_0 \left(|A_1|^2 + 2|A_2|^2 \right) - N_\theta \left(|B_1|^2 + |B_2|^2 \right) \right] A_1 - K_\theta B_1 B_2 A_2^*, \quad (4.102)$$

$$\dot{A}_2 = \left[\gamma_2 - N_0 \left(|A_2|^2 + 2|A_1|^2 \right) - N_\theta \left(|B_1|^2 + |B_2|^2 \right) \right] A_2 - K_\theta B_1 B_2 A_1^*, \quad (4.103)$$

$$\dot{B}_1 = \left[\gamma_2 - N_0 \left(|B_1|^2 + 2|B_2|^2 \right) - N_\theta \left(|A_1|^2 + |A_2|^2 \right) \right] B_1 - K_\theta A_1 A_2 B_2^*, \quad (4.104)$$

$$\dot{B}_2 = \left[\gamma_2 - N_0 \left(|B_2|^2 + 2|B_1|^2 \right) - N_\theta \left(|A_1|^2 + |A_2|^2 \right) \right] B_2 - K_\theta A_1 A_2 B_1^*. \quad (4.105)$$

The dot within the rest of this subsection denotes the derivative with respect to t_{42}.

The analysis of pattern selection for this problem is based on [12] (see Section D.1 of Appendix D). All the branching patterns are supercritical, standing cross rolls (SCRs) do not exist, and alternating rolls (rectangular) (ARs-R) are selected:

$$A_1 = A_2 = iB_1 = iB_2 = a\exp(-i(\gamma_{0i}t_{40} + \omega^{(R)}t_{42})), \quad (4.106)$$

$$a = \sqrt{\frac{\gamma_{2r}}{S_{Rr}}}, \quad \omega^{(R)} = -\gamma_{2i} + \frac{S_{Ri}}{S_{Rr}}\gamma_{2r}, \quad S_R = 3N_0 + 2N_\theta - K_\theta. \quad (4.107)$$

Keeping in mind the definitions of N_θ and K_θ, one can see that $S_{Rr} = \alpha_r(1 + 4\cos^2\theta)$ and $S_{Ri} = (1 + 4\cos^2\theta)\alpha_i + 5\beta_i$.

Recall that during a quarter of the period, ARs-R evolve from the set of rolls oriented normally to $\mathbf{k}_1^{(R)}$ to rectangular pattern and then to the rolls oriented perpendicularly to $\mathbf{k}_2^{(R)}$; see Figure 4.3.

It is worth noting that except for $\theta = \pi/3$ which is a particular case of hexagonal patterns (see below), (4.102)–(4.103) remain valid even for *finite supercriticality* r_2 and arbitrary values of K with the same coefficients and γ being now the total growth rate, (4.93) (dot stands for the derivative with respect to t_4 in that case). Indeed, any superposition of three waves for rhombic patterns does not satisfy the resonant conditions, equations (4.91) and (4.92), i.e., the two pairs of counter-propagating waves with the wavevectors $\pm\mathbf{k}_1^{(R)}$ and $\pm\mathbf{k}_2^{(R)}$ represent a closed set for $\theta \neq \pi/3$.

Fig. 4.3 Snapshots for AR-Rs during a quarter of the time period; in the next quarter of the period, the snapshots are repeated in the reverse order and with inversion of h

Square Lattice

For the problem under consideration, the square lattice is a particular case of the rhombic lattice. However, keeping in mind further complications of this simple case, we consider it separately. Substitution of $\theta = \pi/2$ in (4.102)–(4.106) yields:

$$\dot{A}_1 = \left(\gamma_2 - N_0\left(|A_1|^2 + 2|A_2|^2\right) - N_{\pi/2}\left(|B_1|^2 + |B_2|^2\right)\right)A_1$$
$$-K_{\pi/2}B_1B_2A_2^*, \qquad (4.108)$$

$$\dot{A}_2 = \left(\gamma_2 - N_0\left(|A_2|^2 + 2|A_1|^2\right) - N_{\pi/2}\left(|B_1|^2 + |B_2|^2\right)\right)A_2$$
$$-K_{\pi/2}B_1B_2A_1^*, \qquad (4.109)$$

$$\dot{B}_1 = \left(\gamma_2 - N_0\left(|B_1|^2 + 2|B_2|^2\right) - N_{\pi/2}\left(|A_1|^2 + |A^2|^2\right)\right)B_1$$
$$-K_{\pi/2}A_1A_2B_2^*, \qquad (4.110)$$

$$\dot{B}_2 = \left(\gamma_2 - N_0\left(|B^2|^2 + 2|B^1|^2\right) - N_{\pi/2}\left(|A_1|^2 + |A^2|^2\right)\right)B_2$$
$$-K_{\pi/2}A_1A_2B_1^*. \qquad (4.111)$$

As usual, a point means $\partial_{t_{42}}$. Again, alternating rolls are selected

$$A_1 = A_2 = iB_1 = iB_2 = a\exp(-i(\gamma_{0i}t_{40} + \omega^{(S)}t_{42})), \qquad (4.112)$$

$$a = \sqrt{\frac{\gamma_2 r}{(3N_0 + 2N_{\pi/2} - K_{\pi/2})_r}}, \tag{4.113}$$

$$\omega^{(S)} = -\gamma_{2i} + \frac{(3N_0 + 2N_{\pi/2} - K_{\pi/2})_i}{(3N_0 + 2N_{\pi/2} - K_{\pi/2})_r} \gamma_{2r}, \tag{4.114}$$

whereas the other patterns are supercritical. The combination of the nonlinear coefficients is simplified as follows for $\theta = \pi/2$: $(3N_0 + 2N_{\pi/2} - K_{\pi/2})_r = \alpha_r$ and $(3N_0 + 2N_{\pi/2} - K_{\pi/2})_i = \alpha_i + 5\beta_i$.

Hexagonal Lattice

For a hexagonal lattice, the coefficients of nonlinear interaction $N_{\pi/3} = \frac{1}{2}\alpha + \frac{5}{4}\beta$, $K_{\pi/3} = 2\alpha + \frac{1}{2}\beta$ are such that the complex-valued relation

$$3N_0 = 2N_{\pi/3} + K_{\pi/3} \tag{4.115}$$

holds true.

This allows rewriting the amplitude equations as follows:

$$\dot{A}_1 = \left(\gamma_2 - (N_0 - N_{\pi/3})(|A_1|^2 - |A_2|^2) - 2N_{\pi/3}S_A\right)A_1 - K_{\pi/3}\hat{\Sigma}A_2^*, \tag{4.116}$$

$$\dot{A}_2 = \left(\gamma_2 + (N_0 - N_{\pi/3})(|A_1|^2 - |A_2|^2) - 2N_{\pi/3}S_A\right)A_2 - K_{\pi/3}\hat{\Sigma}A_1^*, \tag{4.117}$$

$$2S_A = |A_1|^2 + |A_2|^2 + |B_1|^2 + |B_2|^2 + |C_1|^2 + |C_2|^2, \tag{4.118}$$

$$\hat{\Sigma} = A_1A_2 + B_1B_2 + C_1C_2; \tag{4.119}$$

two other pairs of equations are obtained from (4.116) and (4.117) by the cyclic permutation of A, B, and C.

For the further analysis, it is convenient to represent the complex amplitudes as

$$A_j = a_j e^{i\phi_{Aj}}, \; B_j = b_j e^{i\phi_{Bj}}, \; C_j = c_j e^{i\phi_{Cj}}. \tag{4.120}$$

An appropriate choice of the origin and reference time provides three additional conditions coupling these phases; hence only three combinations of the phases are independent. The most convenient way to choose these phases is

$$\Phi = \phi_{A1} + \phi_{A2} - \phi_{B1} - \phi_{B2}, \tag{4.121}$$

$$\Psi = \phi_{A1} + \phi_{A2} - \phi_{C1} - \phi_{C2}, \tag{4.122}$$

$$\Delta = \phi_{A1} - \phi_{A2} + \phi_{B1} - \phi_{B2} + \phi_{C1} - \phi_{C2}. \tag{4.123}$$

Substituting these phase variables into (4.116) and (4.117) and separating the real and imaginary parts, one can derive the set of nine real amplitude equations. Below we need only a particular case of this set, see (4.125)–(4.129); the reader, who is interested in the details, can find the full set of equations in Appendix C of [11]. Introduction of the phase differences obviously excludes oscillation in t_{42}, the fixed points in the framework of the abovementioned set of nine real amplitude equation correspond to limit cycles of the initial set of six complex equations.

The structure of the cubic nonlinear terms within Landau equations (4.116) and (4.117) dictates the degeneracy mentioned in [13]. On one hand, (4.116) and (4.117) are invariant under the transformation

$$A_2 \to e^{iv/3}A_2, \; B_2 \to e^{iv/3}B_2, \; C_2 \to e^{iv/3}C_2 \tag{4.124}$$

for any real v and fixed A_1, B_1, and C_1. On the other hand, Δ is replaced with $\Delta - v$ under this transformation. This degeneracy is important for two pairs of patterns (SHs & SRTs and TwRs & WRs2) for which this replacement really changes the pattern. The corresponding patterns are shown in Figures D.11–D.14. The degeneracy can be removed only at the second order with respect to δ^2 with additional quintic nonlinear terms introduced (see Section 4.2.1.4). Within this subsection, we do not discuss this degeneracy any further.

Next, we simplify the analysis reducing the general set of equations to the particular case of standing patterns, such that $a_1 = a_2 = a$, $b_1 = b_2 = b$, and $c_1 = c_2 = c$. The reason of doing that is twofold: first, it can be easily shown on the base of [13] that all asymmetric patterns—traveling rolls (TRs), traveling rectangles 1 and 2 (TRas1 and TRas2, respectively), and oscillating triangles (OTs)—are supercritical and unstable. Those patterns are shown in Figures D.1, D.6, D.7, and D.10. Moreover, as shown in Appendix C of [11], the selected standing patterns are stable with respect to symmetry breaking.

For the standing patterns, the dynamical system can be recast as follows:

$$\dot{S}_A = 2\left(\gamma_r S_A - 2\mathrm{Re}N_{\pi/3}S_A^2 - \mathrm{Re}K_{\pi/3}|\Sigma_A|^2\right), \tag{4.125}$$

$$\dot{\Sigma}_A = 2\left[\gamma_r - (2\mathrm{Re}N_{\pi/3} + K_{\pi/3})S_A + i\mathrm{Im}(K_{\pi_3}\Sigma_A)\right]\Sigma_A, \tag{4.126}$$

$$\dot{\Phi} = 2\mathrm{Im}\left[K_{\pi/3}\Sigma_A\left(e^{i\Phi} - 1\right)\right], \tag{4.127}$$

$$\dot{\Psi} = 2\mathrm{Im}\left[K_{\pi/3}\Sigma_A\left(e^{i\Psi} - 1\right)\right], \tag{4.128}$$

$$\dot{\Delta} = 0, \tag{4.129}$$

where

$$\Sigma_A = a^2 + b^2 e^{-i\Phi} + c^2 e^{-i\Psi} = \hat{\Sigma}e^{-i(\phi_{A1}+\phi_{A2})}. \tag{4.130}$$

Note that Σ_A is complex and thus the second equation is also complex. In fact, this representation is inconvenient for synchronous standing patterns—standing rolls (SRs), standing rectangles (SRas), standing hexagons (SHs), and standing regular triangles (SRTs) shown in Figures D.2, D.8, D.11, and D.12—those with $\Phi = \Psi = 0$ because $S_A = \Sigma_A$ in this case, and two equations become identical. However, those four patterns, although degenerate (have several vanishing eigenvalues), are unstable according to the analysis of [13].

The remaining three patterns (asynchronous hexagons, AHs), ARs-R, TwRs, and WRs2, shown in Figures 4.3, D.13, and D.14, are neutrally stable within (4.116) and (4.117). Moreover, any combination of their amplitudes (a, b, and c) and two phases

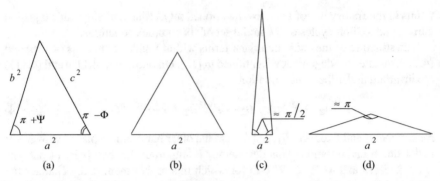

Fig. 4.4 Triangles of a fixed perimeter corresponding to the family of AH: (a) general case; (b) equilateral triangle, $\Phi = -\Psi = 2\pi/3$ or $a = b = c$ corresponding to symmetric patterns; (c) isosceles triangle with a zero angle subtending the base, which corresponds to ARs=R, $\Phi - \Psi = \pi$, or $a = 0$; (d) isosceles triangle with zero base angles, corresponding to the family of OHRs, $\Phi = -\Psi = \pi$ or $a^2 = 2b^2 = 2c^2$

(Φ and Ψ), which satisfies three real conditions $\Sigma_A = 0$ and $S_A = \gamma_{2r}/(2\mathrm{Re}N_{\pi/3})$, solves (4.125)–(4.129). Therefore, the entire two-parameter family

$$a^2 = \mu_S \sin(\Psi - \Phi), \quad b^2 = -\mu_S \sin\Psi, \quad c^2 = \mu_S \sin\Phi, \qquad (4.131)$$

$$\mu_S = \frac{S_A}{\sin(\Psi - \Phi) + \sin\Phi - \sin\Psi}$$

with Φ and Ψ serving as the parameters is selected within the equations with the cubic nonlinearity. In order to find the selected pattern within this two-parameter family, one has to proceed to the second (with respect to δ^2) order accounting for the quintic nonlinearity (see Section 4.2.1.4). The only restriction imposed on the phase differences Φ and Ψ are the positivity conditions for a^2, b^2, and c^2, which yield either

$$0 \le \Phi \le \pi, \quad -\pi \le \Psi \le \Phi - \pi \qquad (4.132)$$

or

$$0 \le \Psi \le \pi, \quad -\pi \le \Phi \le \Psi - \pi. \qquad (4.133)$$

To reiterate, each representative from AHs corresponds to the limit cycle with the complex amplitudes proportional to $\exp\left(-i\omega^{(H)}t_{42}\right)$, where the nonlinear correction to the frequency reads

$$\omega^{(H)} = 2\mathrm{Im}N_{\pi/2}S_A - \gamma_{2i}. \qquad (4.134)$$

Each representatives of AHs corresponds to a superposition of three standing waves oriented according to the wavevectors $\mathbf{k}_j^{(H)}$ and shifted in phase by $\Phi/2$ and $\Psi/2$.

It can be shown from (4.131) that for fixed S_A (or fixed supercriticality γ_2), there exists a one-to-one correspondence between the two-parameter family of AHs and a family of triangles of a fixed perimeter S_A. The sides of each triangle are a^2, b^2, and c^2; they subtend the angles $\Phi - \Psi - \pi$, $\pi + \Psi$, and $\pi - \Phi$, respectively; see Figure 4.4(a).

The important particular case of AHs corresponds to isosceles triangles, which means that at least two out of the amplitudes are equal to each other. Representatives of this case, important for the further analysis (see Section 4.2.1.4), are:

(i) Symmetric patterns, $a^2 = b^2 = c^2$ or $\Phi = -\Psi = \Phi_0 \equiv \pm 2\pi/3$. In terms of the abovementioned one-to-one correspondence, these patterns are described by an equilateral triangle; see Figure 4.4(b). For $\Delta = 0$ and $\Delta = \pi$, symmetric patterns give rise to TwR and WR2, respectively. The temporal evolution of the field h in the fast time τ is shown in Figure D.13(a) for $\Delta = 0$ and Figure D.14(b) for $\Delta = \pi$.

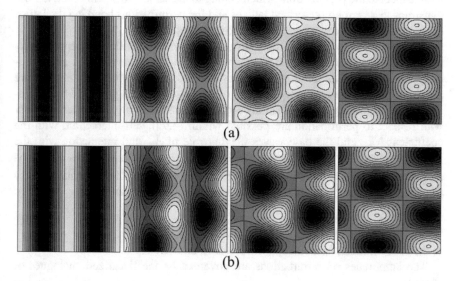

Fig. 4.5 Temporal evolution of h for $\Omega\tau = 0, \pi/8, \pi/4, 3\pi/8, \pi/2$ (left to right) for AH with $a^2 = 2b^2 = 2c^2$. The horizontal and vertical axes are, respectively, X and Y. (a) $\Delta = 0$ (OHR1), (b) $\Delta = \pi$ (OHR2)

(ii) ARs-R are the particular case of AHs with zero amplitude of one of the standing waves. Either AR-A with $a = 0$ and $\Phi - \Psi = \pi$, or AR-B with $b = 0$ and $\Psi = -\pi$, or AR-C with $c = 0$ and $\Phi = \pi$. This pattern corresponds to an isosceles triangle with almost right base angles; see Figure 4.4(c). The temporal evolution of h in this case is shown in Figure 4.3. For this family the degeneracy in Δ is absent because any change in Δ can be eliminated by an appropriate shift in the reference frame. However, for patterns close to ARs-R, Δ plays a decisive role; see Section 4.2.1.4.

(iii) Oscillating hexarolls (OHR) $\Phi = -\Psi = \pi$ correspond to degenerate triangles with the angles $0, 0$, and π depicted in Figure 4.4(d). In this case the amplitude of one standing wave is maximum, whereas two others are equal to each other, i.e., $a^2 = 2b^2 = 2c^2$. One can recast the field h as follows

$$h = 2\left[\sqrt{2}\cos K_c X \cos \Omega \tau - 2\sin\left(\frac{1}{2}K_c X + \frac{3}{4}\Delta\right)\sin\frac{\sqrt{3}}{2}K_c Y \sin \Omega \tau\right], \quad (4.135)$$

the corresponding temporal evolution (in the fast time τ) for $\Delta = 0$ (OHR1) and $\Delta = \pi$ (OHR2) is shown in Figure 4.5.

4.1.1.5 External Perturbations

The stability analysis performed in Section 4.1.1.4, which is based on papers [12, 13], is not complete. Indeed, the cited papers deal with the *internal* stability, with respect to the perturbations which belong to the same lattice. However, in fluid dynamics there are no prescribed lattices and the perturbations breaking the lattice symmetry (*external* perturbations) have to be considered.

Near the stability threshold, only disturbances with the critical wavenumber K_c should be taken into account because waves with other wavelengths decay linearly. (In fact, *modulational* perturbations with the wavenumbers close to K_c are also important.) Therefore, the evolution of the perturbation is governed by the linearized (with respect to perturbation) version of (4.99). Two types of terms linear with respect to perturbation are possible within (4.99): self-interaction, with the coefficient $N_{\theta_{kq}}$ and the base waves playing role of $\tilde{A}_q$, and interaction with the counter-propagating wave with the coefficient $K_{\theta_{kq}}$ and the base waves chosen as $\tilde{A}_{\pm q}$. Consequently, in order to explore the stability with respect to external stability of patterns analyzed in Section 4.1.1.4, we add a small perturbation

$$h' = \delta\left(be^{i\mathbf{p}\cdot\mathbf{R}} + de^{-i\mathbf{p}\cdot\mathbf{R}}\right)e^{-i\Omega\tau} + c.c., \quad \mathbf{p} = K_c\left(\cos\phi, \sin\phi\right) \quad (4.136)$$

to the base solutions h_0 given by (4.77), (4.94), and (4.101).

The amplitudes of perturbations are governed by the linearized analogues of (4.99):

$$\dot{b} = \left(\gamma_2 - \sum_j N_{\theta_j}|A_j|^2\right)b - \sum_{j>0}K_{\theta_j}A_jA_{-j}d^*, \quad (4.137)$$

$$\dot{d} = \left(\gamma_2 - \sum_j N_{\theta_j}|A_j|^2\right)d - \sum_{j>0}K_{\theta_j}A_jA_{-j}b^*, \quad (4.138)$$

where j numerates the base vectors of each lattice (two waves with the opposite j are counter-propagating) and θ_j is the angle between the wavevectors $\mathbf{p}$ and $\mathbf{k}_j$ for the appropriate lattice.

The analysis of this system is performed below for each lattice separately.

Rhombic Lattice

To study the external stability of ARs-R, it is convenient to choose the perturbation wavevector in a slightly different manner, $\mathbf{p} = K_c \left(\cos(\phi + \theta/2), \sin(\phi + \theta/2)\right)$. Substituting this ansatz to (4.137) and (4.138), one arrives at the following set of equations

$$\dot{b} = \left(\gamma_2 - N_b a^2\right) g + K_b A^2 d^*, \qquad (4.139)$$
$$\dot{d} = \left(\gamma_2 - N_b a^2\right) d + K_b A^2 b^*, \qquad (4.140)$$

where $A = A_1^{(R)} = A_2^{(R)}$, $a = |A|$ and

$$N_b = 2N_{\phi+\theta/2} + 2N_{\phi-\theta/2} = 2\left[2\beta + (\beta + 2\alpha)(1 + \cos 2\phi \cos \theta)\right], \quad (4.141)$$
$$K_b = K_{\phi+\theta/2} + K_{\phi-\theta/2} = 2\beta \sin 2\phi \sin \theta. \quad (4.142)$$

Equations (4.139) and (4.140) are linear homogeneous equations with variable coefficients due to oscillations of A^2 with doubled frequency $2\omega^{(R)}$. The standard replacement [14]

$$b = \exp(-i\omega^{(R)} t_{42}) \left(b_+ e^{\lambda t_{42}} + b_- e^{\lambda^* t_{42}}\right)$$

and the similar replacement for d converts this system to the set of equations with constant coefficients.

The growth rate is determined by the following quadratic equation:

$$\lambda^2 + 2\lambda(N_{br} - S_{Rr}) + |N_b - S_R|^2 - |K_b|^2 = 0, \qquad (4.143)$$

where $S_R = 3N_0 + 2N_\theta - K_\theta$, cf. (4.106). According to Vieta's formula, in order to guarantee the stability of the ARs-R, one has to demand the positivity of both coefficients at λ^1 and λ^0.

The former stability condition results in $3 - 4(\cos\theta\cos 2\phi + \cos^2\theta) > 0$ for any ϕ. This is valid for

$$\theta > \pi/3; \qquad (4.144)$$

therefore, ARs-R with the angle between the wavevectors smaller than $\pi/3$ are not analyzed any further.

The latter condition which warrants stability of ARs-R is

$$|S_R - N_b|^2 - |K_b|^2 > 0. \qquad (4.145)$$

and it cannot be simplified further in the general case. As stated above, the Lewis number is rather small for most binary mixtures; this, with the exception discussed below, means that $|\alpha| \gg |\beta|$ and therefore $|N_b - S_R| \gg |K_b|$ except for the case $\theta \approx \pi/3$, where $N_b \approx S_R$ according to (4.144). An accurate analysis of ARs-R with $\theta \approx \pi/3$ is performed in Appendix B of [15] for the Marangoni convection: it is shown there that the external instability takes place for $\theta < \pi/3 + O(L^2)$, so that the instability domain slightly increases in comparison with (4.144).

At extremely small $|\psi|$, $|\psi| = O(L^2)$, where $|\alpha| \approx |\beta|$, there exists another domain of instability. This case also was thoroughly studied in Appendix B of [15] for the surface tension-driven convection. Summarizing, omitting the terms of order of L^2, (4.144) is both necessary and sufficient stability condition for ARs-R.

As mentioned above, see Section 4.1.1.4, the pattern branching and pattern selection can be extended to finite values of r_2 (finite-amplitude regimes). The present stability analysis, (4.144) and (4.145), is also appropriate even at finite supercriticality, but it becomes incomplete since far from the minimum of the neutral stability curve, external perturbations with a wavenumber different from K_c has to be taken into account.

Square Lattice

In the absence of the quadratic nonlinearity in (4.89) and (4.90), the patterns on a square lattice are the particular case $\theta = \pi/2$ of the patterns on a rhombic lattice. Hence, the stability analysis with respect to external perturbations developed in the previous paragraph can be easily reduced to squares.

The first of the stability conditions, (4.144), is valid for squares. The second one, (4.145), is reduced to

$$|3N_0 + 2N_{\pi/2} - K_{\pi/2} - 2(3\beta + 2\alpha)|^2 - 4|\beta|^2 > 0, \qquad (4.146)$$

which is equivalent to $|3\alpha + \beta|^2 - 4|\beta|^2 > 0$. For sufficiently small L, which is the practical case, this inequality holds true except for the extremely small values of the separation number, $|\psi| = O(L^2)$.

Hexagonal Lattice

Finalizing this subsection, let us consider the stability of AHs with respect to external perturbations. Although the set of amplitude equations with cubic nonlinearity is degenerate and higher-order nonlinear terms have to be taken into account, the analysis of external perturbations in t_{42} is important. Indeed, some patterns, which belong to AHs, can be unstable with respect to external perturbations, and, therefore, their delicate study based on quintic nonlinearity is unnecessary. In particular, it has been shown above that ARs-R with $\theta = \pi/3$ are unstable with respect to external perturbations; therefore, AHs close to those patterns are expected to be unstable as well.

The general equations (4.137) and (4.138) are reduced to (4.139) and (4.140) but with the following coefficients of self- and cross-interaction:

$$N_b a^2 = 2(2\alpha + \beta)\left(a^2 \cos^2 \phi + b^2 \cos^2 \phi_- + c^2 \cos^2 \phi_+\right) + 2\beta S_A, \quad (4.147)$$

$$K_b A^2 = 2\beta \left(A^2 \cos^2 \phi + B^2 \cos^2 \phi_- + C^2 \cos^2 \phi_+\right), \quad \phi_\pm = \phi \pm \frac{2\pi}{3}. \quad (4.148)$$

The stability conditions for AHs can be rewritten as

$$W > 0 \text{ and} \tag{4.149}$$
$$|2\alpha + \beta|^2 W^2 - 4|K_b|^2 > 0, \tag{4.150}$$

where

$$W \equiv 4(a^2 \cos^2 \phi + b^2 \cos^2 \phi_- + c^2 \cos^2 \phi_+) - 1.$$

Applying the triangle inequality, one can show that $W \geq 0$ for any value of ϕ and any AHs. The minimum value, $W = 0$, is attained for ARs-R, AHs with zero amplitude of one of the standing waves.

Dealing with the second condition, (4.149), we again take into account that $|\alpha| \gg |\beta|$ for sufficiently small values of the Lewis number. Therefore, (4.150) is met except for (i) either the abovementioned case of small $|W|$ or (ii) extremely small $|\psi|$: $|\psi| = O(L^2)$. Both these cases are analyzed for the Marangoni convection in Appendix D of [11]; it is shown that instability of AHs with respect to external perturbations takes place within $O(L^2)$ vicinity of ARs-R in case (i) or for the whole two-parameter family of AHs in case (ii), which is also $O(L^2)$-vicinity of $\psi = \psi_c$.

4.1.2 Non-Boussinesq Effects: Quadratic Nonlinearity

Equations (4.89) and (4.90) contain only cubic nonlinear terms since all quadratic terms vanish due to the problem symmetry with respect to the plane $z = 0$. Quadratic terms can be generated, e.g., by non-Boussinesq effects, such as dependence of the viscosity and/or other kinetic coefficients on the temperature (solute concentration). For the sake of simplicity, below we deal with the linear dependence of the viscosity on the temperature only [1]:

$$\eta_* = \eta_*^{(0)}(1 + \eta_T(T_* - T_*^{(0)})). \tag{4.151}$$

This assumption results in the replacement of $\nabla^2 \mathbf{v}$ term in (4.9) with $\nabla \cdot ((1 - z + \nu\Theta)(\nabla\mathbf{v} + (\nabla\mathbf{v})^T))$, where the superscript T denotes the transposed tensor and $\nu = \eta_T A d_*$.

In order to simplify the derivation, we assume weak non-Boussinesq effect, $\nu = \varepsilon^2 \nu_2$ (a more general discussion of similar effects is presented in Section 4.2.4 for a different problem). That leads to the additional term

$$-\nu_2 \left[(-z + F)\mathbf{U}_z^{(0)} \right]_z$$

in the right-hand side of (4.69). The first term in the parenthesis, $-z$, does not contribute to the solvability conditions; therefore, only the contribution of the second one, F, is discussed below. This term results in corrections $-\nu_2 r_0 V_\parallel^{(R)} F \nabla h$ and $-\nu_2 r_0 V_\perp^{(R)} \nabla \cdot (F \nabla h)$ in (4.82) and (4.83), for $\mathbf{U}^{(2)}$ and $W^{(2)}$, respectively, which, in turn, leads to the additional terms, $-v_2 r_0 \nabla \cdot (F\nabla h)$ and $v_2 \psi r_0 \nabla \cdot (F\nabla h)$, in the right-hand sides of (4.89) and (4.90), respectively.

The contribution of these terms to the Landau equations has to be discussed in more details. First, the quadratic nonlinear terms contribute to the solvability conditions, if (4.91) and (4.92) are satisfied for $s_3 = 0$, whereas $s_2 = \pm 1$ as before. It can be readily seen that these solvability conditions are satisfied if two of wavevectors are orthogonal; therefore, the quadratic nonlinearity affects only the patterns on a square lattice. (It is worth noting that for the steady patterns generated by a monotonic instability, hexagons are affected by a quadratic nonlinearity.) It can be easily checked that two perpendicular base waves with the wavevectors $\mathbf{k}_1^{(S)}$ and $\mathbf{k}_2^{(S)}$ create the "oblique" wave $K_c(1,1)$ and the resonant conditions are satisfied for $s_2 = 1$. Moreover, this oblique wave interacts with, say, the second base wave giving rise to the first one ($s_2 = -1$ for the base wave in this case). Therefore, for the pattern on a square lattice, h has to be rewritten as follows:

$$h = \delta \left(A_1 e^{iK_c X} + A_2 e^{-iK_c X} + B_1 e^{iK_c Y} + B_2 e^{-iK_c Y} \right) e^{-i\Omega\tau}$$
$$+ \delta^2 \left(C_1 e^{i\varphi_+} + C_2 e^{-i\varphi_-} + C_3 e^{-i\varphi_+} + C_4 e^{i\varphi_-} \right) e^{-2i\Omega\tau} + c.c. \qquad (4.152)$$

where $\varphi_\pm = K_c(X \pm Y)$. The expansions of the time derivative and the growth rate remain unchanged, as well as the first-order problem (4.96).

At the second order, the generation of the oblique waves takes place:

$$\partial_{t_{40}} C_1 = \Gamma_0 C_1 - \mu_2 A_1 B_1, \qquad (4.153)$$

where $\Gamma_0 = \gamma(\sqrt{2}k_c)$ at $r_2 = r_{2c}$ and

$$\mu_2 = v_2 r_0 k_c^2 \left(1 + \frac{iL}{\tilde{\Omega}} \right).$$

Similar three equations are obtained for the remaining oblique waves. In view of the already known dependence of base amplitudes on the time t_{40}, this equation can be integrated easily:

$$C_1 = \frac{\mu_2 A_1 B_1}{\Gamma_0 - 2\gamma_0}. \qquad (4.154)$$

At the third order, interaction of C_1 with B_1 (and C_4 with B_2) produces the wave with the amplitude A_1. This in fact leaves Landau equations (4.108) and (4.109) valid; only one of the coefficients is changed:

$$N_{\pi/2} = \beta + \frac{2\mu_1\mu_2}{\Gamma_0 - 2\gamma_0}. \qquad (4.155)$$

$$\mu_1 = \frac{v_2 r_0 k_c^2}{2(1 + \tilde{\Omega}^2)} \left(2L + 1 - \tilde{\Omega}^2 + \frac{i}{\tilde{\Omega}}(L(1 - \tilde{\Omega}^2) - 2\tilde{\Omega}^2) \right).$$

Pismen [1] predicted the importance of this resonance for quadratic nonlinear terms and described this effect. However, the time derivative for the oblique waves was overlooked, and therefore the term $-2\gamma_0$ was missed.

The conclusions obtained in [12] are still valid. The analysis shows that with an increase in v_2, ARs change the direction of their bifurcation at certain v_{2c}, whereas the other patterns remain unstable and supercritical. With further growth of v_2 StSs and TSs also bifurcate subcritically.

4.1.3 Restitution Procedure

4.1.3.1 The Idea of Restitution Procedure

Mathematical tools for solving nonlocal equations similar to the solvability conditions for (4.89) and (4.90) with (4.74) and (4.75) are not well developed; in contrast, there are numerous methods for solving local partial differential equations. Therefore, it is tempting to restore a set of partial differential equations on the base of (4.74), (4.75), (4.89), and (4.90). We discuss that idea in a general form representing the solution in powers of small ε, which is the "longwave parameter" (in the problem analyzed in Section 4.1.1 $\varepsilon = \sqrt{B}$).

At the leading order, one yields the linear homogeneous problem of the form

$$\mathscr{L}(\mathbf{F}) = 0, \tag{4.156}$$

where $\mathbf{F}$ is the vector of the perturbations and $\mathscr{L}$ is the linear differential operator, which includes only superposition of ∂_τ and ∇^2:

$$\mathscr{L}_j(\mathbf{F}) = \alpha_{jl}\partial_\tau F_l + \beta_{jl}\nabla^2 F_l + r\gamma_{jl}\nabla^2 F_l,$$

where we accept the Einstein rule of summation. The parameter r, named "Rayleigh number" below, is the bifurcating parameter; the matrices α, β, and γ depend on the remaining parameters of the problem.

In the particular case of the buoyancy convection in a binary mixture layer, $\mathscr{L}$ is given by (4.74) and (4.75), so that the matrices are

$$\alpha = \begin{pmatrix} 1 & 0 \\ 0 & 1 \end{pmatrix}, \ \beta = -\begin{pmatrix} 1 & 0 \\ \psi L & L \end{pmatrix}, \ \gamma = \begin{pmatrix} 1 & -1 \\ -\psi & \psi \end{pmatrix}, \tag{4.157}$$

with $\mathbf{F} = (F, G)$, the leading parts of the temperature and solute concentration, respectively.

At the stability boundary $r = r_0$, the solution to (4.156) can be presented as a superposition of traveling waves which neither grow nor decay in τ. Both r_0 and the frequency of neutral disturbances Ω can be readily found from an algebraic eigenvalue problem; the dispersion relation is quadratic $\Omega = \tilde{\Omega}K^2$, where K is the wavenumber of the solution. The amplitudes of traveling waves are yet unknown.

In order to determine their evolution near the stability threshold, we introduce the fast time scale $\tau_0 = \tau$ and the slow time scale $\tau_1 = \varepsilon\tau$ and expand the Rayleigh number $r = r_0 + \varepsilon r_1 + \ldots$ and, hence, the operator $\mathcal{L}$:

$$\mathcal{L} = \mathcal{L}_0 + \varepsilon\mathcal{L}_1 + \ldots. \qquad (4.158)$$

$$(\mathcal{L}_0)_{jl} = \alpha_{jl}\partial_{\tau_0} + \beta_{jl}\nabla^2 + r_0\gamma_{jl}\nabla^2, \quad (\mathcal{L}_1)_{jl} = \alpha_{jl}\partial_{\tau_1} + r_1\gamma_{jl}\nabla^2, \qquad (4.159)$$

With these expansions we proceed to the next order in ε, which results in the following linear nonhomogeneous problem:

$$\mathcal{L}_0(\mathbf{Q}) = \mathbf{N}(\mathbf{F}) - \mathcal{L}_1(\mathbf{F}), \qquad (4.160)$$

where $\mathbf{Q}$ are the analogues of $\mathbf{F}$ at the first order with respect to ε. The vector function $\mathbf{N}(\mathbf{F})$ generally speaking includes linear, quadratic, and cubic terms:

$$\mathbf{N}(\mathbf{F}) = \mathbf{N}^{(l)}(\mathbf{F}) + \mathbf{N}^{(q)}(\mathbf{F};\mathbf{F}) + \mathbf{N}^{(c)}(\mathbf{F};\mathbf{F};\mathbf{F}).$$

The linear operator includes the terms of zeroth and fourth orders with respect to ∇: zeroth-order terms are similar to heat loss in (4.89), and the fourth-order terms are "dispersive" ones. To reiterate, the solvability conditions of this nonhomogeneous linear problem serve as the evolution equations for the abovementioned amplitudes.

The "naive" idea is to combine partial differential equations (4.156) and (4.160), introducing $\mathbf{F}_r = \mathbf{F} + \varepsilon\mathbf{Q}$ and adding (4.160) multiplied by ε to (4.156). This gives the following nonlinear problem:

$$\mathcal{L}(\mathbf{F}_r) = \varepsilon\mathbf{N}(\mathbf{F}_r), \qquad (4.161)$$

where in the right-hand side $\mathbf{F}$ is replaced with $\mathbf{F}_r$; this replacement produces only $O(\varepsilon^2)$ terms, which will be neglected in the following analysis. Equation (4.158) is also taken into account. Nonlinear equations (4.161) provide the desired set of partial differential equations.

However, this restitution (or reconstitution or resummation) procedure mixes different orders of the asymptotically small parameter ε in a single set of equations; therefore the applicability of this approach is questionable. For the finite-amplitude regimes, $|\mathbf{F}_r| = O(1)$, the only asymptotically correct procedure is to expand $\mathbf{F}_r$ in powers of ε, which returns back to the sequential solution of linear homogeneous (4.156) and nonhomogeneous (4.160) problems. Thus, (4.161) is only a compact way of presentation of these two problems. Any attempt to solve numerically (4.161) with finite ε may lead to asymptotically incorrect results.

More intricate situation takes place near the stability threshold, where $|\mathbf{F}| = O(\delta)$ and the question arises, whether or not these two limits ($\delta \to 0$ and $\varepsilon \to 0$) commute. The asymptotically correct way (see Sections 4.1.1 and 4.1.2) is to expand (4.161) first in ε and then to consider the limit $\delta \ll 1$ in the resulting equations. In contrast, the weakly nonlinear analysis within (4.161) requires the opposite sequence of expansions: first, small-amplitude expansion in δ takes place, whereas within each order the solution is presented as a series in ε. In the particular case of the Rayleigh–Bénard convection in a binary liquid, the relationship between regular expansion and the restitution was addressed in [16]. The authors checked the calculations of [1] and then obtained different results within the restitution approach. Thus the main conclusion of [16] is that that two limits do not commute. However, as we shall show below, the accurate expansion within (4.161) provides exactly the same results as the regular expansion does, and therefore the commutation takes place.

In order to prove that, we perform both the regular expansion (Section 4.1.3.2) and the restitution procedure (Section 4.1.3.3) for (4.156), (4.160), and (4.161), correspondingly.

4.1.3.2 Regular Expansion for Small-Amplitude Solution

Dealing with the small-amplitude waves within the regular expansion, we expand both $\mathbf{F}$ and $\mathbf{Q}$ in powers of δ:

$$\mathbf{F} = \delta\mathbf{F}^{(1)} + \delta^2\mathbf{F}^{(2)} + \dots, \quad \mathbf{Q} = \delta\mathbf{Q}^{(1)} + \delta^2\mathbf{Q}^{(2)} + \dots, \quad (4.162)$$

This expansion has to be supplemented by introduction of two time scales $\tau_1^{(0)} = \tau_1$ and $\tau_1^{(2)} = \delta^2\tau_1$ and the expansion of the Rayleigh number r_1 in powers of δ^2: $r_1 = r_1^{(0)} + \delta^2 r_1^{(2)}$. This again results in expansion of the linear operator: $\mathscr{L}_1 = \mathscr{L}_1^{(0)} + \delta^2\mathscr{L}_1^{(2)} + \dots$; the expressions for these operators are evident; $\mathscr{L}_0$ is not affected by the expansion in δ.

The leading (first)-order problem for (4.160) yields:

$$\mathscr{L}_0(\mathbf{Q}^{(1)}) = \mathbf{N}^{(l)}(\mathbf{F}^{(1)}) - \mathscr{L}_1^{(0)}(\mathbf{F}^{(1)}). \quad (4.163)$$

The solvability condition for this linear nonhomogeneous problem gives the neutral stability curve $r_1^{(0)} = r_{1*}(K)$, the critical wavenumber K_c, and the frequency $-\gamma_i^{(0)}$ of neutral perturbations. Similar to Section 4.1.1, γ is the growth rate of perturbations in τ_1, which is expanded in δ^2: $\gamma = \gamma^{(0)} + \delta^2\gamma^{(2)} + \dots$.

At the second order, one obtains

$$\mathscr{L}_0(\mathbf{Q}^{(2)}) = \mathbf{N}^{(l)}(\mathbf{F}^{(2)}) - \mathscr{L}_1^{(0)}(\mathbf{F}^{(2)}) + \mathbf{N}^{(q)}(\mathbf{F}^{(1)}; \mathbf{F}^{(1)}). \quad (4.164)$$

The solvability conditions for this problem serve to determine $\mathbf{F}^{(2)}$; it is nonzero only for patterns on a square lattice, where two perpendicular waves produce the resonant "oblique" one with the wavenumber $\sqrt{2}K_c$; see Section 4.1.2. Keeping in mind that $\mathbf{N}^{(l)} - \mathscr{L}_1^{(0)}$ contains the derivatives with respect to slow time $\tau_1^{(0)}$, one can see that $\mathbf{F}^{(2)}$ are given by the relations similar to (4.154).

For other patterns, the resonant conditions (4.91) and (4.92) are not satisfied, and therefore the problem (4.164) is solvable for $\mathbf{F}^{(2)} = 0$; the corresponding solution $\mathbf{Q}^{(2)}$ is not needed any further.

The third order for (4.160) reads

$$\mathcal{L}_0(\mathbf{Q}^{(3)}) = \mathbf{N}^{(l)}(\mathbf{F}^{(3)}) - \mathcal{L}_1^{(0)}(\mathbf{F}^{(3)}) - \mathcal{L}_1^{(2)}(\mathbf{F}^{(1)})$$
$$+ \mathbf{N}^{(q)}(\mathbf{F}^{(1)}; \mathbf{F}^{(2)}) + \mathbf{N}^{(q)}(\mathbf{F}^{(2)}; \mathbf{F}^{(1)}) + \mathbf{N}^{(c)}(\mathbf{F}^{(1)}; \mathbf{F}^{(1)}; \mathbf{F}^{(1)}). \qquad (4.165)$$

The solvability condition for this problem gives rise to the Landau equations with cubic nonlinearity. It is worth noting that for the waves with the critical wavenumber K_c and critical frequency $-\gamma_i^{(0)}$, the first two terms in the right-hand side do not contribute to the solvability condition since the corresponding problem (4.163) is solvable.

4.1.3.3 Restitution Procedure for Small-Amplitude Solution

Within the restitution procedure, we represent the solution to (4.161) as a power series in δ as follows:

$$\mathbf{F}_s = \delta\mathbf{F}^{(1)} + \delta^2\mathbf{F}^{(2)} + \dots. \qquad (4.166)$$

We also introduce two different time scales and expand the Rayleigh number in δ:

$$\tau^{(0)} = \tau, \ \tau^{(2)} = \delta^2\tau, \ r = r^{(0)} + \delta^2 r^{(2)}. \qquad (4.167)$$

It should be noted that the expansion of r in δ leads to the expansion of the linear operator $\mathcal{L}$:

$$\mathcal{L} = \mathcal{L}^{(0)} + \delta^2\mathcal{L}^{(2)} + \dots.$$

At the first order, the linear stability problem reads

$$\mathcal{L}^{(0)}(\mathbf{F}^{(1)}) = \varepsilon\mathbf{N}^{(l)}(\mathbf{F}^{(1)}). \qquad (4.168)$$

The terms of different order in ε are mixed in this equation. This, in particular, leads to a change in the dispersion relation—in addition to the second-order spatial derivative, there appears $\mathbf{N}^{(l)}$-term containing the fourth-order ones and some terms, which do not bear any gradients. Moreover, the procedure of calculation of the right-hand side is based on the fact that it serves only as a small correction to the terms in the left-hand side. Therefore, one has to draw another expansion, in ε, as well:

$$\mathbf{F}^{(j)} = \mathbf{F}_0^{(j)} + \varepsilon\mathbf{F}_1^{(j)} + \dots, \ \tau_l^{(j)} = \varepsilon^l\tau^{(j)}, \ r^{(j)} = r_0^{(j)} + \varepsilon r_1^{(j)} + \dots \ (j > 0, \ l = 0, 1).$$

$$(4.169)$$

This, again, results in additional decomposition of $\mathcal{L}^{(0,2)}$. In what follows we need only the problems at zeroth and first orders with respect to ε. Most of the slow time scales $\tau_l^{(j)}$ and corrections to the Rayleigh number $r_l^{(j)}$ are not needed; only the necessary ones will be mentioned below.

The leading-order problem coincides with (4.21):

$$\mathcal{L}_0^{(0)}(\mathbf{F}_0^{(1)}) = 0. \qquad (4.170)$$

We are interested in $\mathbf{F}_0^{(1)}$ which is neutral in the corresponding time $\tau_0^{(0)}$; therefore $r_0^{(0)}$ is equal to r_0 found within the asymptotically correct approach, Section 4.1.3.2. The neutral perturbation oscillates with the frequency $\Omega = \tilde{\Omega}K^2$.

The first-order problem

$$\mathscr{L}_0^{(0)}(\mathbf{F}_1^{(1)}) = \mathbf{N}^{(l)}(\mathbf{F}_0^{(1)}) - \mathscr{L}_1^{(0)}(\mathbf{F}_0^{(1)}), \qquad (4.171)$$

defines the neutral stability curve $r_1^{(0)} = r_{1*}(K)$. This curve and, therefore, the critical correction to the Rayleigh number $r_{1c}^{(0)}$ and wavenumber K_c coincide with those obtained from (4.163). Therefore $\mathbf{F}_0^{(1)}$ includes the waves only of the wavenumber K_c. It is also worth noting that the neutral waves oscillate in $\tau_1^{(1)}$ with the frequency $-\gamma_i^{(0)}$.

At the second order, with respect to δ one obtains:

$$\mathscr{L}_0^{(0)}(\mathbf{F}_0^{(2)}) = 0, \qquad (4.172)$$

$$\mathscr{L}_0^{(0)}(\mathbf{F}_1^{(2)}) = \mathbf{N}^{(l)}(\mathbf{F}_0^{(2)}) - \mathscr{L}_1^{(0)}(\mathbf{F}_0^{(2)}) + \mathbf{N}^{(q)}(\mathbf{F}_0^{(1)}, \mathbf{F}_0^{(1)}). \qquad (4.173)$$

Although the first of these equations is homogeneous, its solution can be nontrivial because the resonance takes place in the second equation. Oblique waves with the wavenumber $\sqrt{2}K_c$, which are generated by interaction of two waves with orthogonal wavevectors, have to be included in $\mathbf{F}_0^{(2)}$ with yet unknown amplitudes. Their amplitudes are found from the solvability conditions for (4.173)—the right-hand side must not have a projection onto the oblique waves, which are eigenfunctions for $\mathscr{L}_0^{(0)}$. In fact, this solution is the same as $\mathbf{F}^{(2)}$ found within the regular expansion from (4.164). The solution to (4.173) $\mathbf{F}_1^{(2)}$, which contains nonresonant quadratic waves, is not needed below. Indeed, $\mathbf{F}_1^{(2)}$ has the factor $\delta^2\varepsilon$; its contribution at the third order (via the terms $\mathbf{N}^{(q)}(\mathbf{F}_0^{(1)};\mathbf{F}_1^{(2)})$ and $\mathbf{N}^{(q)}(\mathbf{F}_1^{(2)};\mathbf{F}_0^{(1)})$) is proportional to $\delta^3\varepsilon^2$, which is small in comparison with the terms of order $\delta^3\varepsilon$ produced by either cubic nonlinearity or quadratic terms including $\mathbf{F}_0^{(1)}$ and $\mathbf{F}_0^{(2)}$.

At the third order with respect to δ the problem reads:

$$\mathscr{L}_0^{(0)}(\mathbf{F}_0^{(3)}) = 0, \qquad (4.174)$$

$$\mathscr{L}_0^{(0)}(\mathbf{F}_1^{(3)}) = -\mathscr{L}_1^{(2)}(\mathbf{F}_0^{(1)}) - \mathscr{L}_1^{(0)}(\mathbf{F}_0^{(3)}) + \mathbf{N}^{(l)}(\mathbf{F}_0^{(3)})$$
$$+ \mathbf{N}^{(q)}(\mathbf{F}_0^{(2)}, \mathbf{F}_0^{(1)}) + \mathbf{N}^{(q)}(\mathbf{F}_0^{(1)}, \mathbf{F}_0^{(2)}) + \mathbf{N}^{(c)}(\mathbf{F}_0^{(1)}). \qquad (4.175)$$

The solution $\mathbf{F}_0^{(3)}$ to the first equation contains all the combinations of resonant waves, including those with K_c and the frequency of neutral perturbations $-\gamma_i^{(0)}$. Those waves do not contribute to the solvability conditions for (4.175). The solvability conditions for (4.175) are the same as those calculated for (4.165), and thus the corresponding Landau equations are exactly the same.

Therefore, accurate weakly nonlinear expansion of (4.161) provides the results equivalent to those for the regular expansion. However, this derivation breaks the hopes to use the restitution procedure for finite ε. Indeed, finite values of ε completely change the structure of the solution, and, in particular, the dispersion relation is no more quadratic in K. This is especially important at the second order, since the oblique waves cease to be resonant at any finite ε.

4.2 Marangoni Convection

In the present section, we study the longwave Marangoni convection in the binary liquid layer. Though that problem has still attracted less attention than that in the case of the buoyancy convection, we see a certain advantage in a more detailed description of the main ideas of the longwave approach taking this kind of convection as a basic example. Besides keeping in mind numerous applications of the Marangoni convection in microgravity and microfluidics, we would like to emphasize that this case is somewhat more general that that of the buoyancy convection. Because specific symmetries characteristic for the Boussinesq buoyancy convection are absent, some additional (quadratic) effects appear for the Marangoni convection, and they change the analysis to a great extent.

First, we consider the "longitudinal" monotonic and oscillatory instability modes, which are developed at small Biot number in the case of non-deformable interface, i.e., in the limit of large Galileo number (Section 4.2.1). This case will be studied more thoroughly than the buoyancy one, including the analysis of the fifth-order terms for hexagonal patterns. The deformational instability is discussed in Section 4.2.2. Specific distinguished limits that appear in the case of asymptotically small Lewis number are described in Section 4.2.3. In Section 4.2.4 we include into consideration the dynamics and sorption kinetics of the surfactant adsorbed on the layer surface.

4.2.1 Non-deformable Surface

4.2.1.1 Formulation of the Problem

We consider a layer of incompressible binary liquid atop a heated substrate at $z_* = 0$ with a prescribed normal component of the temperature gradient $-A$ there, which corresponds to the case of low thermal conductivity of the wall. The upper free boundary $z_* = d_*$ is non-deformable under conventional assumption $G \gg 1$, where $G = gd_*^3/v^2$ is the Galileo number. Heat flux from the surface is governed by Newton's law of cooling (3.97). Both thermocapillary effect and solutocapillary effect are taken into account; consequently, the surface tension is assumed linear with respect to both the temperature T_* and solute concentration C_*:

$$\gamma_* = \gamma_*^{(0)} - \gamma_T (T_* - T_*^{(0)}) + \gamma_C (C_* - C_*^{(0)}), \qquad (4.176)$$

where $\gamma_*^{(0)}$, $T_*^{(0)}$, and $C_*^{(0)}$ are the reference values of the surface tension, temperature, and concentration, respectively. The coefficients γ_T and γ_C can be of either sign. Typically, $\gamma_T > 0$ (normal thermocapillary effect), but an anomalous thermocapillary effect, $\gamma_T < 0$, is also observed in some solutions. For aqueous solutions of inorganic salts, γ_C is typically positive; for solutions of surfactants, $\gamma_C < 0$. For experimental data on the dependence of the surface tension on the temperature and concentration, see [17].

In this section we restrict ourselves to the analysis of the Soret-driven convection only; the mass flux $\mathbf{J}_m$ is given by

$$\mathbf{J}_m = -\rho D (\nabla C_* + \alpha \nabla T_*), \qquad (4.177)$$

where ρ, D, and α are the density, mass diffusivity, and the Soret coefficient of the binary mixture, respectively. Therefore, in the motionless fluid, the heat flux across the layer creates the concentration gradient in order to vanish $\mathbf{J}_m$.

The surface tension-driven convection in this system is governed by the following boundary value problem:

$$\rho \left[\partial_{t_*} \mathbf{v}_* + (\mathbf{v}_* \cdot \nabla_*) \mathbf{v}_* \right] = -\nabla_* p_* + \eta \nabla_*^2 \mathbf{v}_*, \ \nabla_* \cdot \mathbf{v}_* = 0, \qquad (4.178)$$

$$\partial_{t_*} T_* + \mathbf{v}_* \cdot \nabla_* T_* = \kappa \nabla_*^2 T_*, \qquad (4.179)$$

$$\partial_{t_*} C_* + \mathbf{v}_* \cdot \nabla_* C_* = D \nabla_*^2 (C_* + \alpha_* T_*); \qquad (4.180)$$

$$\mathbf{v}_* = 0, \ \partial_{z_*} T_* = -a, \ \partial_{z_*} (C_* + \alpha T_*) = 0 \text{ at } z_* = 0, \qquad (4.181)$$

$$w_* = 0, \ k_{th} \partial_{z_*} T_* + q_* T_* = 0, \ \partial_{z_*} (C_* + \alpha_* T_*) = 0,$$

$$\eta_* \partial_{z_*} \mathbf{u}_* = -\nabla_{*2} (\sigma_T T_* - \sigma_C C_*) \text{ at } z_* = d_*. \qquad (4.182)$$

Here $\mathbf{v}_* = \mathbf{u}_* + w_* \mathbf{e}_z$, $\mathbf{u}_*$ is a 2D projection of the velocity onto the $x_* - y_*$ plane, w_* is a z_*-component of the velocity, and $\nabla_{*2} = (\partial_{x*}, \partial_{y*})$; η, ν, κ, k_{th}, and q are the dynamics kinematic viscosities, thermal diffusivity, thermal conductivity, and heat transfer rate according to Newton's law of cooling, respectively. As usual, subscripts denote partial derivatives with respect to the corresponding variables.

Choosing the scales for the time, length, temperature, solute concentration, velocity, and pressure fields as d_*^2/κ, d_*, Ad_*, $A\sigma_t/\sigma_c$, κ/d_*, and $\eta\kappa/d_*^2$, respectively, we arrive at the following dimensionless set of governing equations and boundary conditions:

$$P^{-1} [\mathbf{v}_t + (\mathbf{v} \cdot \nabla) \mathbf{v}] = -\nabla p + \nabla^2 \mathbf{v}, \ \nabla \cdot \mathbf{v} = 0, \quad (4.183)$$

$$T_t + \mathbf{v} \cdot \nabla T = \nabla^2 T, \quad (4.184)$$

$$L^{-1} (C_t + \mathbf{v} \cdot \nabla C) = \nabla^2 (C + \chi T); \quad (4.185)$$

$$\mathbf{v} = 0, \ T_z = -1, \ C_z = \chi \text{ at } z = 0, \quad (4.186)$$

$$w = 0, \ T_z + BT = 0, \ C_z - \chi BT = 0, \ \mathbf{u}_z = -M\nabla_2 (T - C) \text{ at } z = 1. \quad (4.187)$$

The following five dimensionless parameters govern the boundary value problem (4.183)–(4.187):

$$M = \frac{\gamma_T A d_*^2}{\rho \nu \kappa}, \quad B = \frac{q d_*}{k_{th}}, \quad \chi = \frac{\alpha \gamma_C}{\gamma_T}, \quad P = \frac{\nu}{\kappa}, \quad L = \frac{D}{\kappa}, \quad (4.188)$$

which are the Marangoni, Biot, Soret, Prandtl, and Lewis numbers, respectively. We also use the Schmidt number, $S = P/L$. It is worth noting that values of the Lewis number are rather small for most mixtures within the range of $L \leq 10^{-2}$. However, within this chapter, we do not assume L asymptotically small, except for Section 4.2.3. The Biot number is also usually small and in what follows the limiting case $B \ll 1$ is analyzed, where the longwave instability is possible.

The physical system under consideration obviously possesses the conductive state (the base state):

$$\mathbf{v}_0 = 0, \quad T_0 = \frac{B+1}{B} - z, \quad C_0 = \chi \left(z - \frac{1}{2} \right) + \langle C \rangle, \quad p_0 = const. \quad (4.189)$$

At small B, the z-independent term in T_0 is large, which has a clear reason: boundary condition (4.187) for T means that the temperature of the ambient gas far from the layer surface is chosen as a reference value of the temperature. The gas is of a low thermal conductivity, and, therefore, the layer temperature has to be high to sustain the fixed heat flux. This constant part of the temperature (as well as the mean solute concentration $\langle C \rangle$) does not influence the further analysis.

4.2.1.2 Linear Stability Analysis

In order to study the linear stability, we introduce small perturbations to the base state (4.189). In view of the rotational symmetry of the problem in the x–y plane, one can restrict the analysis to 2D perturbations only. Therefore, the velocity field is 2D, $\mathbf{v} = (u, 0, w)$, and the perturbations are independent of y. The coordinates t and x are separated as follows: $f(x, z, t) = \tilde{f}(z) \exp(\lambda t + ikx)$, where f is any perturbation field and λ and k are the complex-valued growth rate and real wavenumber of perturbations. Resulting set of linearized equations reads:

$$ik\tilde{u} + \tilde{w}' = 0, \quad (4.190)$$

$$\lambda P^{-1} \tilde{u} = -ik\tilde{p} + \Delta \tilde{u}, \quad (4.191)$$

$$\lambda P^{-1} \tilde{w} = -\tilde{p}' + \Delta \tilde{w}, \quad (4.192)$$

$$\lambda \tilde{T} - \tilde{w} = \Delta \tilde{T}, \quad (4.193)$$

$$L^{-1} \left(\lambda \tilde{C} + \chi \tilde{w} \right) = \Delta \left(\tilde{C} + \chi \tilde{T} \right); \quad (4.194)$$

$$\tilde{u} = \tilde{w} = \tilde{T}' = \tilde{C}' = 0 \text{ at } z = 0, \quad (4.195)$$

$$\tilde{w} = 0, \quad \tilde{T}' + B\tilde{T} = 0, \quad \left(\tilde{C} + \chi \tilde{T} \right)' = 0, \quad \tilde{u}' = -Mik \left(\tilde{T} - \tilde{C} \right) \text{ at } z = 1. \quad (4.196)$$

Here $\Delta = d^2/dz^2 - k^2$ is the Fourier transform of the Laplace operator, and a prime denotes the z-derivative. For arbitrary k (shortwave mode), this set of equations has to be solved numerically; only the neutral stability curve $M = M_*(k)$ for the monotonic mode ($\lambda = 0$) can be found analytically; see (4.229) below and [18]. In contrast, for the longwave instability, which takes place at small B, further progress is possible [19].

Because the longwave stability analysis in the case of Marangoni convection is quite similar to that carried out in Section 4.1.1.2 for the buoyancy convection, below we present only the results. The details of the derivation can be found in [18].

We apply (4.20)–(4.22) to rescale the wavenumber, growth rate, and velocity components and to present the perturbation and the rescaled growth rate as a power series in ε^2; again, $\varepsilon = B^{1/4} \ll 1$ is used. The expansion of the Marangoni number

$$M = 48(m_0 + \varepsilon^2 m_2 + \ldots) \tag{4.197}$$

is used similar to (4.22) for the Rayleigh number R.

The solution of the zeroth order boundary value problem reads:

$$\tilde{T}_0 = a_0, \quad \tilde{C}_0 = b_0, \tag{4.198}$$

$$\tilde{P}_0 = -72m_0 h_0, \quad \tilde{U}_0 = m_0 i K h_0 V_\|, \quad \tilde{W}_0 = -m_0 K^2 h_0 V_\perp, \tag{4.199}$$

$$V_\|(z) = 12z(2 - 3z), \quad V_\perp(z) = -12z^2(1 - z), \tag{4.200}$$

where the constant $-h_0 = b_0 - a_0$ has the meaning of the perturbation of the surface tension.

At the second order, we obtain:

$$(K^2 + \Lambda_0)\, a_0 - m_0 K^2(a_0 - b_0) = 0, \tag{4.201}$$

$$\chi K^2 a_0 + (K^2 + \Lambda_0 L^{-1})\, b_0 + m_0 \chi L^{-1} K^2(a_0 - b_0) = 0. \tag{4.202}$$

This set of linear algebraic equations coincides with (4.33) and (4.34) with obvious replacements: $r_0 \to m_0$ and $\psi \to \chi$. Therefore, the stability thresholds for monotonic and oscillatory modes can be easily found:

$$m_0^{(m)} = \frac{L}{\chi_L}, \quad \chi_L = L + \chi + \chi L, \tag{4.203}$$

$$m_0^{(o)} = \frac{1 + L}{1 + \chi}, \tag{4.204}$$

the squared frequency of neutral perturbations for the oscillatory mode reads

$$\Omega^2 = K^4 L \left(1 - \frac{m_0^{(o)}}{m_0^{(m)}}\right) = K^4 \left[-\frac{\chi(1 + L + L^2) + L^2}{1 + \chi}\right]. \tag{4.205}$$

The oscillatory mode takes place when this expression is positive; hence,

$$-1 < \chi < \chi_c, \ \chi_c = -\frac{L^2}{1+L+L^2}. \tag{4.206}$$

According to [8], (4.203) can be restored from the classical Pearson's result $m_0 = 1$ [20] for a pure liquid; see the similar explanation for the buoyancy convection below (4.36).

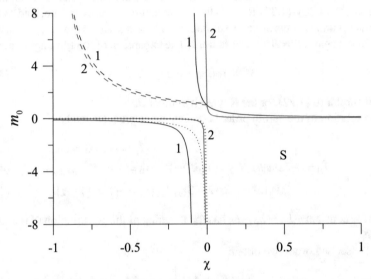

Fig. 4.6 Linear stability for the longwave Marangoni convection in a binary mixture in the plane $\chi - m_0$ for $L = 0.1$ (lines 1) and $L = 0.01$ (lines 2). Solid, dashed, and dotted lines correspond to monotonic mode, (4.203), oscillatory mode, (4.204), and the monotonic mode with $k = O(\sqrt{B})$, (4.228), respectively. Domain of stability is marked "S"

Stability boundaries $m_0^{(m,o)}(\chi)$ for different values of the Lewis number are depicted in Figure 4.6 by solid and dashed lines, respectively; they coincide with corresponding lines in Figure 4.1(a). The oscillatory mode is critical for χ in the interval (4.206). The variation of Λ_0/K^2 within this interval is similar to that shown in Figure 4.1(b) with r replaced with m. The monotonic instability occurs at $\chi > \chi_c$ for heating from below; for heating from above and negative Soret numbers, the competition of two monotonic mode takes place as explained below.

For Λ_0 being an eigenvalue, (4.201) and (4.202) are proportional to each other; hence the dependence between a_0 and b_0 can be found from any of those equations. The proportionality between a_0 and b_0 is taken into account below.

The solution of the second-order problem is

$$\tilde{T}_2 = a_2 - m_0 K^2 h_0 T_{l2}(z), \tag{4.207}$$

$$\tilde{C}_2 = b_2 + m_0 \chi \left(1 + L^{-1}\right) K^2 h_0 T_{l2}(z), \tag{4.208}$$

$$\tilde{p}_2 = 12m_0K^2h_0z(2-3z) + \frac{6}{5}m_0h_0\Gamma - 72\Phi, \quad (4.209)$$

$$\tilde{U}_2 = iKm_0h_0\Gamma U_{l2}(z) + iK\Phi V_{\parallel}, \quad (4.210)$$

$$\tilde{W}_2 = -m_0h_0K^2\Gamma W_{l2}(z) - K^2\Phi V_{\perp}, \quad (4.211)$$

$$\Gamma = 2K^2 + \frac{\Lambda_0}{P}, \quad \Phi = \frac{\chi_L m_0^2 h_0 K^2}{30L} + m_0h_2 + m_2h_0, \quad (4.212)$$

$$T_{l2}(z) = z^4 - \frac{3}{5}z^5 - \frac{1}{2}z^2 + \frac{1}{15}, \quad U_{l2}(z) = \frac{z}{5}(20z^2 - 15z^3 + 3z - 6), \quad (4.213)$$

$$W_{l2}(z) = \frac{z^2}{5}(3z^3 - 5z^2 - z + 3), \quad (4.214)$$

where m_0 is equal to either $m_0^{(m)}$ or $m_0^{(o)}$ for the monotonic and oscillatory modes, respectively; in the former case $\Lambda_0 = 0$, and in the latter one, $\Lambda_0 = -i\Omega$.

In order to obtain Λ_2, one has to proceed to the fourth order; only energy and mass balances are needed for that goal.

Let us discuss the corresponding results separately for each mode.

For the oscillatory mode, the second-order correction is complex:

$$\Lambda_2^{(o)} = -\frac{1}{2}\left(1 + i\frac{L-\chi m_0}{\Omega}\right) + \frac{m_2K^2}{2}\left(1 + \chi + i\frac{\chi_L}{\Omega}\right) + \frac{m_0K^4}{120P}f \quad (4.215)$$

$$f_r = \chi\left[2\left(S^2 - SP + P^2\right) - 3(S+P)\right] + P(-4S + 2P - 3), \quad (4.216)$$

$$f_i = -S\frac{2}{\tilde{\Omega}}\chi_L(3P - Sm_0\chi_L) + 3S(1+\chi)\tilde{\Omega}. \quad (4.217)$$

The condition $\Lambda_{2r} = 0$ yields the neutral stability curve:

$$m_2^{(o)} = m_0^{(o)}\left(\frac{f_\beta^{(o)}}{K^2} + f_K^{(o)}K^2\right), \quad (4.218)$$

$$f_\beta^{(o)} = \frac{1}{1+L}, \quad f_K^{(o)} = -\frac{f_r}{60P(1+\chi)}. \quad (4.219)$$

It is clearly seen that $f_\beta^{(o)}$ is always positive, whereas $f_K^{(o)} > 0$ if $S+P > 3/2$. In that case, the neutral stability curve has the only minimum

$$m_2^{(o)} = 2m_0^{(o)}\sqrt{f_\beta^{(o)}f_K^{(o)}} \text{ at } K_c^4 = \frac{f_\beta^{(o)}}{f_K^{(o)}}. \quad (4.220)$$

Returning back to unrescaled variables, one can readily obtain:

$$M_* \approx 48m_0^{(o)}\left(1 + f_\beta^{(o)}\frac{B}{k^2} + f_K^{(o)}k^2\right). \quad (4.221)$$

Within the unrealistic parameter range $S + P < 3/2$, shortwave perturbations grow faster than the longwave ones; we do not discuss this case any further.

For the monotonic mode we obtain:

$$\Lambda_2^{(m)} = L^2 \frac{-(1+\chi) + m_2^{(m)} K^2 m_0^{-2} - m_0^{-1} K^4 / 15}{\chi(1 + L + L^2) + L^2}. \tag{4.222}$$

Equating Λ_2 to zero, one can find the neutral stability curve:

$$m_2^{(m)} = m_0^{(m)} \left(\frac{f_\beta^{(m)}}{K^2} + f_K^{(m)} K^2 \right), \tag{4.223}$$

$$f_\beta^{(m)} = \frac{L(1+\chi)}{\chi L}, \quad f_K^{(m)} = \frac{1}{15}.$$

Therefore, for $f_\beta^{(m)} > 0$ the neutral stability curve is determined by (4.218); one only has to replace $f_\beta^{(o)}$ ($f_K^{(o)}$) with $f_\beta^{(m)}$ ($f_K^{(m)}$). Returning back to the unscaled parameters, one obtain

$$M_* \approx 48 m_0^{(m)} \left(1 + f_\beta^{(m)} \frac{B}{k^2} + f_K^{(m)} k^2 \right). \tag{4.224}$$

Analysis of (4.223) shows that $f_\beta^{(m)} > 0$ either for $\chi > \chi_\infty^{(m)}$ or for $\chi < -1$, and therefore, the neutral stability curve has the only minimum at K_c given by (4.220) with the abovementioned replacement.

At $-1 < \chi < \chi_\infty^{(m)}$ formally $m_2^{(m)} \to -\infty$ at small K and the analysis in the framework of the conventional asymptotics $k \sim B^{1/4}$ fails. As shown in [21], another asymptotics with $k \sim B^{1/2}$ has to be developed for this mode. In order to elaborate this case, we introduce the small parameter $\bar\varepsilon = \sqrt{B}$ and replace ε with $\bar\varepsilon$ in (4.20)–(4.22). The boundary value problem and, hence, the solution at the zeroth order are unaffected by this replacement; at the second order, the boundary conditions at the free surface are $T_2' = -T_0$ and $C_2' = \chi T_0$. The only correction to (4.201) is the additional term a_0 in the left-hand side, whereas (4.202) remains unchanged.

The resulting quadratic equation for Λ_0 reads:

$$\Lambda_0^2 + [\beta_K + L - m_0(1+\chi)] \Lambda_0 K^2 + [L\beta_K - m_0(L + \chi\beta_K + \chi L)] K^4 = 0, \tag{4.225}$$

where $\beta_K = 1 + K^{-2}$. It can be easily seen that for the oscillatory mode, these asymptotics yield a rather obvious result: additional mechanism of dissipation leads to the increase in the correction to the Marangoni number for any finite K, whereas at $K \gg 1$ (4.218) is reproduced.

In contrast, for the monotonic mode, the result is nontrivial. The neutral stability curve $m_0 = m_0^{(B)}$ is given by

$$m_0^{(B)} = \frac{L\beta_K}{L + \chi\beta_K + \chi L} \tag{4.226}$$

or, in terms of the unrescaled wavenumber,

$$M_* = 48 \frac{L\left(1 + BK^{-2}\right)}{L + \chi\left(1 + BK^{-2}\right) + \chi L}. \tag{4.227}$$

Within the interval $-1 < \chi < \chi_\infty^{(m)}$, $m_0^{(B)}$ decreases monotonically from

$$m_0^{(l)} = \frac{L}{\chi} \tag{4.228}$$

to $m_0^{(m)}$ as K grows from zero to infinity. Thus, the critical wavenumber is $K = 0$, and the threshold value of the Marangoni number is given by $m_0^{(l)}$. Moreover, at $\chi_\infty^{(m)} < \chi < 0$, the maximum (critical) value of $m_0^{(B)}$ is still given by (4.228), whereas at a certain value K_* of the wavenumber $m_0^{(B)}$ diverges and no instability for heating from above exists at $K > K_*$.

Therefore, additional dissipative mechanism, the heat losses from the free surface, leads to the layer destabilization. At first glance, this effect seems to be counterintuitive, but in fact it is not. Indeed, for the parameter range $m_0 < 0$ and $\chi < 0$, the solutal and thermal Marangoni effects compete, and the temperature gradient is stabilizing. Additional heat flux decreases the temperature gradient and, therefore, effectively amplifies the role of the solutal (destabilizing) effect.

Consequently, for $-1 < \chi < 0$ the stability threshold is determined by (4.228), and the critical perturbations are genuine longwave ones, with $K = 0$. The corresponding stability boundaries are shown in Figure 4.6 by the dotted lines. Note, that these nontrivial results retain for small, but nonzero, values of the Biot number. At $B = 0$, the K^{-2}-term in (4.227) vanishes and one returns back to $M_* = 48L/\chi_L$, cf. (4.223). Therefore, at $-1 < \chi < 0$ the neutral stability curves for $B = 0$ and arbitrary small $B \neq 0$ are completely different. In contrast, for the conventional asymptotics, $B = 0$ is not a distinguished case, only critical wavenumber vanishes in that limit.

It should be emphasized that when dealing with the linear stability problem, one has to consider all possible values of the disturbance wavenumber k. Therefore, complete analysis of both asymptotics (i) $k \sim B^{1/4}$ and (ii) $k \sim B^{1/2}$ is necessary; the conventional asymptotics (i) alone does not provide critical perturbations for $-1 < \chi < \chi_\infty^{(m)}$. Moreover, analysis of the mode with $k \sim B^{1/2}$ allows finding longwave instability for heating from below at $\chi_\infty^{(m)} < \chi < 0$, which is absent for the conventional scaling.

The analogues of (4.223) and (4.226) are also valid for the Rayleigh–Bénard convection in a layer with the solid boundaries. In order to derive these neutral stability curves, one just has to replace m with r, χ with ψ, and $1/15$ with $1/231$ in $f_K^{(m)}$. All the other features, including the values of the Soret number separating different regimes, remain the same for the buoyancy convection. Moreover, the similar results are also valid for the mixed convection in a binary liquid layer [22, 23]: one just has to replace m_0 with $m_0 + r_0$ and to add the corresponding threshold values for both. (Note that the linear stability analysis performed in Section 4.1.1.2 is not just a particular case of the cited papers because of the difference of the conditions at the upper boundary.)

Both the asymptotics can be calculated from the general formula for the neutral stability curve $M_*(k)$ derived in [18]:

$$M_* = \frac{4Lkg_3(k)(\sinh 2k - 2k)(B\cosh k + g_3(k))}{\chi B\cosh^2 kg_2(k) + \frac{1}{4}g_3(2k)[\chi_L g_2(k) - \chi Bg_1(k)] - \chi_L g_1(k)g_3^2(k)}, \quad (4.229)$$

where $g_1(k) = 2 + k^2$, $g_2(k) = k^2 + \sinh^2 k$, and $g_3(k) = k\sinh k$. At small k expanding the hyperbolic functions in powers of k and keeping only the lower powers one immediately arrives at:

$$M_* \approx \frac{48L\left[B + \left(1 + \frac{13}{15}B\right)k^2 + \left(\frac{2B}{7} + \frac{8}{15}\right)k^4\right]}{\chi B + \left(\chi + L + \chi L + \frac{4}{5}\chi B\right)k^2 + \left[\frac{7}{15}(\chi + L + \chi L) + \frac{17}{75}\chi B\right]k^4}. \quad (4.230)$$

Assuming (i) $B = O(k^4)$ and (ii) $B = O(k^2)$ one can reproduce (4.224) and (4.227), respectively. Analysis of (4.230) shows that the latter mode is critical at $-1 < \chi < 0$ and $M_* < 0$ as stated above. Numerical tabulation of (4.229) performed in [18] demonstrates that for rather small B, there is no any other critical mode; the neutral stability curve $M_*(k)$ has the only minimum, which is either true longwave ($-1 < \chi < 0$, $M_* < 0$), mode (ii), or takes place at small $k \sim B^{1/4}$, mode (i).

In contrast, at $-1 < \chi < \chi_c$, see (4.206); the competition of the longwave and shortwave oscillatory modes takes place. That competition substantially diminishes the interval of χ where the longwave oscillatory mode is critical: $\chi_1(L) < \chi < \chi_c$, where χ_1 depends on the Lewis number and can be roughly estimated as $\chi_1 \approx -0.1$; see the details in [18].

4.2.1.3 Amplitude Equations for the Monotonic Modes

In the case of conventional mode, the amplitude equation which describes the nonlinear evolution of monotonic perturbations is quite similar to that discussed in Section 3.2.1 for the convection in a pure liquid. The reader can find that amplitude equation, with the coefficients changed due to the Soret-induced solutocapillary contribution, in [19].

Below we discuss only the novel mode (ii) with unusual scaling. The marginal stability curve (4.226) predicts the critical wavenumber $K = 0$; therefore, the typical convective length scale ℓ should be larger than $B^{-1/2}$ typical for this mode. Accurate analysis shows that the scaling with $\ell = O(B^{-1})$ is distinguished: the consideration of this case allows to describe the cases of either smaller or larger ℓ by a certain reduction of equations.

In order to derive the amplitude equation, we start with the following rescalings:

$$\mathbf{u} = \tilde{\varepsilon}\mathbf{U}, \ w = \tilde{\varepsilon}^2 W, \ T = T_0 + \tilde{\varepsilon}\Theta, \ C = C_0 + \Sigma, \tag{4.231}$$

$$(X,Y) = \tilde{\varepsilon}(x,y), \ \tau = \tilde{\varepsilon}^4 t, \ B = \tilde{\varepsilon}\tilde{\beta}. \tag{4.232}$$

supplemented with the expansions in terms of $\tilde{\varepsilon}$,

$$M = 48(m_0 + \tilde{\varepsilon}m_1 + \ldots), \ f = f_0 + \tilde{\varepsilon}f_1 + \ldots, \ f = (\Theta, \Sigma, \mathbf{U}, W, p). \tag{4.233}$$

The boundary value problem at the zeroth order and its solution are obvious:

$$\Sigma_0'' = 0, \ p_0' = 0, \ \mathbf{U}_0'' = \nabla p_0, \ W_0' = -\nabla \cdot \mathbf{U}_0; \tag{4.234}$$

$$\Sigma_0' = \mathbf{U}_0 = W_0 = 0 \text{ at } z = 0, \tag{4.235}$$

$$\Sigma_0' = W_0 = 0, \ \mathbf{U}_0' = 48m_0\nabla\Sigma_0 \text{ at } z = 1; \tag{4.236}$$

$$\Sigma_0 = G(X,Y,\tau), \ p_0 = 72m_0 G,$$

$$\mathbf{U}_0 = -m_0 V_\parallel(z)\nabla G, \ W_0 = -m_0 V_\perp(z)\nabla^2 G. \tag{4.237}$$

The solution (4.237) coincides with that obtained within the linear problem (4.198) and (4.199) with the obvious replacements, $iK \to \nabla$ and $h = -G$ owing to the scaling for the temperature.

At the first order, the boundary value problem is similar, with additional equation for $\Theta_0 = \theta(X,Y,\tau)$. One only has to replace the subscripts $(0 \to 1)$ and to extend the expansion for $m(\Theta - \Sigma)$: $-m_0 G \to m_0 h_1 - m_1 G$, where $h_1 = \theta - G_1$.

The second-order problem reads

$$(\Sigma_2 + \chi\Theta_2)'' = -\nabla^2\Sigma_0 + L^{-1}(\chi W_0 + \mathbf{U}_0 \cdot \nabla\Sigma_0), \tag{4.238}$$

$$\Theta_2'' = -W_0, \tag{4.239}$$

$$\mathbf{U}_2'' - \nabla P_2 = -\nabla^2\mathbf{U}_0 + P^{-1}(\mathbf{U}_0 \cdot \nabla\mathbf{U}_0 + W_0\mathbf{U}_0'), \tag{4.240}$$

$$P_2' = W_0'', \ W_0' = -\nabla\mathbf{U}_0; \tag{4.241}$$

$$\mathbf{U}_0 = W_0 = \Sigma_2' = \Theta_2' = 0 \text{ at } z = 0, \tag{4.242}$$

$$W_2 = (\Sigma_2 + \chi\Theta_2)' = 0, \ \Theta_2' = -\tilde{\beta}\Theta_1,$$

$$\mathbf{U}_2' = -48\nabla[m_0(\Theta_2 - \Sigma_2) + m_1 h_1 - m_2\Sigma_0] \text{ at } z = 1. \tag{4.243}$$

The solvability conditions for mass and heat balance equations are as follows:

$$\left(\frac{\chi}{L}m_0 - 1\right)\nabla^2 G = 0, \tag{4.244}$$

$$m_0\nabla^2 G - \tilde{\beta}\theta = 0, \tag{4.245}$$

which provide both the critical Marangoni number (4.228) and the coupling between the leading part of the perturbations of the temperature and solute concentration:

$$\theta = \frac{m_0\nabla^2 G}{\tilde{\beta}}.$$

The solution to the second-order problem has the following form:

$$T_2 = -\frac{m_0}{10}\nabla^2 G\left(10z^4 - 6z^5 - 1\right) + \theta_1, \quad (4.246)$$

$$C_2 = \nabla^2 G T_{l2}(z) + G_2 - \frac{m_0}{5L}(\nabla G)^2 \left(20z^3 - 15z^4 - 2\right)$$
$$-\chi(T_2 - \theta_1), \quad (4.247)$$

$$p_2 = -72\Gamma + m_0\nabla^2 G\left(V_\perp(z) + \frac{12}{5}\right) + \Pi_2, \quad (4.248)$$

$$\mathbf{U}_2 = V_\parallel(z)\nabla\Gamma - m_0 U_{l2}(z)\nabla\nabla^2 G + \frac{72m_0^2}{P}q_1(z)\nabla\left(|\nabla G|^2\right)$$
$$+\frac{144m_0^2}{P}q_2(z)\nabla^2 G\nabla G + \frac{1}{2}\nabla\Pi_2(z^2 - 2z), \quad (4.249)$$

$$W_2 = V_\perp(z)\nabla^2\Gamma - m_0 W_{l2}(z)\nabla^4 G - \frac{72m_0^2}{P}q_3(z)\nabla^2\left(|\nabla G|^2\right)$$
$$-\frac{144m_0^2}{P}q_4(z)\nabla\cdot\left(\nabla^2 G\nabla G\right), \quad (4.250)$$

$$\Gamma = m_0 h_2 + m_1 h_1 - m_2 G - \frac{m_0}{30}\left[9m_0(1+\chi) - 1\right]\nabla^2 G + \frac{3m_0^2}{5L}(\nabla G)^2, \quad (4.251)$$

$$q_1 = \frac{z}{210}\left(70z^3 - 126z^4 + 63z^5 - 18z + 8\right), \quad (4.252)$$

$$q_2 = \frac{z}{30}\left(12z^4 - 5z^3 - 6z^5 - 4\right), \quad (4.253)$$

$$q_3 = \frac{z^2}{210}\left(14z^3 - 21z^4 + 9z^5 - 6z + 4\right), \quad (4.254)$$

$$q_4 = \frac{z^2}{420}\left(28z^4 - 14z^3 - 12z^5 - 13z + 11\right). \quad (4.255)$$

Recall that the functions $T_{l2}(z)$, $U_{l2}(z)$, and $W_{l2}(z)$ are given by (4.213) and (4.214).

At the third order, we have to demand only the solvability of the heat and mass balance equations with the corresponding boundary conditions:

$$(C_3 + \chi T_3)'' = -\nabla^2(C_1 + \chi T_1) + L^{-1}(\chi W_1 + \mathbf{U}_1\cdot\nabla C_0 + \mathbf{U}_0\cdot\nabla C_1), \quad (4.256)$$

$$T_3'' = -\nabla^2 T_1 - W_1 + \mathbf{U}_0\cdot\nabla T_1, \quad (4.257)$$

$$T_3' = (C_3 + \chi T_3)' = 0 \text{ at } z = 0, \quad (4.258)$$

$$T_3' = -\tilde{\beta}T_1, \quad (C_3 + \chi T_3)' = 0 \text{ at } z = 1. \quad (4.259)$$

The first equation is solvable if

$$m_1\nabla^2 G = \frac{m_0^2}{\tilde{\beta}}(1+\chi)\nabla^4 G. \quad (4.260)$$

The general solution to (4.260) is obvious:

$$G = \sum_\mathbf{k} A_\mathbf{k}(\tau)e^{i\mathbf{k}\cdot\mathbf{R}} + c.c., \quad (4.261)$$

where all the wavevectors contributing to G have the same length k coupled with the correction to the Marangoni number:

$$m_1 = -\frac{m_0^2}{\tilde{\beta}}(1+\chi)k^2. \qquad (4.262)$$

This expression agrees with the expansion of (4.226) at small $K^2/\tilde{\beta}$. Note that at this stage, we have fixed the perturbation wavenumber; from physical point of view, this corresponds to an analysis of patterns in a confined cavity. Indeed, for an infinite layer, the set of wavenumbers is continuous and infinitely small k are possible to minimize the correction m_1.

The solvability condition for the second equation, (4.257), serves to define G_1:

$$\nabla^2 [\theta(m_0 - 1) - m_0 G_1 - m_1 G] + \tilde{\beta}\theta = 0.$$

Solution to (4.256) and (4.257) are not needed; at the fourth order, one needs only the solvability of the mass balance equation:

$$(C_4 + \chi T_4)'' = L^{-1} (\chi W_2 + \mathbf{U}_2 \cdot \nabla C_0 + \mathbf{U}_1 \cdot \nabla C_1 +$$
$$\mathbf{U}_0 \cdot \nabla C_2 + \partial_\tau C_0) \qquad (4.263)$$
$$-\nabla^2 (C_2 + \chi T_2) \qquad (4.264)$$
$$(C_4 + \chi T_4)' = 0 \text{ at } z = 0, 1. \qquad (4.265)$$

Integrating the right-hand side of the equation across the layer, one arrives at

$$L(1+\chi)\left(\nabla^2\theta_1 + k^2\theta_1\right) = G_\tau + \frac{(1+\chi)\chi_L m_0^2}{\tilde{\beta}^2}k^6 G + \chi m_2 \nabla^2 G$$

$$+\gamma_1 L\nabla^4 G - \gamma_2 \frac{L}{\chi}\nabla^2\left(|\nabla G|^2\right) - \gamma_3 \frac{L}{\chi}\nabla \cdot (\nabla^2 G\nabla G)$$

$$-\frac{48L}{35\chi^2}\nabla \cdot \left(|\nabla G|^2 \nabla G\right) + \frac{312L}{35\chi^2 S}\mathbf{J} \cdot \nabla G, \qquad (4.266)$$

where

$$\mathbf{J} = \nabla\nabla^{-2}\nabla \cdot (\nabla^2 G\nabla G) - \nabla^2 G\nabla G, \qquad (4.267)$$

$$\gamma_1 = \frac{1}{30}[9(1+\chi)m_0 + 2], \quad \gamma_2 = \frac{1}{10S} + \frac{5}{3}, \quad \gamma_3 = \frac{1}{5S} + \frac{1}{10} - \frac{L}{2}. \qquad (4.268)$$

The right-hand side of this equation can be derived from the amplitude equation obtained by Shtilman and Sivashinsky [10] with the replacement of the time scale (by the factor L), the amplitude function (by the factor $\theta = -G\chi^{-1}$), and replacement $P \to S$. This is equivalent to the replacement of the thermocapillary convection with the solutocapillary one. There are, however, three exceptions: the second term in the right-hand side, $9(1+\chi)m_0$ contribution in γ_1, and $-L/2$ term in γ_3. These three terms are the only impacts of the heat balance on the solutocapillary convection.

However, despite the abovementioned similarity with the Shtilman–Sivashinsky equation, (4.266) is completely different from the mathematical point of view. Indeed, G solves linear equation (4.260), and therefore it can be represented as (4.261). Moreover, any term with the wavenumber k in the right-hand side of (4.266) is secular. This provides the amplitude nonlinear equations for the Fourier amplitudes A_k:

$$\dot{A}_k = \gamma A_k - \left(K_0|A_k|^2 + \sum_q K_{kq}|A_q|^2 \right) A_k + Q B_k^* C_k^*, \qquad (4.269)$$

$$\gamma = -\frac{(1+\chi)\chi_L m_0^2}{\tilde{\beta}^2} k^6 + m_2 \chi k^2 - \gamma_1 L k^4, \qquad (4.270)$$

$$K_0 = \frac{144L}{35\chi^2} k^4, \; K_{kq} = K(\theta_{kq}) = \frac{96L}{35\chi^2} k^4 \left(1 + 2\cos^2\theta_{kq}\right), \qquad (4.271)$$

$$Q = \frac{L}{2\chi}\left(\frac{1}{5S} - 1 - L\right) k^4. \qquad (4.272)$$

where θ_{kq} is the angle between the wavevectors k and q; B_k and C_k are the amplitudes of the waves tilted by $\pm 2\pi/3$ to k, so that the corresponding three wavevectors are the base vectors for a hexagonal lattice, q are represent all the wavevectors, which enter into (4.261).

The striking feature of (4.269) is that both quadratic and cubic nonlinear terms are present in the same amplitude equations without any additional assumptions. This property takes place because for this mode we do not assume the Fourier amplitudes small; only longwave expansion is used.

Equations (4.269) admit simple pattern selection similar to that carried out in Section 3.2.1.2 (see also [24] and references therein). Because $T_1 = K(\pi/6)/K_0 = 5/3$, $T_2 = K(\pi/3)/K_0 = 1$, and $T_3 = K(\pi/2)/K_0 = 2/3$, one can readily conclude that rolls and quasicrystals are always unstable, hexagons are stable within the whole domain of existence, and squares are stable with respect to hexagons in a certain region of γ. For real binary mixtures, $5S > 1/(1+L)$ and therefore $Q > 0$; it means that up-hexagons are selected.

4.2.1.4 Oscillatory Mode

In this section we present the results on the nonlinear dynamics for the oscillatory mode. This analysis, in fact, has much in common with that for the buoyancy convection; see Sections 4.1.1.3–4.1.1.5. However, because of the lack of the reflection symmetry in the layer midplane, some additional (quadratic) effects appear.

Derivation of the Amplitude Equations

Similar to the buoyancy convection, the nonlinear dynamics is governed by nonlocal equations, which arise from the solvability conditions of the problem at the fourth order in $\varepsilon = B^{1/4}$. The derivation of those solvability conditions is almost verbatim of that performed in Section 4.1.1.3; therefore we omit that cumbersome procedure, which can be found in [15].

At the second order, the evolution of the leading parts of the temperature (F) and solute concentration (G) perturbations is governed by the following linear homogeneous equations:

$$\mathscr{L}_1(F,G) = F_\tau - (1 - m_0)\nabla^2 F - m_0 \nabla^2 G = 0, \qquad (4.273)$$

$$\mathscr{L}_2(F,G) = G_\tau - \chi(L + m_0)\nabla^2 F - (L - \chi m_0)\nabla^2 G = 0. \qquad (4.274)$$

Similar to Section 4.1.1.3, at $m_0 = m_0^{(0)}$ and $\chi_1(L) < \chi < \chi_c$, the longwave perturbations neither grow nor decay but oscillate in the fast time τ with the frequency given by (4.205). Thus, the general solution to (4.273) and (4.274) can be presented by (4.77) for $h = F - G$ and (4.78) for F and G, where again the evolution of the Fourier harmonics $A_\mathbf{k}(t_4)$ is governed by the fourth-order problem. One has only to replace r_0 in (4.78) with m_0; $-h$ now plays role of the surface tension perturbation.

The solvability conditions at the fourth order read

$$\mathscr{L}_1(Q,R) = -\dot{F} - F - m_2\nabla^2 h - \frac{m_0}{60}\nabla^4\left(3\tilde{\Phi} - 2\frac{m_0\chi_L}{L}h\right)$$
$$- \frac{m_0^2}{10}\left(1 + \frac{2}{P}\right)\nabla\cdot(\nabla^2 h\nabla h) - \frac{m_0^2}{10}\nabla^2\left(\frac{1}{P}(\nabla h)^2 + 6\Gamma\right)$$
$$+ \frac{48m_0^2}{35}\nabla\cdot[(\nabla h\cdot\nabla F)\nabla h] - \frac{312m_0^2}{35P}J(F,\tilde{\Psi}), \qquad (4.275)$$

$$\mathscr{L}_2(Q,R) = -\dot{G} + \chi m_2\nabla^2 h + \frac{\chi m_0}{60}\nabla^4\left(3\tilde{\Phi} - 2\frac{m_0\chi_L}{L}h\right)$$
$$+ \frac{\chi m_0^2}{10}\left(1 + L^{-1} + \frac{2}{P}\right)\nabla\cdot(\nabla^2 h\nabla h) + \frac{\chi m_0^2}{10}\nabla^2\left(\frac{1}{P}(\nabla h)^2 + 6\Gamma\right)$$
$$- \frac{48m_0^2}{35}\nabla\cdot\left\{\left[\nabla h\cdot\nabla(\chi F - L^{-1}G)\right]\nabla h\right\} - \frac{312m_0^2}{35P}J(G,\tilde{\Psi}), \qquad (4.276)$$

where $J(f,g) = f_X g_Y - f_Y g_X$ is a Jacobian of the mapping from (X,Y) to (f,g),

$$\tilde{\Phi} = \left[2 + \frac{m_0}{P}(1 + \chi)\right]h - \frac{1 - \chi L}{P}F + S^{-1}G, \ \Gamma = \nabla h\cdot\nabla\left[(1 + \chi)F - L^{-1}G\right].$$

$$(4.277)$$

and the toroidal potential $\tilde{\Psi}$ is defined by (4.85).

Comparing the set of equations (4.273) and (4.274) to (4.89) and (4.90), one can see the similarity of those problems. In both cases of buoyancy and surface tension-driven convection, the nonlinear evolution is governed by the solvability conditions for a certain linear nonhomogeneous problem. Moreover, the linear operators are identical in both cases, whereas the forms of the right-hand sides are similar: they comprise linear terms with even orders of the gradient up to fourth one, the cubic nonlinear terms, and contributions of the toroidal potential Ψ. The only difference is that for the Marangoni convection, quadratic terms are also included into the analysis because of the reason discussed in Chapter 2: there is no reflection symmetry with respect to the middle plane, and, therefore, there is no reason for vanishing quadratic (with respect to F and G) terms. It will be shown below that these quadratic terms make the analysis much more complicated.

Similar to the buoyancy convection, general solution (4.77) and (4.78) contains arbitrary "constants" $A_{\mathbf{k}}$ that depend on the slow time variable t_4 and represent the amplitudes of the waves. The set of equations (4.275) and (4.276) is a linear nonhomogeneous problem with respect to Q and R with a nonhomogeneity depending on the abovementioned wave amplitudes $A_{\mathbf{k}}$. At m_0 given by (4.204), which is assumed below, the corresponding homogeneous problem has a nontrivial solution, and one needs to consider the solvability conditions of (4.275) and (4.276). These solvability conditions provide the following set of equations describing the nonlinear evolution of the amplitudes:

$$\dot{A}_{\mathbf{k}} = \gamma(K)A_{\mathbf{k}} - N^{(1)}_{\mathbf{k}_1\mathbf{k}_2}A_{\mathbf{k}_1}A_{\mathbf{k}_2} - N^{(2)}_{\mathbf{k}_1\mathbf{k}_2}A_{\mathbf{k}_1}A^*_{\mathbf{k}_2}$$
$$- K^{(1)}_{\mathbf{k}_1\mathbf{k}_2\mathbf{k}_3}A_{\mathbf{k}_1}A_{\mathbf{k}_2}A_{\mathbf{k}_3} - K^{(2)}_{\mathbf{k}_1\mathbf{k}_2\mathbf{k}_3}A_{\mathbf{k}_1}A_{\mathbf{k}_2}A^*_{\mathbf{k}_3} - K^{(3)}_{\mathbf{k}_1\mathbf{k}_2\mathbf{k}_3}A_{\mathbf{k}_1}A^*_{\mathbf{k}_2}A^*_{\mathbf{k}_3}, \qquad (4.278)$$

where nonlinear interactions contribute to (4.278) only if resonant conditions (4.91) and (4.92) are satisfied. Similarly to those conditions, $s_{2,3}$ are ±1 for the cubic terms, whereas for the quadratic terms $s_2 = \pm1$ and $s_3 = 0$. (Again, minus corresponds to the complex-conjugate amplitude.) The sum over all possible pairs (triples) of wavevectors $\mathbf{k}_1$ and $\mathbf{k}_2$ ($\mathbf{k}_1$, $\mathbf{k}_2$, and $\mathbf{k}_3$) satisfying (4.91) and (4.92) is assumed in (4.278).

The linear growth rate agrees with (4.215):

$$\gamma(K) = -\frac{1}{2}\left(1 + i\frac{L - \chi m_0}{\tilde{\Omega}}\right) + \frac{m_2 K^2}{2}\left(1 + \chi + i\frac{\chi_L}{\tilde{\Omega}}\right) + \frac{m_0 K^4}{120 PS}f, \quad (4.279)$$

where $f = f_r + if_i$ is given by (4.216) and (4.217).

The coefficients describing the quadratic interaction $N^{(1,2)}_{\mathbf{k}_1\mathbf{k}_2}$ are given by

$$N^{(1)}_{\mathbf{k}_1\mathbf{k}_2} = \tilde{\mu}_2 k_1^2 k_2^2, \quad N^{(2)}_{\mathbf{k}_1\mathbf{k}_2} = \tilde{\mu}_1 k_2^2\left(k_1^2 - k_2^2\right), \qquad (4.280)$$

$$\tilde{\mu}_1 = \mu^{(1)} + \mu^{(2)}, \quad \tilde{\mu}_2 = 2\mu^{(1)} + 2\mu^{(2)}\left(1 + \frac{2}{P}\right), \qquad (4.281)$$

$$\mu^{(1)} = \frac{m_0^2 \chi}{20 L}\left(1 + \frac{i}{\tilde{\Omega}}\right), \quad \mu^{(2)} = \frac{m_0^2}{20}\left(1 + \chi + \frac{i}{\tilde{\Omega}}\chi_L\right). \qquad (4.282)$$

The resonant interaction of the first kind, $N^{(1)}_{\mathbf{k}_1\mathbf{k}_2}$, takes place for orthogonal $\mathbf{k}_1$ and $\mathbf{k}_2$; the second coefficient, $N^{(2)}_{\mathbf{k}_1\mathbf{k}_2}$, is nonzero for perpendicular $\mathbf{k}_2$ and $\mathbf{k} = \mathbf{k}_1 - \mathbf{k}_2$.

The coefficients of the cubic interaction $K^{(1,2,3)}_{\mathbf{k}_1\mathbf{k}_2\mathbf{k}_3}$ are

$$K^{(1)}_{\mathbf{k}_1\mathbf{k}_2\mathbf{k}_3} = -\tilde{\alpha}(\mathbf{k}_1 \cdot \mathbf{k}_2)(\mathbf{k} \cdot \mathbf{k}_3) + \tilde{J}_c k_1^2\frac{(\mathbf{k}_1 \times \mathbf{k}_2) \cdot (\mathbf{k}_3 \times \mathbf{k})}{(\mathbf{k}_1 + \mathbf{k}_2)^2} + \{1,2,3\}, \quad (4.283)$$

$$K^{(2)}_{\mathbf{k}_1\mathbf{k}_2\mathbf{k}_3} = \tilde{\alpha}\,(\mathbf{k}_1\cdot\mathbf{k}_2)\,(\mathbf{k}\cdot\mathbf{k}_3) + \tilde{\beta}\,(\mathbf{k}_1\cdot\mathbf{k})\,(\mathbf{k}_2\cdot\mathbf{k}_3)$$

$$+\tilde{J}_c(k_3^2 - k_1^2)\frac{(\mathbf{k}_1\times\mathbf{k}_3)\cdot(\mathbf{k}_2\times\mathbf{k})}{(\mathbf{k}_1-\mathbf{k}_3)^2} + \{1,2\}, \quad (4.284)$$

$$K^{(3)}_{\mathbf{k}_1\mathbf{k}_2\mathbf{k}_3} = (\tilde{\alpha}-\tilde{\beta})\,(\mathbf{k}_1\cdot\mathbf{k})\,(\mathbf{k}_2\cdot\mathbf{k}_3) - \tilde{\beta}\,(\mathbf{k}_1\cdot\mathbf{k}_2)\,(\mathbf{k}_3\cdot\mathbf{k})$$

$$+\tilde{J}_c k_2^2 \frac{(\mathbf{k}_2\times\mathbf{k}_3)\cdot(\mathbf{k}_1\times\mathbf{k})}{(\mathbf{k}_2+\mathbf{k}_3)^2} + \{2,3\}, \quad (4.285)$$

$$\tilde{\alpha} = \frac{24m_0^2}{35L}\left[L+1+\frac{i}{\tilde{\Omega}}\left(L^2+1-m_0\right)\right],\ \tilde{\beta} = -\frac{48m_0^2}{35}\frac{i}{\tilde{\Omega}},\ \tilde{J}_c = \frac{312m_0^2}{35P}. \quad (4.286)$$

Here the symbol $\{n,m\}$ ($\{n,m,l\}$) denotes all the remaining terms produced by permutation of n and m (n, m, and l). The permutations of two identical elements must not be taken into account; three identical wavevectors in $K^{(1)}_{\mathbf{k}_1\mathbf{k}_2\mathbf{k}_3}$ are obviously impossible in view of the resonant conditions, (4.91) and (4.92) with $s_2 = s_3 = 1$.

Below we analyze these equations for three different lattices: rhombic, square, and hexagonal; except for the latter case, both small-amplitude and finite-amplitude perturbations are considered.

(Hereafter the subscripts r and i denote the real and imaginary parts of corresponding expressions, respectively.) The rest of the coefficients of nonlinear coupling $C^{(1,3)}_{jlm}$ can be readily derived from $K^{(1)}_{\mathbf{k}_1\mathbf{k}_2\mathbf{k}_3}$ and $K^{(3)}_{\mathbf{k}_1\mathbf{k}_2\mathbf{k}_3}$; see (4.283) and (4.285); they are nonzero only if the resonant conditions are satisfied.

Rhombic Lattice

First, we discuss the weakly nonlinear analysis for a rhombic lattice [15], thus dealing with a small supercriticality: $\delta^2 = m_2 - m_2^{(o)} \ll 1$, where $m_2 = m_2^{(o)}(K)$ is the neutral stability curve given by (4.218). The growth rate is also expanded as $\gamma = \gamma_0 + \gamma_2\delta^2$, where $\gamma_0 = i\gamma_{0i}$ is the growth rate at the neutral stability curve, whereas $\gamma_2 = \partial\gamma/\partial m_2$ similar to (4.95). Finally, we introduce two time scales t_{40} and t_{42} following (4.95). Near the stability threshold, (4.102) and (4.103) are valid with the coefficients of nonlinear interaction given by (4.97); one only has to replace the coefficients $\alpha = \tilde{\alpha}K^4$ and $\beta = \tilde{\beta}K^4$ with those for Marangoni convection; see (4.286).

The results of pattern selection coincide with those presented in Section 4.1.1.4 for the Rayleigh–Bénard problem: ARs-R are selected on the rhombic lattices. This coincidence is expected, because of the same structure of the coefficients of nonlinear interaction; see (4.97). Stability conditions (D.9) are satisfied because $\alpha_r > 0$ and $\beta_r = 0$; the last condition (D.11) is met since $|\alpha/\beta| \sim L^{-1}$, which is numerically large for most of mixtures.

The pattern selection discussed is valid at $m_2 \approx m_2^{(o)}(K)$ for any K; dealing with the external perturbations we, however, have to fix K equal to the wavenumber of critical perturbations K_c, (4.220). The stability of small-amplitude ARs-R with respect to the external perturbations is guaranteed if the inequalities (4.144) and

(4.145) hold true. Similar to the buoyancy convection, the instability of ARs-R with $\theta > \pi/3$ takes place only in two cases: for $\theta - \pi/3 = O(L^2)$ and for $|\chi| = O(L^2)$, which are not considered below (see [15] for the details).

However, in contrast to the buoyancy convection, quadratic nonlinear terms enter (4.278), and those terms make the analysis of external perturbations for ARs-R more complicated. At small supercriticality, only the cross-interaction coefficients for the wave with perpendicular wavevectors are changed similar to the case analyzed in Section 4.1.2. Therefore, the perturbation of the form (4.136) with $\mathbf{p} = K_c(0,1)$ has to be considered separately. This analysis performed in [15] results in the following:

$$\text{Re}\left(\frac{2\mu_1\mu_2}{\Gamma - 2i\gamma_{0i}}\right) - \alpha_r(1 + 4\cos 2\theta) > 0, \qquad (4.287)$$

$$\left|\frac{2\mu_1\mu_2}{\Gamma - 2i\gamma_{0i}} - \alpha(1 + 4\cos 2\theta) - \beta\cos 2\theta\right|^2 - 4|\beta|^2\sin^4\theta > 0, \qquad (4.288)$$

where $\mu_{1,2} = \tilde{\mu}_{1,2}K^4$ are defined by (4.281). Here and below $\Gamma = \gamma\left(\sqrt{2}K\right)$ is the growth rate for the oblique wave at $m_2 = m_2^{(o)}(K)$. Computations show that both the conditions are met except for the limit of extremely small $|\chi| < L^2$.

The striking feature of the rhombic lattice is that (4.102) and (4.103) remain valid even for finite supercriticality; one only needs to restore the full growth rate γ instead of γ_2 and full time scale t_4 instead of t_{42} (the terms with t_{40} should be omitted). Of course, the rescaling of the amplitudes is not needed for finite supercriticality; therefore tildes for $A_{\pm\mathbf{k}_{1,2}^{(R)}}$ must be omitted in (4.101). Thus, ARs-R given by (4.106) (with the abovementioned replacements) are stable on the rhombic lattice.

However, analysis of external perturbations for finite-amplitude regimes changes significantly. Indeed, far from the stability threshold, there is no reason to consider perturbations of the fixed wavenumber K_c only; disturbances with all possible wavevectors must be analyzed. These calculations were carried out first in [25]; the details of the analysis are presented in Appendix E. In Section E.1 we perform the general analysis of perturbations; they are classified as either simple (see *Class IVa* there) or resonant ones. The former type of perturbations originates from the nonlinear term $|A_\mathbf{q}|^2 A_\mathbf{k}$, which enters (4.278) with the coefficients $K_{\mathbf{kqq}}^{(2)}$; therefore, each wave of the base solution contributes to the evolution of a single-wave perturbation. The development of resonant perturbations, which comprise at least two interacting waves, is more complex. Each wave of the base solution gives rise to two classes (*Classes I* and *II*) of resonant perturbations, which stem from the quadratic terms in (4.278); each pair of the base waves generates up to four classes (*Classes III–VI*) of perturbations. In Section E.2 further simplification of the resonant classes for the stability analysis of ARs-R is performed; seven resonant families are considered.

To analyze the evolution of simple perturbations, we add a small perturbation in the form

$$h' = b\exp\left[i\left(K\mathbf{p}\cdot\mathbf{R} - \tilde{\Omega}l^2 T\right)\right], \qquad (4.289)$$

to the base solution. Here $\mathbf{p} = (p, q)$ and $l^2 = p^2 + q^2$.

The self-interaction of this perturbation via the four base waves, which form ARs-R, results in

$$\dot{b} = \left(\gamma_1 - K_{pq} a^2 \right) b, \quad (4.290)$$

$$K_{pq} = 2(2\alpha + \beta) \left[p^2 + (p\cos\theta + q\sin\theta)^2 \right] + 4\beta l^2 + 2J_c F(p,q), \quad (4.291)$$

$$F = (l^4 - 1) \left[\frac{q^2}{(1+l^2)^2 - 4p^2} + \frac{l^2 - (p\cos\theta + q\sin\theta)^2}{(l^2+1)^2 - 4(p\cos\theta + q\sin\theta)^2} \right].$$

Here $J_c = \tilde{J}_c K^4$ and $\tilde{J}_c$ is defined by (4.286); K_{pq} is a sum of coefficients $K_{pqq}^{(2)}$ (see (4.284)), where $\mathbf{q}$ is equal to each of the base wavevectors $\pm\mathbf{k}_1^{(R)}$ and $\pm\mathbf{k}_2^{(R)}$.

It is worth noting that $F(p,q)$ is finite at $(p,q) = \pm(1,0)$ and $(p,q) = \pm(\cos\theta, \sin\theta)$, where $l^2 = p^2 + q^2 = 1$ and both numerator and denominator vanish for the corresponding term in $F(p,q)$. Therefore, the growth rate λ remains finite at any p and q. Its real part λ_r reads

$$\lambda_r = -4\alpha_r \left[p^2 + (p\cos\theta + q\sin\theta)^2 \right] a^2 - 2J_c F(p,q) a^2 + \mathrm{Re}\gamma_1. \quad (4.292)$$

We first analyze λ_r for large m_2; in this case $\gamma = \gamma^{(2)} m_2$, $\gamma_1 = \gamma^{(2)} m_2 l^2$, where $\gamma^{(2)} \approx \partial\gamma/\partial m_2$. It can be shown that ARs-R are stable at large m_2, if the Schmidt number exceeds a certain critical value:

$$S > S_c = \frac{26}{3g(\sin^2\theta)(1+L)}, \quad g(x) = \frac{3 - \sqrt{9 + 15x - 8x^2 - 16x^3}}{3x}. \quad (4.293)$$

The dependence of $S_c(1+L)$ on $\sin^2\theta$ is demonstrated in Figure 4.7. The threshold value grows monotonically as θ decreases from $\pi/2$ approaching the infinite value at $\theta = \pi/3$. For smaller θ ARs-R are unstable at any S. (Recall that they are also unstable to the perturbations of the wavenumber K; see (4.144).) This instability mode is caused by the emergence of a mean toroidal flow; indeed, J_c-term describes the influence of the toroidal potential $\tilde{\Psi}$ on the dynamics of perturbations.

The critical perturbations are longwave with $p \to 0$ and $q \to 0$, whereas λ_r depends on the ratio p/q only. The maximum real part of the growth rate is attained at

$$\left(\frac{p}{q} \right)_c = -\frac{\sin 2\theta \left[4 + 3g\left(\sin^2\theta \right) \right]}{6 \left[1 - \sin^2\theta\, g\left(\sin^2\theta \right) \right]}. \quad (4.294)$$

Because of the obvious symmetry reasons, the perturbations with counter-propagating waves have the same growth rate; therefore the angle between the X-axis and the critical wavevector can be chosen from the interval $[0, \pi)$. This angle increases from $2\pi/3$ $((p/q)_c = -1/\sqrt{3})$ to $3\pi/4$ $((p/q)_c = -1)$ as θ grows from $\pi/3$ to $\pi/2$. (In fact, at $\theta = \pi/2$ the longwave minimum does not depend on p/q any more.)

The single-wave ansatz (4.289) for the perturbation (and, hence, stability conditions (4.292)) is valid for *simple* perturbations only. There also exist seven single-parameter families of *resonant* perturbations, which are studied in detail in Ap-

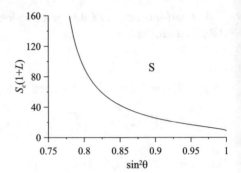

Fig. 4.7 The domain of stability for ARs-R (marked by "S") at large m_2; see (4.293)

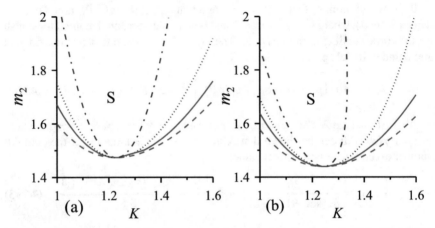

Fig. 4.8 Stability domains for ARs-R at $\chi = -0.1$ and $L = 0.01$ (marked with "S"). (a): $S = 100$. (b): $S = 20$. The neutral stability curves are shown by the dashed curves. The solid, dotted, and dashed–dotted curves correspond to $\theta = \pi/2$, $\theta = 0.4\pi$, and $\theta = 0.35\pi$, respectively

pendix E.2. At large m_2 these *resonant* modes are unimportant; they either decay or lead to the instability at the Schmidt number smaller than S_c given by (4.293).

At finite m_2, the stability conditions for the *simple* and *resonant* perturbations are investigated numerically. Numerical simulations show that the only resonant family which can compete with the *simple* perturbations is family (i). The results of computations are shown in Figure 4.8 for two values of S, $S = 100$ and $S = 20$, and three angles θ. The maximum value, $\theta = \pi/2$, should be thought of as the limit $\theta \to \pi/2$ in this figure. For $\theta = 0.35\pi$ the critical perturbations are *simple* along the whole corresponding curve. For larger θ the switching of the critical perturbations is observed at $K > K_c$: the right branches (at $K > K_c$) still correspond to *simple* perturbations, whereas the left branches are for the *resonant* perturbations that belong to family (i).

Figure 4.8 demonstrates the tendency, which is clear from (4.293): ARs-R become more stable as θ increases. It is seen that for $S = 100$ and $\theta = 0.4\pi$, 0.5π, ARs-R with any K become stable as m_2 grows; the nonlinear solution with $K = K_c$ is stable within the whole interval of its existence. For $\theta = 0.35\pi$, (4.293) predicts $S_c = 104.1$; therefore, this pattern is unstable at large m_2. However, this instability takes place at sufficiently large m_2 and cannot be shown on the scale of Figure 4.8(a). A similar situation takes place for $\theta = 0.4\pi$ at $S = 20$: (4.293) yields $S_c = 24.38$, but for $S = 20$ the right branch of the dotted line in Figure 4.8(b) turns to the left only at $K = 1.848$ and $m_2 = 6.657$.

To summarize, for values of the Schmidt number typical for most of binary liquids ($S > 100$), ARs-R are stable even at large reduced Marangoni number m_2, except for patterns with θ close to $\pi/3$. For sufficiently small values of S, ARs-R are stable only within a certain window of m_2; the stability region is adjacent to the minimum of the marginal stability curve.

Square Lattice

For a square lattice, the wavevectors, which enter the general solution (4.77) for h, can be enumerated as

$$\mathbf{k}_{nj} = n\mathbf{k}_1^{(S)} + j\mathbf{k}_2^{(S)},$$

where $\mathbf{k}_{1,2}^{(S)}$ are the base vectors for the square lattice; see Figure 4.2. The same renaming is applied to the amplitudes: $A_{\mathbf{k}_{nj}} \to A_{nj}$. The corresponding rewriting of (4.278) is obvious:

$$\dot{A}_{nj} = H_{nj}, \; H_{nj} = \gamma_{nj}A_{nj} - \sum_{g,l,p,q} A_{gl}\left(D_{glpq}^{(1)}A_{pq} + D_{glpq}^{(2)}A_{pq}^*\right)$$

$$- \sum_{g,l,p,q,r,s} A_{gl}\left(C_{glpqrs}^{(1)}A_{pq}A_{rs} + C_{glpqrs}^{(2)}A_{pq}A_{rs}^* + C_{glpqrs}^{(3)}A_{pq}^*A_{rs}^*\right), \qquad (4.295)$$

where $\gamma_{nj} \equiv \gamma(k_{nj})$ is the growth rate for the harmonics with the wavenumber of $k_{nj} = K\sqrt{n^2 + j^2}$; K is the wavevector of the entire pattern. We do not present the coefficients of nonlinear interaction, which can be easily extracted from (4.280)–(4.285); see [26] for the details.

The reduced form of resonant conditions (4.91) and (4.92) is also evident:

$$n = g + s_2p + s_3r, \; j = l + s_2q + s_3s, \qquad (4.296)$$

$$n^2 + j^2 = g^2 + l^2 + s_2\left(p^2 + q^2\right) + s_3\left(r^2 + s^2\right) \qquad (4.297)$$

with the same rules for the coefficients $s_{2,3}$ as explained below (4.278).

The weakly nonlinear analysis for a square lattice has been carried out in [15]; technically the procedure is similar to that discussed for patterns on a rhombic lattice with introduction of two time scales and expansion of the growth rate with respect to small $\delta^2 = m_2 - m_2^{(o)}$. Near the stability threshold, four amplitudes are the leading ones,

$$A_{\pm 1,0} = \delta A_{1,2}, \; A_{0,\pm 1} = \delta B_{1,2}, \tag{4.298}$$

and the other Fourier harmonics are small, $A_{jn} = O(\delta^2)$ for $|j| + |n| > 1$. Most of them are unimportant near the stability threshold, but the oblique waves play role (see Section 4.1.2, where the influence of the oblique waves is elucidated). Those amplitudes are determined by

$$A_{jn} = \frac{\mu_2 A_{j0} A_{0n}}{\Gamma - 2i\gamma_i^{(0)}} \sim \delta^2, \; |j| = |n| = 1, \tag{4.299}$$

and they provide the only distinction from the limiting case $\theta = \pi/2$ of the rhombic lattice. (Here and below $\mu_{1,2} = \tilde{\mu}_{1,2} K^4$ and $\Gamma \equiv \gamma_{11}$ as in (4.287) and (4.288).)

Truncated set of equations (4.108) and (4.109) is valid with the coefficients given by (4.97), and, similar to rhombic patterns, $\alpha = \tilde{\alpha} K^4$ and $\beta = \tilde{\beta} K^4$ (see (4.286)). The only replacement is for the coefficient $N_{\pi/2}$, which is now given by

$$N_{\pi/2} = \beta + \frac{\mu_1 \mu_2}{\Gamma - 2i\gamma_i^{(0)}}. \tag{4.300}$$

Similar to Section 4.1.2, the additional μ-term describes generation of the oblique waves and their interaction with the base waves. This is the only term caused by the quadratic interaction near the stability threshold.

Like in the case of the buoyancy convection, ARs are selected; this pattern is given by (4.112). This pattern and the remaining patterns from the list in Appendix D.1 emerge through the direct bifurcation. The stability of the selected pattern with respect to external perturbations is determined by (4.146) too; the computations show that ARs are stable.

In contrast to rhombic patterns, for squares at finite supercriticality, four-wave representation (4.298) is not valid anymore; all higher harmonics have to be taken into account. Hence, at finite $m_2 - m_2^{(o)}$ (4.295) can be solved only numerically; to that end we truncate the solution for h as follows: $A_{nj} = 0$ for $|n| + |j| > N$. Therefore, the order of the resulting complex-valued dynamical system is equal to $2N(N+1)$. (No interaction of the spatially uniform part of the solution A_{00} with the other Fourier harmonics occurs; this Fourier mode decays exponentially, and it is omitted below.) The convergence of the series for h is very fast, which allows setting $N = 10$ in the numerical simulations. Tests with N up to 30 were performed to control the accuracy, and they indicated no difference in the abovementioned results for $N > 10$. Estimation of A_{nj} and large n and j is presented in [26].

The definition of ARs should be revised because of the influence of higher harmonics; they are not a superposition of two standing waves shifted in phase by $\pi/2$ any further. The pattern which evolves continuously from ARs as $m_2 - m_2^{(o)}(K)$ grows has the following symmetry properties:

$$A_{nj} = A_{-nj} = A_{n,-j} = A_{-n,-j}, \tag{4.301}$$
$$A_{nj} = A_{jn} \text{ at } n + j = 2l, \tag{4.302}$$

$$A_{nj} = iA_{jn} \text{ at } j = 2l, \ n = 2g + 1, \tag{4.303}$$

where l and g are arbitrary integer numbers. In addition to this family, there also exist ARs with $A_{nj} = -iA_{jn}$ for even j and odd n; they give rise to the same patterns but in the inverted fast time τ.

It is remarkable that even for ARs at finite supercriticality, each Fourier harmonics oscillates in t_4 with a constant amplitude and the phase linearly growing with the time:

$$A_{nj} = a_{nj} \exp(i\phi_{nj}) \exp\left[-i\omega_{AR}\left(n^2 + j^2\right) t_4\right]. \tag{4.304}$$

The frequency ω_{AR} depends on m_2 when the remaining parameters are kept fixed; near the stability threshold $\omega_{AR} \approx -\gamma_{0i} + \omega^{(S)}\delta^2$ is linear in $\delta^2 = m_2 - m_2^{(o)}(K)$; see (4.114). By appropriate choice of the reference time, one of the phases can be set to zero; below we assume that $\phi_{01} = 0$.

Equation (4.304) means that the time variable t_4 can be easily excluded and ARs correspond to a fixed point within a certain reduced dynamical system. Indeed, comparing (4.77) with $\Omega = \tilde{\Omega}K^2$ and (4.304) one can readily see that ω_{AR} has a meaning of the nonlinear correction to the frequency Ω. Therefore the substitution of $\Omega + \varepsilon^2\omega_{AR}$ instead of Ω in (4.77) excludes the dependence of the amplitudes A_{nj} on the slow time variable t_4 for ARs. This is equivalent to replacement of the phases ϕ_{nj} with some linear combinations, for example, $\arg(A_{nj}) - (n^2 + j^2)\arg(A_{01})$. The resulting reduced real-valued dynamical system of order $4N(N+1) - 1$ can be derived from (4.295) by separation of the real and imaginary parts with A_{nj} replaced by $a_{nj}\exp(i\phi_{nj})$.

In fact, there exists a simpler way to study the fixed points corresponding to ARs. To that end, we substitute (4.304) into (4.295) and end up with a set of $2N(N+1)$ complex-valued equations which allows us to find the real amplitude a_{01}, the remaining $2N(N+1) - 1$ complex amplitudes $a_{nj}\exp(i\phi_{nj})$, and the frequency ω_{AR}. Solution to those transcendental complex-valued equations instead of a direct numerical integration of the evolutionary problem significantly speeds up the numerical computations. Taking into account the symmetry properties (4.301)–(4.303), we reduce this problem to a set of $N(N+4)/4$ (for even N) complex equations

$$-i\omega_{AR}(n^2 + j^2)A_{nj} = H_{nj}, \ n \geq j \geq 0, \ 0 < n + j \leq N. \tag{4.305}$$

In the computation of H_{nj} given by (4.295), we take into account the symmetry properties (4.301)–(4.303).

In addition to these low-cost computations, numerical solution of (4.295) is also performed in order to evaluate the stability of the obtained solutions on the fixed lattice. Tests with the wavenumber K replaced by K/p with an integer p are carried out to control the influence of subharmonics. These calculations have resulted in the emergence of ARs with a basic wavenumber Kq/p; the integer number q depends on both the set of problem parameters and initial condition.

An example of computations is presented in Figures 4.9 and 4.10. The former plot presents the evolution of the field h during a quarter of a period in the fast time τ. (The leading part of the longitudinal velocity $\mathbf{U}^{(0)}$ at any fixed Z is normal to isolines of h.) The snapshots are similar to that shown in Figure D.5 but do not coincide exactly: on one hand, at $\tau = 0$ ($\Omega\tau = \pi/2$), the solution is close to the system of rolls oriented along the X- (Y-) axis, whereas at $\Omega\tau = \pi/4$, it resembles a square pattern. On the other hand, owing to the presence of higher harmonics shifted in phase, the patterns do not exactly coincide with rolls or squares.

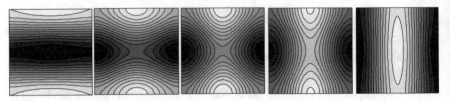

Fig. 4.9 Evolution of the field h for ARs in the fast time τ: $\Omega\tau = 0, \pi/8, \pi/4, 3\pi/8, \pi/2$ (from left to right). $S = 200$, $\chi = -0.01$, $K = 1.8$, $m_2 = 1.5$

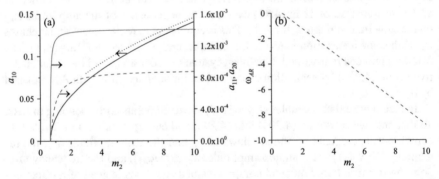

Fig. 4.10 Amplitude curves for ARs with $S = 200$, $\chi = -0.01$, $K = 1.8$. Panel (a): the moduli of the lowest Fourier amplitudes vs. the reduced Marangoni number m_2. The left vertical axis: $a = a_{10}$, the solid and dotted lines show the numerical results and the asymptotic behavior for large m_2, respectively. The right vertical axis: $c = a_{11}$ (the solid line) and a_{20} (the dashed line). Panel (b): The nonlinear correction ω_{AR} to the frequency vs. m_2

The dependence of the moduli of the lowest Fourier harmonics a_{10}, a_{11}, and a_{20} on the reduced Marangoni number m_2 is depicted in Figure 4.10(a). For the sake of brevity, here and below we use the notation $a \equiv a_{10}$ and $c \equiv a_{11}$ keeping in mind the symmetry properties of ARs, (4.301)–(4.303). It is clear that a_{nj} rapidly diminishes with the increase in $|n| + |j|$. At small supercriticality, the curves agree well with the results of the weakly nonlinear analysis, cf. (4.112).

The nonlinear correction $\omega_{AR}(m_2)$ to the frequency is displayed in Figure 4.10(b). It is clear that ω_{AR} is nearly linear in m_2 even far from the linear stability threshold.

It is noteworthy that (4.295) allows for the further analytical progress in the opposite limiting case, $m_2 \gg 1$. Simple observation yields

$$a_{nj} = O\left(\sqrt{m_2}\right) \text{ for } |n| + |j| = 1,$$

whereas the amplitudes of higher harmonics are either finite or decay as the inverse integer powers of $\sqrt{m_2}$. With this in mind, one can arrive at a simple set of ordinary differential equations for the leading complex amplitudes

$$\dot{A}_{10} = A_{10}\left[\gamma^{(1)}m_2 - N_0\left(|A_{10}|^2 + 2|A_{-1,0}|^2\right)\right.$$
$$\left. - \tilde{N}_{\pi/2}\left(|A_{01}|^2 + |A_{0,-1}|^2\right)\right] - K_{\pi/2}A_{01}A_{0,-1}A^*_{-1,0}, \qquad (4.306)$$

$$\dot{A}_{-1,0} = A_{-1,0}\left[\gamma^{(1)}m_2 - N_0\left(|A_{-1,0}|^2 + 2|A_{10}|^2\right)\right.$$
$$\left. - \tilde{N}_{\pi/2}\left(|A_{01}|^2 + |A_{0,-1}|^2\right)\right] - K_{\pi/2}A_{01}A_{0,-1}A^*_{10}, \qquad (4.307)$$

and a similar pair of equations with permutation of the first and the second subscripts. Here again $\gamma^{(1)} = \partial\gamma/\partial m_2$, whereas $\tilde{N}_{\pi/2} = \beta$. The difference between $N_{\pi/2}$ and $\tilde{N}_{\pi/2}$ is the only difference with the set of amplitude equations appropriate for small amplitudes.

Thus, the influence of the quadratic terms is absent at large m_2, and we return to the limit $\theta = \pi/2$ of rhombic pattern or to the square patterns for the Rayleigh–Bénard convection; see Section 4.1.1.4. In contrast, near the stability threshold, no such a reduction takes place since the quadratic terms are important.

The limiting solution for large m_2 is shown in Figure 4.10(a) by the dotted line. The discrepancy between the numerical and the analytical results is quite large, but it is mainly caused by the difference between γ_r and $\gamma_r^{(1)}m_2$. Substituting the full expression of the growth rate γ_r instead of $\gamma_r^{(1)}m_2$ in the analytical results for large m_2 yields a curve which cannot be distinguished from the solid line on the scale of Figure 4.10(a).

As shown for patterns on a rhombic lattice, the remaining patterns, TSs, StSs, TRs, and SRs (see Appendix D.1) are "supercritical" and unstable within (4.306)–(4.307). This means that no one amplitude grows infinitely in $\tilde{t}_4$, which verifies the truncation of the full dynamical system (4.295). Thus, these patterns are unstable both near the stability threshold and at large m_2.

However, as shown in [26], stable TSs exist at finite values of $m_2 - m_2^{(o)}$ and extremely small values of the Schmidt number, $S < 10$. It should be noted that TSs form a close solution based on three waves only, say, A_{10}, A_{01}, and A_{11}. These three complex amplitudes solve a set of transcendental equations, and they can be found analytically either at small or at large $m_2 - m_2^{(o)}$ only. Another interesting feature is that the stability exchange between ARs and TSs occurs via emergence of the so-called asymmetric pattern, a solution which demonstrates features of both standing and traveling patterns. (This solution can be roughly thought as a superposition of two counter-propagating TSs with different amplitudes and frequencies.) We do not discuss these patterns and bifurcations in detail just referring to the cited paper.

We finalize the analysis of square patterns discussing stability of ARs with respect to external perturbations. At large m_2 ARs represent just a particular case $\theta = \pi/2$ of ARs-R; moreover, the same is valid for the stability analysis with just a few exceptions. In particular, ARs are stable with respect to *simple* perturbations at

$$S > S_c = \frac{26}{3(1+L)},$$ (4.308)

which is a particular case when $\theta \to \pi/2$ of (4.293).

However, the critical perturbations differ for ARs on rhombic and square lattices, although in both cases they are longwave, $l^2 = p^2 + q^2 \ll 1$. For a rhombic lattice, the growth rate at small l depends on p/q, so that there exists a critical ratio $(p/q)_c$; see (4.294). In contrast, for on a square lattice, the leading term in λ_r is proportional to l^2 and, there exists no critical ratio p/q.

Most of the *resonant* perturbations are the same as those found on a rhombic lattice (see Appendix E.2) with only three exceptions. The detailed analysis of these perturbations is performed in Appendix E.3. A stability analysis with respect to these resonant perturbations does not provide instability conditions stronger than (4.308).

Returning to finite values of m_2, we restrict ourselves to the stability analysis of ARs with respect to *simple* perturbations only since a complete study of the resonant perturbations seems to be unrealistic. Indeed, even in the limit of large m_2, when the base solution comprises four waves only, there are many resonant perturbations. At finite m_2, the situation is complicated drastically: there exist two families of quadratic resonances for each wave in (4.77) and up to six families of cubic resonances for each pair of waves in the base solution.

For simple disturbances the perturbed field h has the following form

$$h = h_0 + h', \; h' = b(t_4)\exp\left[iK\left(pX + qY\right) - i\Omega l^2 \tau\right],$$ (4.309)

with h_0 corresponding to ARs, we obtain

$$\dot{b} = \left(\gamma_{pq} - \sum_{n,j} K_{nj}(p,q)a_{nj}^2\right) b,$$ (4.310)

$$K_{nj}(p,q) = (2\alpha + \beta)(np + jq)^2 + \beta(n^2 + j^2)l^2$$
$$+ J_c\frac{(pj - qn)^2\left(l^2 - n^2 - j^2\right)}{(n-p)^2 + (j-q)^2}.$$

Here $l^2 = p^2 + q^2$.

The real part of the growth rate λ_r

$$\lambda_r = \text{Re}\left(\gamma_{pq}\right) - \sum_{n,j}\text{Re}\left[K_{nj}(p,q)\right]a_{nj}^2,$$ (4.311)

$$\text{Re}\left[K_{nj}(p,q)\right] = 2\alpha_r(np + jq)^2 + J_c\frac{(pj - qn)^2\left(l^2 - n^2 - j^2\right)}{(n-p)^2 + (j-q)^2}$$

is calculated numerically; the equation $\lambda_r^{(max)} = 0$ is then solved, where $\lambda_r^{(max)}$ is the maximal value of λ_r with respect to the wavevector of perturbation $K(p,q)$.

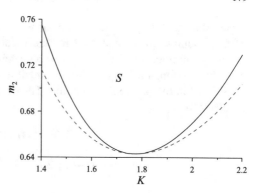

Fig. 4.11 Domain of stability of ARs (marked by "S") for $\chi = -0.01$, $S = 200$. The dashed line displays the neutral stability curve for the conductive state $m_2 = m_2^{(o)}(K)$. The solid line corresponds to simple external perturbations

A typical Busse balloon for ARs is presented in Figure 4.11; it is qualitatively similar to the domains shown in Figure 4.8 for ARs-R. In a wide parameter range, ARs are stable with respect to external perturbations starting from a certain critical value of m_2. A solution with the critical wavenumber K_c is stable for $m_2 > m_c$ in this case.

Hexagonal Lattice. Inclusion of Quintic Nonlinearity

As discussed in Section 4.1, for the Rayleigh–Bénard convection, the weakly nonlinear analysis on a hexagonal lattice is much more complicated than for either rhombic or square lattices. The entire three-parameter family of asynchronous hexagons (AHs) is selected within the standard equations with cubic nonlinearity. Quintic nonlinear terms have to be included to select the stable pattern within this family; such analysis, first conducted in [11], is the goal of the present subsection. The analysis of hexagonal patterns at finite supercriticality $m_2 - m_2^{(0)}$ has not been performed yet. Fortunately, quadratic nonlinearities are unessential for small-amplitude hexagonal patterns with the only exception; an additional external perturbation has to be considered as we shall explain below. Therefore, the results of Section 4.1.1.4 can be applied directly to the Bénard–Marangoni convection; one has only to replace α and β with those given by (4.286).

According to the results of Section 4.1.1.4, the selected family AHs is given by (4.131). The phase differences Φ, Ψ, and Δ, defined by (4.121)–(4.123), do not vary in the slow time t_{42}, but they can depend in the slower one $t_{44} = \delta^4 t_4$. Equations (4.275) and (4.276) are not used any further; therefore no conflict of notations is possible for Φ and Ψ; within the current section, they state for the phase differences only.

The stability analysis with respect to external perturbations carried out in Section 4.1.1.4 also remains valid, although for the Marangoni convection it has to be extended. Inequalities (4.149) and (4.150) are necessary but not sufficient conditions, because the external perturbation with the wavevector orthogonal to one of the base wavevectors, say to $\mathbf{k}_1^{(H)}$, has to be considered separately. Indeed, for such a perturbation, the quadratic nonlinearity becomes important and leads to generation of oblique waves and their influence back on the perturbation. This can be taken into account by the appropriate replacement of the coefficient of nonlinear cross-interaction, similarly to that for small-amplitude patterns on a square lattice considered in the previous subsection. The corresponding additional stability conditions read:

$$\text{Re}W_\perp > 0, \ |W_\perp| > 3|\beta|a^2, \tag{4.312}$$

$$W_\perp = (2\alpha + \beta)(2S_A - 3a^2) + \frac{4\mu_1\mu_2 a^2}{\Gamma - 2i\gamma_{0i}},$$

where $\Gamma = \gamma\left(\sqrt{2}k\right)$ similar to that for square lattice, $S_A = a^2 + b^2 + c^2$. It is clear that the same inequalities should be valid with a^2 replaced by either b^2 or c^2. It should be noted that conditions (4.312) guarantee stability of AHs on a superlattice, the superposition of two hexagonal lattices with the angle $\pi/2$ between the base wavevectors.

It can be shown analytically that (4.149) is always met; the first condition in (4.312) is satisfied if ARs (on a square lattice) branch supercritically, which is the case. Condition (4.150) and the second inequality in (4.312) were studied numerically in [11]; it was shown that a three-parameter family of AHs is stable with the following two exceptions:

(i) AH close to ARs-R (either $a = 0$ or $b = 0$ or $c = 0$);
(ii) All AHs are unstable for small values of the Soret number $|\chi| = O(L^2)$.

To finalize the discussion of the amplitude equation at the third order in δ, we describe the emergence of $O\left(\delta^3\right)$ nonlinear terms in h. There are two types of such contributions. The first one takes place for any lattice: the solution to the third-order problem includes $A_{1,2}^{(2)}$ ($B_{1,2}^{(2)}$ and $C_{1,2}^{(2)}$), which are $O(\delta^2)$-corrections to the base amplitudes $A_{1,2}$ ($B_{1,2}$ and $C_{1,2}$), respectively. (Recall that the base amplitudes contribute to h with the small factor δ; therefore $A_{1,2}^{(2)}$ and the others provide δ^3 terms in h.) These corrections solve the equations

$$\partial_{t40}A_{1,2}^{(2)} = \gamma_0 A_{1,2}^{(2)}$$

and the similar equations for the remaining four amplitudes. Since γ_0 is purely imaginary, they oscillate in the fastest time like the base waves do.

The second type of solution is intrinsic to a hexagonal lattice only. It is clear that the wave set with the wavevectors $\mathbf{k}_1 = \mathbf{k}_2 = \mathbf{k}_2^{(H)}$, $\mathbf{k}_3 = \mathbf{k}_1^{(H)}$, and $\mathbf{k} = \sqrt{3}K_c(0,1) = \mathbf{k}_2^{(H)} - \mathbf{k}_3^{(H)}$ satisfies the resonant conditions (4.91) and (4.92), with $s_2 = s_3 = 1$. (The base vectors of the lattice are determined by (4.100); see Figure 4.2.) The same wave is also generated via the interaction of waves with $\mathbf{k}_1 = \mathbf{k}_2 = -\mathbf{k}_3^{(H)}$, $\mathbf{k}_3 = -\mathbf{k}_1^{(H)}$ for the same s_2 and s_3.

Totally, at the third order, six additional waves are generated in such a way; they are shown in Figure 4.12. Those waves do not contribute to the evolution of leading amplitudes $A_{1,2}$, $B_{1,2}$, and $C_{1,2}$ within the cubic truncation, but they are important for generation of quintic nonlinear terms.

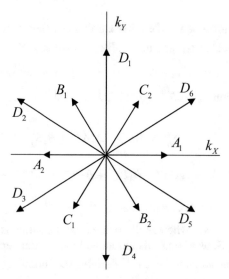

Fig. 4.12 Wavevectors of interacting perturbations on the hexagonal lattice

The evolution of D_1 is governed by the equation

$$\partial_{t40}D_1 = \Gamma_3 D_1 - K_{D1}\left(A_1 B_1^2 + A_2 C_2^2\right),$$

where $\Gamma_3 = \gamma_0\left(\sqrt{3}k\right)$ and $K_{D1} = 3\alpha/2$. The steady solution for D_1 is obvious

$$D_1 = \frac{K_{D1}\left(A_1 B_1^2 + A_2 C_2^2\right)}{\Gamma_3 - 3\gamma^{(0)}}. \tag{4.313}$$

The similar equations for the remaining five amplitudes can be easily obtained by obvious symmetry transformations.

Proceeding to the fifth order in δ, we take into account the interactions of those six additional waves with the base ones, which contribute to evolution of the base waves. As an example, one can mention the interaction of the waves with the amplitudes D_1, B_1, and B_1, which contribute in the slow evolution of A_1; therefore $\mathbf{k}_1 = \sqrt{3}K_c(0,1)$, $\mathbf{k}_{2,3} = \mathbf{k}_2^{(H)}$, and $\mathbf{k} = \mathbf{k}_1^{(H)}$ with $s_2 = s_3 = -1$ in (4.91) and (4.92). Accounting for all interactions of such kind (with the coefficient of nonlinear interaction being $K_{D2} = -3\beta/4$), we arrive at the following set of equations in the fifth order:

$$\partial_{t42}A_1^{(2)} = \hat{L}_{A1} - \dot{A}_1 - \kappa\left(c^4 + b^4\right)A_1 - \kappa\left[(C_1^* B_2)^2 + (B_1^* C_2)^2\right]A_2, \tag{4.314}$$

$$\partial_{t42}A_2^{(2)} = \hat{L}_{A2} - \dot{A}_2 - \kappa\left(c^4 + b^4\right)A_2 - \kappa\left[(C_2^* B_1)^2 + (B_2^* C_1)^2\right]A_1, \tag{4.315}$$

$$\hat{L}_{A1} \equiv -2A_1 \left[\sigma \mathrm{Re} \left(A_1^* A_1^{(2)} - A_2^* A_2^{(2)} \right) \right] - 2K_1 S_A^{(2)} A_1 - K_2 A_2^* \hat{\Sigma}^{(2)},$$

$$\hat{L}_{A2} \equiv 2A_2 \left[\sigma \mathrm{Re} \left(A_1^* A_1^{(2)} - A_2^* A_2^{(2)} \right) \right] - 2K_1 S_A^{(2)} A_2 - K_2 A_1^* \hat{\Sigma}^{(2)},$$

with similar four equations for the other amplitudes. Henceforth a dot denotes the derivative with respect to t_{44}, and the coefficients are $\kappa \equiv K_{D1} K_{D2} / \left(\Gamma_3 - 3\gamma^{(0)} \right) = -\frac{9}{8} \alpha \beta / \left(\Gamma_3 - 3\gamma^{(0)} \right)$, $\sigma \equiv N_0 - N_{\pi/3} = \frac{1}{2}\alpha - \frac{1}{4}\beta$. For the sake of simplicity, we also use the expansions $S_A = S_A^{(0)} + \delta^2 S_A^{(2)}$, $\hat{\Sigma} = \delta^2 \hat{\Sigma}^{(2)}$, where

$$S_A^{(2)} = \mathrm{Re} \sum_{j=1}^{2} \left[A_j^* A_j^{(2)} + B_j^* B_j^{(2)} + C_j^* C_j^{(2)} \right], \qquad (4.316)$$

$$\hat{\Sigma}^{(2)} = A_1 A_2^{(2)} + A_2 A_1^{(2)} + B_1 B_2^{(2)} + B_2 B_1^{(2)} + C_1 C_2^{(2)} + C_2 C_1^{(2)}. \qquad (4.317)$$

We represent the complex amplitudes as $A_j^{(2)} = a_j^{(2)} \exp \left(i\phi_{Aj} \right)$ $(j = 1, 2)$, where the $a_j^{(2)}$ are yet unknown, whereas the phases are the same as for the cubic truncation; see (4.120). Substituting this ansatz and the similar representations for the other amplitudes, one can eliminate $a_j^{(2)}$ deriving the following set of equations for the phases

$$\dot{\Phi} = 2 \frac{c^2}{S_A^{(0)}} \mathrm{Im} \left(F e^{-i\Psi} \right) (1 - 2\cos\Delta), \quad (4.318)$$

$$\dot{\Psi} = -2 \frac{b^2}{S_A^{(0)}} \mathrm{Im} \left(F e^{-i\Phi} \right) (1 - 2\cos\Delta), \quad (4.319)$$

$$\dot{\Delta} = -2Q \left(a^4 + b^4 + c^4 \right) \sin\Delta, \quad (4.320)$$

$$F = \kappa \left[\left(b^4 - a^4 \right) \left(1 - e^{i\Psi} \right) - \left(c^4 - a^4 \right) \left(1 - e^{i\Phi} \right) \right], \quad Q = \frac{\mathrm{Re}(\sigma \kappa^*)}{\sigma_r}.$$

The amplitudes a, b, and c depend on the slowest time t_{44} only via the phases according to (4.131) because the relaxation of the amplitudes in t_{42} is "fast" and (4.131) is valid at any value of t_{44}. It is important that (4.318)–(4.320) are valid only if the amplitudes a, b, and c are finite. The cases of ARs-R and close patterns with, e.g., $a \ll \min(b,c)$ should be analyzed separately. To shorten the formulas, we set $S_A^{(0)} = 1$ below, which can be achieved by rescaling of the parameter δ: $\delta^2 S_A^{(0)} \to \delta^2$. Therefore, it is assumed $a^2 + b^2 + c^2 = 1$ below.

Dynamical system (4.318)–(4.320) has five fixed points, which correspond to limit cycles within the full set of equations. These fixed points are:

(i) and (ii) symmetric patterns, TwRs and WRs2 (see Appendix D.2), with $\Phi = -\Psi = \Phi_0 = \pm 2\pi/3$, i.e., $a = b = c$. Recall that $\Delta = 0$ for TwRs and $\Delta = \pi$ for WRs2.

(iii) Alternating rolls (rhombic) (ARs-R): $\Phi - \Psi = \pi$, i.e., $a = 0$, $b = c \neq 0$ (ARs-A); see Appendix D.2 as well. There exist two similar patterns with either $b = 0$ (ARs-B) or $c = 0$ (ARs-C).

(iv) and (v) oscillating hexarolls (OHRs) with $\Phi = -\Psi \to \pi$, i.e., $a^2 = 2b^2 = 2c^2$ and either $\Delta = 0$ (OHR1) or $\Delta = \pi$ (OHR2). The snapshots for these patterns are shown in Figure 4.5. Two similar pairs of patterns exist with either b or c being the dominant amplitude.

Linear stability analysis within (4.318)–(4.320) yields the following growth rates for these patterns:

(i) TwRs:

$$-\frac{2}{9}\kappa, \quad -\frac{2}{9}\kappa^*, \quad -\frac{2}{3}Q;$$

(ii) WRs2:

$$\frac{2}{3}\kappa, \quad \frac{2}{3}\kappa^*, \quad \frac{2}{3}Q;$$

(iii) ARs-R:

$$\frac{1}{2}\kappa_r, \quad -\frac{3}{2}\kappa_r, \quad \pm Q;$$

(iv) OHR1:

$$-\frac{3}{16}\kappa_r, \quad 0, \quad -\frac{3}{4}Q;$$

(v) OHR2:

$$\frac{9}{16}\kappa_r, \quad 0, \quad \frac{3}{4}Q.$$

The growth rates for TwRs and WRs2 conform with the stability conditions derived in [13]. (Recall that in the cited paper, the only degeneracy was that with respect to Δ.) Nonlinear stability analysis can be performed for zero growth rates for OHR1 and OHR2; both those patterns are unstable, and they are not discussed any further.

Note, that ARs-R are not even a fixed point within (4.318)–(4.320) because one or two phases of Φ and Ψ grow linearly with the time. However, combining these equations, one can derive the conditions presented. In addition, genuine ARs-R do not depend on Δ at all, but a pattern close to ARs-R does; $\Delta = 0$ and $\Delta = \pi$ correspond to stable and unstable manifolds for ARs-R as we shall discuss later.

The stability of the patterns is determined by the signs of κ_r and Q only. If $\kappa_r Q > 0$, one of the symmetric patterns is stable, and the system approaches the sole stable solution. For $\kappa_r < 0$ and $Q < 0$, WRs2 are stable, whereas for $\kappa_r > 0$ and $Q > 0$, TwRs are stable. For Marangoni convection in a binary liquid, $\kappa_r < 0$, whereas Q changes its sign; consequently, WRs2 are stable within a certain domain of the problem parameters; see Figure 4.13.

The dynamics is more interesting if $\kappa_r Q < 0$ (see Figure 4.13), when no stable solution exists. It can be easily shown that for $\kappa_r < 0$ and $Q > 0$ the stable manifolds for ARs-R, WRs2, and TwRs are $\Delta = 0$, $\Delta = \pi$, and $\Phi = -\Psi = \Phi_0$, respectively, whereas the unstable manifolds are $\Delta = \pi$, $\Phi = -\Psi = \Phi_0$, and $\Delta = 0$, respectively. Therefore, the heteroclinic connection ARs-R $\to$ WRs2 $\to$ TwRs $\to$ ARs-R,

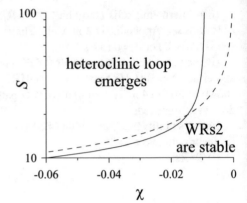

Fig. 4.13 Stability boundaries S vs. χ for WRs2 ($Q < 0$) and heteroclinic loop ($Q > 0$); $L = 0.01$ (solid line) and $L = 10^{-5}$ (dashed line)

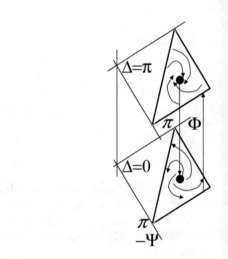

Fig. 4.14 Scheme of the heteroclinic connection. Condition (4.132) is met inside the triangles. Solid thick lines (sides of the triangles) correspond to ARs-R, the lower central point to TwRs, the upper one to WRs2

sketched in Figure 4.14, is possible since for each fixed point the stable manifolds coincide with the unstable manifold of the preceding fixed point. Hence, the necessary stability condition formulated in [27] is valid for this heteroclinic cycle.

To apply the sufficient condition of the stability of a heteroclinic cycle [28], we calculate the product of the real parts of the positive (unstable) eigenvalues over all saddle points of the heteroclinic cycle and the similar product of the leading negative (stable) eigenvalues. These two products are equal to each other, so that the system is exactly at the stability boundary, and a more delicate analysis is needed. This coincidence means that perturbations grow/decay with the number of the cycle slower than superexponentially.

Numerical simulations show that this heteroclinic loop is indeed attracting; however, we have to describe briefly the evolution of patterns close to ARs-R before discussing the numerical results. As stated above, set (4.318)–(4.320) becomes incomplete, when one of the amplitudes, say a, becomes small and therefore $\Phi - \Psi \approx \pi$. According to these equations, the evolution toward ARs-A is governed by formulas

$$a^2 = a_-^2 \exp\left[\frac{1}{2}\kappa_r(\tau_2 - \tau_-)\right], \tag{4.321}$$

$$\Delta = \Delta_- \exp[-Q(\tau_2 - \tau_-)], \tag{4.322}$$

$$\Phi = \frac{1}{2}\kappa_i(\tau_2 - \tau_-) + \Phi_-, \tag{4.323}$$

whereas the dynamics along the unstable manifold is determined by formulas

$$a^2 = a_+^2 \exp\left[-\frac{3}{2}\kappa_r(\tau_2 - \tau_+)\right], \tag{4.324}$$

$$\Delta = \pi + \Delta_+ \exp[Q(\tau_2 - \tau_+)], \tag{4.325}$$

$$\Phi = -\frac{3}{2}\kappa_i(\tau_2 - \tau_+) + \Phi_+. \tag{4.326}$$

Here $a_\pm$, $\Delta_\pm$, $\Phi_\pm$, and the time moments $\tau_\pm$ are arbitrary at this stage.

A separate analysis is needed to describe, how one of the asymptotics matches another one. Referring to [11] for the details, we present here only the idea of such analysis. In order to describe the "small" waves, the rescaled amplitudes $\tilde{A}_j = \delta^{-1}A_j \exp(i\omega_1 t_{42} + i\omega_2 t_{44})$ ($j = 1, 2$) are introduced; those amplitudes evolve in the slow time according to the following set of equations:

$$\dot{\tilde{A}}_1 = -\frac{\kappa}{4}(\tilde{A}_1 + 2\tilde{A}_2) + \sigma\left(|\tilde{A}_2|^2 - |\tilde{A}_1|^2\right)\tilde{A}_1, \tag{4.327}$$

$$\dot{\tilde{A}}_2 = -\frac{\kappa}{4}(\tilde{A}_2 + 2\tilde{A}_1) + \sigma\left(|\tilde{A}_1|^2 - |\tilde{A}_2|^2\right)\tilde{A}_2. \tag{4.328}$$

The solution of this set of equations can be matched with both asymptotics, (4.321)–(4.323) and (4.324)–(4.326). Thus, $|\tilde{A}_{1,2}|$ decay with $|\Delta| \ll 1$, which is equivalent to $\tilde{A}_1 \approx \tilde{A}_2$, then after a certain relaxation process, both amplitudes start growing with $\Delta \approx \pi$ (or $\tilde{A}_1 \approx -\tilde{A}_2$). In fact, the evolution within (4.327) and (4.328) can be tracked solving five sets of linear ordinary differential equations, which are matched one by one. This matching allows one to construct the mapping

$$\{a_-, \Delta_-, \Phi_-\} \to \{a_+, \Delta_+, \Phi_+\}$$

and, hence, it provides the initial condition for (4.318)–(4.320) in form (4.324)–(4.326) at $\tau = \tau_+$ for known a_-, Δ_-, and Φ_- at $\tau = \tau_-$. The analysis carried out in [11] shows that

$$a_+ = a_-, \quad \Delta_+ = \eta\Delta_-^{1/3}, \quad \Phi_+ = \Phi_- + \varphi, \tag{4.329}$$

where η and φ are constants found numerically. It is interesting, that these constants depend on $\ln \delta$; although δ is asymptotically small and the solution is sought as a series in δ, its logarithm also determines the evolution. Numerical solution shows that typical value of η is less than unity within the parameter domain, where the heteroclinic loop takes place. This characteristic fluctuates with from one cycle to another depending on subtle details, but the fluctuations remain much smaller than the mean value.

We are now at the point to discuss the numerical results for (4.318)–(4.320) matched with (4.327) and (4.328). An example of elementary loop of the heteroclinic cycle WRs2 $\rightarrow$ TwRs $\rightarrow$ ARs-R $\rightarrow$ WRs2 is demonstrated in Figure 4.15.

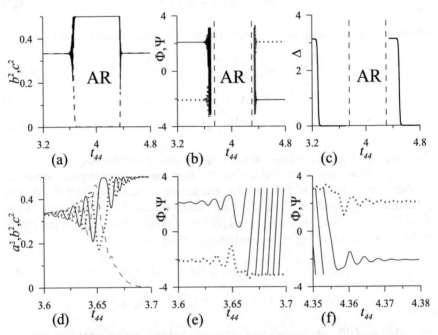

Fig. 4.15 Evolution within a single loop of the heteroclinic connection for $\chi = -0.05, L = 0.01, S = 200$. During the period marked with "AR," the system evolves according to (4.327) and (4.328). (a) Dashed and solid lines correspond to $b^2(t_{44})$ and $c^2(t_{44})$, respectively; (b) $\Phi(t_{44})$ and $\Psi(t_{44})$ shown by the solid and dashed lines, respectively; (c) $\Delta(t_{44})$. Details of the evolution near ARs-B are shown in panels (d), (e), and (f): (d) the squared amplitudes a^2, b^2, and c^2 shown by the dotted, dashed, and solid lines, respectively; (e) and (f) phase differences Φ and Ψ shown by the solid and dotted lines, respectively

The system is alternately attracted to and then repelled from each of three unstable fixed points. Starting from a pattern close to WRs2 at $t_{44} = 3.2$, the system approaches TwRs: Δ decreases suddenly; see Figure 4.15(c). After that, the differences in the amplitudes a^2, b^2, and c^2 start growing as seen in Figure 4.15(a) and (d). At the final stage of this process, b becomes small, and one of the phases is almost constant, $\Psi \approx -\pi$, whereas another one, Φ, varies almost linearly with t_{44}; see Figure 4.15(e). Hence, the system visits ARs-B, where the phases Φ and Δ and the smallest amplitude b^2 evolve as it has been described above: b diminishes at $\Delta \ll 1$, and then Δ approaches π fast, and b starts to grow. The system leaves the vicinity of ARs-R along its unstable manifold; at this stage Φ varies almost linearly with t_{44} (Figure 4.15(f)). At $t_{44} \approx 4.37$, the vicinity of WRs2 is reached: $a^2 \approx b^2 \approx c^2$ ($\Psi = -\Phi = 2\pi/3$) and $\Delta \approx \pi$, which finalizes the loop.

Such evolution, however, does not form a regular cycle because of two reasons: first, any ARs-R pattern among the three (ARs-A, ARs-B, or ARs-C) can be reached at the current loop. Second, ARs-R and WRs2 are the saddle-foci, i.e., the details of the attraction to these points are different from one cycle to another. Moreover, the observed dynamics at large times strongly depends on the accuracy of computations, and the model must be further improved to guarantee the needed accuracy. To that end a new variable, $\xi \equiv \ln |\tan(\Delta/2)|$ is introduced, which allows transforming (4.320) into

$$\dot{\xi} = -2Q\left(a^4 + b^4 + c^4\right). \tag{4.330}$$

Additionally, close to symmetric AHs, a complex variable $v \equiv \ln \zeta$ is introduced, where $\zeta \equiv \Phi - \Phi_0 + e^{i2\pi/3}(\Psi + \Phi_0)$. In terms of ζ (4.318) and (4.319) read

$$\dot{v} = \frac{2}{9}\kappa(1 + 2\tanh\xi). \tag{4.331}$$

The system is interpreted to be close to symmetric AHs at $\tau_-^{(s)} < \tau_2 < \tau_+^{(s)}$, when $|\zeta| < \varepsilon$ ($\varepsilon = 10^{-6}$ in the computations); otherwise, (4.318) and (4.319) are integrated along with (4.330). An introduction of these two variables considerably increases the accuracy of computations because an exponential decay of Δ (or $\pi - \Delta$) and $\Phi - \Phi_0$, $\Psi + \Phi_0$ with t_{44} corresponds to the linear variation of ξ and v, respectively.

The results of computations are presented in Figure 4.16. The figure and many other numerical tests, which are not presented here, demonstrate that at large t_{44}, only one of ARs-R is observed. For instance, under initial conditions chosen, ARs-A is found, i.e., $a \ll 1$ at the ARs-R stages of the heteroclinic loop. It is also worth noting that the duration of each cycle grows. (As shown in [11], both these features are lost because of numerical noise, when untransformed equations (4.318)–(4.320) are solved.) In Figure 4.17(a) the duration of stay τ_N in a vicinity of an unstable fixed points vs. the cycle number N is depicted; τ_N increases almost linearly with N for both ARs-R and symmetric hexagons. As we have noted above, the heteroclinic

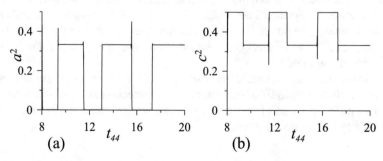

Fig. 4.16 Evolution of the squared amplitudes in slow time τ_2 for $\chi = -0.05$, $L = 0.01$, $S = 200$: $a^2(t_{44})$ and $c^2(t_{44})$ are shown in panels (a) and (b), respectively

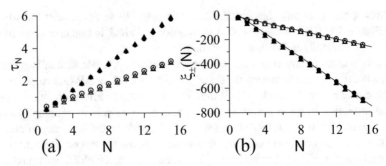

Fig. 4.17 Characteristics of the Nth cycle for $\chi = -0.05$, $L = 0.01$, $S = 200$. Circles and triangles correspond to $\delta = 10^{-6}$ and $\delta = 10^{-4}$, respectively. (a): Duration of an ARs-R state, $\tau_+(N) - \tau_-(N)$ (open symbols), and symmetric AHs $\tau_+^{(s)}(N) - \tau_-^{(s)}(N)$ (filled symbols) depending on N. (b): Representation $\xi_-(N)$ (filled symbols) and $\xi_+(N)$ (open symbols). The lines depict the asymptotic expression corresponding to the global maps given by equation (4.332)

loop under consideration realizes the stability boundary according to [28]; therefore, the period grows slower than exponentially.

In Figure 4.17(b) the variation of $\xi_\pm(N) = \mp\xi(\tau_\pm)$ is also depicted. It is clear from this figure that $\Delta_\pm$ becomes very small after first several cycles; therefore, during each loop, the system is closer to the unstable fixed points. This shows (along with the increase in τ_N) that the vector flow corresponding to (4.318)–(4.320), (4.327), and (4.328) is contracting in the vicinity of the heteroclinic cycle. To prove this more rigorously, the global map for the heteroclinic connection was constructed in [11]. It was shown that $\Delta_\pm$ at $(N+1)$th loop could be expressed via the same quantities at Nth loop as follows:

$$\Delta_-^{(N+1)} = \frac{\eta^3}{4}\Delta_-^{(N)}, \ \Delta_+^{(N+1)} = \frac{\eta}{4^{1/3}}\Delta_+^{(N)}, \tag{4.332}$$

where η is the factor which enters (4.329); recall that this value is rather small, and, therefore, the heteroclinic loop is stable (attracting). Mapping (4.332) is juxtaposed with the numerical results in Figure 4.17(b) and a perfect agreement is found.

Finally, the effect of δ is demonstrated in Figure 4.17. (Recall, that $\delta \ll 1$, but $\ln\delta$ determines the dynamics near ARs-R.) Variation of δ changes drastically each loop of evolution (the duration of the cycle increases for smaller δ), but the linear dependencies of τ_N and $\xi_\pm(N)$ on N remain almost the same, while δ increases by two orders of magnitude.

The particular problem considered above provides an excellent example demonstrating methods of analysis of oscillatory pattern selection and the difficulties that appear on different stages of that analysis. Below we consider some generalizations and modifications of that problem.

4.2.2 Deformable Surface

4.2.2.1 Finite *Bi*

Marangoni convection in binary mixtures in the presence of a surface deformation was first addressed by Castillo and Velarde [29, 30]. Similar to the case of a pure liquid, in both cases of independent temperature and concentration gradients [31] and of the concentration gradients induced by the Soret effect [32, 33], it was found that for sufficiently small values of the Bond number $Bo = Ga/Ca$ (i.e., in very thin layers or under reduced gravity) the neutral stability curve has an additional minimum at zero wavenumber. A similar result was obtained in the general case of a multicomponent fluid [34].

Depending on the system parameters, both monotonic and oscillatory instabilities are possible. For instance, Bhattacharjee [32] considered the Marangoni instability in binary liquid mixtures when the temperature of the substrate is fixed. In the longwave limit, the monotonic instability was found in the case of small Biot number. The oscillatory instability was found in the limit of a small Lewis number in a definite interval of the Soret number but only for finite wavelength. Joo [33] analyzed the stability of binary liquid mixture heated from above, when the instability is driven by solutocapillary effect, while the thermocapillary effect is stabilizing. In a contradistinction to the opposite case (when thermocapillarity is destabilizing and solutocapillarity is stabilizing), only monotonic instability is possible.

Generally, for finite Ga, Ca, and Bi, the longwave instability in a binary liquid layer is controlled by two conservation laws: (i) conservation of the liquid volume and (ii) conservation of the solute mass. Mixing of the corresponding slow modes may create a longwave oscillatory instability.

Podolny et al. [35, 36] developed a strongly nonlinear theory of the longwave instability in the case of a prescribed heat flux on the bottom rigid plate and an arbitrary Bi at the free surface. The following rescaling of parameters was employed:

$$\xi = \varepsilon x, \ \tau = \varepsilon^4 t, \ G = \varepsilon^2 g, \ L = \varepsilon^2 l, \ M = \varepsilon^2 m. \tag{4.333}$$

This rescaling allows to derive a set of longwave evolution equations in terms of the film thickness h and the disturbance of the total concentration Ψ defined as

$$\Psi = \int_0^h (C - \bar{C}) dz. \tag{4.334}$$

Here the local film thickness h is nonnegative, while Ψ can be of either sign. The technique of longwave expansions allows to derive a closed system of strongly nonlinear equations governing the spatiotemporal evolution of the fields h and Ψ,

$$\frac{\partial h}{\partial \tau} + \frac{\partial}{\partial \xi} Q_1(h, \Psi) = 0, \ \frac{\partial \Psi}{\partial \tau} + \frac{\partial}{\partial \xi} Q_2(h, \Psi) = 0, \tag{4.335}$$

where

$$Q_1(h, \Psi) = -4h^3 \left(g\frac{\partial h}{\partial \xi} - Ca\frac{\partial^3 h}{\partial \xi^3} \right) + 3m\chi h^2 \frac{\partial h}{\partial \xi} - 6m\Psi \frac{\partial h}{\partial \xi} + 6mh\frac{\partial \Psi}{\partial \xi}, \quad (4.336)$$

$$Q_2(h, \Psi) = \frac{1}{24Ph} \left\{ -h^3(8\Psi + \chi h^2) \left(g\frac{\partial h}{\partial \xi} - Ca\frac{\partial^3 h}{\partial \xi^3} \right) + \right.$$

$$\left[-12m\Psi^2 + 4(6l + m\chi h^2)\Psi + m\chi^2 h^4 - 12l\chi h^2 \right] \frac{\partial h}{\partial \xi} - \quad (4.337)$$

$$\left. 2h(-6m\Psi - m\chi h^2 + 12l)\frac{\partial \Psi}{\partial \xi} \right\}$$

are the flow rate and the solute flux, respectively. The identical equations are derived when G, L, and M are $O(1)$, but $Ca = O(\varepsilon^{-2})$.

Predictions of the weakly nonlinear theory developed in [35] have been verified in [37] by means of numerical simulations. The latter paper contains also numerical simulations of finite-amplitude flow regimes, which include traveling waves, standing waves, quasiperiodic regimes, and chaotic regimes.

A generalization of equations (4.336), (4.337) in the case where the Biot number at the rigid plane is also arbitrary, was done in [38]. Regions of stability of different patterns and those of subcritical instability were determined.

4.2.2.2 Small Bi

The case of deformable surface and small Bi is quite complex even in the framework of the linear stability theory. The analysis fulfilled for a purely Marangoni instability [21] has shown that there are two distinguished regions of wavenumbers, $k \sim Bi^{1/4}$ and $k \sim Bi^{1/2}$, which have to be investigated separately. In the former region, both thermocapillary and solutocapillary effects are significant, while in the latter region, the solutocapillary effect prevails. That circumstance creates a rather complex picture of neutral curves for monotonic and oscillatory instabilities. A generalization of the theory for combined Marangoni and Rayleigh instabilities, which was carried out in [22], revealed a diversity of instability types in the longwave region.

4.2.3 *The Limit of Small Lewis Number*

The theory presented above was developed for *finite* values of the Lewis number L. In fact, that is equivalent to the assumption of $Bi \ll L$. However, most of liquid binary mixtures are characterized by small values of the Lewis number, i.e., the typical characteristic time of the diffusion is larger than that of the heat transfer processes. An analysis of the case of *asymptotically small* Lewis number can be especially important in the context of particle-laden fluids. A binary mixture is the simplest model of a nanofluid, with Soret effect being an analogue of the thermophoresis [39]. The diffusion coefficient and hence the Lewis number are remarkably small for nanoparticles. Some experiments [40–43] show that the surface tension depends on the concentration of particles; therefore an analogue of a solutocapillary effect takes place.

The analysis of both longwave and shortwave instability modes for asymptotically small L has been carried out in [18]. For $L = O(Bi) = O(k^2)$ and $\chi = O(1)$, $\chi > 0$, one has found the boundary of a monotonic instability,

$$M_m(k) = \frac{48L}{\chi} \left[1 + \frac{k^2}{5} \left(1 - \frac{Bi_*}{Bi + k^2} \right) \right], \quad Bi_* = \frac{15(\chi + 1)L}{\chi}. \tag{4.338}$$

For $Bi > Bi_*$, the minimum of $M_m(k)$, which takes place at $k_c = 0$, corresponds to the standard threshold of solutocapillary instability. For $Bi < Bi_*$, the minimum takes place at $k_c^2 = \sqrt{BiBi_*} - Bi$. Equation (4.338) matches both asymptotics of small B [19] and finite B [35]. For small $\chi = O(L)$, one obtains the neutral curve

$$M_m(k) = \frac{48L}{L + \chi} \left[1 + \frac{k^2}{15} + \frac{LBi}{(L + \chi)k^2} \right], \tag{4.339}$$

which has the minimum at $k_c^4 = 15LB/(L + \chi)$.

For the oscillatory mode, three qualitatively different cases have been revealed: (i) $Bi \ll L \ll 1$; (ii) $Bi^3 \ll L \ll Bi$; (iii) $L \ll Bi^3$.

In the case (i), it was found that the convential longwave mode with $k = O(Bi^{1/4})$ studied in Section 4.2.1 is not critical. A novel longwave oscillatory mode with $k_c \sim \sqrt{L}$ has been revealed.

In the case (ii), which corresponds to the formation of a diffusive boundary layer near the free surface, the critical parameters are:

$$M_c = 48 \left[1 - \frac{4}{5} \left(\frac{72\chi^2 L}{5Bi} \right)^{1/4} \right], \quad \omega_c = 5Bi, \, k_c = \left[\frac{(5Bi)^5}{72\chi^2 L} \right]^{1/8}. \tag{4.340}$$

The mechanism of the oscillatory instability emerging in this case can be explained as follows. Because the Marangoni number is less than 48, an initial disturbance triggers a slowly decaying thermocapillary flow. The solute concentration at this stage behaves as a passive scalar advected by the flow. This advection leads to accumulation of the solute near the stagnation point at the free surface and to the slow diffusion of the solute directly to the surface. This, in turn, gives rise to an intensive solutocapillary flow sweeping the spot of a high solute concentration. Recall that the concentration gradient is potentially stable for negative χ. At the last stage, the pure thermocapillary decay of the flow occurs, and the cycle is repeated.

In the case (iii), one finds

$$M_c = 48 \left(1 + 2\sqrt{\frac{Bi}{15}} \right), \quad k_c = (15Bi)^{1/4}, \tag{4.341}$$

which coincides with the known result for the *monotonic* Marangoni instability in a pure liquid [44, 45]. However, for the binary mixture, this mode is oscillatory with the frequency

$$\omega_c = (16200\chi^2 Bi^2 L)^{1/5}.$$

The case of an asymptotically small Lewis number for the Rayleigh–Marangoni convection in a binary liquid in a layer with a deformable surface was studied in [23].

4.2.4 Surfactant Dynamics

In this section we consider how the sorption kinetics influences the longwave Marangoni convection in a layer of binary liquid. In fact, this consideration for fast adsorption (Section 4.2.4.3) is resembling the analysis of non-Boussinesq effects in Section 4.1.2: the key effect is the appearance of the terms, which depend on the amplitude function itself, rather than on their gradients only. However, in spite of similar mathematics, the effect of sorption kinetics is different from that of the temperature dependence of the viscosity. Moreover, in this section nonlinear regimes are tracked even for finite sorption rates (see Section 4.2.4.4), although the corresponding set of amplitude equations is ill-posed. Also, the convection in the presence of an insoluble surfactant is considered.

4.2.4.1 Governing Equations

The problem statement is similar to that formulated in Section 4.2.1: a layer of a binary liquid atop of a substrate of low heat conductivity is heated from below; the Soret effects create the vertical solute stratification. The upper free surface is non-deformable; the heat flux from it is governed by Newton's law of cooling.

The only distinction is that the solute is a soluble surfactant; therefore it forms both the volume (bulk) and surface phases with an adsorption–desorption exchange between them. In order to describe this, we introduce the surface concentration Γ_* (in addition to bulk concentration C_*) and consider the simplest (linear) sorption kinetics within the following relations valid at the free surface:

$$D_* (C_* + \alpha T_*)_{z_*} = -k_a C_* + k_d \Gamma_*, \tag{4.342}$$

$$\partial_{t_*} \Gamma_* + \nabla_{*2} \cdot (\Gamma_* \mathbf{u}_*) = D_s \nabla_{*2}^2 \Gamma_* + k_a C_* - k_d \Gamma_*, \tag{4.343}$$

where D_s, k_a, and k_d are the surface diffusion coefficient for the surfactant, adsorption, and desorption rates, respectively.

Instead of (4.176) we use

$$\gamma_* = \gamma_*^{(0)} - \gamma_T (T_* - T_*^{(0)}) - \gamma_\Gamma (\Gamma_* - \Gamma_*^{(0)}); \tag{4.344}$$

therefore, the surface tension linearly depends on the surface concentration Γ_* rather than on C_*, but the linearity of γ_* in the temperature is retained. The reference values $\gamma_*^{(0)}$ and $T_*^{(0)}$ are unimportant for the further analysis, whereas $\Gamma_*^{(0)}$ is discussed below. Substituting (4.344) into tangential stress balance, we end up with the following expression:

$$\eta_* \partial_{z_*} \mathbf{u}_* = -\nabla_{*2} \left(\gamma_T T_* + \gamma_\Gamma \Gamma_* \right). \tag{4.345}$$

Equations (4.342) and (4.345) replace the last two equations in (4.182); the boundary value problem (4.178)–(4.182) should be also supplemented by (4.343).

The scales for the velocity, length, temperature, time, and pressure are chosen the same as in Section 4.2.1, whereas for the surface and bulk concentration, they are $\gamma_\Gamma^{-1} \gamma_T A d_*$ and $k_d \gamma_T A d_* (k_a \sigma_\Gamma)^{-1}$, respectively. The bulk concentration C_* introduced as the mass fraction is dimensionless by definition, but this rescaling simplifies the comparison with the results of Section 4.2.1. The resulting dimensionless boundary value problem reads

$$\nabla \cdot \mathbf{v} = 0, \tag{4.346}$$

$$\frac{1}{P} \left(\mathbf{v}_t + \mathbf{v} \cdot \nabla \mathbf{v} \right) = -\nabla p + \nabla^2 \mathbf{v}, \tag{4.347}$$

$$T_t + \mathbf{v} \cdot \nabla T = \nabla^2 T, \tag{4.348}$$

$$\frac{1}{L} \left(C_t + \mathbf{v} \cdot \nabla C \right) = \nabla^2 C + \chi \nabla^2 T; \tag{4.349}$$

$$\mathbf{v} = 0, \ T_z = -1, \ C_z + \chi T_z = 0 \text{ at } z = 0, \tag{4.350}$$

$$w = 0, \ C_z + \chi T_z + \frac{K_a}{L} \left(C - \Gamma \right) = 0, \ \mathbf{u}_z = -M \nabla_2 \left(T + \Gamma \right),$$

$$C - \Gamma = K_d^{-1} \left[\Gamma_t + \nabla_2 \cdot (\Gamma \mathbf{u}) - L_s \nabla_2^2 \Gamma \right], \ T_z = -BT \text{ at } z = 1. \tag{4.351}$$

The boundary value problem (4.346)–(4.351) is characterized by eight dimensionless parameters. Five of them are given by (4.188) with the only correction for the Soret number:

$$\chi = \frac{\alpha k_a \gamma_\Gamma}{k_d \gamma_T}.$$

The sign of χ is chosen in the opposite way to (4.188); there positive χ_C stands for the aqueous solution of inorganic salt, when the surface tension increases with growth of C. For a surfactant, in contrast, $\gamma_\Gamma > 0$, which means that the surface tension decreases with the growth of Γ.

The additional three parameters characterize the sorption kinetics and two-dimensional diffusion of the surface phase:

$$K_a = \frac{k_a d_*}{\kappa}, \ K_d = \frac{k_d d_*^2}{\kappa}, \ L_s = \frac{D_s}{\kappa}.$$

These parameters represent the dimensionless adsorption and desorption coefficients and the surface Lewis number, respectively. The nondimensional time to reach the equilibrium between the phases is K_d^{-1}. At $K_d^{-1} \to 0$ the bulk and surface phases are in the local equilibrium, $C(z = 1) = \Gamma$, i.e., the problem reduces to that studied in Section 4.2.1. (Henceforth we refer to that limiting case as "infinitely fast kinetics" or "absence of the adsorption.") In the opposite case, $K_d \ll 1$, the surfactant is insoluble, and, therefore, the bulk concentration C is irrelevant. That limiting case was investigated in [46–48].

Substituting $C - \Gamma$ from the fourth to the second condition of (4.351), one can obtain:

$$(C + \chi T)_z = -\frac{K_{ad}}{L} \left[\Gamma_t + \nabla_2 \cdot (\Gamma \mathbf{u}) - L_s \nabla^2 \Gamma \right], \tag{4.352}$$

where the Langmuir number $K_{ad} = K_a K_d^{-1} = k_a / (k_d d)$ is introduced. This parameter characterizes the ratio of characteristic time scales of the desorption and adsorption.

The equilibrium base state within (4.346)–(4.351) corresponds to the conductive distributions of both T and C in a fluid at rest:

$$\mathbf{v}_0 = 0, \ T_0 = -z + \frac{B+1}{B}, \ C_0 = \Gamma_0 + \chi (z - 1). \tag{4.353}$$

The equilibrium value of the surface concentration Γ_0—the dimensionless value of Γ_* appearing in (4.344)—provides the exact balance between absorption and desorption. Either Γ_0 or the mean bulk concentration $\langle C \rangle = \Gamma_0 - \chi/2$ is the ninth dimensionless parameter, which governs the dynamics of the system. (Recall that for $K_d \gg 1$ only the gradient of concentration is relevant, whereas the similar constant $\langle C \rangle$ in (4.189) is not.) Both Γ_0 and $\langle C \rangle$ include the temperature gradient a_* via the scales for the surface and volume concentrations; therefore in experiments it is tricky to keep them fixed varying M. In what follows, we use the so-called elasticity number:

$$N = M \langle C \rangle = \frac{\gamma_\Gamma k_a \langle C_* \rangle d}{k_d \eta \kappa},$$

which does not contain a_*; $\langle C_* \rangle$ is the unscaled mean value of the bulk concentration. However, to shorten the formulas, we also make use of $\Gamma_0 = N/M + \chi/2$.

The presence of Γ_0 in the model results in some restrictions on the dimensionless parameters. Indeed, one has to provide positive values of both the surface and bulk concentrations of the surfactant C_0 sign$M$ and Γ_0 sign$M$. (The scales for both bulk and surface concentration are proportional to $\gamma_\Gamma a_*$, and therefore they are of the same signs with the Marangoni number.) Then, the unscaled concentration $C_{0*} = C_0 k_d \gamma_\Gamma a_* d / (k_a \gamma_\Gamma)$ has to be less than unity across the layer. In fact, even stronger restriction is needed: the unscaled equilibrium concentration C_{0*} must be far from both zero and unity, because we have neglected the dependence of the Soret coefficient on the concentration. Otherwise, keeping in mind that $\alpha_* = D_T C_* (1 - C_*)$, where D_T is the thermal diffusion coefficient, one obtains nonlinear variation of C_0 with z [49, 50]. Demanding, for instance, $1/4 < C_{0*}(z) < 3/4$ for any z, i.e., the maximum value of C_{0*}, does not exceed three times of its minimum value, we end up with the restriction:

$$N > |M\chi|. \tag{4.354}$$

Below this inequality is taken into account.

Finalizing this section, we estimate the typical values of parameters K_a and K_d governing the sorption kinetics. For 1-decanol the desorption time is small enough, $k_d = 10 \div 100$ s^{-1} depending on the concentration of the surfactant. This provides $K_d = 10^4 \div 10^5$ for 1 cm layer, i.e., the sorption kinetics serves only as a weak effect (however, it can be still important; see Section 4.2.4.3). For polyethoxylated surfactants the adsorption and desorption are very slow processes, $k_d \sim 10^{-4}$ s^{-1}, $k_a \sim 10^{-3}$ cm s^{-1} [51], which results in $K_d \sim 0.1$ and $K_a \sim 1$. Therefore, below we analyze both cases of finite K_a and K_d and the limit of fast kinetics, $K_{ad} \ll 1$.

For most binary liquids, the Lewis number is rather small, although we do not treat it to be asymptotically small, setting $L = 0.01$ in calculations. Analysis of asymptotically small L can be found in [18] and [52] for infinite and finite values of the Langmuir number K_d, respectively. The available results of measurements for D_s are somewhat controversial. The majority of authors (see, e.g., [53]) found that the surface diffusion was smaller than the bulk one; an opposite conclusion was reported in [54]. Hereafter we assume $L_s = 0$ in numerical calculations, though retain it in analytical formulas. In fact, setting $L_s \sim L$ does not change the numerical results substantially.

4.2.4.2 Linear Stability Analysis

We start the analysis of the surfactant dynamics with the linear stability problem for base state (4.353). Technically, the procedure is similar to that carried out in Section 4.2.1.2; therefore, we only briefly describe it. Two-dimensional normal perturbations of the form $f(x,z,t) = \tilde{f}(z) \exp(ikx + \lambda t)$ are introduced; they are governed by the following boundary value problem:

$$ik\tilde{u} + \tilde{w}' = 0, \quad (4.355)$$

$$\frac{\lambda}{P}\tilde{u} = -ik\tilde{p} + \Delta\tilde{u}, \quad (4.356)$$

$$\frac{\lambda}{P}\tilde{w} = -\tilde{p}' + \Delta\tilde{w}, \quad (4.357)$$

$$\lambda\tilde{T} - \tilde{w} = \Delta\tilde{T}, \quad (4.358)$$

$$\lambda\tilde{C} + \chi\tilde{w} = L\Delta\left(\tilde{C} + \chi\tilde{T}\right); \quad (4.359)$$

$$\tilde{u} = \tilde{w} = \tilde{T}' = \tilde{C}' = 0 \text{ at } z = 0, \quad (4.360)$$

$$\tilde{w} = \tilde{T}' + B\tilde{T} = 0, \quad \left(\tilde{C} + \chi\tilde{T}\right)' = -\frac{K_{ad}}{L}\left(\lambda\tilde{\Gamma} + \Gamma_0 ik\tilde{u} + L_s k^2\tilde{\Gamma}\right),$$

$$\tilde{u}' = -Mik\left(\tilde{T} + \tilde{\Gamma}\right), \quad \tilde{C} - \tilde{\Gamma} = K_d^{-1}\left(\lambda\tilde{\Gamma} + \Gamma_0 ik\tilde{u} + L_s k^2\tilde{\Gamma}\right) \text{ at } z = 1. \quad (4.361)$$

This problem is an obvious extension of (4.190)–(4.196) for $K_d^{-1} \neq 0$. In what follows we deal with the longwave perturbations ($k \ll 1$) at small B for the conventional asymptotics, $k \sim \varepsilon = B^{1/4}$, only. The problem for finite k at finite, but numerically small B, typical for an experimental situation, was solved in [55]. The longwave disturbances were shown to be critical in a wide range of the problem parameters. The monotonic mode with $k \sim \sqrt{B}$ was also analyzed in the cited paper.

The perturbation fields, growth rate, and wavenumber are expanded using (4.20)–(4.22) with the expansion for the Marangoni number (4.197) rather than the Rayleigh number. Additionally, the disturbance of the surface concentration reads $\tilde{\Gamma} = \Gamma_0 + \varepsilon^2 \Gamma_2 + \ldots$.

The asymptotic analysis similar to that applied in the previous sections leads to following results. The stability threshold for the monotonic mode is determined by

$$m_*^{(m)} = \frac{L + K_{ad}(L_s + N/4)}{L - \chi - \chi L + K_{ad}(L_s - 6\chi)}, \tag{4.362}$$

whereas for the oscillatory mode it reads

$$m_*^{(o)} = \frac{1 + L + K_{ad}(1 + L_s + N/4)}{1 - \chi + K_{ad}(1 - 6\chi)}. \tag{4.363}$$

This mode exists for positive values of the squared frequency of perturbations:

$$\Omega^2 = K^4 \left(1 - \frac{m_*^{(o)}}{m_*^{(m)}} \right) \frac{L + K_{ad}(L_s + N/4)}{1 + K_{ad}}. \tag{4.364}$$

This condition can be presented as follows:

$$\chi_{0\infty} > \chi > \chi_0, \tag{4.365}$$

where

$$\chi_0 = \frac{L^2 + K_{ad}(2LL_s - (1-L)N/4) + K_{ad}^2(L_s^2 - (1-L_s)N/4)}{1 + L + L^2 + K_{ad}(7 + L(1 + L_s + N/4)) + 6K_{ad}^2}$$

and $\chi = \chi_{0\infty}$ corresponds to the pole for $m_*^{(o)}(\chi)$:

$$\chi_{0\infty} = \frac{1 + K_{ad}}{1 + 6K_{ad}}. \tag{4.366}$$

The stability thresholds (4.362) and (4.363) match both known limiting cases: for infinitely fast kinetics, $K_{ad} = 0$, they reduce to (4.203) and (4.204), respectively; at $K_{ad} \gg 1$, $\chi = 0$ the results for insoluble surfactant are reproduced [46].

Results of computation are presented in Figure 4.18; they are qualitatively similar to that for $K_{ad} = 0$, Figure 4.6. To simplify the comparison, we plot $m_0(-\chi)$, since the sign of the solutocapillary effect differs from that used in Section 4.2.1.2. The sorption kinetics results in a layer stabilization; $|m_*|$ increases for both modes as N and K_{ad} grow.

We do not analyze monotonic mode with $k \sim \sqrt{B}$, referring a reader to [55]. Again, the adsorption/desorption kinetics does not influence this mode much; qualitatively the stability map remains similar to Figure 4.6.

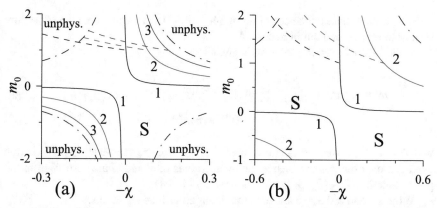

Fig. 4.18 Critical Marangoni number m_0 for the longwave modes vs. χ at $L_s = 0, L = 0.01$; (a): $N = 10$ (b): $N = 40$. Curves 1, 2, and 3 correspond to $K_{ad} = 0, 0.04, 0.1$, respectively. The solid lines show the stability boundaries for the monotonic mode, (4.362); the dashed lines are for the oscillatory mode (4.363). Domains of stability are marked by "S." For positive (negative) m_0, the condition (4.354) is violated above (below) the dashed–dotted lines; these unphysical domains are marked by "unphys"

4.2.4.3 Small K_{ad}. Oscillatory Mode

We start the analysis of nonlinear dynamics with the limit of small K_{ad} (fast adsorption). It will be shown that even in this case the nonlinear dynamics changes substantially. For the monotonic mode the analysis of this limiting case is not presented; as shown in [55] the corresponding amplitude equation resembles the equation for the Marangoni convection in a pure liquid (see Section 3.2). The only feature to be mentioned for this case is that stable rolls are found. For the oscillatory mode, there are two distinguished limits, $K_{ad} \sim \varepsilon^2$ (Case I) and $K_{ad} \sim \varepsilon$ (Case II).

Case I

We start the nonlinear analysis of the oscillatory mode with the limiting case, $K_{ad} = O(\varepsilon^2)$, dealing with the limit of fast desorption, $K_d \sim \varepsilon^{-2}$, $K_a = O(1)$. The derivation of the amplitude equation in this limit is similar to that for small non-Boussinesq effects addressed in Section 4.1.2, although, as will be shown below, the pattern selection differs. To get the amplitude equations, we use expansions (4.61) and (4.62) with the series for Rayleigh number replaced by (4.197) and two additional expansions for the Langmuir number and the surface concentration:

$$K_{ad} = \varepsilon^2 k_{ad}, \; k_{ad} = O(1), \; \Gamma = \Gamma_0 + \Gamma^{(0)} + \varepsilon^2 \Gamma^{(2)} + \ldots \qquad (4.367)$$

(The fields Θ and Σ represent the perturbations of the temperature and solute concentration, so that $T = T_0 + \Theta, C = C_0 + \Sigma$.)

Substituting these expansions into (4.346)–(4.351), one can repeat the procedure similar to that carried out in Section 4.1.1.3.

At the zeroth order, the solution is given by

$$\Theta^{(0)} = F(X,Y,\tau,t_4) + \langle \Theta^{(0)}(t_4) \rangle, \quad \sum^{(0)} = \Gamma^{(0)} = G(X,Y,\tau,t_4), \quad (4.368)$$

$$\mathbf{U}^{(0)} = m_0 V_\parallel \nabla h, \ W^{(0)} = m_0 V_\perp \nabla^2 h, \quad (4.369)$$

with $h = F + G$. The second term in $\Theta^{(0)}$ stands for the temperature perturbation averaged over the domain; hence the mean values of F and G are zero. Within the present section $\nabla = (\partial_X, \partial_Y)$, m_0 is defined by (4.204).

At the second order, we arrive at the linear amplitude equations:

$$\mathscr{L}_1(F,G) = \mathscr{L}_2(F,G) = 0, \quad (4.370)$$

where $\mathscr{L}_{1,2}$ are given by (4.273) and (4.274). Hence, adsorption does not contribute at the fast dynamics. At m_0 given by (4.204), the solution oscillates in τ with the frequency Ω, (4.205).

Proceeding to the fourth order and demanding the solvability of the corresponding boundary value problem, we arrive at (4.275) and (4.276) with the additional term $-k_{ad}\sigma_n$ in the right-hand side of (4.276), where

$$\sigma_n = G_\tau - 12m_0 \nabla \cdot ((\Gamma_0 + G)\nabla h) - L_s \nabla^2 G. \quad (4.371)$$

(G_τ can be expressed via $\nabla^2 F$ and $\nabla^2 G$ using relations (4.274).)

The new term $k_{ad}\sigma_n$, which stems from the adsorption kinetics, includes both linear and quadratic contributions. Linear one results in a correction to the growth rate, proportional to k_{ad}:

$$\gamma^{(ads)} = \gamma - \frac{k_{ad}K^2}{2}\left[1 + L_s + \frac{N}{4} - m_0(1 - 6\chi)\right]$$
$$- \frac{ik_{ad}K^2}{2\Omega}\left[L_s + \frac{N}{4} - \Omega^2 - m_0(L_s - 6\chi)\right], \quad (4.372)$$

where γ is given by (4.279).

The additional term in $\gamma^{(ads)}$ agrees with the expansion of $\gamma(K_{ad})$ at small K_{ad}. This contribution is proportional to K^2 similar to that containing m_2 in γ. Therefore, at $k_{ad} \neq 0$ the critical value of m_2 contains a correction independent of K, whereas the critical wavenumber K_c does not dependent on k_{ad}, N, and L_s. Below we fix $K = K_c$ and $m_2 = m_{2c}$, corresponding to the minimum of the neutral stability curve $m_2(K)$, for the oscillatory mode.

Quadratic nonlinearity in σ_n affects pattern selection on a square lattice only (in a way similar to Section 4.1.2, see also Section 4.2.1.4 for square patterns), whereas branching of rolls, rhombic patterns, and hexagons (Section 4.2.1.4) remains unchanged. Recall that all patterns on these lattices bifurcate supercritically.

The computations based on Appendix D.1 show that all patterns on a square lattice bifurcate supercritically. As mentioned above, for one-dimensional patterns no oblique waves are generated, and the directions of branching for TRs and SRs do not depend on k_{ad}. In contrast, the stability of these patterns is affected by the rescaled Langmuir number as well. Applying the stability conditions, one can easily show that ARs lose their stability as k_{ad} grows, whereas TRs become stable. Within a certain interval of k_{ad}, shown in Figure 4.19, bistability takes place. Both the abovementioned patterns are stable, and the system evolves toward one of them depending on the initial conditions.

It is noteworthy that in spite of similar contributions of adsorption kinetics and non-Boussinesq effects (Section 4.1.2) to the amplitude equations, their influence on the branching and pattern selection is quite different. Recall, that in the former case, the direct Hopf bifurcation is replaced with the inverse one, as non-Boussinesq effects amplify.

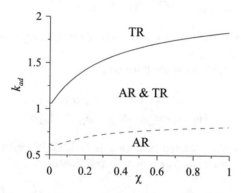

Fig. 4.19 Case I: domains of stability for patterns on a square lattice; $P = 2, L = 0.01$. Within the domains marked by "AR" and "TR" alternating folls and traveling rolls are stable, respectively. In the domain marked "AR & TR" multistability occurs

Case II

As we have just shown, in the case $K_{ad} = O(\varepsilon^2)$, the amplitude equations are similar to those with $K_{ad} = 0$, except for several additional terms. With further growth of K_{ad}, the structure of the amplitude equations changes dramatically. The next distinguished limit for the oscillatory mode is

$$K_{ad} = \varepsilon \tilde{k}_{ad}. \tag{4.373}$$

At large k_{ad} TRs are selected, the same should be valid at small $\tilde{k}_{ad}$; therefore, only one-dimensional solutions are analyzed below. To that end, in addition to the expansions given by (4.61), we introduce the first slow time $\tau_1 = \varepsilon^3 t$ and include terms of odd orders in ε in (4.197) and (4.62):

$$f = f^{(0)} + \varepsilon f^{(1)} + \ldots, \ f = (\Theta, \Gamma, \Sigma, U, W, P), \tag{4.374}$$

$$M = 48(m_0 + \varepsilon m_1 + \ldots). \tag{4.375}$$

The boundary value problems at the zeroth and second orders and their solutions do not change; the solution at the first order reads:

$$\Theta^{(1)} = F_1(X, Y, \tau, \tau_1, t_4), \ \Gamma^{(1)} = G_1(X, Y, \tau, \tau_1, t_4), \ h_1 = F_1 + G_1, \tag{4.376}$$

$$U^{(1)} = V_\parallel \partial_X (m_0 h_1 + m_1 h), \ W^{(1)} = V_\perp \partial_X^2 (m_0 h_1 + m_1 h). \tag{4.377}$$

This solution resembles (4.368) and (4.369).

The solvability conditions at the second order remains unchanged (4.370); its one-dimensional solution reads

$$h = \delta \left(A_1 e^{iKX} + A_2 e^{-iKX} \right) e^{-i\Omega K^2 \tau} + c.c., \tag{4.378}$$

where the complex amplitudes of two counter-propagating waves depend on the slow times τ_1, t_4, etc.

The solvability conditions at the third order are:

$$\mathscr{L}_1(F_1, G_1) = -m_1 h_{XX} - \partial_{\tau_1} F, \tag{4.379}$$

$$\mathscr{L}_2(F_1, G_1) = \chi m_1 h_{XX} - \partial_{\tau_1} G - \tilde{k}_{ad} \tilde{\sigma}_n, \tag{4.380}$$

where $\tilde{\sigma}_n$ is the one-dimensional reduction of (4.371). We split $\tilde{\sigma}_n$ into two parts, linear ($\tilde{\sigma}_{nl}$) and quadratic ($\tilde{\sigma}_{nn}$) in F and G, as follows:

$$\tilde{\sigma}_{nl} = G_\tau - 12m_0\Gamma_0 h_{XX} - L_s G_{XX}, \ \Gamma_0 = \frac{N}{48m_0} + \frac{\chi}{2}, \tag{4.381}$$

$$\tilde{\sigma}_{nn} = -12m_0 (Gh_X)_X. \tag{4.382}$$

It is clear that $\tilde{\sigma}_{nl}$ does not produce terms important for the further analysis; the nonlinear part of the solution is

$$h_n^{(1)} = \frac{4m_0\tilde{k}_{ad}}{\tilde{\Omega}^2} \left[\xi \left(A_1^2 e^{2iKX} + A_2^2 e^{-2iKX} \right) e^{-2i\Omega\tau} + 3(1 - \xi_r - m_0) A_1 A_2^* e^{2iKX} \right] + c.c.,$$

where $\xi = \xi_r + i\xi_i = (m_0 - 1 - i\tilde{\Omega})(2 + i\tilde{\Omega})/(1 + i\tilde{\Omega})$. The corresponding part of the surface concentration $\Gamma_n^{(1)}$ can be easily expressed using $h_n^{(1)}$.

The solvability conditions at the fourth order represent the one-dimensional reduction of (4.275) and (4.276) with the additional term $12m_0\tilde{k}_{ad} \left(\Gamma_n^{(1)} h_X + Gh_{nX}^{(1)} \right)_X$ in the right-hand side of (4.276).

The solvability conditions for (4.275) and (4.276) with the above additional term yield the following set of amplitude equations:

$$\dot{A}_1 = \left(\gamma_2 - \tilde{N}_0|A_1|^2 - \tilde{N}_\pi|A_2|^2\right)A_1, \tag{4.383}$$

$$\dot{A}_2 = \left(\gamma_2 - \tilde{N}_0|A_2|^2 - \tilde{N}_\pi|A_1|^2\right)A_2. \tag{4.384}$$

The bifurcations within (4.383) and (4.384) are determined by the real parts of the self- and cross-interaction coefficients

$$\tilde{N}_{0r} = \alpha_r - \frac{72m_0^2\tilde{k}_{ad}^2K^2}{\tilde{\Omega}^2(1+\tilde{\Omega}^2)}\left[(2m_0-1)(m_0-1)+\tilde{\Omega}^2(1+m_0)\right], \tag{4.385}$$

$$\tilde{N}_{\pi r} = 2\alpha_r + \frac{72m_0^2\tilde{k}_{ad}^2K^2}{\tilde{\Omega}^2(1+\tilde{\Omega}^2)}(m_0+1)(L+L^2), \tag{4.386}$$

where α is given by (4.286). At small $\tilde{k}_{ad}$ ($K_{ad} \sim \varepsilon^2$) the last term in $\tilde{N}_{0r}$ vanishes, and the formula $\tilde{N}_{0r} = \alpha_r = N_{0r}$ is restored, as we have discussed in the previous paragraph. At rather large $\tilde{k}_{ad}$, the real part of the self-interaction coefficient becomes negative, since the contribution of the sorption kinetics into $\tilde{N}_{0r}$ is less than zero. In contrast, $\tilde{N}_\pi$ grows as $\tilde{k}_{ad}$ increases; therefore SRs do not play any role within this limit.

The curve, which corresponds to $\tilde{N}_{0r} = 0$ and therefore separates domains of supercritical and subcritical Hopf bifurcations for TRs, is shown in Figure 4.20. This plot corresponds to the critical perturbations with $K = K_c$. Similar to Case I, both L_s and N contribute to $\tilde{N}_{0,\pi}$ neither directly nor via selection of K_c; hence, Figure 4.20 is valid irrespectively of these two parameters.

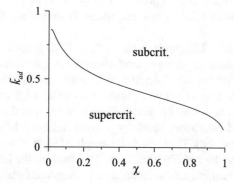

Fig. 4.20 Case II: domains of supercritical and subcritical Hopf bifurcations for TRs within (4.383)–(4.386); $P = 2, L = 0.01$

4.2.4.4 Nonlinear Evolution at Finite K_{ad}

At finite values of the Langmuir number, we apply expansions (4.61) and (4.374) with odd orders omitted in the former one. The Marangoni number is rescaled via the relation $M = 48m$, but it is not expanded in ε.

At the leading (second) order, the solvability conditions read

$$F_\tau = \nabla^2(F - mh), \tag{4.387}$$

$$G_\tau = \nabla^2 \frac{(L + K_{ad}L_s)G + L\chi F + m\chi h}{1 + K_{ad}} + \frac{12mK_{ad}}{1 + K_{ad}}\nabla \cdot (G\nabla h), \tag{4.388}$$

where $F(X, Y, \tau) = \Theta^{(0)}$, $G = \Gamma_0 + \Gamma^{(0)}$ and $h = F + G$. For the sake of brevity, the mean value Γ_0 is included into G.

When linearized about $G = \Gamma_0$ and $F = 0$, (4.387) and (4.388) reproduce the linear stability problem studied in Section 4.2.4.2. In the limiting case of insoluble surfactant ($K_{ad} \gg 1$ and $\chi = 0$), (4.387) and (4.388) reduce to the corresponding amplitude equations derived in [47]. A similar set of equations was also analyzed in [56] for axisymmetric regimes of buoyancy convection in a binary mixture layer with a source of solute at the origin. Both the axial symmetry and the presence of the source are of crucial importance, and the results obtained in [56] cannot be used for the problem under consideration.

It is clear that in contrast to the amplitude equations obtained in the absence of adsorption, (4.273) and (4.274), equations (4.387) and (4.388) are nonlinear; see the last term in (4.388). Indeed, for $K_{ad} = 0$ only gradients of the amplitude functions F and G (perturbations of the temperature and concentration, respectively) contribute to the nonlinear evolution, rather than F and G themselves. As a consequence, the nonlinear terms in the amplitude equations bear additional differentiations with respect to spatial variables, and they do not appear at the second order in ε. Thus, at $K_{ad} = 0$ (or $K_{ad} = O(\varepsilon)$; see Section 4.2.4.3), (4.387) and (4.388) are linear; they determine the threshold value of $m = m_*$, where the longwave disturbances neither grow nor decay.

In the case of finite Langmuir number, $K_{ad} = O(1)$, such value m_* does not exist: the full nonlinear set of equations describes coupled nonlinear diffusion of two scalars whose "diffusivities" depend on G. Above the stability threshold, given by (4.362) or (4.363), an effective diffusivity becomes negative and the problem is ill-posed. Moreover, it can be readily shown in the one-dimensional case that both monotonic and oscillatory spatially periodic solutions bifurcate subcritically within (4.387) and (4.388). Thus, for finite-amplitude perturbations, the problem is ill-posed even below the stability threshold. This leads to the infinitely fast growth of shortwave perturbations, whose characteristic length scale is smaller than ε^{-1}.

The well-posedness of (4.387) and (4.388) can be restored only by adding the terms with fourth derivatives. This procedure was carried in Section 4.2.4.3 for sufficiently fast sorption kinetics. Another example of such a restoring is an assumption that both F and $G - \Gamma_0$ are $O(\varepsilon^2)$. This analysis for the monotonic mode results in the Sivashinsky equation (1.81), which is known to be ill-posed as well. Thus, a longwave weakly nonlinear analysis based on scalings (4.61) and (4.374) is not self-consistent at $K_{ad} = O(1)$.

4.2.4.5 Deformable Interface

The modification of nonlinear evolution equations (4.335) in the case of adsorption and desorption of the solute on the free deformable surface has been carried out in [57]. One obtains:

$$\frac{\partial h}{\partial \tau} + \nabla \cdot \mathbf{Q}_1 = 0, \quad \frac{\partial \Psi}{\partial \tau} + \nabla \cdot (\mathbf{Q}_2 + K_{ad}\mathbf{Q}_3) = 0; \quad (4.389)$$

explicit expression for $\mathbf{Q}_1$, $\mathbf{Q}_2$, and $\mathbf{Q}_3$ are given in [57]. Weakly nonlinear asymptotic expansions were applied to (4.389) in order to investigate the stability of convective patterns emerging in the vicinity of the oscillatory instability threshold: single traveling and standing waves, superpositions of two traveling or two standing waves, and superpositions of three traveling waves [57, 58]. The theoretical predictions have been confirmed and extended by the numerical solution of evolution equations (4.389).

4.2.4.6 Instability in the Presence of an Insoluble Surfactant

In conclusion to the present section, let us consider a heated liquid layer in the presence of an *insoluble* surfactant adsorbed at the non-deformable ($Ga \gg 1$) free surface. It should be emphasized that the liquid is not a binary one; thus the mechanism of the oscillatory instability characteristic for binary liquids is not relevant. Still, the presence of a surfactant with the surface concentration Γ, which diminishes the surface tension γ according to the law $\gamma_\Gamma = -d\gamma/d\Gamma > 0$, may lead to the suppression of stationary convective flows and to the generation of a specific kind of oscillations [59, 60]. Those phenomena can be explained in the following manner.

The transfer of the surfactant along the interface is governed by the equation [61]

$$\frac{\partial \Gamma}{\partial t} + \nabla_\perp \cdot (\mathbf{v}_\perp \Gamma) = D_0 \Delta_\perp \Gamma. \quad (4.390)$$

Here, $\mathbf{v}_\perp = \mathbf{v} - v_n \mathbf{n}$ is the tangential component of the fluid velocity at the interface, $\nabla_\perp = \nabla - \mathbf{n}(\mathbf{n} \cdot \nabla)$, $\Delta_\perp = \nabla_\perp^2$, and D_0 is the surface diffusion coefficient. Let us consider a flow in the bulk that generates at the interface a velocity field with a nonvanishing interfacial divergence $q = \nabla_\perp \cdot \mathbf{v}_\perp$. Assume that the surface diffusion coefficient is small, so that the diffusion of the surfactant is negligible compared to the advection of the surfactant by the flow. Assume also that the initial distribution of the surfactant is homogeneous. Obviously, because of the advection of the surfactant, the surface concentration decreases in the region of a divergent interfacial flow where $q > 0$, and it increases in the region of a convergent flow where $q < 0$. The tangential stresses generated by the inhomogeneity of the surfactant concentration are directed opposite to the fluid motion. Thus, they will prevent the development of a monotonic instability and suppress stationary flows. At the same time, such "negative feedback" can lead to oscillatory instability ("overstability").

Note that the arguments given above do not depend on the physical nature of the flow. Thus, we can expect that for both buoyancy-driven convection and surface tension-driven convection, monotonic instability will be suppressed and replaced by oscillatory instability.

Linear Stability Theory

The effect of the surfactant on the Marangoni convection can be described analytically [46], if we assume that a constant heat flux $-\lambda A$ is applied at the bottom plate (λ is the conductivity coefficient, $-A$ is the vertical temperature gradient), and the heat transfer coefficient at the upper surface q is small: the Biot number

$$Bi = qd/\lambda = O(\varepsilon^4), \ \varepsilon \ll 1. \tag{4.391}$$

In that case, the minimum of the neutral curve is in the longwave region, $k = O(\varepsilon)$. The relevant nondimensional parameters, in addition to the Marangoni number, are the elasticity number of the surfactant, $N = \alpha_\Gamma d\Gamma_0/\eta\chi$ (Γ_0 is the mean surfactant concentration), and the surfactant Lewis number, $L = D_0/\chi$, which is typically small.

Using the longwave expansions, one can find that the instability of the quiescent state is monotonic, if $N < N_* = 4L^2/(1-L)$; the critical Marangoni number is, at the leading order,

$$M_m = 48 + \frac{12N}{L}. \tag{4.392}$$

For $N > N_*$, the instability becomes oscillatory, with the threshold

$$M_o = 48 + 12(N+4L) \tag{4.393}$$

and the frequency

$$\omega = \omega_0 k^2, \ \omega_0 = \frac{1}{2}[(1-L)(N-N_*)]^{1/2}. \tag{4.394}$$

More detailed expressions, obtained in the case of a deformable interface (for finite Ga), can be found in [46].

Note that the influence of surfactants on different kinds of convective instabilities in two-layer systems between the plates with fixed temperature was studied in [62–67].

Nonlinear Development of an Oscillatory Instability Mode: Non-deformable Surface

Let us describe the methodology of the investigation of nonlinear wavy pattern near the oscillatory instability threshold, at $M - M_0 = M_2\varepsilon^2$.

The problem under consideration has two conservation laws: the exact conservation law for the amount of the surfactant and the approximate conservation law for the mean temperature, due to the smallness of the Bio number $Bi = \beta \varepsilon^4$, $\varepsilon \ll 1$. The growth rate is described by formula (1.25) (with R replaced by M); hence $\sigma_r = O(\varepsilon^4)$, while the frequency has a quadratic dispersion law; hence $\sigma_i = O(\varepsilon^2)$. Thus, for the description of oscillatory instability, we need two slow time scales, $t_2 = \varepsilon^2 t$, which is characteristic for the period of oscillations, and $t_4 = \varepsilon^4 t$, which characterizes the growth of oscillations. Also, we rescale the horizontal coordinates as $X = \varepsilon x$ and $Y = \varepsilon y$.

Applying the technique of multiscale expansions [47], one obtains, at the zeroth order of expansions, two coupled linear equations which govern the temporal evolution of the mean temperature distribution, $T^{(0)} = F_0(X, Y, t_2, t_4)$, and the disturbance of the nondimensional surfactant concentration, $\Gamma^{(0)} = 1 + G_0(X, Y, t_2, t_4)$:

$$L_1(F_0, G_0) \equiv \frac{\partial F_0}{\partial t_2} - \nabla^2 F_0 + \frac{1}{48} \nabla^2 (M_0 F_0 + N G_0) = 0, \qquad (4.395)$$

$$L_2(F_0, G_0) \equiv \frac{\partial G_0}{\partial t_2} - L \nabla^2 G_0 - \frac{1}{4} \nabla^2 (M_0 F_0 + N G_0) = 0, \qquad (4.396)$$

where $\nabla = (\partial/\partial X, \partial/\partial Y)$. Relation (4.392)–(4.394) can be easily obtained from (4.395), (4.396). M_0 is taken at the stability boundary. In a contradistinction to the case of the dispersion law $\omega \sim k^3$, where the dispersion of waves is balanced by the nonlinearity forming nonlinear (cnoidal) waves (see Section 2.1), the stronger dispersion, $\omega \sim k^2$, dominates at the leading order; therefore the leading-order wavy pattern is just a superposition of harmonic waves.

At the first order, one obtains the following system of equations:

$$L_1(F_1, G_1) = 0, \qquad (4.397)$$

$$L_2(F_1, G_1) = \frac{1}{4} \nabla [G_0(M_0 \nabla F_0 + N \nabla G_0)]. \qquad (4.398)$$

At the second order, one finds:

$$L_1(F_2, G_2) = -\frac{\partial F_0}{\partial t_4} - \beta F_0 - \frac{1}{48} M_2 \nabla^2 F_0$$

$$- \left(\frac{4 + 3P^{-1}}{60} \right) \frac{\partial F_0}{\partial t_2^2} + \left(\frac{112 + 24P^{-1} - M_0}{480} \right) \nabla^2 \frac{\partial F_0}{\partial t_2} + \left(\frac{M_0 - 80}{480} \right) \nabla^4 F_0, \qquad (4.399)$$

$$L_2(F_2, G_2) = \frac{1}{4} \nabla [G_0(M_0 \nabla F_1 + N \nabla G_1) + G_1(M_0 \nabla F_0 + N \nabla G_0)] - \frac{\partial G_0}{\partial t_4} + \frac{1}{4} M_2 \nabla^2 F_0$$

$$- \frac{1}{40} \left\{ (M_0 - 32) \nabla^4 F_0 + (32 + 16P^{-1} - M_0) \nabla^2 \frac{\partial F_0}{\partial t_2} - 16P^{-1} \frac{\partial F_0}{\partial t_2^2} \right\}, \qquad (4.400)$$

where $M_2 = (M - M_0)/\varepsilon^2$, $P = \nu/\chi$ is the Prandtl number.

The obtained system of equations can be used for the investigation of the non-linear evolution of finite-amplitude disturbances. However, that is done in a way very different from the case of waves governed by the CGLE. In the zeroth order, our evolution system is linear, and that means that the solution can be taken as a superposition of harmonic waves with arbitrary wavevectors (though the amplitudes of waves are finite). Below we explain the methodology of the analysis taking the simplest solution of equations (4.395) and (4.396) in the form of a one-dimensional, spatially periodic traveling wave:

$$F_0 = A_0(t_4)e^{i(KX-\omega_0 K^2 t_2)} + c.c., \quad G_0 = B_0(t_4)e^{i(KX-\omega_0 K^2 t_2)} + c.c., \tag{4.401}$$

where

$$B_0 = \frac{M_0 A_0}{4i\omega_0 - (N+4L)}. \tag{4.402}$$

At the next order, one finds:

$$F_1 = A_1(t_4)e^{2i(KX-\omega_0 K^2 t_2)} + c.c., \quad G_1 = B_1(t_4)e^{2i(KX-\omega_0 K^2 t_2)} + c.c., \tag{4.403}$$

where

$$A_1 = (a_{1,r} + ia_{1,i})A_0^2, \tag{4.404}$$

$$a_{1,r} = \frac{8[4(L-1)L + (L-2)N]}{4L^2 + (L-1)N}, \quad a_{1,i} = \frac{8[4(L-1)+N]\omega_0}{4L^2 + (L-1)N};$$

$$B_1 = (b_{1,r} + ib_{1,i})A_0^2, \tag{4.405}$$

$$b_{1,r} = \frac{-48\{4N + (4L+N)[12L(L-1)+(3L-5)N]\}}{N[4L^2 + (L-1)N]},$$

$$b_{1,i} = \frac{-96[24L(L-1)+2(5L-4)N+N^2]\omega_0}{N[4L^2 + (L-1)N]}.$$

In the second order of expansion, the solvability condition gives an evolution equation for $A(t_4)$,

$$\frac{dA_0}{dt_4} = \sigma A_0 + \kappa|A_0|^2 A_0, \tag{4.406}$$

which is the well-known complex Landau equation. Here, $\sigma = \sigma_r + i\sigma_i$ is the linear complex growth rate, and $\kappa = \kappa_r + i\kappa_i$ is the complex Landau constant. For the real part of the Landau constant, one finds:

$$\kappa_r = \frac{6K^2(4+4L+N)[12L(L-1)-(389+377L)N-95N^2]}{4L^2 + (L-1)N}, \tag{4.407}$$

which is positive in the realistic case when $L < 1$. Therefore, the traveling wave solution bifurcates into the subcritical region, where it is unstable. In the super-critical region, the amplitude of a traveling wave grows without a saturation. That circumstance makes it unnecessary to analyze any other nonlinear patterns: in the considered problem, the weakly nonlinear analysis is unable to find stable wavy regimes.

Nonlinear Evolution of a Deformable Free Surface

The deformability of the boundary (finite values of G) involves an additional conservation law: the conservation of the liquid volume by surface deformation. If the Biot number is small, one obtains *three* amplitude equations for three active variables, averaged temperature F, surfactant concentration Γ, and local layer thickness h. As we know, in the case of an instability mode essentially connected with the surface deformation, one can significantly modify the approach while extending it from the near critical region, $M - M_0 = M_2 \varepsilon^2 \ll 1$, to arbitrary values of $M - M_0$. That is possible if the surface tension is strong, i.e., the inverse capillary number Ca is large; therefore shortwave disturbances are suppressed. Already at the leading order of expansions, one obtains a closed system of nonlinear equations governing the evolution of the active variables. Note that in the case $Ca = O(1)$, the application of the longwave expansions for $M - M_0 = O(1)$ gives an ill-posed system of equations [68], which contains no stabilizing terms with fourth spatial derivatives. The assumption $C = O(\varepsilon^{-2})$ shifts those stabilizing terms into the leading-order equations.

If the Biot number is finite, the averaged temperature is not conserved at the leading order, hence one gets two equations for active variables Γ and h. Such equations have been derived for longwave Marangoni convection in a liquid layer with insoluble surfactant in the presence of evaporation, for a constant kinetic resistance to evaporation [69] and for a kinetic resistance depending on the surfactant concentration [70].

References

1. L.M. Pismen, Phys. Rev. A **38**, 2564 (1988)
2. F.S. Gaeta, G. Perna, G. Scala, F. Bellucci, J. Phys. Chem. **86**, 2967 (1982)
3. P. Kolodner, H. Williams, C. Moe, J. Chem. Phys. **88**, 6512 (1988)
4. G.W.T. Lee, P. Lucas, A. Tyler, J. Fluid Mech. **135**, 235 (1983)
5. W. Hort, S.J. Linz, M. Lücke, Phys. Rev. A **45**, 3737 (1992)
6. J. Liu, G. Ahlers, Phys. Rev. E **55**, 6950 (1997)
7. S. Hollinger, M. Lücke, H.W. Muller, Phys. Rev. E **57**, 4250 (1998)
8. A. Bergeon, D. Henry, H. BenHadid, Microgravity Q. **5**, 123 (1995)
9. D.A. Nield, J. Fluid Mech. **71**, 411 (1975)
10. L. Shtilman, G.I. Sivashinsky, Physica D **52**, 477 (1991)
11. S. Shklyaev, A.A. Nepomnyashchy, A. Oron, Phys. Rev. E **84**, 056327 (2011)
12. M. Silber, E. Knobloch, Nonlinearity **4**, 1063 (1991)
13. M. Roberts, J.W. Swift, D.H. Wagner, Contemp. Math. **56**, 283 (1986)
14. A.H. Nayfeh, *Perturbation Methods* (Wiley, New York, 1973)
15. S. Shklyaev, A.A. Nepomnyashchy, A. Oron, Phys. Fluids **19**, 072105 (2007)
16. T. Clune, M.C. Depassier, E. Knobloch, *Instabilities and Nonequilibrium Structures* (Kluwer, Dordrecht, 1996)

17. J.J. Jasper, J. Phys. Chem. Ref. Data **1**, 841 (1972)
18. S. Shklyaev, A.A. Nepomnyashchy, A. Oron, Phys. Fluids **21**, 054101 (2009)
19. A. Oron, A.A. Nepomnyashchy, Phys. Rev. E **69**, 016313 (2004)
20. J.R.A. Pearson, J. Fluid Mech. **4**, 489 (1958)
21. A. Podolny, A. Oron, A.A. Nepomnyashchy, Phys. Fluids **17**, 104104 (2005)
22. A. Podolny, A. Oron, A.A. Nepomnyashchy, Math. Model. Nat. Phenom. **3**(1), 1 (2008)
23. A. Podolny, A. Oron, A.A. Nepomnyashchy, Fluid Dyn. Mater. Process. **6**, 13 (2010)
24. A.A. Golovin, A.A. Nepomnyashchy, L. Pismen, Physica D **81**, 117 (1995)
25. S. Shklyaev, A.A. Nepomnyashchy, A. Oron, SIAM J. Appl. Math. **73**, 2203 (2013)
26. S. Shklyaev, A.A. Nepomnyashchy, A. Oron, SIAM J. Appl. Math. **74**, 1005 (2014)
27. M. Krupa, I. Melbourne, Ergodic Theory Dyn. Syst. **15**, 121 (1995)
28. I. Melbourne, P. Chossat, M. Golubitsky, Proc. R. Soc. Edinb. Sec. A Math. **113**, 315 (1989)
29. J.L. Castillo, M.G. Velarde, J. Fluid Mech. **125**, 463 (1982)
30. J.L. Castillo, M.G. Velarde, J. Colloid Interface Sci. **108**, 264 (1985)
31. M.-I. Char, K.-T. Chiang, Int. J. Heat Mass Transf. **39**, 407 (1996)
32. J.K. Bhattacharjee, Phys. Rev. E **50**, 1198 (1994)
33. S.W. Joo, J. Fluid Mech. **293**, 127 (1995)
34. J.R.L. Skarda, F.E. McCaughan, Int. J. Heat Mass Transf. **42**, 2387 (1999)
35. A. Podolny, A. Oron, A.A. Nepomnyashchy, Phys. Fluids **18**, 054104 (2006)
36. A. Podolny, A. Oron, A.A. Nepomnyashchy, J. Non-Equilib. Thermodyn. **32**, 1 (2007)
37. M. Morozov, A. Oron, A.A. Nepomnyashchy, Phys. Fluids **25**, 052107 (2013)
38. A. Podolny, A. Oron, A.A. Nepomnyashchy, Phys. Rev. E **76**, 026309 (2007)
39. J. Buongiorno, ASME Trans. J. Heat Transf. **128**, 240 (2006)
40. L. Dong, D. Johnson, Langmuir **19**, 10205 (2003)
41. L. Dong, D. Johnson, Adv. Space Res. **32**, 149 (2003)
42. H. Xue, J. Fan, Y. Hu, R. Hong, K. Cen, J. Appl. Phys. **100**, 104909 (2006)
43. F. Ravera, M. Ferrari, L. Liggieri, G. Loglio, E. Santini, A. Zanobini, Colloids Surf. A **323**, 99 (2008)
44. G.I. Sivashinsky, Physica D **4**, 227 (1982)
45. E. Knobloch, Physica D **41**, 450 (1990)
46. A. Mikishev, A.A. Nepomnyashchy, Microgravity Sci. Technol. **22**, 415 (2010)
47. A. Mikishev, A.A. Nepomnyashchy, Phys. Rev. E **82**, 046306 (2010)
48. A.S.A. Hamid, S.A. Kechil, A.S.A. Aziz, World Acad. Sci. Eng. Technol. **58**, 24 (2011)
49. S. Slavtchev, P. Kalitzova-Kurteva, A. Oron, J. Theor. Appl. Mech. (Sofia) **39**, 63 (2009)
50. P. Kalitzova-Kurteva, S. Slavtchev, A. Oron, J. Theor. Appl. Mech. (Sofia) **39**, 31 (2009)
51. R. Pan, J. Green, C. Maldarelli, J. Colloid Interface Sci. **205**, 213 (1998)
52. S. Shklyaev, A.A. Nepomnyashchy, Eur. Phys. J. Spec. Top. **192**, 155 (2011)
53. M.L. Agrawal, R.D. Neuman, J. Colloid Interface Sci. **121**, 366 (1988)
54. D.S. Valkovska, K.D. Danov, J. Colloid Interface Sci. **223**, 314 (2000)
55. S. Shklyaev, A.A. Nepomnyashchy, J. Fluid Mech. **718**, 428 (2013)
56. D.S. Goldobin, D.V. Lyubimov, Sov. Phys. JETP **104**, 830 (2007)
57. M. Morozov, A. Oron, A.A. Nepomnyashchy, Phys. Fluids **26**, 112101 (2014)
58. M. Morozov, A. Oron, A.A. Nepomnyashchy, Phys. Fluids **27**, 082107 (2015)
59. J.C. Berg, A.A. Acrivos, Chem. Eng. Sci. **20**, 737 (1965)
60. H.J. Palmer, J.C. Berg, J. Fluid Mech. **51**, 38 (1972)
61. V.G. Levich, *Physicochemical Hydrodynamics* (Prentice Hall, Englewood Cliffs, 1962)
62. A.Y. Gilev, A.A. Nepomnyashchy, I.B. Simanovskii, Zh. Prikl. Mekh. Tekhn. Fiz. **5**, 76 (1986, in Russian)
63. A.A. Nepomnyashchy, I.B. Simanovskii, Fluid Dyn. **21**, 169 (1986)
64. A.A. Nepomnyashchy, I.B. Simanovskii, Fluid Dyn. **23**, 302 (1988)
65. A.A. Nepomnyashchy, I.B. Simanovskii, Zh. Prikl. Mekh. Tekhn. Fiz. **1**, 146 (1989, in Russian)
66. A.A. Nepomnyashchy, I.B. Simanovskii, Sov. Phys. Dokl. **34**, 420 (1989)
67. A.A. Nepomnyashchy, I.B. Simanovskii, Zh. Prikl. Mekh. Tekhn. Fiz. **3** 73 (1992, in Russian)
68. A. Mikishev, A.A. Nepomnyashchy, Microgravity Sci. Technol. **23**(Suppl 1), S59 (2011)
69. A. Mikishev, A.A. Nepomnyashchy, Phys. Fluids **25**, 054109 (2013)
70. A. Mikishev, A.A. Nepomnyashchy, Fluid Dyn. Res. **46**, 041420 (2014)

Chapter 5
Instabilities of Parallel Flows

The stability theory of viscous parallel flows is a traditional part of the fluid mechanics that has a long history (see [1–4]). As well as the Rayleigh–Benard problem, the abovementioned problem, which has important applications like laminar-turbulent transition in channel flows and boundary layers, is a touchstone for different approaches of the nonlinear science. The nonlinear evolution of disturbances on the background of a parallel flow in a channel was the first physical problem that was considered by means of the complex Ginzburg–Landau equation [5, 6]. The most obvious difference between the problems connected with parallel flows and the convection problems is the anisotropy and (as a rule) lack of the reflection symmetry characteristic for the former problem. The typical instability of a parallel flow, predicted by the linear theory, is the *non-degenerated oscillatory instability* rather than a stationary instability or twofold degenerated oscillatory instability of convective problems.

In the overwhelming majority of examples, the instability of parallel flows is a *short-wavelength instability*. Usually, the characteristic longitudinal scale of disturbances generating the instability coincides with the transverse scale of the flow itself. However, there are several examples of *long-wavelength instabilities* of parallel flows, connected with some *conservation laws*.

In the present chapter, we consider two typical problems of this kind. In Section 4.1, the flows with free boundaries are considered. For such flows, the crucial circumstance is the conservation of the fluid volume. We shall concentrate on the classical problem of waves generation in a liquid film flowing down an inclined plane that will give us a possibility to consider several qualitatively different types of nonlinear waves. In Section 4.2, we consider a flow generated by an external spatially periodic parallel force. In this case the long-wavelength instability is connected with the conservation of transverse components of the momentum.

In both cases, the existence of the conservation law gives us the possibility to expect that the instabilities are of the type II, and the nonlinear waves generated by those instabilities are governed by equations listed in Section 1.4.2 and by their multidimensional modifications.

© Springer Science+Business Media, LLC 2017
S. Shklyaev, A. Nepomnyashchy, *Longwave Instabilities and Patterns in Fluids*,
Advances in Mathematical Fluid Mechanics,
https://doi.org/10.1007/978-1-4939-7590-7_5

5.1 Waves on an Inclined Plane

In this section, we consider the classical problem of wave motions in a liquid film flowing down an inclined plane originating in the papers of P.L. Kapitsa [7, 8]. It is necessary to emphasize from the very beginning that the weakly nonlinear analysis performed in this section has a quite restricted applicability region. Even for moderate Reynolds numbers, the characteristic amplitude of waves becomes comparable with the mean thickness of the film, and a *strongly nonlinear* (though longwave) approach is necessary (see, e.g., [9, 10]). We shall not apply here that approach, limiting ourselves to consideration of simple (but generic) nonlinear equations.

Let us consider a viscous incompressible liquid flowing under the action of the gravity force (gravity acceleration is g) along the inclined rigid surface (see Figure 5.1.).

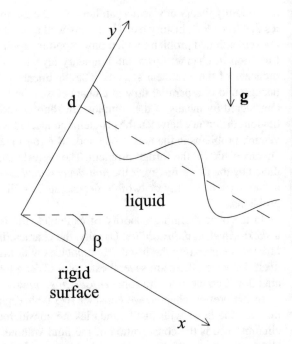

Fig. 5.1 Schematic of liquid flow on inclined plane

The density of the liquid is ρ, the coefficient of kinematic viscosity is v, and the surface tension coefficient is σ; the angle between the surface and the horizontal direction is β; the mean thickness of the layer is equal to d. We use d, v/d, d^2/v, and $\rho(v/d)^2$ as units of length, velocity, time, and pressure. The two-dimensional motion of the fluid is governed by the system of equations

$$\frac{\partial u}{\partial t} + u\frac{\partial u}{\partial x} + v\frac{\partial u}{\partial y} = -\frac{\partial p}{\partial x} + \Delta u + G\sin\beta, \tag{5.1}$$

$$\frac{\partial v}{\partial t} + u\frac{\partial v}{\partial x} + v\frac{\partial v}{\partial y} = -\frac{\partial p}{\partial y} + \Delta v - G\cos\beta, \tag{5.2}$$

$$\frac{\partial u}{\partial x} + \frac{\partial v}{\partial y} = 0; \quad \Delta = \frac{\partial^2}{\partial x^2} + \frac{\partial^2}{\partial y^2}; \tag{5.3}$$

the axes x and y and the origin are chosen in such a way that the rigid boundary of the layer is described by the equation $y = 0$; u and v are x- and y-components of velocity, p is pressure, and $G = gd^3/v^2$ is Galileo number. The boundary conditions on the rigid boundary are:

$$y = 0: \quad u = v = 0; \tag{5.4}$$

on the free boundary (h is the deflection of the surface):

$$y = 1 + h: \quad \frac{\partial u}{\partial y} + \frac{\partial v}{\partial x} + \frac{2\partial h/\partial x(\partial v/\partial y - \partial u/\partial x)}{\left[1 - (\partial h/\partial x)^2\right]} = 0, \tag{5.5}$$

$$p + \frac{\gamma G^{1/3}\partial^2 h/\partial x^2}{\left[1 + (\partial h/\partial x)^2\right]^{1/2}} - 2\frac{\partial v}{\partial y} + \frac{\partial h}{\partial x}\left(\frac{\partial u}{\partial y} + \frac{\partial v}{\partial x}\right) = p_a, \tag{5.6}$$

$$\frac{\partial h}{\partial t} + u\frac{\partial h}{\partial x} = v, \tag{5.7}$$

where

$$\gamma = \frac{\gamma_0}{\rho(v^4 g)^{1/3}}$$

p_a is the atmospheric pressure.

The boundary value problem (5.1)–(5.7) has always a solution describing a stationary parallel flow with a flat surface:

$$h_0 = 0, \; v_0 = 0, \; u_0 = G\sin\beta\, U_0(y), \; p_0 = p_a + G\cos\beta\, P_0(y), \tag{5.8}$$

where

$$U_0(y) = y - \frac{y^2}{2}, \; P_0(y) = 1 - y.$$

5.1.1 Linear Stability Theory

Let us impose a two-dimensional disturbance on the motion (5.8):

$$u = u_0 + \tilde{u}, \; v = v_0 + \tilde{v}, \; h = h_0 + \tilde{h}, \; p = p_0 + \tilde{p}. \tag{5.9}$$

We shall use the stream function:

$$\tilde{u} = \frac{\partial \tilde{\psi}}{\partial y}, \; \tilde{v} = -\frac{\partial \tilde{\psi}}{\partial x}. \tag{5.10}$$

Let us substitute (5.9) and (5.10) into (5.1)–(5.7) and linearize the equations with respect to functions describing disturbances. Assuming

$$(\tilde{h}, \tilde{\psi}, \tilde{p}) = (1, \Psi(y), P(y)) \exp(ikx + \sigma t),$$

we obtain the following eigenvalue problem determining the spectrum of growth rates $\sigma(\alpha, G, \gamma)$:

$$\sigma \Psi' + ikG \sin\beta\, U_0 \Psi' - ikG \sin\beta\, U_0' \Psi = -ikP - k^2 \Psi' + \Psi''', \tag{5.11}$$

$$- ik\sigma \Psi + ik^2 G \sin\beta\, U_0 \Psi = -P' + ik^3 \Psi - ik \Psi''; \tag{5.12}$$

$$\text{at } y = 0: \ \Psi = \Psi' = 0; \tag{5.13}$$

$$\text{at } y = 1: \ U_0'' G \sin\beta + \Psi'' + \alpha^2 \Psi = 0, \tag{5.14}$$

$$P_0' G \cos\beta + P - \gamma G^{1/3} \alpha^2 + 2i\alpha \Psi' = 0, \tag{5.15}$$

$$\sigma + i\alpha G \sin\beta\, U_0 + i\alpha \Psi = 0 \tag{5.16}$$

(prime denotes a differentiation with respect to y).

For long waves, the solution of the problem (5.11)–(5.16) can be obtained by means of expansions in powers of the small parameter k:

$$\Psi = \sum_{n=0}^{\infty} k^n \Psi_n, \ P = \sum_{n=0}^{\infty} k^n P_n, \ \sigma = \sum_{n=0}^{\infty} k^n \sigma_n. \tag{5.17}$$

The expansions (5.17) are substituted into the equations and boundary conditions (5.11)–(5.16), and the terms of the same order in k are collected. Omitting the calculations, we present the expressions for σ_n up to the fourth order:

$$\sigma_1 = -iG \sin\beta; \tag{5.18}$$

$$\sigma_2 = \frac{2}{15}(G \sin\beta)^2 - \frac{1}{3} G \cos\beta; \tag{5.19}$$

$$\sigma_3 = i \left[G \sin\beta + \frac{4}{63}(G \sin\beta)^3 - \frac{10}{63} G \cos\beta\, G \sin\beta \right]; \tag{5.20}$$

$$\sigma_4 = -\frac{1}{3}\gamma G^{1/3} - \frac{75872}{202705}(G \sin\beta)^4 + \frac{17363}{155925}(G \sin\beta)^2 G \cos\beta \tag{5.21}$$

$$- \frac{157}{224}(G \sin\beta)^2 - \frac{2}{45}(G \cos\beta)^2 + \frac{3}{5} G \sin\beta.$$

The stability of the parallel flow at small k is determined by the sign of σ_2. One can see that the flow (5.8) is unstable ($\sigma > 0$) with respect to long-wavelength ($k \to 0$) disturbances, if $G > G_0$, where

$$G_0 = \frac{5 \cos\beta}{2 \sin^2 \beta}. \tag{5.22}$$

The shape of the neutral curve at small k depends on the sign of σ^4. For many fluids the coefficient γ (*the Kapitsa number*) is large, e.g., for water at 15°C $\gamma = 2850$; in this case, for moderate G, the first term in the expression (5.21) prevails, and we have indeed a long-wavelength instability.

However, if the inclination angle β is small, the threshold value G_0 is large; that is why even at $G \sim G_0$, the additional terms in (5.21) are significant. Indeed, using the expressions for σ_2 and σ_4, we find that the neutral curve has the following form at small k:

$$G = G_0 \left[1 + H_0(\gamma, \beta) k^2 \right],$$

where

$$H_0(\gamma, \beta) = \gamma \left(\frac{2}{5} \right)^{2/3} \frac{\sin^{4/3} \beta}{\cos^{5/3} \beta} + \frac{7743}{2240} - \cot^2 \beta \frac{1}{36036} \qquad (5.23)$$

(see Figure 5.2.)

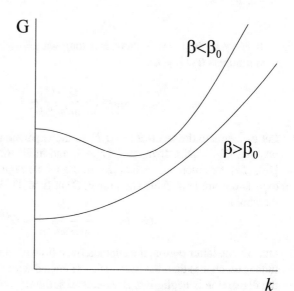

Fig. 5.2 Neutral curves for wave instability

The function $H_0(\gamma, \beta)$ decreases monotonically when β decreases. If $\beta < \beta_0$, where β_0 is defined by the equation $H_0(\gamma, \beta_0) = 0$, the point $k = 0$ becomes a point of the maximum of the neutral curve. The angle β_0 is very small ($\beta_0 \approx 9'45''$ in the limit $\gamma \to 0$; $\beta_0 \approx 9'7''$ for water, $\gamma = 2850$); see line 1 in Figure 5.3. Let us note however that the critical Reynolds number that can be calculated as

$$Re_0 = \frac{1}{3} G_0 \sin \beta$$

is about 300 for such inclination angles; hence the shortwave Tollmien–Schlichting instability is still irrelevant.

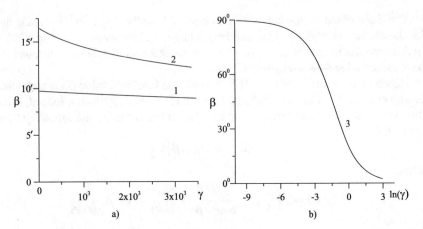

Fig. 5.3 Critical inclination angles; (a) line 1, $H_0(\gamma,\beta) = 0$, and line 2, $H_0(\gamma,\beta) = 2.25$; (b) line 3, $H_0(\gamma,\beta) = 3.6$

If $H_0(\gamma,\beta)$ is positive, there is a long-wavelength instability in the interval of wavenumbers $0 < k < k_m$,

$$k_m = \left(\frac{G - G_0}{G_0 H_0(\gamma,\beta)} \right)^{1/2}. \tag{5.24}$$

Let us estimate the terms $\Sigma_n = \sigma_n k^n$ in the expansion (5.17) for $k \sim k_m$. It is obvious that $|\Sigma_2| \sim |\Sigma_4|$. If $H_0 \gg 1$ ($\gamma \gg 1$, and inclination angles are moderate), then $|\Sigma_2|, |\Sigma_4|$ are much larger than $|\Sigma_3|$ in the whole region where the long-wavelength expansions are applicable (i.e., for $k_m G \sin \beta \ll 1$), except a small region near the threshold:

$$G - G_0 \sim \frac{1}{G_0 H_0 \sin^2 \beta} \sim G_0^{2/3} \gamma^{-1}. \tag{5.25}$$

Outside the latter region, the characteristic time of the evolution of the wave amplitude is much less than the time scale connected with the dispersion of waves; hence the dispersion is negligible. If γ decreases, the region (5.25) extends. For small inclination angles ($H_0 \sim 1$), in the whole region where long-wavelength expansions are applicable, we obtain the opposite inequality:

$$|\Sigma_3| \gg |\Sigma_2|, |\Sigma_4|,$$

and the dispersion of waves is essential. Certainly, there is an intermediate parameters region where $|\Sigma_3| \sim |\Sigma_2|, |\Sigma_4|$.

We consider each region separately.

5.1.2 Small Inclination Angles

5.1.2.1 One-Dimensional Amplitude Equations and Traveling Wave Solutions

For finite-amplitude motions, we shall use long-wavelength expansions. Let us introduce a new variable

$$x_1 = \varepsilon x;$$

that means that we substitute

$$\frac{\partial}{\partial x} = \varepsilon \frac{\partial}{\partial x_1}$$

into (5.1)–(5.7). There are several possibilities to choose the parameter ε (as the width of the instability interval, as a wavenumber of an actual disturbance, etc.); this question is discussed below.

Though the initial disturbance of the fluid interface can be arbitrary, the amplitude of stationary waves in a viscous fluid, arising spontaneously because of an instability of a parallel flow, is quite definite. This is the crucial difference between waves generated by instability in a viscous fluid and waves on the interface of an ideal fluid that can have arbitrary amplitudes (from a certain interval of allowed values). This circumstance is very important, because it turns out that the amplitude of waves tends to zero as $\varepsilon \to 0$. Hence the function h, characterizing the deflection of the interface, itself should be expanded into series in powers of ε:

$$h = \varepsilon^N \sum_n \varepsilon^n h_n. \tag{5.26}$$

Similarly, we expand the finite-amplitude disturbances of pressure and stream function:

$$p = \varepsilon^N \sum_n \varepsilon^n p_n, \quad \psi = \varepsilon^N \sum_n \varepsilon^n \psi_n \tag{5.27}$$

($u = \partial \psi / \partial y$, $v = -\varepsilon \partial \psi / \partial x_1$). The quantity N should be determined by solving the problem.

Taking into account that phenomena connected with propagation, dispersion, and growth of waves, corresponding to different terms in the expansion for σ, have different characteristic time scales, we introduce a set of time variables

$$\tau_n = G_0 \sin \beta \, \varepsilon^n t, \tag{5.28}$$

hence

$$\frac{\partial}{\partial t} = G_0 \sin \beta \sum_n \varepsilon^n \frac{\partial}{\partial \tau_n}, \quad n = 1, 2, \dots \tag{5.29}$$

(we introduced the multiplier $G_0 \sin \beta$ in order to simplify the equations that will be obtained below). Finally, the connection between G and ε will be represented in the form

$$G = G_0 + G_2 \varepsilon^2. \tag{5.30}$$

Substituting expansions (5.26)–(5.30) into (5.1)–(5.6), we obtain at the nth order in ε a system of equations that allows to express the solutions u_n, v_n, and p_n in terms of functions $h_l(x_1, \tau_1, \tau_2, \ldots)$, $0 \leq l \leq n$. Substituting these solutions into (5.7), we find at each order an equation that determines the terms in the expansion (5.26) for h.

In the present section, we consider the case $H_0 \sim 1$, $G - G_0 \sim \varepsilon^2$ that describes, as shown above, the motions near the instability threshold at small inclination angles (or at arbitrary inclination angles, if the constant γ is not large). In this case, the stabilizing contribution of the surface tension into σ_4 does not prevail and should be considered together with other terms of the same order in ε.

One can show that we should take $N = 2$ in the expansions (5.26), (5.27); otherwise we do not obtain the solutions in the form of stationary traveling waves.

The equation of zeroth order gives:

$$\frac{\partial h_0}{\partial \tau_1} + \frac{\partial h_0}{\partial x_1} = 0, \tag{5.31}$$

hence

$$h_0 = h_0(\xi), \quad \xi = x_1 - \tau_1. \tag{5.32}$$

The solutions ψ_0, p_0 are expressed by means of the function $h_0(\xi)$ and its derivative with respect to ξ.

At the first order we obtain:

$$\frac{\partial h_0}{\partial \tau_2} + \frac{\partial^2 h_0}{\partial \xi^2} \left(\frac{2}{15} G_0 \sin \beta - \frac{1}{3} \cot \beta \right) = 0. \tag{5.33}$$

Because of (5.22), h_0 does not depend on τ_2. The solutions ψ_1, p_1 depend on the function $h_0(\xi)$ and its derivatives, and on a new unknown function $h_1(\xi)$, that should be calculated in the next orders.

At the second order we obtain the KdV equation for the function h_0:

$$\frac{\partial h_0}{\partial \tau_3} + \frac{\partial^3 h_0}{\partial \xi^3} + \frac{\partial}{\partial \xi}(h_0^2) = 0. \tag{5.34}$$

This equation has (among others) a four-parametric family of solutions

$$h_0 = h_0(\xi - c_0 \tau_3 - \xi_0)$$

corresponding to traveling periodic (cnoidal) and solitary waves. Imposing the periodicity condition

$$h_0(\xi + 2\pi/q) = h_0(\xi)$$

and the condition

$$\int_0^{2\pi/q} h_0(\xi)d\xi = 0$$

(h_0 is a deflection of the thickness of the surface from its mean value, by definition), we obtain a two-parametric set of solutions (see Section 3.3.1.2):

$$h_0(\xi_1;q;s;\xi_0) = \frac{6q^2K^2(s)}{\pi^2}\left[1 - \frac{E(s)}{K(s)} - s^2 sn^2\frac{qK(s)(\xi_1 - \xi_0)}{\pi}\right], \tag{5.35}$$

$$c_0 = \frac{4q^2K^2(s)}{\pi^2}\left[2 - s^2 - 3\frac{E(s)}{K(s)}\right], \tag{5.36}$$

$$\xi_1 = \xi - c_0\tau_3,$$

where ξ_0 is an arbitrary constant, s is a modulus of elliptic functions, and $K(s)$, $E(s)$ are complete elliptic integrals. The function h_0 satisfies the ordinary differential equation

$$\frac{d^3h_0}{d\xi_1^2} - c_0\frac{dh_0}{d\xi_1} + \frac{d}{d\xi_1}(h_0^2) = 0, \; h_0(\xi_1 + 2\pi/q) = h_0(\xi_1). \tag{5.37}$$

Let us note that $h_0(\xi_1)$ is an even function of $\xi_1 - \xi_0$. In the limit where $s \to 1$ and $q \to 0$ simultaneously in such a way that $qK(s)/\pi$ tends to a finite constant q_1, we get a two-parametric (with parameters q_1 and ξ_0) family of solitary waves,

$$\lim_{\xi_1 \to \pm\infty} h_0(\xi_1) = 0.$$

The parameter $s = s(q)$ and the doubled amplitude of the wave

$$A = (h_0)_{max} - (h_0)_{min} = \frac{6q^2K^2(s)}{\pi^2}$$

of the stationary wave with the fixed wavenumber q are determined from the solvability condition for the equation governing h_1, which is obtained in the third order:

$$\frac{\partial h_1}{\partial \tau_3} + \frac{\partial^3 h_1}{\partial \xi^3} + 2\frac{\partial}{\partial \xi}(h_0 h_1)$$

$$+ \frac{\partial h_0}{\partial \tau_4} + \frac{1}{3}\cot\beta\left[\frac{\partial^4 h_0}{\partial \xi^4}\cdot H_0 + \frac{\partial^2 h_0}{\partial \xi^2}\cdot\frac{G_2}{G_0} + \frac{3}{2}\frac{\partial^2}{\partial \xi^2}(h_0^2)\right] = 0, \tag{5.38}$$

where H_0 is determined by the expression (5.23).

To determine G_2, we have to define the small parameter ε that is used still formally. Let us choose $\varepsilon = k_m$ according to (5.24); in this case, $G_2/G_0 = H_0$. Certainly this choice is possible only in the supercritical region, if $H_0 > 0$, and in the subcritical region, if $H_0 < 0$.

For a stationary traveling periodic wave, we have to take into account the fact that the phase velocity of the traveling wave solution can be a function of G. This means that in the definition of ξ_1 we have to put

$$\xi_1 = \xi - c\tau_3,$$

where

$$c = c_0 + \varepsilon c_1 + \dots$$

We obtain the following ordinary differential equation:

$$\frac{d^3h_1}{d\xi_1^3} - c_0\frac{dh_1}{d\xi_1} + 2\frac{d}{d\xi_1}(h_0h_1) = c_1\frac{dh_1}{d\xi_1} - \frac{1}{3}\cot\beta\left[\frac{d^4h_0}{d\xi_1^4}\cdot H_0 + \right. \tag{5.39}$$

$$\left.\frac{d^2h_0}{d\xi_1^2}\cdot\frac{G_2}{G_0} + \frac{3}{2}\frac{d^2}{d\xi_1^2}(h_0^2)\right],$$

$$h_1(\xi_1 + 2\pi/q) = h_1(\xi_1),\quad G_2/G_0 = H_0.$$

The solvability condition to (5.39) is the orthogonality of the right-hand side of this equation to solutions of the adjoint problem

$$-\frac{d^3\Psi}{d\xi_1^3} + c_0\frac{d\Psi}{d\xi_1} - 2h_0\frac{d\Psi}{d\xi_1} = 0,\quad \Psi(\xi_1 + 2\pi/q) = \Psi(\xi_1). \tag{5.40}$$

The problem (5.40) has two linearly independent solutions: $\Psi = 1$ and $\Psi = h_0$. The condition of the orthogonality of the right-hand side of (5.39) to the first solution is satisfied identically, but the orthogonality condition to the second solution gives the following equation:

$$H_0\int_0^{2\pi/q}\left[\left(\frac{d^2h_0}{d\xi_1^2}\right)^2 - \left(\frac{dh_0}{d\xi_1}\right)^2\right]d\xi_1 - 3\int_0^{2\pi/q}h_0\left(\frac{dh_0}{d\xi_1}\right)^2 d\xi_1 = 0. \tag{5.41}$$

Using (5.37), we can rewrite (5.41) in the form:

$$\left(H_0 - \frac{3}{2}\right)\int_0^{2\pi/q}\left(\frac{d^2h_0}{d\xi_1^2}\right)^2 d\xi_1 - \left(H_0 + \frac{3}{2}c_0\right)\int_0^{2\pi/q}\left(\frac{dh_0}{d\xi_1}\right)^2 d\xi_1 = 0. \tag{5.42}$$

Substituting (5.35) and (5.36) into (5.42), we find the connection between parameters s and q:

$$q^2(s) = H_0\left[\frac{\pi}{K(s)}\right]^2$$

$$\left[\left(H_0 - \frac{3}{2}\right)\frac{20}{7}\frac{(2 - 4s^2 + s^4 + s^6) + (-2 + 3s^2 + 3s^4 - 2s^6)E(s)/K(s)}{(-2 + 3s^2 - s^4) + 2(1 - s^2 + s^4)E(s)/K(s)}\right. \tag{5.43}$$

$$\left. - 6[2 - s^2 - 3E(s)/K(s)]\right]^{-1}.$$

Formulas (5.35), (5.36), and (5.43) together describe the two-parametric (parameters s and ξ_0) family of stationary traveling waves in a viscous fluid layer. These equations were obtained first in [11].

Let us assume that $H_0 > 0$, and consider some limit cases. The limit $s \to 0$ corresponds to small amplitude waves. Using the asymptotics of functions $E(s)$ and $K(s)$ as $s \to 0$, we find:

$$q^2(s) = H_0 \left[H_0 + \frac{3}{16}s^4(1+s^2)\left(H_0 - \frac{9}{4}\right) + \frac{1365}{8192}s^8\left(H_0 - \frac{207}{91}\right) + O(s^{10}) \right]^{-1}.$$

(5.44)

One can see that for $0 < H_0 < 2.25$, the small amplitude wave motions exist for $q^2 > 1$; that corresponds to a finite-amplitude (subcritical) instability of the parallel flow (see Figure 5.4(a)). The inclination angle corresponding to $H_0(\gamma, \beta) = 2.25$ is indicated in Figure 5.3 (line 2). For $H_0 > 2.25$ the wave motions with sufficiently small amplitudes have $q^2 < 1$ (supercritical instability). However, expression (5.44) shows that if $H_0 - 2.25$ is small, for larger values of s q^2 starts to grow and becomes larger than 1. Hence, there is an interval of values of H_0, where the stationary wave solutions with sufficiently large amplitudes exist in the stability region $q^2 > 1$ of the parallel flow, though the bifurcation of small amplitude solitons is supercritical. In this case there exists a region $q_m < q < 1$ where there are two solutions with different amplitudes for any value of q.

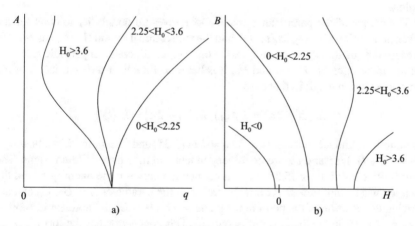

Fig. 5.4 Amplitide curves: (a) on the plane (q, A); (b) on the plane (H, B)

In order to determine the upper boundary of the interval of H_0, where the amplitude curve has features described above, we consider the opposite limit $s \to 1$. In the latter case

$$q^2 \approx \left(\frac{\pi}{\Lambda}\right)^2 \frac{H_0}{(20/7)(H_0 - 3.6) + 18/\Lambda},$$

(5.45)

where

$$\Lambda = \ln \frac{4}{\sqrt{1-s^2}}.$$

One can see that for s large enough the solutions exist in the region $q^2 > 1$, if $H_0 < 3.6$; in that case $q \to \infty$ for a certain finite s. If $H_0 > 3.6$, then $q \to 0$ as $s \to 1$ (see Figure 5.4(a)), and the solution h_0 tends to a finite-amplitude solitary wave:

$$ho = 6q_1^2 \cosh^{-2}(q_1\xi_1), \quad q_1 = \left(\frac{0.35H_0}{H_0 - 3.6}\right)^{1/2}. \qquad (5.46)$$

The inclination angle corresponding to $H_0 = 3.5$ is shown in Figure 5.3(b). Let us note that the expressions (5.45) and (5.46) cease to be justified as $q \to \infty$, $q_1 \to \infty$, because in this case the main assumption, used by derivation of the amplitude equations, is violated.

Thus, when H_0 decreases, first (at $H_0 = 3.6$) the finite amplitude solutions arise in the stability region of the parallel flow, and then (at $H_0 = 2.25$) the type of bifurcation on the neutral curve is changed.

The results of a numerical tabulation of formula (5.43) for intermediate values of s give no new qualitative features compared with the predictions obtained by consideration of limit cases.

If $H_0 < 0$ (e.g., the point $G = G_0$ corresponds to a maximum of the neutral curve), $q^2 < 1$ for any s, and this corresponds to a subcritical instability. As $q \to 0$, the periodic solution tends to the solitary wave solution (5.46) that exists in the subcritical region.

The choice of the parameter ε made above gives no possibility to consider the subcritical region $G < G_0$, as $H_0 > 0$, and the supercritical region $G > G_0$, as $H_0 < 0$, because the quantity k_m is meaningless in these cases. To consider periodic solutions for arbitrary signs of $G - G_0$ and H_0, it is better to fix ε as a wavenumber k of the periodic solution itself; in that case

$$h_0(\xi_1 + 2\pi) = h_0(\xi_1), \quad h_1(\xi_1 + 2\pi) = h_1(\xi_1)$$

by definition, and it is necessary to fix $q = 1$ in (5.35) and (5.36). In (5.39), however, $G_2 = (G - G_0)/\varepsilon^2$ and H_0 are now independent and may have different signs. The solvability condition for (5.39) gives a connection between the quantity s (and the doubled wave amplitude $B = 6K^2(s)s^2/\pi^2$) and the parameter $H = G_2/G_0$ characterizing the position of the point in the plane (k, G) ($H > H_0$ in the linear instability region, $H < H_0$ in the linear stability region). This connection has the form

$$H = \frac{H_0}{q^2(s)},$$

where $q^2(s)$ is the expression (5.43) that is now not necessary positive. One can see that if $H_0 > 3.6$, then $H > H_0$ for any B; if $s \to 1$, then $H \to +\infty$. In the interval $2.25 < H_0 < 3.6$, as B grows, first H grows and reaches a certain maximal value H_m, and then it starts to decrease; as $s \to 1$, $H \to -\infty$. For $H_0 < 2.25$, H decreases monotonously, as B grows. The dependences described above are shown in Figure 5.4(b).

Thus, the behavior of nonlinear wave motions near the threshold $G = G_0$ is determined by the quantity H_0 defined by (5.23). As $H_0 > 3.6$, the solution exists only in the supercritical region. For $2.25 < H_0 < 3.6$ the small-amplitude solutions bifurcate into the supercritical region; in the region $G_0(1 + H_0k^2) < G < G_0(1 + H_mk^2)$, there exist two solutions with different amplitudes in each point (k, G); in the subcritical region $G < G_0(1 + H_0 k^2)$, there exists only one solution. As $H_0 < 2.25$, the solution exists only in the subcritical region.

Let us discuss now the results obtained in this section from the point of view of the general arguments presented in Section 1.4.2. Applying the restitution procedure, we introduce

$$h = \varepsilon h_0 + \varepsilon^2 h_1, \quad \frac{\partial}{\partial \tau} = G_0 \sin \beta \left(\varepsilon^3 \frac{\partial}{\partial \tau_3} + \varepsilon^4 \frac{\partial}{\partial \tau_4} \right), \quad \frac{\partial}{\partial X} = \varepsilon \frac{\partial}{\partial \xi},$$

and write equations (5.34) and (5.38) together in the form

$$\frac{\partial h}{\partial \tau} = \sigma_3 \frac{\partial^3 h}{\partial X^3} + \kappa_1 \frac{\partial (h^2)}{\partial X} + \varepsilon \left[\sigma_2' \frac{\partial^2 h}{\partial X^2} + \sigma_4 \frac{\partial^4 h}{\partial X^4} + \kappa_2 \frac{\partial^2 (h^2)}{\partial X^2} \right], \tag{5.47}$$

where σ_3, $\sigma_2' \equiv (d\sigma_2/dG)(G - G_0)$ and σ_4 are defined by expressions (5.19)–(5.21) calculated at $G - G_0$, $\kappa_1 = G_0 \sin \beta$, $\kappa_2 = (G_0/2) \cos \beta$. The equation (5.47) is identical to the *dissipation-modified Korteweg-de Vries equation* (1.77) that is the most general amplitude equation for the nonlinear waves generated by a II_o instability in the generic case. Hence, the analysis performed in the present section is actually the investigation of traveling wave solutions for the dissipation-modified Korteweg-de Vries equation. Recall that this equation was already discussed in the context of Marangoni waves (see Section 3.3.1.2).

In conclusion of this subsection, let us present the structure of higher-order terms in the right-hand side of the equation (5.47):

$$\frac{\partial h}{\partial \tau} = \ldots + \varepsilon^2 \frac{\partial}{\partial X} \left[\sigma_3' \frac{\partial^2 h}{\partial X^2} + \sigma_5 \frac{\partial^4 h}{\partial X^4} + \kappa_1' h^2 + \kappa_3 \frac{\partial^2 (h^2)}{\partial X^2} + \kappa_4 \left(\frac{\partial h}{\partial X} \right)^2 + \kappa_5 h^3 \right]$$

$$+ \varepsilon^3 \frac{\partial}{\partial X} \left[\sigma_2'' \frac{\partial h}{\partial X} + \sigma_4' \frac{\partial^3 h}{\partial X^3} + \sigma_6 \frac{\partial^5 h}{\partial X^5} + \kappa_2' \frac{\partial}{\partial X} (h^2) + \kappa_6 h \frac{\partial^3 h}{\partial X^3} + \right. \tag{5.48}$$

$$\left. \kappa_7 \frac{\partial h}{\partial X} \frac{\partial^2 h}{\partial X^2} + \kappa_8 \frac{\partial}{\partial X} (h^3) \right],$$

where all the coefficients are real constants. The expression of order ε^2 is necessary, for instance, for calculation of the correction to the phase velocity c_1 that is not determined from the solvability condition (5.41) to (5.39). This correction is found from the solvability condition to the equation for h_2 that contains the terms with coefficient $\sigma_3', \ldots, \kappa_5$ in the right-hand side. Fortunately, using the parity property of h_0, one can show that all these terms do not contribute actually to this solvability condition, which is always satisfied if $c_1 = 0$. The solution h_2 itself, however, depends on all these coefficients. The first nonzero correction to the phase velocity which is obtained from the solvability condition to the equation for h_3 depends on all 13 coefficients in the expression (5.48).

5.1.2.2 Stability of Traveling Waves

The traveling wave solutions bifurcating into the stable region as $H_0 < 2.25$ and the solutions on the upper branch of the amplitude curve for $2.25 < H_0 < 3.6$ are definitely unstable. However, even in the case $H_0 > 3.6$ we cannot expect that the stationary traveling waves are stable for any $0 < q < 1$. Indeed, if $q = 1 - \delta^2, \delta \ll 1$, the amplitude of the traveling wave is of order of δ. It is clear that the wave with the amplitude small enough is unstable, because it is close to the unstable trivial solution $h = 0$. On the other side, the solution with small q ($s \to 1$) which looks like a rare periodic chain of solitary waves is nearly zero everywhere except relatively in small regions of peaks. That is why the waves with small q should be unstable, too. We can expect, however, that the stationary traveling waves are stable if their wavenumbers are inside a certain interval.

In the present subsection, we shal use a more standard and universal scaling for the dissipative-modified KdV equation than one used in (5.34) and (5.39), in order to use the obtained results for other physical problems different from film flows, specifically, for that considered in Section 3.3.1.2. Starting from the general equation (1.77), by means of the scaling transformation

$$a = \alpha U, \ T = \beta t, \ X = \gamma x, \tag{5.49}$$

where

$$\alpha = \frac{3\sigma_3\sigma_2(R - R_c)}{\kappa_1\sigma_4}, \beta = |\sigma_3| \left(\frac{\sigma_2(R - R_c)}{\sigma_4} \right)^{3/2}, \gamma = -sign(\sigma_3)\sqrt{\frac{\sigma_2(R - R_c)}{\sigma_4}}, \tag{5.50}$$

we obtain a universal equation with only two parameters:

$$U_T + 6UU_X + U_{XXX} + \varepsilon \left(U_{XX} + U_{XXXX} + D(U^2)_{XX} \right) = 0, \tag{5.51}$$

where

$$\varepsilon = \frac{\sqrt{\sigma_2(R - R_c)\sigma_4}}{|\sigma_3|}; \ D = 3\frac{\sigma_3\kappa_2}{\kappa_1\sigma_4}. \tag{5.52}$$

The connection between (5.51) and equations (5.34) and (5.39) of the previous subsection is given by formulas:

$$\varepsilon_{new} = \varepsilon_{old} \cdot \frac{1}{3}\cot\beta \cdot H_0, \ h_0 + \varepsilon_{old}h_1 = 3U,$$

$$\tau = T, \ \xi = X, \ D = \frac{9}{2H_0},$$

where ε_{old} and ε_{new} denote the coefficients ε used in the previous and in the present sections.

Let us list the main results of the previous subsection using new notations (see Section 3.3.1.2):

$$u(\xi) = \frac{2q^2 K^2}{\pi^2} \left[dn^2 \frac{qK(\xi - \xi_0)}{\pi} - \frac{E}{K} \right] + O(\varepsilon), \tag{5.53}$$

$$\xi = X - cT,$$

$$c = \frac{4q^2 K^2}{\pi^2} \left(2 - s^2 - \frac{3E}{K} \right) + O(\varepsilon^2),$$

$$A = u_{max} - u_{min} = \frac{2q^2 K^2}{\pi^2},$$

$$q^2(s) = \frac{\pi^2}{4K^2} \left[-\left(2 - s^2 - \frac{3E}{K}\right) - \frac{D - 3}{7} \right.$$

$$\left. \frac{-14(1 - s^2 + s^4)E^2/K^2 + 2(10 - 15s^2 + 13s^4 - 4s^6)E/K - 2(3 - 6s^2 + 5s^4 - 2s^6)}{2(1 - s^2 + s)E/K} \right]^{-1},$$

where $E \equiv E(s)$ and $K \equiv K(s)$; three types of behavior of $A(q)$ are realized now in regions $D < 5/4(H_0 > 3.6)$, $5/4 < D < 2(2.25 < H_0 < 3.6)$, and $D > 2(H_0 < 2.25)$.

Also, another representation of the solution (5.53) can be used:

$$u_0 = u_0(\xi - \xi_0; \omega, i\omega') = \frac{c_0}{2} - 2\wp(\xi - \xi_0 + i\omega'; \omega, i\omega') \tag{5.54}$$

where $\wp(z)$ is the Weierstrass elliptic function with periods ω and $i\omega'$,

$$\omega = \pi/q, \quad \omega' = K'\pi/Kq; \quad K' = K(\sqrt{1 - s^2}). \tag{5.55}$$

Now, let us consider the evolution of small disturbances $\tilde{u}$ imposed on the stationary traveling wave u. Let us substitute $U = u(\xi) + \tilde{u}$ into (5.51) and linearize the equation with respect to $\tilde{u}$. In the reference frame moving together with the traveling wave, we obtain the following equation for the disturbance $\tilde{u}$:

$$\tilde{u}_t - c\tilde{u}_\xi + 6(u\tilde{u})_\xi + \tilde{u}_{\xi\xi\xi} + \varepsilon(\tilde{u}_{\xi\xi} + \tilde{u}_{\xi\xi\xi\xi} + 2D(u\tilde{u})_{\xi\xi}) = 0. \tag{5.56}$$

We put $\tilde{u} = \tilde{\Phi}'(\xi)\exp(\lambda t)$ and integrate once:

$$\lambda\tilde{\Phi} - c\tilde{\Phi}' + 6u\tilde{\Phi}' + \tilde{\Phi}''' + \varepsilon[\tilde{\Phi}'' + \tilde{\Phi}'''' + 2D(u\tilde{\Phi}')'] = \tilde{a}. \tag{5.57}$$

To eliminate the constant $\tilde{a}$, we put $\tilde{\Phi} = \Phi + \tilde{a}/\lambda$, which is possible if $\lambda \neq 0$. Finally, we obtain the following equation:

$$\lambda\Phi - c\Phi' + 6u\Phi' + \Phi''' + \varepsilon[\Phi'' + \Phi'''' + 2D(u\Phi')'] = 0. \tag{5.58}$$

The solution is stable if $Re\lambda \leq 0$ for any disturbances.

We assume asymptotic expansion as before:

$$\Phi = \Phi_0 + \varepsilon\Phi_1 + \ldots \tag{5.59}$$

$$\lambda = \lambda_0 + \varepsilon\lambda_1 + \ldots, \tag{5.60}$$

and we get the following system of equations:

$$O(1): L_0(\Phi_0) = 0 \tag{5.61}$$
$$O(\varepsilon): L_0(\Phi_1) = -\Omega_1 \Phi_0 - L_1(\Phi_0), \tag{5.62}$$

where:

$$L_0(\Phi) = \lambda_0 \Phi - c_0 \Phi' + 6u_0 \Phi' + \Phi''',$$
$$L_1(\Phi) = 6u_1 \Phi' + (\Phi'' + \Phi'''' + 2D(u_0 \Phi')')$$

We construct solutions to the system, according to Floquet's theory, in the form:

$$\Phi_n(\xi) = f_n(\xi)e^{i\tilde{q}\xi} \tag{5.63}$$
$$f_n\left(\xi + \frac{2\pi}{q}\right) = f_n(\xi) \tag{5.64}$$

where $\tilde{q}$ is a real number (quasi-wavenumber) which can be chosen inside the Brillouin zone:

$$|\tilde{q}| \le q/2. \tag{5.65}$$

The solutions of the problem for Φ_0 were found by Spektor [12]:

$$\Phi_0 = \frac{\sigma^2(\xi + i\omega' + \alpha)}{\sigma^2(\xi + i\omega')\sigma^2(\alpha)} \exp\left[-2(\xi + i\omega')\zeta(\alpha)\right], \tag{5.66}$$

$$\lambda_0 = -4\wp'(\alpha), \tag{5.67}$$

$$\tilde{q} = 2i\left[\zeta(\alpha) - \frac{\alpha}{\omega}\zeta(\omega)\right], \tag{5.68}$$

where σ and ζ are Weierstrass elliptic functions. Let us note that the formula (5.68) gives $\tilde{q}$ in the representation of extended Brillouin zone and $\tilde{q}$ does not satisfy the condition (5.65). The function f_0 corresponding to Φ_0 has the form

$$f_0 = f_{even} + i f_{odd} \tag{5.69}$$

where f_{even} and f_{odd} are real functions. The quasi-wavenumber $\tilde{q}$ is a real number, when $Re\,\alpha$ is equal to $n\omega$. It can be shown that λ_0 is an imaginary number. Thus, in order to find stability domains, we have to consider the next order of the asymptotic expansion.

The solvability condition for the periodic solution $f_1(\xi)$ can be written by means of the solution of the adjoint problem

$$L_0^*(\Psi) = \lambda_0 \Psi + c_0 \Psi' - 6(u_0\Psi)' - \Psi''' = 0 \tag{5.70}$$

corresponding to the quasi-wavenumber $\tilde{q}$. Obviously, this solution is

$$\Psi = \Phi_0'. \tag{5.71}$$

It can be shown that this solvability condition can be written in the form

$$(\Phi_0', \lambda_1 \Phi_0 + 6u_1 \Phi_0' + \Phi_0'' + \Phi_0'''' + 2D(u_0 \Phi_0')') = 0 \tag{5.72}$$

where we use the notation

$$(f,g) \equiv \int_0^{\frac{2\pi}{q}} f^*(\xi) g(\xi) d\xi. \tag{5.73}$$

(though the functions f, g are not periodic in our case, the product f^*g is periodic). Taking into account the parity properties of functions u_0 and u_1 and the property (5.69), we find that λ_1 is a real number:

$$\lambda_1 = -\frac{[(1 + Dc_0/3)(\Phi_0', \Phi_0'') + (1 - D/3)(\Phi_0', \Phi_0'''') - (D\Omega_0/3)(\Phi_0', \Phi_0')]}{(\Phi_0', \Phi_0)}. \tag{5.74}$$

Though in the film stability problem the coefficient D is always positive, for sake of completeness of the mathematical analysis, we shall consider here arbitrary signs of D.

Let us calculate the dependences $\lambda_1(\tilde{q})$ for fixed values of parameters (q, D) by means of the formula (5.74). Our goal is to find the domain in (q, D) plane where $\lambda_1 \leq 0$ for all the possible values of $\tilde{q}$, corresponding to the stability region. These calculations were performed numerically in [13]. Actually, variables (k, D, α) were used. Recall that $Re\alpha = n\omega$, where n is integer. It was sufficient to fulfill the calculations only for two values of n: $n = 0$ and $n = 1$. Other values of n do not define new instability modes. For each (q, D), we found the points $\tilde{q}_m(q, D)$ where the functions $\lambda_1(\tilde{q})$ take their maximal values $\lambda_m(q, D) = max \, \lambda_1(\tilde{q})$. The curves $\lambda_m(q, D) = 0$ define the boundaries of the stability region, and the dependences $\tilde{q}_m(q, Q)$ determine the type of the instability.

Three types of instabilities have been found. The right boundary of the stability region has been obtained for $Re\alpha = \omega$. For $D > D_c$ ($D_c = -6.8 \pm 0.1$) the right boundary has been calculated in the limit $\alpha \to \omega + i\omega'$, $\tilde{q} \to q$ (let us remind that for the reduced quasi-wavenumber $\tilde{q} - q$, we have still $|\tilde{q} - q| \gg \varepsilon$). This instability corresponds to a weakly inhomogeneous translation of the original wave. The other part of the right boundary of the stability region ($D < D_c$) corresponds to finite values of $|\tilde{q} - q|$. On this boundary, a small-amplitude wave with an incommensurable wavelength starts to grow on the background of the periodic wave. The left boundary has been obtained for $Re\alpha = 0$ and corresponds to finite values of the reduced quasi-wavenumber. On the left boundary, the corresponding disturbance generates a wave with smaller wavelength between the peaks of the original wave. For $D \geq 1.25$ the stability region disappears.

The case $D \to -\infty$ can be considered analytically. In this case the problem has two proper small parameters:

$$\tilde{\varepsilon} = |D|^{1/2}\varepsilon, \quad \delta = |D|^{-1/2}. \tag{5.75}$$

The stationary traveling wave solution $u = \delta v$ is governed by the equation

$$-cv' + v''' + 3\delta(v^2)' + \tilde{\varepsilon}(-(v^2)'' + \delta(v'' + v''')) = 0.$$

Assuming that $q = O(1)$, we represent the solution in the form of a series in $\tilde{\varepsilon}$ and δ:

$$v = \sum_{n=0}^{\infty} \tilde{\varepsilon}^n \sum_{m=0}^{\infty} \delta^m v_n^{(m)}, \quad c = \sum_{n=0}^{\infty} \tilde{\varepsilon}^n \sum_{m=0}^{\infty} \delta^m c_n^{(m)}. \tag{5.76}$$

Using the abovementioned expansions, we find:

$$v = A\cos q\xi + \delta\frac{A^2}{2q^2}\cos 2q\xi + \tilde{\varepsilon}\frac{A^2}{3q}\sin 2q\xi + \ldots, \tag{5.77}$$

$$c = -q^2 + \ldots, \tag{5.78}$$

$$A^2(q) = \frac{2}{3}q^2(1 - q^2). \tag{5.79}$$

For the disturbance $\tilde{v}exp(\lambda T)$, we substitute the expansions

$$\tilde{v} = \sum_{n=0}^{\infty} \tilde{\varepsilon}^n \sum_{m=0}^{\infty} \delta^m \tilde{v}_n^{(m)}, \quad \lambda = \sum_{n=0}^{\infty} \tilde{\varepsilon}^n \sum_{m=0}^{\infty} \delta^m \lambda_n^{(m)} \tag{5.80}$$

into the equation:

$$\lambda\tilde{v} - c\tilde{v}' + \tilde{v}''' + 6\delta(v\tilde{v})' + \tilde{\varepsilon}(-2(v\tilde{v})'' + \delta(\tilde{v}'' + \tilde{v}''')) = 0.$$

We obtain:

$$\tilde{v} = e^{i\tilde{q}\xi}\left(1 + \delta\frac{2iA(q)}{q\tilde{q}}\sin(q\xi) + \tilde{\varepsilon}\frac{2A(q)}{3q\tilde{q}}[\tilde{q}\sin(q\xi) - iq\cos(q\xi)] + \ldots\right), \tag{5.81}$$

$$\lambda = i(-q^2\tilde{q} + \tilde{q}^3) + \tilde{\varepsilon}\delta(\tilde{q}^2 - \tilde{q}^4 - 2A^2(q)) + \ldots. \tag{5.82}$$

Using the expression (5.79), we find that $Re\lambda < 0$ when $1/2 < q < \sqrt{3}/2$.

It should be noted that the expression (5.74) is valid only if the reduced quasi-wavenumber $\tilde{q}$ inside the Brillouin zone (i.e., $\tilde{q}\,mod\,q$) satisfies the condition $\tilde{q} \gg \varepsilon$. In the region $\tilde{q} = O(\varepsilon)$, the perturbation theory should be constructed in a different way. It is known that the nonperturbed KdV equation has three acoustic modes [14]. Putting $\tilde{q} = Q\varepsilon$, $Q = O(1)$ and performing the asymptotic expansions for the solution of the equation (5.56), we obtain a cubic dispersion relation determining the growth rate ωQ for three resonant modes. The expressions (5.67) and (5.74) for $\tilde{q} = 0$, $\pm q$ correspond to the limit $Q \to \infty$ for the roots of this cubic dispersion relation (see [13]). However, the disturbances with finite Q which have actually wavenumbers of order $O(\varepsilon^2)$ (because of the scaling transformation (5.49)) are impossible if the length of the physical system is less than $O(\varepsilon^{-2})$, and their physical relevance is unclear. That is why we shall not discuss this case in detail here.

5.1.2.3 Two-Dimensional Waves

For three-dimensional disturbances with the wavevector (k_x, k_y) (x is the direction of the parallel flow, and y is the transverse direction), the linear stability theory gives the expression for the growth rate

$$\sigma = \sum_{n,m=0}^{\infty} k_x^n k_y^m \sigma_{n,m}, \qquad (5.83)$$

where $\sigma_{n,0} = \sigma_n$ (see expressions (5.18)–(5.21)),

$$\sigma_{02} = -\frac{1}{3} G \cos\beta; \quad \sigma_{12} = iG \sin\beta \left(1 - \frac{10}{63} G \cos\beta\right);$$

$$\sigma_{22} = -\frac{2}{3}\gamma G^{1/3} + \frac{17363}{155925}(G \sin\beta)^2 G \cos\beta - \frac{157}{224}(G \sin\beta)^2$$
$$-\frac{4}{45}(G \cos\beta)^2 + \frac{6}{5} G \cos\beta;$$

$$\sigma_{04} = -\frac{1}{3}\gamma G^{1/3} - \frac{2}{45}(G \cos\beta)^2 + \frac{3}{5} G \cos\beta.$$

For an inclined plane ($\beta \neq \pi/2$), $\sigma_{02} \neq 0$. In that case, the characteristic scale of the transverse disturbances is $y_2 = \varepsilon^2 y$. The only modification of the leading order amplitude equation (5.47) is the additional term $\varepsilon\sigma_{02}\partial^2 h/\partial y_2^2$ in the right-hand side of the equation. By means of a scale transformation and a change of the reference frame, one can rewrite the amplitude equation in the following standard form:

$$U_T + 6UU_X + U_{XXX} + \varepsilon\left(U_{XX} + U_{XXXX} - U_{YY} + D(U^2)_{XX}\right) = O(\varepsilon^2), \qquad (5.84)$$

where

$$\varepsilon = \varepsilon\frac{(\sigma_2'\sigma_4)^{1/2}}{|\sigma_3|}, \quad D = \frac{3\sigma_3\kappa_2}{\kappa_1\sigma_4} = \frac{9}{2H_0(\gamma,\beta)}. \qquad (5.85)$$

Thus, we obtain an *anisotropic dissipation-modified Korteweg-de Vries equation*. It allows to consider a class of oblique waves

$$U = u(\eta), \quad \eta = X + \alpha Y - cT, \quad |\alpha| < 1. \qquad (5.86)$$

The analysis of solutions and their stability in the framework of equation (5.85) has been done in [15, 16].

5.1.3 Vertical Plane, Moderate Surface Tension

The derivation of the amplitude equation has to be modified in the case of a vertical plane ($\beta = \pi/2$), where $\sigma_{02} = 0$ and $G_0 = 0$. In the present section, we assume $G \ll$

1, $\gamma = O(1)$ (i.e., the surface tension is moderate). Taking G as the small parameter, we find

$$\sigma_{20} \sim \frac{2}{15}G^2, \quad \sigma_{40} \sim -\frac{1}{3}\gamma G^{1/3},$$

thus the scale of the instability region in the k_x-direction is

$$k_m = \left(-\frac{\sigma_{20}}{\sigma_{40}}\right)^{1/2} \sim (2/5\gamma)^{1/2}G^{5/6}.$$

Because of the equality $\sigma_{02} = 0$, the scale of the instability region in k_y-direction is determined by terms

$$\sigma_{22} \sim 2\sigma_{40}, \quad \sigma_{04} \sim \sigma_{40},$$

hence it is the same as in the k_x-direction. Estimating different terms in the amplitude equations, we find that the terms

$$\sigma_{30}\frac{\partial^3 h}{\partial x^3}, \quad \sigma_{12}\frac{\partial^3 h}{\partial x \partial y^2}, \quad \kappa_1\frac{\partial}{\partial x}(h^2) = O(G^{31/6}),$$

and

$$\sigma_{20}\frac{\partial^2 h}{\partial x^2}, \quad \sigma_{40}\frac{\partial^4 h}{\partial x^4}, \quad \sigma_{22}\frac{\partial^4 h}{\partial x^2 \partial y^2}, \quad \sigma_{04}\frac{\partial^4 h}{\partial y^4} = O(G^{16/3}),$$

while all other terms are much smaller. By rescaling, we obtain a *dissipation-modified Zakharov–Kuznetsov equation*,

$$U_T + UU_X + \Delta U_X + \varepsilon(U_{XX} + \Delta^2 U) = o(\varepsilon), \tag{5.87}$$

where $\varepsilon \sim G^{1/6}$ and $\Delta = \partial^2/\partial X^2 + \partial^2/\partial Y^2$.

It is known that one-dimensional solitary and cnoidal waves governed by the nonperturbed Zakharov–Kuznetsov equation are unstable with respect to transverse disturbances [12, 17, 18]. In the case of a primary one-dimensional soliton, the instability leads to formation of a chain of two-dimensional solitons [17, 19]. Numerical simulations of the perturbed Zakharov–Kuznetsov equation [20–22] show the existence of oblique V-shaped chains of two-dimensional solitons on the background of weak chaotic waves.

5.1.4 Strong Surface Tension

Another kind of expansions is used in the problem of film flows of a liquid with large surface tension [23]. Assume that $G - G_0 = O(1)$, but $\gamma \gg 1$. Then the width of the instability interval,

$$k_m = \left[\frac{2\sin^2\beta G^{2/3}(G - G_0)}{5\gamma}\right]^{1/2}, \tag{5.88}$$

is narrow, and it can be used as a small parameter ε. In the reference frame moving with the phase velocity of waves, one obtains an *anisotropic Kuramoto-Sivashinsky equation* [24, 25] (cf. (1.79)),

$$\frac{\partial h}{\partial \tau} = \sigma_{20}\frac{\partial^2 h}{\partial \xi^2} + \sigma_{02}\frac{\partial^2 h}{\partial y_1^2} + \sigma_{40}\frac{\partial^4 h}{\partial \xi^4} + \sigma_{22}\frac{\partial^4 h}{\partial \xi^2 \partial y_1^2} + \sigma_{04}\frac{\partial^4 h}{\partial y_1^4} + \kappa_1\frac{\partial(h^2)}{\partial \xi}. \tag{5.89}$$

By rescaling, that equation is transformed to a standard form:

$$U_T + U_{XX} - nU_{YY} + \Delta_{\perp}^2 U + (U^2)_X = 0, \tag{5.90}$$

where

$$n = G_0/(G - G_0).$$

The most characteristic feature of the Kuramoto–Sivashinsky equation is the generation of the spatio-temporal chaos [26] that manifests itself as spontaneous creation and annihilation of pulses, with the typical mean distance between pulses corresponding to the wavenumber $k = 0.71$ [27–30]. The properties of the spatio-temporal chaos are described in detail elsewhere (see, e.g., [31]). Here we discuss only the stability of stationary spatially periodic waves with the plane front,

$$U = U_0(X); \ U_0(X + 2\pi/q) = U_0(X), \tag{5.91}$$

with respect to disturbances

$$\tilde{U}(X,Y,T) = F(X)\exp(i\tilde{q}_x X + i\tilde{q}_y Y)\exp\sigma T, \tag{5.92}$$

where $F(X)$ is a $2\pi/q$-periodic function, which satisfies equation

$$\mu F + \left(\frac{d}{dX} + i\tilde{q}_x\right)^4 F + (1 - 2\tilde{q}_y^2)\left(\frac{d}{dX} + i\tilde{q}_x\right)^2 F + 2\left(\frac{d}{dX} + i\tilde{q}_x\right)(U_0 F) = 0. \tag{5.93}$$

Here $\mu = \sigma + n\tilde{q}_y^2 + \tilde{q}_y^4$.

For longwave disturbances, one can use the expansions

$$F = \sum_{n=0}^{\infty} F_n \delta^n, \ \mu = \sum_{n=0}^{\infty} \mu_n \delta^n, \tag{5.94}$$

where

$$\delta = (\tilde{q}_x^2 + \tilde{q}_y^4)^{1/2}, \ \tilde{q}_x = \delta\cos\phi, \ \tilde{q}_y^2 = \delta\sin\phi.$$

If only one-dimensional disturbances are taken into account ($\phi = 0$), solutions (5.91) are stable within the interval of wavenumbers $q_2 < q < q_1$, where $q_1 \approx 0.837$ and $q_2 \approx 0.768$ [23, 32, 33]. Both stability boundaries are created by longwave disturbances. If $q > q_1$, then $\mu_1^2 > 0$; in that case, there are two real roots of opposite sign, one of which is always positive. If $q < q_1$, then $\mu_1^2 < 0$ (two imaginary roots), while the coefficient μ_2 is negative if $q > q_2$ and positive otherwise. Thus the instability of a periodic solution is monotonic for $q > q_1$ and oscillatory for $q < q_2$.

The stability of solutions (5.91) with respect to two-dimensional disturbances was investigated in [24, 34]. The oscillatory instability boundary, $q = q_2$, is not changed. The monotonic instability boundary depends on n.

Let us discuss first the case $n = 0$. Near the boundary $q = q_1$, where μ_1^2 changes its sign, one finds that in the region $q < q_1$ (where $\mu_1^2 < 0$) there exists a region of the instability with respect to two-dimensional disturbances; its boundary is determined by the formula

$$\tilde{q}_x = \frac{\tilde{q}_y^6}{\tilde{q}_y^2 + \mu_1^2/2\alpha}, \tag{5.95}$$

where α is a positive number. According to (5.95), the region of wavevectors of growing disturbances is separated from the axis $\tilde{q}_x$ by a "gap", $(\tilde{q}_y^2)_{min} = -\mu_1^2/2\alpha$. The monotonic instability region disappears only for $q < q_3 \approx 0.74$, i.e., $q_3 < q_2$; thus the intervals of monotonic and oscillatory instability overlap, and hence all the one-dimensional waves are unstable with respect to transverse corrugation of the front. The most unstable disturbance has $\tilde{q}_y = q/2$, i.e., the three-dimensional disturbance doubles the longitudinal spatial period of the wave.

The term with $n > 0$ is a stabilizing one. A full stabilization with respect to monotonic three-dimensional disturbances takes place for $n > n_*(q)$. The stability interval appears for $n > n_c = 0.12$, which corresponds to $G < 9.3G_0$. The stability diagram on the plane (q, n) is shown in Figure 5.5.

The nonlinear development of the three-dimensional instability does not create stable three-dimensional waves; instead, an attracting heteroclinic loop appears [25].

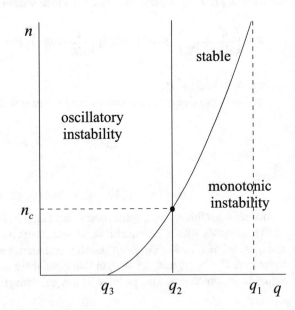

Fig. 5.5 Stability of plane waves with respect to three-dimensional disturbances

5.1.5 Intermediate Parameter Region

In the intermediate case, $|\Sigma_3| \sim |\Sigma_2|, |\Sigma_4|$, the long small-amplitude waves are governed by the *Kawahara equation* [35, 36],

$$\frac{\partial h}{\partial \tau} = \sigma_2' \frac{\partial^2 h}{\partial X^2} + \sigma_3 \frac{\partial^3 h}{\partial X^3} + \sigma_4 \frac{\partial^4 h}{\partial X^4} + \kappa_1 \frac{\partial (h^2)}{\partial X}. \tag{5.96}$$

The scaled equation,

$$U_T + U U_X + U_{XX} + \delta U_{XXX} + U_{XXXX} = 0, \tag{5.97}$$

includes just one non-dimensional parameter, δ, characterizing the relative strength of dispersion. It has been found that dispersion changes drastically the behavior of waves: the spatio-temporal chaos is suppressed, and nearly periodic wave trains, which consist of identical one-hump pulses, are developed [37–39]. With the growth of δ, the numerous families of stationary periodic waves, each ending in a solitary wave, disappear, and the only family of one-hump pulse trains remains [40]. The solutions of that family are stable in a wide interval of wavenumbers [40]. More details can be found in the book [41].

5.2 Kolmogorov Flow

In the present section, we consider the longwave instabilities of a short-scale spatially periodic unidirectional flows in an unbounded space which is created by a spatially periodic external force. Though this formulation of the problem looks artificial, the obtained results may be important for understanding the phenomenon of spontaneous generation of large-scale structures by small-scale flows which is observed in different physical contexts (coherent structures in turbulent flows, atmospheric cyclones, etc.) That is why the investigation of a particular relatively simple case may be useful for solving that intriguing problem.

5.2.1 Derivation of the Amplitude Equation

Assume that a viscous incompressible fluid of density ρ and kinematic viscosity ν moves under the action of an external force parallel to the x-axis and varying in the direction of the y-axis,

$$\mathbf{F} = A \sin(y/d) \mathbf{e}_x, \tag{5.98}$$

where $\mathbf{e}_x$ is a unit vector in the direction of the x-axis. Choosing d, ν/d and $\rho \nu^2/d^2$ as the length, velocity, and pressure scales, respectively, we write the equation of motion in the non-dimensional form as follows:

$$\frac{\partial \mathbf{v}}{\partial t} + (\mathbf{v} \cdot \nabla)\mathbf{v} = -\nabla p + \Delta \mathbf{v} + R \sin y \mathbf{e}_x; \ \nabla \cdot \mathbf{v} = 0, \tag{5.99}$$

and $R = Ad^3/\rho \nu^2$ is the Reynolds number. Later on, we consider two-dimensional flows ($v_z = 0$), and use the notations $u \equiv v_x$, $v \equiv v_y$. We also require that the mean flux of the fluid vanishes,

$$\lim_{l \to \infty} \frac{1}{l} \int_{-l}^{l} u(x,y)dy = 0, \ \lim_{l \to \infty} \frac{1}{l} \int_{-l}^{l} v(x,y)dx = 0, \tag{5.100}$$

and the functions $\mathbf{v}$ and ∇p are bounded. The system (5.99) has a solution which satisfies conditions (5.100) and the conditions of boundedness,

$$\mathbf{v} = \mathbf{V} = (R \sin y, 0), \ p = P = \text{const.} \tag{5.101}$$

Flow (5.101) is the famous *Kolmogorov flow* suggested by Kolmogorov as the object for studying the role of instabilities in the laminar-turbulent transition. The linear stability theory of that flow was developed by Meshalkin and Sinai [42] who found that the Kolmogorov flow is unstable with respect to long-wavelength disturbances, unlike the flows in channels and boundary layers which are subject to short-wavelength instabilities. Expansion in terms of the wave number k yields the following expression for the growth rate σ:

$$\sigma(k,R) = (R^2/2 - 1)k^2 - R^2(1 + R^2/4)k^4 + O(k^6). \tag{5.102}$$

One can see that the instability appears as $R > R_0 = \sqrt{2}$ in the interval of wavenumbers $0 < k < k_m$, where

$$R = \sqrt{2}[1 + 3k_m^2/2 + O(k_m^4)]. \tag{5.103}$$

As expected, in the instability region

$$\sigma(k,R) = O(k_m^4).$$

In order to construct long-scale two-dimensional solutions, we introduce variables $X = k_m x$, $T = k_m^4 t$. The appropriate scales of the disturbances of the velocity and the pressure are as follows [43]:

$$\mathbf{v} = \mathbf{V} + k_m \mathbf{v}', \ p = P + k_m^2 p'.$$

Omitting primes, we obtain the following system of equation for the rescaled disturbances:

$$\frac{\partial v}{\partial y} = -k_m \frac{\partial u}{\partial X}, \tag{5.104}$$

$$\frac{\partial^2 u}{\partial y^2} = R \cos y v + k_m \left(R \sin y \frac{\partial u}{\partial X} + v \frac{\partial u}{\partial y} \right) +$$

$$k_m^2 \left(u \frac{\partial u}{\partial X} - \frac{\partial^2 u}{\partial X^2} + \frac{\partial p}{\partial X} \right) + k_m^4 \frac{\partial u}{\partial T}, \tag{5.105}$$

$$\frac{\partial p}{\partial y} = -\frac{\partial^2 u}{\partial X \partial y} - R \sin y \frac{\partial v}{\partial X} - v \frac{\partial v}{\partial y} +$$

$$k_m \left(-u \frac{\partial v}{\partial X} + \frac{\partial^2 v}{\partial X^2} \right) - k_m^3 \frac{\partial v}{\partial T}; \tag{5.106}$$

in the last equation, the continuity equation (5.104) has been used.

Let us introduce expansions for the rescaled disturbances,

$$\mathbf{v} = \sum_{n=0}^{\infty} \mathbf{v}_n k_m^n, \quad p = \sum_{n=0}^{\infty} p_n k_m^n \tag{5.107}$$

and describe the relation between k_m and R as

$$R = \sum_{n=0}^{\infty} R_{2n} k_m^{2n}. \tag{5.108}$$

Substitute expansions (5.107) and (5.108) into equations (5.104)–(5.106), and equate the terms of the same order.

The zeroth order system of equations is as follows:

$$\frac{\partial v_0}{\partial y} = 0, \quad \frac{\partial^2 u_0}{\partial y^2} = R_0 \cos y v_0, \quad \frac{\partial p_0}{\partial y} = -\frac{\partial^2 u_0}{\partial X \partial y} - R_0 \sin y \frac{\partial v_0}{\partial X}.$$

Its bounded solution is

$$v_0 = v_0(X, T), \quad u_0 = -R_0 \cos y v_0 + u_0^{(1)}(X, T),$$

$$p_0 = 2 R_0 \cos y \frac{\partial v_0}{\partial X} + p_0^{(1)}(X, T).$$

From the first order equations we obtain

$$v_1 = R_0 \sin y \frac{\partial v_0}{\partial X} + v_1^{(1)}(X, T),$$

$$u_1 = -R_0 \sin y v_0^2 - R_0 \cos y v_1^{(1)} + u_1^{(1)}(X, T),$$

$$p_1 = R_0 \sin y \frac{\partial}{\partial X} \left(v_0^2 \right) + \frac{1}{2} \sin 2y \frac{\partial^2 v_0}{\partial X^2} +$$

$$2 R_0 \cos y \frac{\partial v_1^{(1)}}{\partial X} + p_1^{(1)}(X, T).$$

From the solvability conditions for v_1 and p_1 and the boundedness condition, with (5.100) taken into account, we find:

$$u_0^{(1)} = 0, \; R_0 = \sqrt{2}.$$

In the second order we have

$$v_2 = -R_0 \cos y \frac{\partial}{\partial X} \left(v_0^2 \right) + \ldots,$$

$$u_2 = -R_2 \cos y v_0 - R_0 \cos y \left(3 \frac{\partial^2 v_0}{\partial X^2} - v_0^3 \right) + \ldots.$$

We denote by ellipsis the terms related to the solutions of homogeneous equations that are not used in the derivation of the leading order amplitude equation. The solvability condition of the equation for u_2 yields

$$p_0^{(1)} = v_0^2 + C,$$

where C is an arbitrary constant.

Finally, in the third order the solvability condition of the equation for p_3 gives a closed nonlinear equation for the function v_0 [44]:

$$\frac{\partial v_0}{\partial T} + 3 \frac{\partial^4 v_0}{\partial X^4} + R_2 R_0 \frac{\partial^2 v_0}{\partial X^2} - \frac{2}{3} \frac{\partial^2}{\partial X^2} \left(v_0^3 \right) = 0. \tag{5.109}$$

Because k_m is defined as a wavenumber of a neutral disturbance, the linearized equation (5.109) has a time-independent solution $v_0 = \exp(iX)$. That implies $R_2 = 3\sqrt{2}/2$, which agrees with (5.103). Equations (5.109) is transformed to the standard form of the Cahn–Hilliard equation,

$$\frac{\partial V}{\partial \tau} + \frac{\partial^4 V}{\partial X^4} + \frac{\partial^2 V}{\partial X^2} - \frac{\partial^2}{\partial X^2} \left(V^3 \right) = 0 \tag{5.110}$$

by the scaling transformation

$$v_0 = \frac{3}{2} \sqrt{2} V, \; \tau = 3T.$$

Using the properties of the solutions of the Cahn–Hilliard equation discussed in detail in Section 2.2.1, we come to the conclusion that all the spatially periodic motions are unstable with respect to disturbances violating the periodicity. The coarsening process leads to the permanent growth of the characteristic scale of the motion [45].

The analysis of the spatiotemporal dynamics created by the instability of the Kolmogorov flow beyond the weakly nonlinear approach has been carried out in [46–50].

5.2.2 Generalizations

The fully two-dimensional formulation of the problem (5.99) ignores the finite depth
of the liquid layer unavoidable in experiments [51]. The friction on the bottom surface leads to an additional dissipative term $+\alpha V$, $\alpha > 0$ in the left-hand side of
(5.110) [52].

In [53], the instability of an arbitrary unidirectional spatially periodic flow,

$$\mathbf{V} = Ru(y)\mathbf{e}_x, \ u(y+L) = u(y), \ \int_0^L u(y)dy = 0,$$

was considered. It was shown that generally an *oscillatory* instability takes place
with the growth of the Reynolds number R, and the dispersion law in the longwave
limit is $\omega(k) \sim \omega_3 k^3 + o(k^3)$. If $\omega_3 \neq 0$, the leading order amplitude equation is
linear, and one obtains harmonic waves with amplitudes determined by the Lan-
dau equations obtained from the solvability conditions at the next order. In the
case $\omega_3 = 0$ (specifically, that happens if $u(y)$ is an odd function), the leading
order amplitude equation is similar to (5.110), but it contains an additional term
$+A\partial(V\partial^2 V/\partial X^2)/\partial X$. The quantity A characterizes the asymmetry of the velocity
profile. If the velocity profile is even with respect to a certain point (as it takes place
for the profile (5.101), $A = 0$.

Longwave instability of unidirectional flows periodic in two coordinates,

$$\mathbf{V} = u(y,z)\mathbf{e}_x, \ u(y+L_1,z) = u(y,z+L_2) = u(y,z)$$

was studied in [54–56].

It should be noted that the abovementioned investigations are incomplete in the
following aspect: periodic boundary conditions in the directions normal to the direction of the basic flow were postulated from the very beginning. However, there
is a crucial difference between flows in a channel, where the disturbances certainly
satisfy the same boundary conditions as the basic flow, and flows in an unbounded
space generated by an external periodic force. In the latter case, the flow is not necessarily periodic, and the only physical restriction for disturbances is boundness. That
means that the disturbances can be considered in the form of Floquet functions, e.g.,
in the two-dimensional case

$$\mathbf{v}' = \mathbf{w}(y)\exp[i(k_x x + k_y y) + \sigma t], \ \mathbf{w}(y+L) = \mathbf{w}(y). \tag{5.111}$$

Gotoh, Yamada, and Mizushima [57] were the first investigators who considered the
stability of flows with respect to disturbances in the form (5.111). However, only in
the remarkable paper of Dubrulle and Frisch [58], a really non-trivial result was
obtained: in the generic case, the wavevector of the most dangerous disturbance of
the parallel spatially periodic flow is inclined to the direction of the basic flow. In
that case, the frequency of longwave oscillations $\omega \sim k^3$ and the nonlinear regimes
are governed by a perturbed Korteweg-de Vries equation.

Further generalizations of the longwave linear stability theory of spatially periodic flows have been done in [59–63]. For modeling the nonlinear development of large-scale instabilities of spatially periodic flows, the *Navier–Stokes–Kuramoto–Sivashinsky* equation, with the viscous term $\Delta\mathbf{u}$ replaced by the terms $-\Delta\mathbf{u} - \mu\Delta^2\mathbf{u}$, has been suggested [64].

References

1. C.C. Lin, *Theory of Hydrodynamic Stability* (Cambridge University Press, Cambridge, 1967)
2. R. Betchov, W.O. Criminale, *Stability of Parallel Flows* (Academic, New York/London, 1967)
3. D.D. Joseph, *Stability of Fluid Motions*, vol. I, II (Springer, New York, 1976)
4. P.G. Drazin, W.H. Reid, *Hydrodynamic Stability* (Cambridge University Press, Cambridge, 2004)
5. K. Stewartson, J.T. Stuart, J. Fluid Mech. **48**, 529 (1971)
6. A. Davey, L.M. Hocking, K. Stewartson, J. Fluid Mech. **63**, 529 (1974)
7. P.L. Kapitsa, Zh. Exper. Teor. Fiz. **18**, 3 (1948, in Russian)
8. P.L. Kapitsa, S.P. Kapitsa, Zh. Exper. Teor. Fiz. **19**, 105 (1949, in Russian)
9. Y.Y. Trifonov, O.Y. Tsvelodub, J. Fluid Mech. **229**, 531 (1991)
10. H.-C. Chang, E.A. Demekhin, D.E. Kopelevich, J. Fluid Mech. **63**, 299 (1993)
11. A.A. Nepomnyashchy, Proc. Perm State Univ. **362**, 114 (1976, in Russian)
12. M.D. Spektor, Sov. Phys. JETP **61**, 104 (1988)
13. D.E. Bar, A.A. Nepomnyashchy, Physica D **86**, 586 (1995)
14. G.B. Whitham, *Linear and Nonlinear Waves* (Wiley, New York, 1974)
15. A.A. Nepomnyashchy, *Fluid Dynamics at Interfaces*, ed. by W. Shyy, R. Narayanan (Cambridge University Press, Cambridge, 1999), p. 85
16. D.E. Bar, A.A. Nepomnyashchy, Physica D **132**, 411 (1999)
17. V.E. Zakharov, E.A. Kuznetsov, Sov. Phys. JETP **39**, 285 (1974)
18. M.A. Allen, G. Rowlands, J. Plasma Phys. **53**, 63 (1995)
19. P. Frycz, E. Infeld, J. Plasma Phys. **41**, 441 (1989)
20. S. Toh, H. Iwasaki, T. Kawahara, Phys. Rev. A **40**, 5472 (1989)
21. K. Indireshkumar, A.L. Frenkel, Phys. Rev. E **55**, 1174 (1997)
22. T. Ogawa, C. Liu, Physica D **108**, 277 (1997)
23. A.A. Nepomnyashchy, Fluid Dyn. **9**, 354 (1974)
24. A.A. Nepomnyashchy, Proc. Perm State Univ. **316**, 91 (1974, in Russian)
25. A.A. Nepomnyashchy, *Fluid Dynamics*, Pt. 7 (Perm State Pedagogical Institute, Perm, 1974, in Russian), p. 43
26. T. Yamada, Y. Kuramoto, Prog. Theor. Phys. **56**, 681 (1976)
27. P. Manneville, Phys. Lett. A **84**, 129 (1981)
28. Y. Pomeau, A. Pumir, P. Pelce, J. Stat. Phys. **37**, 39 (1984)
29. B.I. Shraiman, Phys. Rev. Lett. **57**, 325 (1986)
30. S. Toh, J. Phys. Soc. Jpn. **56**, 949 (1987)
31. T. Bohr, M.H. Jensen, G. Paladin, A. Vulpiani, *Dynamical Systems Approach to Turbulence* (Cambridge, Cambridge University Press, 1998)
32. B.I. Cohen, J.A. Krommes, W.M. Tang, M.N. Rosenbluth, Nucl. Fusion **16**, 971 (1976)
33. U. Frisch, Z.S. She, O. Thual, J. Fluid Mech. **168**, 221 (1986)
34. H.-C. Chang, M. Cheng, E.A. Demekhin, D. Kopelevich, J. Fluid Mech. **270**, 251 (1994)
35. G.M. Homsy, Lect. Appl. Math. **15**, 191 (1976)
36. J. Topper, T. Kawahara, J. Phys. Soc. Jpn. **44**, 663 (1978)
37. T. Kawahara, Phys. Rev. Lett. **51**, 381 (1983)
38. T. Kawahara, S. Toh, Phys. Fluids **28**, 1636 (1985)
39. T. Kawahara, S. Toh, Phys. Fluids **31**, 2103 (1988)

40. H.-C. Chang, A. Demekhin, D. Kopelevich, Physica D **63**, 299 (1993)
41. S. Kalliadasis, C. Ruyer-Quil, B. Scheid, M.G. Velarde, *Falling Liquid Films* (Springer, Berlin/Heidelberg, 2012)
42. L.D. Meshalkin, I.B. Sinai, J. Appl. Math. Mech. **25**, 1700 (1961)
43. J.S.A. Green, J. Fluid Mech. **62**, 273 (1974)
44. A.A. Nepomnyashchy, J. Appl. Math. Mech. **40**, 836 (1976)
45. Z.S. She, Phys. Lett. A **124**, 161 (1987)
46. W.E, C.-W. Shu, Phys. Fluids **5**, 998 (1993)
47. V. Borue, S.A. Orszag, J. Fluid Mech. **306**, 293 (1996)
48. G.J. Chandler, R.R. Kerswell, J. Fluid Mech. **722**, 554 (2013)
49. D. Lucas, R. Kerswell, J. Fluid Mech. **750**, 518 (2014)
50. N.M. Evstigneev, N.A. Magnitskii, D.A. Silaev, Differ. Equ. **51**, 1292 (2015)
51. N.F. Bondarenko, M.Z. Gak, F.V. Dolzhansky, Atmos. Oceanic Phys. **15**, 711 (1979)
52. G.I. Sivashinsky, Physica D **17**, 243 (1985)
53. A.A. Nepomnyashchy, Proc. Perm State Pedagog. Inst. **152**, 77 (1976, in Russian)
54. L. Shtilman, G. Sivashinsky, J. Phys. (Paris) **47**, 1137 (1986)
55. V. Yakhot, G. Sivashinsky, Phys. Rev. A **35**, 815 (1987)
56. M.A. Brutyan, P.L. Krapivsky, Phys. Lett. A **152**, 211 (1991)
57. K. Gotoh, M. Yamada, J. Mizushima, J. Fluid Mech. **127**, 45 (1983)
58. B. Dubrulle, U. Frisch, Phys. Rev. A **43**, 5355 (1991)
59. A. Wirth, S. Gama, U. Frisch, J. Fluid Mech. **288**, 249 (1995)
60. Y.S. Khazan, A.A. Nepomnyashchy, J. Fluid Mech. **325**, 283 (1996)
61. A. Novikov, G. Papanicolaou, J. Fluid Mech. **446**, 173 (2001)
62. A. Novikov, Nonlinearity **16**, 1607 (2003)
63. A. Novikov, J. Comput. Phys. **195**, 341 (2004)
64. S. Gama, U. Frisch, Comput. Syst. Eng. **6**, 325 (1995)

Chapter 6
Instabilities of Fronts

In the previous chapters, we considered problems where patterns appeared due to an instability of a *motionless* state or a *steady* flow. In both cases, the base state of the system was characterized by a set of *time-independent* fields of variables $u(\mathbf{r})$. In the present chapter, we consider instabilities of moving *fronts* described by *traveling waves* $u(\mathbf{r} - \mathbf{v}t)$.

There are numerous examples of moving fronts in physics, chemistry, and biology. Front dynamics is a well-developed branch of nonlinear science which stems from seminal works of Fisher [1] and Kolmogorov, Petrovsky, and Piskunov [2] and from the mathematical theory of combustion [3]. The reader can find a description of the mathematical theory of front dynamics and its diverse applications in [4–6].

We have selected two typical examples. In Section 6.1 we consider propagating *combustion* fronts, which correspond to a transition from an unstable state (mixture of the fuel and an oxidant) and a stable state (products of combustion). In Section 6.2 we consider the propagation of *solidification fronts* between two locally stable phases, liquid and solid. In both cases, we consider only *transverse* instabilities of fronts leading to formation of patterns at the interface between phases. The considered cases by no means exhaust the phenomena characteristic for the front propagation, but they allow to demonstrate their diversity.

6.1 Combustion Fronts

6.1.1 Formulation of the Problem

Let us consider an exothermic chemical reaction that transforms the "fuel" into the "product" with the reaction rate

$$f(T) = A\exp(-E/RT) \tag{6.1}$$

© Springer Science+Business Media, LLC 2017
S. Shklyaev, A. Nepomnyashchy, *Longwave Instabilities and Patterns in Fluids*,
Advances in Mathematical Fluid Mechanics,
https://doi.org/10.1007/978-1-4939-7590-7_6

(*Arrhenius law*), where T is the temperature, R is the universal constant, E is the activation energy, and A is a certain constant. The problem is described by two variables, temperature T and fuel concentration C:

$$\frac{\partial C}{\partial t} = D\Delta C - f(T)C, \tag{6.2}$$

$$\frac{\partial T}{\partial t} = \chi\Delta T + qf(T)C \tag{6.3}$$

(we do not consider the product concentration, because it does not influence the evolution of the temperature and the fuel concentration field). We impose the boundary conditions

$$x \to -\infty: T = T_-, C = C_-; \; x \to \infty: T = T_+, C = 0$$

(it is assumed that the fuel is fully consumed by the reaction). Formally, the boundary condition $T(x \to -\infty) = T_-$ is incompatible with the law (6.1), because $f(T_-) \neq 0$. However, if $RT_- \ll E$, then $f(T_-)$ is exponentially small and its nonzero value can be disregarded. T_+ is not an independent parameter: it is determined by the heat balance; the corresponding expression will be given below (equation (6.4)).

In the analysis presented below, we follow the works by Sivashinsky, Matkowsky, and Margolis (see, e.g., [7–9]).

As a first step, we consider the one-dimensional plane front solution

$$T = T(z), C = C(z), z = x + vt$$

moving to the left ($v > 0$). We find:

$$vC_z = DC_{zz} - f(T)C; \; vT_z = \chi T_{zz} + qf(T)C.$$

Integrating both equations from $z = -\infty$ to $z = \infty$ and using the boundary conditions, we find:

$$-vC_- = -\int_{-\infty}^{\infty} f(T)Cdx, \; v(T_+ - T_-) = q\int_{-\infty}^{\infty} f(T)Cdx.$$

Thus,

$$T_+ = T_- + qC_-. \tag{6.4}$$

Let us transform the one-dimensional version of equations (6.2), (6.3) to a nondimensional form:

$$T = T_- + \Theta T_+(1 - \sigma); \; \sigma = T_-/T_+ < 1; \; C = YC_-; \; z = \xi\chi/v; \; t = \tau\chi/v^2.'$$

We obtain:

$$\frac{\partial Y}{\partial \tau} + \frac{\partial Y}{\partial \xi} = L^{-1}\frac{\partial^2 Y}{\partial \xi^2} - \Lambda Y \exp\left[\frac{Z(\Theta - 1)}{\sigma + (1 - \sigma)\Theta}\right], \tag{6.5}$$

$$\frac{\partial \Theta}{\partial \tau} + \frac{\partial \Theta}{\partial \xi} = \frac{\partial^2 \Theta}{\partial \xi^2} + \Lambda Y \exp\left[\frac{Z(\Theta - 1)}{\sigma + (1 - \sigma)\Theta}\right]. \tag{6.6}$$

Here

$$L = \frac{\chi}{D} \, (\text{ Lewis number }), \, Z = \frac{E(1-\sigma)}{RT_+} \, (\text{ Zeldovich number }),$$

$$\Lambda = A \exp\left(-\frac{E}{RT_+}\right) \frac{\chi}{v^2}.$$

The boundary conditions are:

$$Y(-\infty) = 1, \, \Theta(-\infty) = 0; \, Y(\infty) = 0, \, \Theta(\infty) = 1. \tag{6.7}$$

Later on, we assume that $Z \gg 1$, which is indeed a typical situation.

As mentioned above, the boundary condition $\Theta(-\infty) = 0$ cannot be actually satisfied, because according to the equation

$$\frac{\partial \Theta}{\partial \tau}(-\infty) \sim \exp\left(-\frac{Z}{\sigma}\right) \neq 0.$$

However, the violation of this boundary condition is exponentially small for $Z \gg 1$, and we disregard it.

6.1.2 Plane Stationary Front

The plane stationary front is described by the problem (6.5)–(6.7) with the time derivative $\partial Y / \partial \tau$ and $\partial \Theta / \partial \tau$ omitted.

For large Z, the reaction rate grows very fast with the growth of temperature. Because of that, the reaction takes place in a thin layer called *combustion front* where the temperature is already close to the maximum one, $\Theta \approx 1$, while the concentration of the fuel C is still different from zero. The thickness of the combustion front is $O(\varepsilon)$, where $\varepsilon \equiv Z^{-1}$. Below we present the asymptotic theory of the combustion front propagation. We find solutions outside the front (*outer solutions*), where there is no reaction, and the solution inside the front (*inner solutions*), where the reaction takes place. Matching of these solutions will give us the value of Λ (which can be called "eigenvalue" of the nonlinear problem), i.e., the front velocity.

In the outer regions, the reaction term can be neglected: in the region $\xi < 0$, because the temperature is too low, and in the region $\xi > 0$, because the fuel concentration is too low. The solutions can be presented in the form of a series in ε,

$$Y = Y_0 + \varepsilon Y_1 + \ldots, \, \Theta = \Theta_0 + \varepsilon \Theta_1 + \ldots$$

Actually, we are interested only in the leading-order terms governed by the equations

$$\frac{dY_0}{d\xi} = L^{-1} \frac{d^2 Y_0}{d\xi^2}, \, \frac{d\Theta_0}{d\xi} = \frac{d^2 \Theta_0}{d\xi^2}. \tag{6.8}$$

In the region $\xi < 0$, the solution has to satisfy the boundary conditions

$$Y_0(-\infty) = 1, \; \Theta_0(-\infty) = 0.$$

The solution is

$$Y_0(\xi) = 1 + A_- e^{L\xi}, \; \Theta_0(\xi) = B_- e^{\xi}, \tag{6.9}$$

where A_- and B_- are constants. In the region $\xi > 0$, equations (6.8) are solved with the boundary conditions

$$Y_0(\infty) = 0, \; \Theta_0(\infty) = 1.$$

The solution is just

$$Y_0(\xi) = 0, \; \Theta_0(\xi) = 1. \tag{6.10}$$

The intensive fuel consumption and heat release lead to jumps of mass flux and heat flux across the front, i.e., discontinuity of the spatial derivatives of the concentration and temperature fields. However, the diffusion and heat diffusion equations do not allow jumps of the concentration and temperature themselves. Hence, solutions (6.9) and (6.10) should be matched in the point $\xi = 0$:

$$Y_0(0^-) = Y_0(0^+) = 0, \; \Theta_0(0^-) = \Theta_0(0^+) = 1. \tag{6.11}$$

That gives the following expressions for the outer solution in the region $\xi < 0$:

$$Y_0(\xi) = 1 - e^{L\xi}, \; \Theta_0(\xi) = e^{\xi}. \tag{6.12}$$

Note that from the point of view of the outer expansion, the reaction takes place in the infinitely thin layer, i.e., the outer solution can be considered as the solution of the following system of equations:

$$\frac{dY_0}{d\xi} = L^{-1}\frac{d^2Y_0}{d\xi^2} - H_0\delta(\xi), \; \frac{d\Theta_0}{d\xi} = \frac{d^2\Theta_0}{d\xi^2} + H_0\delta(\xi), \tag{6.13}$$

where H_0 is the integral of the reaction term across the front which determines the jumps of the fluxes:

$$H_0 = L^{-1}\left[\frac{dY_0}{d\xi}\right] = -\left[\frac{d\Theta_0}{d\xi}\right],$$

where

$$[f] \equiv f(0^+) - f(0^-). \tag{6.14}$$

Solution (6.12) is obtained with $H_0 = 1$.

Let us consider now the fine structure of the combustion front in the region $\xi = O(\varepsilon)$. For this goal, let us introduce the "fast" variable $X = \varepsilon^{-1}\xi$ and assume that the solution is the function of the variable X:

$$Y = y_0(X) + \varepsilon y_1(X) + \varepsilon^2 y_2(X) + \ldots, \; \Theta = \theta_0(X) + \varepsilon\theta_1(X) + \varepsilon^2\theta_2(X) + \ldots$$

Because of the relations (6.11), $y_0(X) = 0$, $\theta_0(X) = 1$. The exponential expression in the reaction rate is just $\exp(\theta_1)$ at the leading order. The balance between the reaction term and the diffusion term is reached if $\Lambda = O(\varepsilon^{-2})$; therefore, we will use the expansion

$$\Lambda = \varepsilon^{-2}\Lambda_{-2} + \varepsilon^{-1}\Lambda_{-1} + \ldots$$

Finally, we get the following system of equations for $y_1(X)$, $\theta_1(X)$:

$$L^{-1}\frac{d^2y_1}{dX^2} - \Lambda_{-2}y_1\exp(\theta_1) = 0, \tag{6.15}$$

$$\frac{d^2\theta_1}{dX^2} + \Lambda_{-2}y_1\exp(\theta_1) = 0. \tag{6.16}$$

The boundary conditions for functions $y_1(X)$, $\theta_1(X)$ are obtained by matching with the solution (6.12), which can be written for small $\xi = \varepsilon X$ as

$$Y_0 = -\varepsilon LX + \ldots, \quad \Theta_0 = 1 + \varepsilon X + \ldots,$$

and with the solution (6.10). We find:

$$X \to -\infty: \; y_1(X) \sim -LX, \; \theta_1(X) \sim X; \; X \to \infty: \; y_1(X) \to 0, \; \theta_1(X) \to 0.$$

Adding equations (6.15) and (6.16), we find

$$\frac{d^2}{dX^2}(\theta_1 + L^{-1}y_1) = 0.$$

Taking into account the boundary conditions, we find:

$$y_1 = -L\theta_1. \tag{6.17}$$

Substituting (6.17) into (6.16), we obtain the following nonlinear eigenvalue problem:

$$\frac{d^2\theta_1}{dX^2} - L\Lambda_{-2}\theta_1\exp(\theta_1) = 0, \tag{6.18}$$

$$X \to -\infty: \; \theta_1(X) \sim X; \; X \to \infty: \; \theta_1(X) \to 0.$$

The first integral of the equation (6.18) is:

$$\frac{1}{2}(\theta_1')^2 - \Lambda_{-2}L(\theta_1 - 1)\exp(\theta_1) = C, \tag{6.19}$$

where C is a constant which is obtained from the boundary condition at $X \to \infty$:

$$C = \Lambda_{-2}L. \tag{6.20}$$

Evaluating both sides of (6.19) at $X \to -\infty$, we find that

$$\Lambda_{-2} = \frac{1}{2L}.$$

Returning to the original variables, we find the following formula for the combustion front velocity:

$$v = \sqrt{\frac{2A}{D} \frac{\chi RT_+^2}{E(T_+ - T_-)}} \exp\left(-\frac{E}{2RT_+}\right).$$

6.1.3 Dynamics of Curved Fronts

Let us generalize now the theory described in the previous subsections. Let fuel concentration and temperature be functions of three spatial variables, $C(x_1,x_2,x_3,t)$ and $T(x_1,x_2,x_3,t)$. Assume that the combustion front is a two-dimensional surface, $x_1 = \Phi(x_2,x_3,t)$. Define the new system of variables, $x = x_1 - \Phi(x_2,x_3,t)$, $y = x_2$, $z = x_3$, $\tau = t$. Using the same transformation to nondimensional variables, as in the previous subsections (i.e., using the velocity v of the plane front for non-dimensionalization), we obtain the following system of equations:

$$\frac{\partial Y}{\partial \tau} - \Phi_\tau \frac{\partial Y}{\partial x} = L^{-1}\Delta Y - w, \quad \frac{\partial \Theta}{\partial \tau} - \Phi_\tau \frac{\partial \Theta}{\partial x} = \Delta\Theta + w,$$

$$w = \Lambda Y \exp\left[\varepsilon^{-1}\frac{\Theta - 1}{\sigma + (1-\sigma)\Theta}\right].$$

The expression for the Laplacian in the coordinates (x,y,z) is

$$\Delta f = (1 + \Phi_y^2 + \Phi_z^2)\frac{\partial^2 f}{\partial x^2} + \frac{\partial^2 f}{\partial y^2} + \frac{\partial^2 f}{\partial z^2} - (\Phi_{yy} + \Phi_{zz})\frac{\partial f}{\partial x} - 2\left(\Phi_y \frac{\partial^2 f}{\partial x \partial y} + \Phi_z \frac{\partial^2 f}{\partial x \partial z}\right).$$

When considering the outer solution, we can replace w by $H\delta(x)$, where H is still an unknown function which is equal to the integral of w over the combustion front.

For the sake of simplicity, we present the theory in the case where the Lewis number is close to L, i.e., $L = 1 + \varepsilon\beta$, hence, $L^{-1} = 1 - \varepsilon\beta + \dots$. We construct the outer solution in the form

$$Y = Y_0(x,y,z,\tau) + \varepsilon Y_1(x,y,z,\tau) + \dots; \quad \Theta = \Theta_0(x,y,z,\tau) + \varepsilon\Theta_1(x,y,z,\tau) + \dots.$$

Note that according to our assumption of the full consumption of the fuel at the reaction front, the outer solution for Y is zero in all the order in the region $x > 0$. The inner solution is presented in the form

$$Y = y_0(X,y,z,\tau) + \varepsilon y_1(X,y,z,\tau) + \dots; \quad \Theta = \theta_0(X,y,z,\tau) + \varepsilon\theta_1(X,y,z,\tau) + \dots,$$

where $X = \varepsilon^{-1}x$. Also, we use the expansions

$$\Phi = \Phi_0 + \varepsilon\Phi_1 + \ldots, \quad H = H_0 + \varepsilon H_1 + \ldots.$$

For Y_0 and Θ_0, we obtain:

$$\frac{\partial Y_0}{\partial \tau} - (\Phi_0)_\tau \frac{\partial Y_0}{\partial x} = \Delta Y_0 - H_0 \delta(x), \tag{6.21}$$

$$\frac{\partial \Theta_0}{\partial \tau} - (\Phi_0)_\tau \frac{\partial \Theta_0}{\partial x} = \Delta \Theta_0 + H_0 \delta(x); \tag{6.22}$$

$$x = -\infty: \ Y_0 = 1, \ \Theta_0 = 0; \ x = \infty: \ Y_0 = 0, \ \Theta_0 = 1.$$

Let us add equations (6.21) and (6.22). We find that the quantity $S_0 = Y_0 + \Theta_0$ satisfies the equation

$$\frac{\partial S_0}{\partial \tau} - (\Phi_0)_\tau \frac{\partial S_0}{\partial x} = \Delta S_0$$

with boundary conditions

$$x = -\infty: \ S_0 = 1; \ x = \infty: \ S_0 = 1.$$

Choose solution $S_0 = 1$ (assuming that it is compatible with the initial conditions). Then $\Theta_0 = 1$ in the region $x > 0$.

Let us consider now the inner solution. Because of the continuity of the fields of Y_0, Θ_0, $Y_0 = 0$ and $\Theta_0 = 1$ as $x = 0$. Therefore, the inner expansion can be written as

$$Y = \varepsilon y_1(X, y, z, \tau) + \ldots; \quad \Theta = 1 + \varepsilon\theta_1(x, y, z, \tau) + \ldots.$$

Because we used the velocity of the plane front for non-dimensionalization, the expansion $\Lambda = \Lambda_{-2}\varepsilon^{-2} + \ldots$ coincides with that used in the previous section. Thus, in our case $\Lambda_{-2} = 1/2$. We obtain the following system of equations in the leading order:

$$[1 + (\Phi_0)_y^2 + (\Phi_0)_z^2]\frac{\partial^2 y_1}{\partial X^2} - \frac{1}{2}y_1 \exp(\theta_1) = 0,$$

$$[1 + (\Phi_0)_y^2 + (\Phi_0)_z^2]\frac{\partial^2 \theta_1}{\partial X^2} + \frac{1}{2}y_1 \exp(\theta_1) = 0.$$

Adding both equations, we find that for the quantity $s_1 = y_1 + \theta_1$,

$$\frac{\partial^2 s_1}{\partial X^2} = 0.$$

Matching with the outer solution for $x > 0$, we find that $s_1 = \theta_1(\infty)$, $y_1 = \theta_1(\infty) - \theta_1$, hence

$$[1 + (\Phi_0)_y^2 + (\Phi_0)_z^2](\theta_1)_{XX} + \frac{1}{2}[\theta_1(\infty) - \theta_1]\exp(\theta_1) = 0 \tag{6.23}$$

(here and below $f(a)$ means $f(a,y,z,\tau)$). Multiplying both sides of equation (6.23) by $(\theta_1)_X$ and integrating from $-\infty$ to ∞, we find:

$$(\theta_1)_X(-\infty) = \frac{\exp\left(\frac{1}{2}\theta_1(\infty)\right)}{\sqrt{1+(\Phi_0)_y^2+(\Phi_0)_z^2}}.$$

Because $(\theta_1)_X(\infty) = 0$, we find that the change of the derivative $(\theta_1)_X$ across the reaction zone,

$$(\theta_1)_X(\infty) - (\theta_1)_X(-\infty) = -\frac{\exp\left(\frac{1}{2}\theta_1(\infty)\right)}{\sqrt{1+(\Phi_0)_y^2+(\Phi_0)_z^2}}. \tag{6.24}$$

The matching conditions

$$\theta_1(\infty) = \Theta_1(0^+), \quad \frac{\partial\theta_1}{\partial X}(\infty) = \frac{\partial\Theta_0}{\partial x}(0^+), \quad \frac{\partial\theta_1}{\partial X}(-\infty) = \frac{\partial\Theta_0}{\partial x}(0^-)$$

allow us to represent the relation (6.24) as the jump of the derivative for the outer solution:

$$[(\Theta_0)_x] = -\frac{\exp(\Theta_1(0^+)/2)}{\sqrt{1+(\Phi_0)_y^2+(\Phi_0)_z^2}}$$

(notation (6.14) is used).

Now we can determine the function H_0 in (6.21), (6.22). Let us integrate (6.22) over x from $x = -\delta$ to $x = \delta$ and take the limit $\delta \to 0$. We find:

$$(1+(\Phi_0)_y^2+(\Phi_0)_z^2)[(\Theta_0^+)_x] + H_0 = 0.$$

Thus,

$$H_0 = \sqrt{1+(\Phi_0)_y^2+(\Phi_0)_z^2}\exp(\Theta_1(0^+)/2),$$

so that equation (6.22) can be rewritten as

$$\frac{\partial\Theta_0}{\partial\tau} - (\Phi_0)_\tau\frac{\partial\Theta_0}{\partial x} = \Delta\Theta_0 + \sqrt{1+(\Phi_0)_y^2+(\Phi_0)_z^2}\exp(\Theta_1(0^+)/2)\delta(x). \tag{6.25}$$

We come to the conclusion that the leading order equation does not allow to get a closed formulation of the outer problem: equation for Θ_0 contains $\Theta_1(0^+)$. Thus, we have to consider the next order of the outer problem.

In the next order, we obtain equations:

$$\frac{\partial Y_1}{\partial\tau} - (\Phi_0)_\tau\frac{\partial Y_1}{\partial x} - (\Phi_1)_\tau\frac{\partial Y_0}{\partial x} = \Delta Y_1 - \beta\Delta Y_0 - w_1,$$

$$\frac{\partial\Theta_1}{\partial\tau} - (\Phi_0)_\tau\frac{\partial\Theta_1}{\partial x} - (\Phi_1)_\tau\frac{\partial\Theta_0}{\partial x} = \Delta\Theta_1 + w_1,$$

where w_1 is the next-order reaction term. Adding both equations and taking into account that $S_0 = Y_0 + \Theta_0 = 1$, we find the following equation for S_1:

$$\frac{\partial S_1}{\partial \tau} - (\Phi_0)_\tau \frac{\partial S_1}{\partial x} = \Delta S_1 + \beta \Delta \Theta_0. \tag{6.26}$$

In the region $x > 0$, $Y_1 = 0$, hence $\Theta_1(0^+) = S_1(0^+)$. Thus, equation (6.25) can be written as

$$\frac{\partial \Theta_0}{\partial \tau} - (\Phi_0)_\tau \frac{\partial \Theta_0}{\partial x} = \Delta \Theta_0 + \sqrt{1 + (\Phi_0)_y^2 + (\Phi_0)_z^2} \exp(S_1(0^+)/2) \delta(x) \tag{6.27}$$

(because S_1 is a continuous function, we can write just $S_1(0)$ instead of $S_1(0^+)$). Equations (6.26) and (6.27) with boundary conditions

$$x \to -\infty : \Theta_0 = 0, \ S_1 = 0; \ x \to \infty : \Theta_0 = 1, \ S_1 = 0$$

give a closed description of the problem.

Instead of using δ-function, we can consider non-smooth solutions of homogeneous equations

$$\frac{\partial \Theta_0}{\partial \tau} - (\Phi_0)_\tau \frac{\partial \Theta_0}{\partial x} = \Delta \Theta_0, \ \frac{\partial S_1}{\partial \tau} - (\Phi_0)_\tau \frac{\partial S_1}{\partial x} = \Delta S_1 + \beta \Delta \Theta_0 \tag{6.28}$$

defined in regions $x < 0$ and $x > 0$, which satisfy the following conditions at $x = 0$:

$$[\Theta_0] = [S_1] = 0; \tag{6.29}$$

$$[(\Theta_0)_x] = -\frac{\exp(S_1(0)/2)}{\sqrt{1 + (\Phi_0)_y^2 + (\Phi_0)_z^2}}; \tag{6.30}$$

$$[(S_1)_x] + \beta[(\Theta_0)_x] = 0. \tag{6.31}$$

The simplest solution of the problem (6.28)–(6.31), which describes the planar front, looks as follows:

$$\bar{\Phi}_0 = -\tau, \ \bar{\Theta}_0(x) = e^x, \ x < 0; \bar{\Theta}_0(x) = 1, \ x > 0; \tag{6.32}$$

$$\bar{S}_1 = -\beta x e^x, \ x < 0; \bar{S}_1 = 0, \ x > 0. \tag{6.33}$$

6.1.4 Linear Stability Theory of the Planar Front

Now we consider the stability of the planar front solution. For the sake of simplicity, we drop the subscripts 0 and 1 and denote the planar front solution (6.32), (6.33) as $(\bar{\Phi}, \bar{\Theta}, \bar{S})$. Impose a small disturbance on the solution (6.32), (6.33). It is convenient to define the disturbances in the following way:

$$\Phi = \bar{\Phi}(\tau) + \phi(y, z, \tau); \tag{6.34}$$

$$\Theta = \bar{\Theta}(x) + \phi \frac{d\bar{\Theta}(x)}{dx} + \theta(x,y,z,\tau); \; x < 0, \, x > 0; \tag{6.35}$$

$$S = \bar{S}(x) + \phi \frac{d\bar{S}(x)}{dx} + s(x,y,z,\tau); \; x < 0, \, x > 0. \tag{6.36}$$

Substituting expressions (6.34)–(6.36) into equation (6.28) and linearizing it with respect to disturbances, we obtain the following system of equations:

$$\frac{\partial \theta}{\partial \tau} + \frac{\partial \theta}{\partial x} = \frac{\partial^2 \theta}{\partial x^2} + \frac{\partial^2 \theta}{\partial y^2} + \frac{\partial^2 \theta}{\partial z^2}, \, x \neq 0;$$

$$\frac{\partial s}{\partial \tau} + \frac{\partial s}{\partial x} = \frac{\partial^2 s}{\partial x^2} + \frac{\partial^2 s}{\partial y^2} + \frac{\partial^2 s}{\partial z^2} + \beta \left(\frac{\partial^2 \theta}{\partial x^2} + \frac{\partial^2 \theta}{\partial y^2} + \frac{\partial^2 \theta}{\partial z^2} \right), \, x \neq 0.$$

The boundary conditions at infinity are:

$$x \to -\infty: \; \theta = 0, \, s = 0; \; x \to \infty: \; \theta = 0, \, s = 0$$

(actually, $\theta = 0$ for all $x > 0$). The jump conditions across the front $x = 0$ are as follows:

$$[\theta] = \phi; \; [s] + \beta \phi = 0; \; [\theta_x] - \phi + \frac{1}{2} s(0^+) = 0; \; [s_x] + \beta \phi + \beta [\theta_x] = 0.$$

Look for solution in the form

$$(\theta, s) = e^{\sigma \tau} e^{i(k_y y + k_z z)} (f(x), g(x)),$$

$$\phi = e^{\sigma \tau} e^{i(k_y y + k_z z)}.$$

We obtain the following eigenvalue problem:

$$\sigma f + f' = f'' - k^2 f, \, x < 0; \; f = 0, \, x > 0; \tag{6.37}$$

$$\sigma g + g' = g'' - k^2 g + \beta (f'' - k^2 f), \, x < 0, \, x > 0, \tag{6.38}$$

where $k^2 = k_y^2 + k_z^2$. Boundary conditions:

$$x \to -\infty: \; f = g = 0; \; x \to \infty: \; f = g = 0;$$

$$x = 0: \; [f] = 1, \; [g] + \beta = 0, \; [f_x] - 1 + \frac{1}{2} g(0^+) = 0, \; [g_x] + \beta + \beta [f_x] = 0.$$

6.1.4.1 Monotonic Instability

Let us find the boundary of the monotonic instability, $\sigma = 0$. The general solution of equation (6.37) is

$$f = Ae^{r+x} + Be^{r-x},$$

where

$$r_{\pm} = \frac{1}{2} \pm \sqrt{\frac{1}{4} + k^2}.$$

Using the boundary conditions, we find:

$$f(x) = -e^{r_+ x}, \; x < 0; f(x) = 0, \; x > 0.$$

Solution of equation (6.38) is

$$g = C_- e^{r_+ x} + D_- e^{r_- x} + \beta \frac{r_+}{2r_+ - 1} x e^{r_+ x}, \; x < 0; g = C_+ e^{r_+ x} + D_+ e^{r_- x}, \; x > 0. \quad (6.39)$$

Using the boundary conditions on the infinity, we find that $D_- = C_+ = 0$. Substituting expressions (6.39) into the boundary conditions at $x = 0$, we obtained three equations for three variables D_+, C_-, and β. First we find that $D_+ = 2r_-$, $C_- = 2r_- + \beta$. Eliminating D_+ and C_-, we obtain the following compatibility condition:

$$r_- \left[2(r_+ - r_-) + \frac{\beta}{r_+ - r_-} \right] = 0. \quad (6.40)$$

Relation (6.40) is satisfied in two cases. The first case is $r_- = 0$, i.e., $k = 0$. Indeed, there always exists a neutral disturbance corresponding to a shift of the position of the front. Another case is

$$\beta = -2(r_+ - r_-)^2 = -2(1 + 4k^2). \quad (6.41)$$

Relation (6.41) determines the *monotonic instability boundary* of the planar front. When $\beta > -2(1 + 4k^2)$, there is no monotonic instability. When $\beta < -2(1 + 4k^2)$, the front is unstable with respect to disturbances with wavenumbers $0 < k^2 < k_m^2$, where

$$k_m^2 = -\frac{\beta + 2}{8}.$$

Thus, we obtained a *monotonic longwave instability*.

6.1.4.2 Oscillatory Instability

Let us consider now the oscillatory instability, $\sigma = i\omega$. The eigenfunctions are:

$$f(x) = -e^{r_+ x}, \; x < 0; f(x) = 0, \; x > 0,$$

$$g(x) = (2r_- + \beta)e^{r_+ x} - \frac{\beta(k^2 - r_+^2)}{2r_+ - 1} x e^{r_+ x}, \; x < 0; g(x) = 2r_- e^{r_- x}, \; x > 0,$$

where

$$r_{\pm} = \frac{1}{2} \pm \sqrt{\frac{1}{4} + k^2 + i\omega}.$$

The dispersion relation is

$$64(i\omega)^3 + (i\omega)^2[8(\beta + 2 + 8k^2) + 16(1 + 8k^2) - \beta^2] +$$

$$2i\omega(\beta + 2 + 8k^2)(1 + 12k^2) + k^2(\beta + 2 + 8k^2)^2 = 0. \qquad (6.42)$$

Taking the real and imaginary parts of equation (6.42), we find:

$$\omega^2 = \frac{1}{32}(\beta + 2 + 8k^2)(1 + 12k^2);$$

$$\beta^2(1 + 12k^2) - 8\beta(1 + 8k^2) - 32(1 + 8k^2)^2 = 0. \qquad (6.43)$$

Equation (6.43) determines the *oscillatory instability boundary*. In the contradistinction to the monotonic instability boundary, which has a threshold $\beta_c = -2$ corresponding to the wavenumber $k = 0$, the minimum of the curve $\beta(k^2)$ determined by (6.43) is located at finite value of k^2, $k^2 = k_c^2 = 1/24$. The threshold value of β, corresponding to the oscillatory instability, is $\beta_c = \beta(k_c^2) = 32/3$. Note that the longwave limit of $\beta(k^2)$, $\beta(0) = 4(1 + \sqrt{3})$, is higher than β_c. Thus, for $L > 1$ a *shortwave oscillatory instability* takes place.

6.1.5 Nonlinear Development of Front Instabilities

6.1.5.1 Monotonic Longwave Instability

Let us consider the nonlinear evolution of an unstable monotonic mode near the instability threshold, i.e., for $\beta = -2(1 + \delta^2)$, $\delta \ll 1$.

The analysis of the linear dispersion relation shows that for such value of β, the wavenumbers of growing disturbances have $k = O(\delta)$, while the growth rate is $O(\delta^4)$. That allows us to expect that the the curved front is characterized by the following scales: $Y = \delta y$, $Z = \delta z$, $T = \delta^4 \tau$. We construct solutions in the following form:

$$\Phi = \bar{\Phi} + \delta^2\phi(Y,Z,T), \; \Theta = \bar{\Theta} + \delta^4\theta(x,Y,Z,T), \; S = \bar{S} + \delta^4 s(x,Y,Z,T).$$

The disturbances are expanded into series:

$$\phi = \phi_0 + \delta^2\phi_1 + \ldots; \; \theta = \theta_0 + \delta^2\theta_1 + \ldots; \; s = s0 + \delta^2 s_1 + \ldots.$$

The leading order problem can be presented in the following form:

$$\frac{\partial\theta_0}{\partial x} = \frac{\partial^2\theta_0}{\partial x^2} - \left(\frac{\partial^2\phi_0}{\partial Y^2} + \frac{\partial^2\phi_0}{\partial Z^2}\right) \cdot \left(\begin{array}{l} e^x, \, x < 0 \\ 0, \, x > 0 \end{array}\right) + \frac{1}{2}s0\,\delta(x),$$

$$\frac{\partial s0}{\partial x} = \frac{\partial^2 s0}{\partial x^2} - 2\frac{\partial^2\theta_0}{\partial x^2} - 2\left(\frac{\partial^2\phi_0}{\partial Y^2} + \frac{\partial^2\phi_0}{\partial Z^2}\right) \cdot \left(\begin{array}{l} xe^x, \, x < 0 \\ 0, \, x > 0 \end{array}\right).$$

The solution is:

$$\theta_0 = \begin{pmatrix} \left(\frac{\partial^2 \phi_0}{\partial Y^2} + \frac{\partial^2 \phi_0}{\partial Z^2} \right) x e^x, \; x < 0 \\ 0, \; x > 0 \end{pmatrix}.$$

$$s_0 = \begin{pmatrix} 2 \left(\frac{\partial^2 \phi_0}{\partial Y^2} + \frac{\partial^2 \phi_0}{\partial Z^2} \right) (1 + x^2) e^x, \; x < 0 \\ 2 \left(\frac{\partial^2 \phi_0}{\partial Y^2} + \frac{\partial^2 \phi_0}{\partial Z^2} \right), \; x > 0 \end{pmatrix}.$$

Note the obtained solution does not satisfy the boundary condition $s(x = \infty, Y, Z, T) = 0$. Actually, because of the longwave nature of the disturbance, it decays on a large spatial scale, $X = \delta x$. The transition between the value of s_0 obtained in the region $x = O(1)$ and the zero value is governed by the equation

$$\frac{\partial s_0}{\partial Z} = \frac{\partial^2 s_0}{\partial Y^2} + \frac{\partial^2 s_0}{\partial Z^2}$$

with the boundary conditions

$$s_0 = \left(\frac{\partial^2 \phi_0}{\partial Y^2} + \frac{\partial^2 \phi_0}{\partial Z^2} \right), \; Z \to 0^+; \; s_0 \to 0, \; Z \to \infty.$$

The solution is

$$s_0 = \frac{1}{2\pi X} \int_{-\infty}^{\infty} \int_{-\infty}^{\infty} dY_1 dZ_1 \exp\{-[(Y - Y_1)^2 + (Z - Z_1)^2]/4X\} \left(\frac{\partial^2 \phi_0}{\partial Y_1^2} + \frac{\partial^2 \phi_0}{\partial Z_1^2} \right).$$

In the next order, the following inhomogeneous linear problem is obtained:

$$-\frac{\partial \phi_0}{\partial T} \cdot \begin{pmatrix} e^x, \; x < 0 \\ 0, \; x > 0 \end{pmatrix} + \frac{\partial \theta_1}{\partial x} = \frac{\partial^2 \theta_0}{\partial Y^2} + \frac{\partial^2 \theta_0}{\partial Z^2} + \frac{\partial^2 \theta_1}{\partial x^2}$$

$$- \left[\left(\frac{\partial \phi_0}{\partial Y} \right)^2 + \left(\frac{\partial \phi_0}{\partial Z} \right)^2 \right] \delta(x) + \left[\left(\frac{\partial \phi_0}{\partial Y} \right)^2 + \left(\frac{\partial \phi_0}{\partial Z} \right)^2 \right] \cdot \begin{pmatrix} e^x, \; x < 0 \\ 0, \; x > 0 \end{pmatrix}$$

$$- \left(\frac{\partial^2 \phi_1}{\partial Y^2} + \frac{\partial^2 \phi_1}{\partial Z^2} \right) \cdot \begin{pmatrix} e^x, \; x < 0 \\ 0, \; x > 0 \end{pmatrix} + \left[\frac{1}{2} s_1 + \frac{1}{2} \left(\frac{\partial \phi_0}{\partial Y} \right)^2 + \left(\frac{\partial \phi_0}{\partial Z} \right)^2 \right] \delta(x);$$

$$-\frac{\partial \phi_0}{\partial T} \cdot \begin{pmatrix} 2(x+1) e^x, \; x < 0 \\ 0, \; x > 0 \end{pmatrix} + \frac{\partial s_1}{\partial x} = \frac{\partial^2}{\partial Y^2} (s_0 - 2\theta_0) + \frac{\partial^2}{\partial Z^2} (s_0 - 2\theta_0)$$

$$+ \frac{\partial^2}{\partial x^2} (s_1 - 2\theta_1 - 2\theta_0) + \left[\left(\frac{\partial \phi_0}{\partial Y} \right)^2 + \left(\frac{\partial \phi_0}{\partial Z} \right)^2 \right] \cdot \begin{pmatrix} 2(x+1) e^x, \; x < 0 \\ 0, \; x > 0 \end{pmatrix}$$

$$- \left(\frac{\partial^2 \phi_1}{\partial Y^2} + \frac{\partial^2 \phi_1}{\partial Z^2} \right) \cdot \begin{pmatrix} 2x e^x, \; x < 0 \\ 0, \; x > 0 \end{pmatrix} - \left(\frac{\partial^2 \phi_0}{\partial Y^2} + \frac{\partial^2 \phi_0}{\partial Z^2} \right) \cdot \begin{pmatrix} 2x e^x, \; x < 0 \\ 0, \; x > 0 \end{pmatrix}.$$

One can show that the solvability conditions of the obtained system of equation are

$$(\phi_0)_T + \frac{1}{2}(\nabla\phi_0)^2 + \Delta\phi_0 + 4\Delta^2\phi_0 = 0. \tag{6.44}$$

Here

$$\nabla = \mathbf{e}_Y \frac{\partial}{\partial Y} + \mathbf{e}_Z \frac{\partial}{\partial Z}, \ \Delta = \nabla^2. \tag{6.45}$$

Equation (6.44) is the famous *Kuramoto–Sivashinsky (KS) equation* (see Section 1.4.2.2), which is a paradigmatic example of the dynamical system which creates spatiotemporal chaotic regimes. It has been found that there exists a certain kind of energy cascade from *short* to *long* scales which determines the large-scale properties of the spatiotemporal chaos. More precisely, the large-scale behavior of the system is governed by the Burgers equation with a certain effective positive *turbulent viscosity*, while the instability manifests itself as an uncorrelated short-scale ("white") Gaussian noise. In other words, the large-scale behavior of the KS chaos can be modeled by the universal Kardar–Parisi–Zhang equation, and it is characterized by corresponding scaling properties [10].

6.1.5.2 Oscillatory Shortwave Instability

The case of the oscillatory shortwave instability is more difficult for the theoretical analysis. In the supercritical region, waves with $k_y^2 + k_z^2 \approx k_c^2$ moving in arbitrary direction along the front are spontaneously generated. Near the instability threshold, $\beta = \beta_c + \delta^2$, each of these waves can be presented as a plane wave slowly modulated in time and space. In the particular case of one-dimensional modulations, the envelope function $A(Y, Z, T)$, where $Y = \delta y$, $Z = \delta z$, and $T = \delta^2 t$, is governed by the *complex Ginzburg–Landau equation* that can be transformed to the standard form,

$$A_t = A + (1 + i\alpha)\nabla^2 A - (s + i\beta)|A|^2 A. \tag{6.46}$$

Here $s = 1$ in the case of a supercritical Hopf bifurcation and $s = -1$ in the case of a subcritical Hopf bifurcation. Despite the apparent simplicity of equation (6.46), the dynamics governed by this equation is incredibly reach. The basic features of this equation have been summarized in the review paper by [11]. It includes spontaneous destruction of waves periodicity, appearance of defects, and different kinds of spatiotemporal chaos. Two-dimensional modulations of plane waves have been hardly investigated. One can expect development of topological defects and spiral-like waves. Waves propagating in different directions interact with each other in a nonlocal way [12, 13].

6.1.6 Sequential Chemical Reaction

In the example considered in the previous sections, the front is subject to either a longwave monotonic instability (if $L < 1$) or a shortwave oscillatory instability (if $L > 1$), but not both kinds of instabilities simultaneously. The situation is different in the case of a two-stage sequential chemical reaction $A \to B \to C$, which creates *two flame fronts*, one at $z = 0$ and another one at $z = H$ [14]. Because the Lewis numbers L_A, L_B of substances A and B are not equal, different combinations of front instabilities are possible.

Later on, we assume that the Zeldovich number $Z \gg 1$ and the Lewis numbers of substances are close to 1, $L_{A,B} = 1 + l_{A,B} Z^{-1}$. The problem is described by the reduced temperature Θ and *two* mass fractions of the components, A and B, which are expanded as

$$(A, B, \Theta) = \sum_{j=0}^{\infty} (A^{(j)}, B^{(j)}, \Theta^{(j)}) Z^{-j}.$$

The leading-order flame dynamics is governed by a closed system of equations for the leading terms $A^{(0)}$ and $B^{(0)}$, and for the enthalpy S, which is a linear combination of the $O(Z^{-1})$ terms,

$$S = A^{(1)} + (1 - \alpha) B^{(1)} + \Theta^{(1)},$$

where α is the ratio of the heat released in the first reaction to the total heat released in the two reactions. Note that the leading term in the reduced temperature expansion, $\Theta^{(0)}$ is linearly related to $A^{(0)}$ and $B^{(0)}$,

$$\Theta^{(0)} = 1 - A^{(0)} - (1 - \alpha) B^{(0)}.$$

The system of two planar fronts moving with the same constant velocity (chosen as a velocity scale) is described by the following expressions for nondimensional variables (see Figure 6.1):

$$A^{(0)} = 1 - e^z, \ B^{(0)} = (1 - e^{-H}) e^z, \ S = [\bar{l}_B (1 - e^{-H}) - l_A] z e^z + \bar{l}_B H e^{z-H}, \ z < 0,$$

$$A^{(0)} = 0, \ B^{(0)} = 1 - e^{z-H}, \ S = \bar{l}_B (H - z) e^{z-H}, \ 0 < z < H, \qquad (6.47)$$

$$A^{(0)} = B^{(0)} = S = 0, \ z > H,$$

where $\bar{l}_B = (1 - \alpha) l_B$. The reduced temperature at the first front, Θ_c, and the distance between the fronts, H, are related to each other and to the heat releases of the reactions,

$$\frac{1 - \Theta_c}{1 - \alpha} = 1 - e^{-H}.$$

A linear stability analysis of the solution (6.47) gives the following relation between the growth rate σ and the wavenumber k [14],

$$2\Gamma^2 (\Gamma - 1) e^{H(\Gamma - 1)} + (\Gamma - 1 - 2\sigma)(l_A + \bar{l}_B (e^{H(\Gamma - 1)} - 1)) = 0, \ \Gamma = (1 + 4\sigma + 4k^2)^{1/2}.$$

$$(6.48)$$

There can be both monotonic and oscillatory instabilities of the planar fronts, which can occur either at a finite perturbation wavenumber or at zero wavenumber.

Consider the nonlinear evolution of perturbations of the uniform solution (6.47), for l_A slightly above the threshold of the long-scale oscillatory instability, in the case where there are overlapping regions of monotonic and oscillatory longwave instabilities. It follows from the dispersion relation (6.48) that near the threshold of the *oscillatory instability*, the complex linear growth rates of the oscillatory and monotonic modes, σ_o and σ_m, are [14]

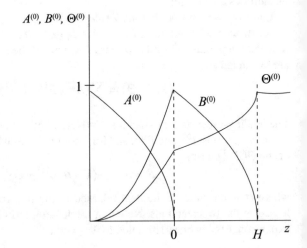

Fig. 6.1 Mass fractions and reduced temperature for the system of two planar fronts

$$\sigma_o \sim i\omega_0 + \gamma(l_A - l_A^o) - \delta k^2 + \dots, \quad \sigma_m \sim -\frac{1}{2}(l_A - l_A^m)k^2 - \chi k^4 + \dots, \quad (6.49)$$

where k is the wavenumber, ω_0 is the frequency, $l_A^o = l_A^o(H, \bar{l}_B)$ and $l_A^m = -2$ are the thresholds of the oscillatory and monotonic instabilities, respectively, and

$$\gamma = \left(\frac{\partial \sigma_o}{\partial l_A}\right)_{l_A = l_A^o, k=0}, \quad \delta = -\frac{1}{2}\left(\frac{\partial^2 \sigma_o}{\partial k^2}\right)_{l_A = l_A^o, k=0}, \quad (6.50)$$

$$\chi = -\frac{1}{24}\left(\frac{\partial^4 \sigma_m}{\partial k^4}\right)_{l_A = l_A^o, k=0} = \frac{l_A^o}{8}[l_A^o(6 - l_A^o) + 4H(l_A^o - \bar{l}_B)]. \quad (6.51)$$

Here γ and δ are complex and ω_0 and χ are real, with $\chi > 0$, $\mathrm{Re}(\delta) > 0$. Typical neutral stability boundaries are shown in Figure 6.2.

Near the oscillatory instability threshold, $l_A = l_A^o + \varepsilon^2$, $\varepsilon \ll 1$, the wavenumbers k of the excited oscillatory perturbations are $O(\varepsilon)$. If χ is numerically large, $\chi = \kappa/\varepsilon^2 = O(\varepsilon^{-2})$, the excited monotonic wavenumbers are also $O(\varepsilon)$ even far from the monotonic instability threshold, for $|l_A^o - l_A^m| = O(1)$. In that case, one obtains a system of coupled equations, the complex Ginzburg–Landau equation for the rescaled complex amplitude of the flame oscillations, $P = \rho \exp(i\theta)$, and the Kuramoto–Sivashinsky equation for the deformation of the first front, Q [15]:

$$\partial_\tau P + \nabla_\perp P \cdot \nabla_\perp Q = \xi P + (1 + iu)\Delta_\perp P - (1 + iv)|P|^2 P - r\Delta_\perp QP, \quad (6.52)$$

$$\partial_\tau Q = -m\Delta_\perp Q - g\Delta_\perp^2 Q - \frac{1}{2}|\nabla_\perp Q|^2 - w|P|^2. \quad (6.53)$$

That system is used when both modes are excited, i.e., when $m > 0$.

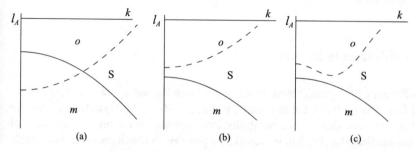

Fig. 6.2 Monotonic (solid line) and oscillatory (dashed line) stability boundaries; (a) overlap between longwave instability regions: (6.52), (6.53) with $m > 0$; (b) no overlap between longwave instability regions: (6.52), (6.53) with $m < 0$, $g = 0$; (c) no overlap between shortwave oscillatory instability region and longwave monotonic instability region: (6.54), (6.55)

If the monotonic mode is linearly damped, $m < 0$, one can set $g = 0$ in equation (6.53); thus, one obtains a system of coupled complex Ginzburg–Landau equation and the Burgers equation. Diverse dynamic regimes, described by both kinds of equations, are described in [15, 16]. In the longwave limit, the evolution of the amplitude disturbances, ρ, are enslaved to the phase disturbances, θ (see Section 7.2). Then the problem is reduced to the system of coupled Burgers equation that was studied in [17].

If coefficient $\mathrm{Re}(\delta)$ in (6.49) changes its sign, a transition between longwave and shortwave oscillatory instabilities takes place. In that case, a stabilizing term with $\Delta_\perp^2$ has to be added in the right-hand side of equation (6.52), which transforms the complex Ginzburg–Landau equation into the *complex Swift–Hohenberg equation*. The estimates of the orders of the terms show that this equation includes terms of different orders:

$$\partial_\tau P = -i\Delta_\perp P + \varepsilon[\gamma_0 P - (\delta_r + i\delta_i)\Delta_\perp P - (1 + iu)\Delta_\perp^2 P - (1 + iv)|P|^2 P -$$

$$\nabla_\perp Q \cdot \nabla_\perp P + (s_r + is_i)P\Delta_\perp^2 Q]. \quad (6.54)$$

The equation for the stable monotonic mode is reduced to

$$\partial_\tau Q = |m|\Delta_\perp Q - w|P|^2. \quad (6.55)$$

The observed dynamics is even more richer that in the case discussed above (see [18]).

In the present section, we have considered the simplest, analytically treatable example of combustion front propagation. Because of the translational invariance of the problem, which is the origin of a neutral disturbance with a zero wavenumber, the monotonic transverse instability of the front is of type II, and it is governed by the universal Kuramoto–Sivashinsky equation. The modulation of shortwave patterns created by an oscillatory instability is governed by the complex Ginzburg–Landau equation. A more intricate situation appears in the case of a sequential chemical reaction, where both modes can coexist and be coupled, which creates a rich dynamics.

6.2 Solidification Fronts

The efficiency of crystal growth processes is often limited by various kinds of instabilities. Among them, the morphological instability of the crystal-melt interface is the best known. This type of instability, first explained in the pioneering works of Mullins and Sekerka [19, 20], is caused by a positive feedback between the solidification front distortion and the heat or solute flux disturbance at the front of the phase transformation. That paradigmatic example of the pattern formation in nonequilibrium systems has been studied for several decades (for a comprehensive description, see [21]).

The nonlinear development of the morphological instability needs a special consideration because it can be *subcritical*. The most interesting case is the *long-scale* morphological instability that appears during directional solidification of systems with small segregation coefficients, typical of many binary semiconductor alloys [21, 22]. Among the basic achievements of the nonlinear theory of the morphological instability, let us mention the derivation of the longwave evolution equation for front distortions known as the Sivashinsky equation [23, 24]. It has been found that in a contradistinction to another basic example of pattern formation, the Rayleigh–Bénard convection, where the instability leads to formation of steady patterns, in the case of morphological instability an unbounded growth of disturbances takes place [25], which leads to formation of deep cells, fingers, and dendrites. We consider that kind of instabilities in Section 6.2.1.

Another mechanism of longwave surface instability, which may be significant by growth of a monocrystal, is the anisotropy of the surface tension of a crystal (Section 6.2.2). The development of the anisotropy-induced instability of the crystal growth front is described by the *convective Cahn–Hilliard equation* (CCHE). The CCHE provides a bridge between the Cahn–Hilliard equation, which has a potential structure, and the Kuramoto–Sivashinsky equation, which gives a paradigmatic example of a spatiotemporal chaos. That equation describes also spinodal decomposition of phase-separating systems in an external field [26–28] and spatiotemporal evolution of the morphology of steps on crystal surfaces [29].

6.2.1 Longwave Morphological Instability

6.2.1.1 Formulation of the Problem

Consider directional solidification of a binary alloy pulled with a constant speed V_0 in the direction of negative z through an applied thermal gradient $G_0 > 0$. We use the *frozen-temperature approximation*, i.e., the temperature field is prescribed as $T(z) = T_i + G_0 z$. The system is described by the diffusion equation for the solute concentration in the melt, C, with the boundary conditions at the crystal-melt interface, $z = h(\mathbf{x}, t)$, and at infinity [21]:

$$C_t = V_0 C_z + D(C_{zz} + \Delta_\perp^2 C), \ z > h(\mathbf{x}, t), \tag{6.56}$$

$$z = h(\mathbf{x}, t) : m\left(\frac{C - C_\infty}{K}\right) = \frac{T_m \gamma}{L_v} 2H + G_0 h, \tag{6.57}$$

$$D\frac{\partial C}{\partial n} = \frac{V_0 + h_t}{\sqrt{1 + (\nabla_\perp h)^2}}(K - 1)C, \tag{6.58}$$

$$z \to \infty : C \to C_\infty \tag{6.59}$$

Here $\mathbf{x} = (x, y)$ are the coordinates along the planar interface, $\nabla_\perp = (\partial_x, \partial_y)$, $\Delta_\perp = \partial_x^2 + \partial_y^2$, $T_i = T_m + mC_\infty/K$ is the temperature at the planar interface in the presence of the solute, T_m is the melting temperature of the pure melt, $m < 0$ is the liquidus slope (for small C), K is the segregation coefficient, γ is the crystal-melt surface tension, L_v is the latent heat per unit volume, H is the mean curvature of the interface, D is the solute diffusion coefficient, and C_∞ is the solute concentration in the melt, far from the interface. The *one-sided model* is employed in which one neglects the impurity diffusion in the solid, thus assuming that it is constant and equal to C_∞ due to the conservation of mass. The boundary condition (6.57) describes the thermodynamic equilibrium at the interface that takes into account the Gibbs–Thomson relation and the constitutional undercooling. The boundary condition (6.58) determines the solute flux balance at the interface. We neglect the heat generation at the interface and assume that thermal diffusivities of the liquid and the solid are equal, so that the interface perturbations do not disturb the thermal field.

Following [23], we choose D/V_0 as the length scale, D/V_0^2 as the time scale, and C_∞/K as the concentration scale and define two dimensionless parameters:

$$W = \frac{G_0 DK}{mV_0 C_\infty (K - 1)}, \ \beta = \frac{\gamma V_0 T_m K}{mL_v DC_\infty (K - 1)}. \tag{6.60}$$

It is convenient to go over to the frame of reference in which the solid–liquid interface corresponds to $z = 0$, by making the coordinate change $z \to z - h(\mathbf{x}, t)$. In this frame of reference, the system (6.56)–(6.59) reads:

$$C_t = (1 + h_t)C_z + \Delta_\perp C + [1 + (\nabla_\perp h)^2]C_{zz} - C_z \Delta_\perp h - 2\nabla_\perp h \cdot \nabla_\perp C_z, \tag{6.61}$$

$$C = 1 - \beta(1-K)H - W(1-K)h, \; z = 0, \tag{6.62}$$

$$(1-K)(1+h_t)C = \nabla_\perp h \cdot \nabla_\perp C - [1 + (\nabla_\perp h)^2]C_z, \; z = 0, \tag{6.63}$$

$$C \to K, \; z \to \infty, \tag{6.64}$$

where

$$H = (-\Delta_\perp h + h_{xx}h_y^2 + 2h_{xy}h_xh_y + h_{yy}h_x^2)[1 + (\nabla_\perp h)^2]^{-3/2}.$$

6.2.1.2 Amplitude Equation

The problem (6.61)–(6.64) has a stationary solution corresponding to a planar solidification front,

$$h^{(s)} = 0, \; C^{(s)} = K + (1-K)\exp(-z). \tag{6.65}$$

In the case of a small segregation coefficient, $K \ll 1$, solution (6.65) is unstable for $W < W_c$ with respect to longwave perturbations [22, 23].

Let us take $W = W_c(1 - s\varepsilon^2)$, where $0 < \varepsilon \ll 1$ is a small parameter, $s = \mathrm{sign}(W_c - W)$. Assume $K = \varepsilon^4 \kappa$, $\kappa = O(1)$, and introduce the long-scale variables $\mathbf{X} = (X, Y) = \varepsilon \mathbf{x}$, $\tau = \varepsilon^4 t$. We construct the solution of the problem (6.61)–(6.64) in the form of the expansions

$$C(\mathbf{X}, z, \tau, \varepsilon) = \sum_{j=0}^{\infty} \varepsilon^{2j} C^{(j)}, \; h(\mathbf{X}, z, \tau, \varepsilon) = \varepsilon^2 \sum_{j=0}^{\infty} \varepsilon^{2j} h^{(j)}. \tag{6.66}$$

and substitute expansions (6.66) in (6.61)–(6.64) in order to obtain the sequence of problems in successive orders of ε^2.

The solvability condition for the problem in the second order gives $W_c = 1$. In the third order, the solvability condition gives the following evolution equation for h(0):

$$h_\tau^{(0)} + \beta \Delta_\perp^2 h^{(0)} + s\Delta_\perp h^{(0)} + \kappa h^{(0)} - \frac{1}{2}\Delta_\perp (h^{(0)})^2 = 0. \tag{6.67}$$

Equation (6.67) is the *damped Sivashinsky equation* [23].

By rescaling of variables, we rewrite equation (6.67) in the form

$$h_T + \Delta_\perp^2 h + s\Delta_\perp h + \alpha h - \frac{1}{2}\Delta_\perp (h^2) = 0, \tag{6.68}$$

where $\alpha = \kappa\beta$, and the operator $\Delta_\perp$ acts on the rescaled long coordinate $\mathbf{X}$.

The parameter α characterizes the *stabilizing* effect of small but finite segregation coefficient (*incomplete* solute rejection). The planar front is linearly stable for $\alpha \geq 1/4$, while for $0 \leq \alpha < 1/4$, $s = 1$, it is unstable with respect to perturbations with the wavenumbers in the interval

$$k_-^2 < k^2 < k_+^2, \; k_\pm^2 = \frac{1 \pm \sqrt{1 - 4\alpha}}{2}. \tag{6.69}$$

In the case of a complete solute rejection, corresponding to a zero segregation coefficient, $\alpha = 0$, we obtain equation which is equivalent to (3.158) up to a scaling of variables. We do not repeat here the discussion of its solutions. Recall only that all the stationary solutions of this equation are unstable, and the instability leads to blow-up.

Near the instability threshold, $\alpha = 1/4 - \delta^2$, the instability takes place in a narrow interval, $k_+ - k_- = O(\delta)$, around $k_c = \sqrt{2}/2$. A one-dimensional solution can be presented as a slowly modulated pattern [30, 31]

$$h(X,T) = \sum_{n=1}^{\infty} \delta^n h^{(n)}(X_0, X_1, T_2), \quad X_0 = X, \quad X_1 = \delta X, \quad T_2 = \delta^2 T, \tag{6.70}$$

where

$$h^{(1)} = A(X_1, T_2)e^{ik_c X_0} + A^*(X_1, T_2)e^{-ik_c X_0}. \tag{6.71}$$

The complex amplitude $A(X_1, T_2)$ is governed by the Ginzburg–Landau equation with real coefficients,

$$A_{T_2} = A + 2A_{X_1 X_1} + \frac{2}{9}|A|^2 A, \tag{6.72}$$

which describes an unbounded growth of disturbances. Thus, the weakly nonlinear approach is not efficient in that case.

6.2.2 Anisotropy-Induced Instability: Convective Cahn–Hilliard Equation

Let us consider now a kinetically controlled crystal growth under unconstrained conditions. Examples include vapor deposition, epitaxial growth, and solidification of hypercooled melt (for review, see [32]). The crystal growth can be subject to different kinds of instabilities [21, 33]. Here we consider only one type of crystal growth instabilities, which is observed in the course of a crystal growth in the case of a strong dependence of the crystal surface tension on the orientation of the surface. That instability is developed if the front propagates in such a way that the surface energy can be diminished by the change of the surface normal direction. In that case, the flat surface with an energetically undesirable orientation is decomposed into a system of domains with preferable orientations. Actually, that is just *formation of facets*. Typically, the transformation of an unstable phase into two stable phases under the restrictions caused by the mass conservation is governed by the Cahn–Hilliard equation. We will see, however, that because of the front propagation, the Cahn–Hilliard equation should be modified.

6.2.2.1 Formulation of the Problem

Let us consider a plane solidification front between a cubic crystal and a hyper-cooled melt. The crystal growth with (as yet unknown) velocity V along the direction at the angle θ with respect to the principal crystalline direction [01]; see Figure 6.3.

The surface tension is given by the formula [34]

$$\gamma = \gamma_0(1 + \varepsilon_\gamma \cos 4\theta). \tag{6.73}$$

The rate of solidification is controlled by kinetic undercooling law

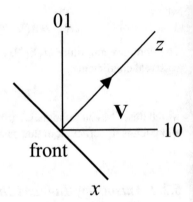

Fig. 6.3 Schematic of the front propagation

$$v_n = \mu_0^{-1}(T_e - T_i), \tag{6.74}$$

where v_n is the normal velocity of the phase transition front, T_i is the local surface temperature, and T_e is the local equilibrium temperature. The dependence of the equilibrium temperature on the surface tension γ is given by the Gibbs–Thomson relation,

$$T_e = T_m\left(1 - \frac{GK}{L}\right), \tag{6.75}$$

where K is the local mean curvature (positive for a convex surface), L is the latent heat of melting per unit volume, and G is the *surface stiffness*,

$$G = \gamma + \frac{\partial^2 \gamma}{\partial \theta^2}. \tag{6.76}$$

Assume now that the front is deformed. In the reference frame moving with the velocity V, its shape is described as $z = h(x,t)$, and the solidification process is governed by the following system of dimensionless equations and boundary conditions in the liquid phase (the heat transfer in the solid phase is disregarded) [35–37]:

$$\Theta_t = \Theta_{zz} + V\Theta_z + \Theta_{xx}, \quad -\infty < x < \infty, \ h(x,t) < z < \infty, \tag{6.77}$$

$$z \to \infty, \; \Theta = 0, \tag{6.78}$$

$$z = h(x,t), \; S(V + h_t) = -\Theta_z + h_x d\Theta_x, \tag{6.79}$$

$$\Theta = 1 + \Gamma E(h_x)h_{xx} - \frac{V + h_t}{(1 + h_x^2)^{1/2}}, \tag{6.80}$$

where $\Theta = (T - T_\infty)/\Delta T$ is the dimensionless temperature, $S = L/(\rho c_p \Delta T)$ is the Stefan number, and $\Gamma = \gamma_0 T_m/(\kappa \mu_0 L)$. Here ΔT is the difference between the melting temperature at the planar front growing in the given direction, T_m, and the temperature at infinity T_∞, L is the latent heat of melting, ρ is the density of the liquid (assumed to be equal to that of the crystal), κ is the melt thermal diffusivity, c_p is the specific heat which we assume to be equal for both crystal and melt, and

$$E(h_x) = \frac{\partial^2 (\tilde{\gamma}\sqrt{1 + h_x^2})}{\partial h_x^2}, \tag{6.81}$$

where $\tilde{\gamma}$ is the dimensionless surface tension [38, 39], $\tilde{\gamma} = \gamma/\gamma_0 = (1 + \varepsilon_\gamma \cos 4\theta)$. For small h_x, one then gets the following expansion for $E(h_x)$:

$$E = f_0(\theta) + f_1(\theta)h_x + f_2(\theta)h_x^2 + O(h_x^3),$$

$$f_0 = 1 - 15\varepsilon_\gamma \cos 4\theta, \; f_1 = 60\varepsilon_\gamma \sin 4\theta, \; f_2 = \frac{3}{2}(95\varepsilon_\gamma \cos 4\theta - 1). \tag{6.82}$$

It can be easily seen that if $|\varepsilon_\gamma| > 1/15$, then there exist crystal growth directions such that $f_0(\theta) < 0$ (see Figure 6.4).

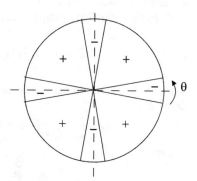

Fig. 6.4 Sign of $f_0(\theta)$ for $\varepsilon_\gamma > 1/15$

A stationary solution of system (6.77)–(6.80) corresponding to a planar front propagating with a constant velocity is [35]

$$\Theta^{(0)} = S exp(-Vz), \; V = 1 - S. \tag{6.83}$$

It exists only at $S < 1$, i.e., if $\Delta T > L/(\rho c_p)$ (the *hypercooled* melt). The *dimensional* velocity of the crystallization front is thus $v_n = \mu_0^{-1}[\Delta T - L/(\rho c_p)]$.

6.2.2.2 Instability of a Planar Front

The stability of solution (6.83) is determined by S and $f_0(\theta)$. The dispersion relation for the perturbations with the wavenumber k and the growth rate σ is [40, 41]

$$S\sigma + SV^2 + \left(V/2 + \sqrt{V^2/4 + \sigma + a^2}\right)(-SV + \sigma + \Gamma f_0 a^2) = 0. \qquad (6.84)$$

In the longwave limit, $k \ll 1$,

$$\sigma = s_2 k^2 + s_4 k^4 + O(k^6), \ s_2 = \frac{S}{V} - \Gamma f_0, \ s_4 = \frac{S}{V^5}[1 + V(1 - \Gamma f_0)](V\Gamma f_0 - 1). \qquad (6.85)$$

In the shortwave limit, $k \to \infty$,

$$\sigma = s'_2 k^2 + s'_1 k + O(1), \ s'_2 = -\Gamma f_0, \ s'_1 = \frac{S\Gamma f_0}{\sqrt{1 - \Gamma f_0}}. \qquad (6.86)$$

In the case where the front moves in the direction with θ preferable from the point of view of the surface energy ($f_0(\theta) > 0$), a longwave instability appears when $S > S_c = \Gamma f_0 / (1 + \Gamma f_0)$. The nonlinear development of that instability is described by the Kuramoto–Sivashinsky equation [40–43], which is typical also for monotonic instabilities of reaction fronts (see Section 6.1).

If $f_0(\theta) < 0$, then both s_2 and s'_2 are positive; actually, the plane crystallization front moving in that direction with a constant velocity is unstable with respect to any wavenumbers. In order to obtain a well-posed boundary value problem, it is necessary to take the surface tension dependence on curvature into account. That leads to the following modification of the Gibbs–Thomson relation (6.80) [37]:

$$\Theta = 1 + \Gamma E(h_x)h_{xx} - Kh_{xxxx} - \frac{V + h_t}{(1 + h_x^2)^{1/2}}, \qquad (6.87)$$

where K is a positive parameter. In that case, the following evolution equation is derived:

$$h_t - \frac{1}{2}Vh_x^2 = (-\mu + \alpha h_x + \beta h_x^2)h_{xx} - (|s_4| + K)h_{xxxx}, \qquad (6.88)$$

where

$$\mu = \frac{S}{V} - \Gamma(1 - 15\varepsilon_\gamma \cos 4\theta),$$

$$\alpha = 60\Gamma \varepsilon_\gamma \sin 4\theta, \ \beta = \frac{S}{V} + \Gamma(135\varepsilon_\gamma \cos 4\theta - 1). \qquad (6.89)$$

By rescaling, one obtains *the convective Cahn–Hilliard equation* for *the slope* $q \sim h_x$,

$$q_t - Dqq_x = (-q + bq^2 + q^3 - q_{xx})_{xx}, \qquad (6.90)$$

which combines the Cahn–Hilliard-like terms characteristic for phase separation (faceting) and the nonlinearity $(q^2)_x$ typical for moving front instabilities.

6.2.2.3 Convective Cahn–Hilliard Equation

Coefficient b can be eliminated by the transformation $u = q + b/3$. Using a moving frame in order to eliminate the linear term with u_x, we arrive at the following equation, which is called *convective Cahn–Hilliard equation*:

$$u_t - Duu_x + (u - u^3 + u_{xx})_{xx} = 0. \tag{6.91}$$

For small driving force $D \to 0$, equation (6.91) is reduced to the well-known Cahn–Hilliard (CH) equation describing spinodal decomposition in phase-separating systems [44] and exhibiting the coarsening dynamics [45] of which faceting of thermodynamically unstable surfaces is a well-known example [46]. With the growth of the driving force, there must be a transition from the coarsening dynamics to a chaotic spatiotemporal behavior, since for $D \to \infty$ the transformation $u \to u/D$ reduces equation (6.91) to the well-known Kuramoto–Sivashinsky (KS) equation. Below we discuss the dynamics of the convective Cahn–Hilliard equation for different values of D.

Stationary Solutions

Let us start with the consideration of stationary solutions $u(x,t) = U(x)$ of equation (6.91) which satisfy the following problem (for more detail, see [47]):

$$U_{xxx} + (U - U^3)_x - \frac{D}{2}U^2 = -\frac{D}{2}A, \quad -\infty < x < \infty; \tag{6.92}$$

$$x \to \pm\infty, \ |U| < \infty. \tag{6.93}$$

Recall that for $D = 0$, that corresponds to the CH equation, all stationary solutions of equation (6.92) can be found analytically [48]. In the opposite limit, $D \to \infty$ (KS equation), the set of stationary solution is much more complicated than in the case of the CH equation. Bifurcations of periodic stationary and traveling wave solutions of the KS equation were studied in [49, 50]. Later, without loss of generality, we assume $D > 0$.

It is convenient to investigate the bifurcations of stationary states in terms of the dynamical system defined by equation (6.92) with parameters D and A. Notably, this dynamical system is measure-preserving. It is also reversible, i.e., invariant with respect to inversion, $x \to -x$, $U \to -U$, as well as invariant with respect to translation $x \to x + C$. Accordingly, all stationary solutions are either invariant with respect to these transformations or (up to an arbitrary shift along x) exist in pairs: if $U(x+C)$ is a family of solutions, then $-U(C-x)$ is also a family of solutions. Both families may coincide.

Constant and Heteroclinic Solutions

For $A=0$, the only bounded solution of the problem (6.92), (6.93) is the trivial equilibrium solution $U = 0$. The linearized problem possesses three eigenvalues on the imaginary axis: $\lambda_1 = 0$ and $\lambda_{2,3} = \pm i$; accordingly, the equilibrium state is structurally unstable. Increasing A removes the degeneracy: the equilibrium splits into two constant solutions, $U = U^{\pm} = \pm\sqrt{A}$; besides, the imaginary eigenvalues are responsible for the creation of the periodic (with respect to x) orbit with period $\approx 2\pi$ whose amplitude locally grows as $\sim \sqrt{A}$ (for small A); below, this orbit will be referred to as the "main family."

Let us start the analysis of the problem (6.92), (6.93) with the consideration of the equilibrium points $U = U^{\pm} = \pm\sqrt{A}$. Consider perturbations of the constant solution, $U = U^{\pm} + \tilde{U}(x)$, $\tilde{U}(x) \sim e^{\lambda x}$, and linearize (6.92) to obtain the following equation for the eigenvalues:

$$\lambda^3 + (1 - 3A)\lambda \mp D\sqrt{A} = 0. \tag{6.94}$$

All eigenvalues are real if

$$D^2 \le D_*^2 = \frac{4(3A - 1)^3}{27A}, \ A > 1/3. \tag{6.95}$$

Otherwise, two eigenvalues are complex conjugate and the third one is real. For the equilibrium point $U = U^+ = \sqrt{A}$, one of the eigenvalues is positive and the other two eigenvalues are either negative or have negative real parts. The unstable manifold $W^u(U^+)$ is one-dimensional, while the stable manifold $W^s(U^+)$ is two-dimensional. For the symmetric counterpart of U^+, the solution $U^- = -\sqrt{A}$, the eigenvalues have opposite signs; hence, its stable manifold $W^s(U^-)$ is one-dimensional and its unstable manifold $W^u(U^-)$ is two-dimensional.

Heteroclinic solutions ("kinks") joining the equilibrium points $U = U^+$ and $U = U^-$ correspond to trajectories on $M_+ = W^u(U^-) \cap W^s(U^+)$ ("positive kinks") and $M_- = W^u(U^+) \cap W^s(U^-)$ ("negative kinks"); see Figure 6.5. As a matter of fact, there are *exact* solutions of this kind [26], one for a positive kink with $A = A_+ = 1 + D/\sqrt{2}$,

$$U = U_+(x) = U_+^0 \tanh \frac{U_+^0}{\sqrt{2}}(x - x_0), \tag{6.96}$$

$$U_+^0 = \sqrt{1 + D/\sqrt{2}}, \ x_0 = const,$$

and the other for a negative kink with $A = A_- = 1 - D/\sqrt{2}$, $D < \sqrt{2}$,

$$U = U_-(x) = -U_-^0 \tanh \frac{U_-^0}{\sqrt{2}}(x - x_0), \tag{6.97}$$

$$U_-^0 = \sqrt{1 - D/\sqrt{2}}, \ x_0 = const.$$

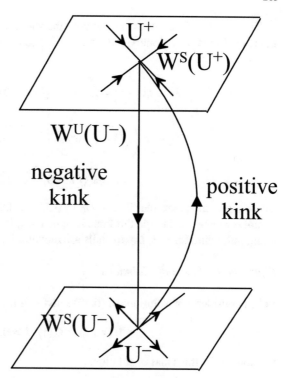

Fig. 6.5 Positive and negative kinks

Since the manifolds $W^s(U^+)$ and $W^u(U^-)$ are two-dimensional for any $A > 0$, their intersection in the three-dimensional phase space is generic, and therefore the solution (6.96) is a representative of a *family* of positive kinks $U_+(x;A)$. A negative kink requires matching of two one-dimensional manifolds in the three-dimensional space, which is not generic, but a codimension-2 event. The symmetry reduces the codimension of this event to 1. Accordingly, we can expect that for a given D, negative kinks exist only for isolated values of A. Solution (6.97) appears to have the largest value of A and the simplest spatial profile among the possible negative kinks: only that solution is monotonic, the other ones have additional humps.

In the phase space, a negative kink provides a connection from $U = U^+ = U^0_-$ to $U = U^- = -U^0_-$; since the positive kink is generic, the "backward" link is always there; hence, the presence of the negative kink implies the existence of a heteroclinic contour that connects the points U^+ and U^-. Accordingly, in the parameter plane, each point that represents a negative kink also corresponds to a heteroclinic contour.

The heteroclinic contour that includes the solution (6.97) plays the principal role in the evolution of stationary patterns. For this contour, the eigenvalues of the points $U^{\pm} = \pm\sqrt{A}$ are

$$\lambda_1 = \mp\left(\sqrt{1 - D/\sqrt{2}} + \sqrt{1 - 3D/\sqrt{2}}\right)/\sqrt{2},$$

$$\lambda_2 = \mp\left(\sqrt{1 - D/\sqrt{2}} - \sqrt{1 - 3D/\sqrt{2}}\right)/\sqrt{2},$$

$$\lambda_3 = \pm\sqrt{1 - D/\sqrt{2}} \cdot \sqrt{2}.$$

These eigenvalues are real if $D \leq D_0 = \sqrt{2}/3 \approx 0.47$; otherwise two of them are complex conjugate and the third one is real. Thus, if $D > D_0$, the kink has an oscillating tail, while for $D < D_0$, the tails are monotonic.

Bifurcations of Periodic Solutions

Let us consider now solutions of (6.92) satisfying the periodicity condition,

$$U(x + l) = U(x), \ l > 0. \tag{6.98}$$

Obviously, for this class of solutions,

$$A = \langle U^2 \rangle = \frac{1}{l}\int_0^l U^2(x)dx. \tag{6.99}$$

Since the dynamical system (6.92) is volume-preserving, the product of Floquet multipliers (eigenvalues of the monodromy matrix) for every periodic solution equals 1. Accordingly, such solutions have either two complex conjugate multipliers on the unit circle (periodic orbit of *elliptic* type) or two real eigenvalues, one of them outside of the unit circle (*hyperbolic* type of the periodic orbit). For an elliptic orbit, variation of parameters lets multipliers $\mu_{1,2} = \exp(\pm 2\pi i\phi)$ wander along the circle; passage of ϕ through each rational number p/q is accompanied by the "period q-tupling": branching of a periodic solution whose period is q times larger than the period of the original orbit. The details of the branching depend on q [51, 52]. For $q = 2$ and $q \geq 5$, the bifurcation is "one-sided": in the parameter space, the newborn periodic solutions exist only on one side of the branching point. For $q = 3$ and $q = 4$, on the contrary, the bifurcation is "two-sided" (transcritical): new branches exist on both sides of the critical parameter value. On one of these branches, the periodic solutions are of elliptic type (and, therefore, give rise to secondary sequences of similar bifurcations); the other branch corresponds to solutions of hyperbolic type.

One can see that, due to the symmetry properties of the equation, if $U(x)$ is a periodic solution of equation (6.92) with a certain value of A, and

$$\langle U \rangle = \frac{1}{l}\int_0^l U(x)dx = C,$$

then $\tilde{U}(x) = -U(2x_0 - x)$, $x_0 = const$ is also a solution of equation (6.92) for the same value of A; obviously, $\langle \tilde{U} \rangle = -C$. The solutions $U(x)$ and $\tilde{U}(x)$ can coincide. In this case, the solution $U(x)$ is odd (antisymmetric to reflections) with respect to the point $x = x_0$, and $C = 0$.

The "main" family of odd stationary periodic solutions, which bifurcates from the trivial solution $U(x) = 0$ with the increase of A from zero, is of a major interest. Recall that in the case of the Cahn–Hilliard equation, $D = 0$, this family exhausts the set of odd stationary solutions that can be calculated analytically (using elliptic Jacobi functions). The wavelength l grows monotonically with A. In the limit $A \to 1$, the wavelength tends to infinity, and the periodic solution is transformed into heteroclinic (kink) solutions $U(x) = \pm \tanh((x - x_0)/\sqrt{2})$.

For $D \neq 0$, the main family with finite values of A was calculated numerically. For small values of A, the periodic orbit is elliptic; with the increase of A, it undergoes a countable set of bifurcations that create new periodic orbits (see [51, 52]).

Depending on D, one can distinguish three different types of interrelation between A and the wavelength of the periodic solution (see Figure 6.6).

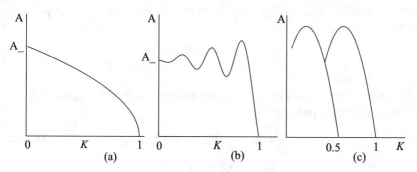

Fig. 6.6 Dependence between amplitude A and wavenumber K: (a) $D < D_0$; (b) $D_0 < D < D_1$; (c) $D > D_1$

In the region $0 < D < D_0 = \sqrt{2}/3$, the main family exists for $0 < A < A_- = 1 - D/\sqrt{2}$, i.e., just for all A below the value corresponding to the principal heteroclinic contour discussed above. The dependence of the wavelength l on the parameter A (or the dependence of A on the wavenumber $K = 2\pi/l$) is monotonic, and the wavelength tends to infinity as A approaches A_-.

In the region $D_0 < D < D_1 \approx 0.8254$, the solutions which belong to the main family exist for any $l > 2\pi$ but the dependence of A on l becomes non-monotonic; thus, the inverse dependence of l on A is not univalued. There exists a countable set of values of l for which the function $A(l)$ has maxima and minima. At the extrema of $A(l)$ that can be called *turning points*, a pair of periodic solutions appears/disappears through a saddle-center bifurcation.

In the case $D > D_1$, the configuration of the main branch of periodic solutions changes again: there remains only one turning point. With the growth of l, all coefficients with odd values of n in the Fourier series

$$U(x;A) = \sum_{n=1}^{\infty} a_n(A) \sin \frac{2\pi n}{l}(x - x_0) \tag{6.100}$$

tend to zero. Thereby, the solution with the period l tends to the solution with the period $2l$. That phenomenon was described in the context of the Kuramoto–Sivashinsky equation [49, 50, 53] where it was named the *noose bifurcation*.

Stability of Stationary Periodic Solutions

Let us investigate the linear stability of the main family of spatially periodic stationary solutions with respect to arbitrary infinitesimal perturbations in the infinite region.

Eigenvalue Problem

Let $U(x)$ be a stationary periodic solution satisfying equation (6.91) with periodic boundary conditions, $U(x) = U(x+l)$. Recall that for the main family of periodic solutions,

$$\int_0^l U(x)dx = 0,$$

because $U(x)$ is odd.

Add a small perturbation $\widetilde{u}$ to the stationary solution and linearize equation (6.91) to obtain the following problem for $\widetilde{u}$:

$$\widetilde{u}_t = -\widetilde{u}_{xx} - \widetilde{u}_{xxxx} + D(U\widetilde{u})_x + 3(U^2\widetilde{u})_{xx}, \tag{6.101}$$

$$|\widetilde{u}| < \infty \text{ for } |x| \to \infty.$$

Following the standard approach of the linear stability theory, consider normal-mode solutions of (6.101),

$$\widetilde{u}(x,t) = e^{t\sigma}V(x;\sigma). \tag{6.102}$$

Substitute (6.102) into (6.101) to obtain the equation:

$$V(x)\sigma = -V_{xx} - V_{xxxx} + D(UV)_x + 3(U^2V)_{xx}. \tag{6.103}$$

Generally, equation (6.103) has four linearly independent solutions for any value of x, $V_j(x)$, $j = 1, 2, 3, 4$. Because of the periodicity of $u(x)$, the functions $V_j(x)$ and $V_j(x+L)$ are simultaneously the solutions of equation (6.103). Therefore,

$$V_j(x+L) = \sum_{k=1}^{4} B_{jk}(\sigma)V_k(x), \tag{6.104}$$

where $\hat{B}(\sigma) = \{B_{jk}(\sigma)\}$ is the monodromy matrix. We are interested only in bounded solutions of equation (6.103),

$$V(x) = \sum_{k=1}^{4} C_j V_j(x), \qquad |V(x)|, \infty \text{ for } x \to \pm\infty, \tag{6.105}$$

that correspond to the eigenvectors of the monodromy matrix,

$$\sum_{k=1}^{4} B_{jk}(\sigma)C_k = B(\sigma)C_j, \tag{6.106}$$

with the eigenvalues satisfying the additional condition,

$$|B(\sigma)| = 1. \tag{6.107}$$

Hence, one can represent the eigenfunctions $V(x)$ in the form of the Floquet–Bloch function [54]

$$V(x) = v(x)e^{iqx}, \tag{6.108}$$

where $v(x)$ is a periodic function, $v(x+L) = v(x)$, and $q = \arg(B(\sigma))/L$ is a real number ("quasi-wavenumber").

Substituting (6.108) into equation (6.103), one obtains the following eigenvalue problem:

$$\sigma(q)v(x) = \mathscr{D}_q^2[(3U^2(x) - 1)v(x)]$$
$$- \mathscr{D}_q^4 v(x) + D\mathscr{D}_q[U(x)v(x)], \tag{6.109}$$
$$v(x+l) = v(x),$$

where $\mathscr{D}_q = d/dx + iq$. If there exists an eigenvalue $\sigma(q)$ with $\mathrm{Re}\,\sigma(q) > 0$, the solution $U(x)$ is unstable.

Longwave Asymptotics

For $q = 0$, equation (6.109) always has a solution $v(x) = U_x(x)$ corresponding to $\sigma(0) = 0$. Indeed, differentiate (6.92) twice with respect to x to find:

$$- v_{xxxx} - v_{xx} + D(Uv)_x + 3(U^2 v)_{xx} = 0. \tag{6.110}$$

The disturbance $v(x) = U_x(x)$ corresponds to an infinitesimal homogeneous translation of the stationary solution $U(x)$. One can expect that $\mathrm{Re}\,\sigma(q)$ can become positive at small q. In the present subsection, we investigate the eigenvalue problem for longwave disturbances that correspond to nonhomogeneous translations applied to the stationary solution.

First, let us introduce the variable $X = Kx$, where $K = 2\pi/l$ is the wavenumber of the stationary regime, and rewrite the problem for the main-family stationary solutions as

$$-K^4 U'''' - K^2 U'' + DKUU' + K^2(U^3)'' = 0, \qquad (6.111)$$

$$U(X + 2\pi) = U(X),$$

where a prime means differentiation with respect to X. Note that

$$\int_0^{2\pi} U(X)dX = 0. \qquad (6.112)$$

Later on, we shift the origin $X = 0$ in such a way that the stationary solution is an odd function of X; therefore, its Fourier series is

$$U = \sum_{n=1}^{N} a_n(K) \sin nX. \qquad (6.113)$$

Equation (6.109) can be written as

$$\sigma(Q)v(X) = K^2 \mathscr{D}_Q^2 [(3U^2(X) - 1)v(X)]$$
$$- K^4 \mathscr{D}_Q^4 v(X) + DK\mathscr{D}_Q[U(X)v(X)], \qquad (6.114)$$
$$v(X + 2\pi) = v(X),$$

where $\mathscr{D}_Q = d/dX + iQ$, $Q \equiv q/K$. One can choose $|Q| \le 1/2$.

Let Q be small. We are looking for a solution in the form

$$v = \sum_{n=0}^{\infty} v_n Q^n, \qquad \sigma = \sum_{n=0}^{\infty} \sigma^{(n)} Q^n, \qquad (6.115)$$

where v_n are periodic functions of X with period 2π. Substituting (6.115) into (6.114), one obtains the sequence of problems in successive orders of Q.

In the zeroth order of Q, one has

$$-K^4 v_0'''' + K^2[(3U^2 - 1)v_0]'' + DK(Uv_0)' - \sigma^{(0)} v_0 = 0, \qquad (6.116)$$

$$v_0(X + 2\pi) = v_0(X),$$

where prime denotes differentiation with respect to X. As shown above, the solution $v_0 = U'$, $\sigma^{(0)} = 0$ always exists. Later on, we assume that there are no other 2π-periodic solutions with $\sigma^{(0)} = 0$ for the chosen values of K and D. This assumption fails for the values of parameters corresponding to the symmetry-breaking bifurcation.

For every integer $i > 0$, the equation for v_i has the form

$$-k^4 v_i'''' + k^2[(3U^2 - 1)v_i]'' + Dk(Uv_i)' = R_i, \quad v_i(X + 2\pi) = v_i(X), \qquad (6.117)$$

where R_i is some linear combination of the functions $v_j(X)$, $0 \leq j < i$, and their derivatives. One can show that the left-hand side of (6.117) is a derivative of a periodic function. Therefore, the integral of the right-hand side, R_i, over X from 0 to 2π must be zero.

At the first order in Q, one obtains

$$-K^4 v_1'''' + K^2[(3U^2 - 1)v_1]'' + DK(Uv_1)' =$$
$$\sigma^{(1)} v_0 - iDU v_0 - 2iK^2[(3U^2 - 1)v_0]' + 4iK^4 v_0''', \quad (6.118)$$

$$v_1(X + 2\pi) = v_1(X).$$

In order to find the solution of equation (6.118), let us differentiate the equation (6.111) with respect to the parameter K:

$$-K^4[\partial_K U]'''' + K^2[(3U^2 - 1)\partial_K U]'' + DK[U\partial_K U]'$$
$$-4K^3 U'''' - 2KU'' + DUU' + 6K[U^2 U']' = 0. \quad (6.119)$$

Comparing equation (6.118) with equation (6.119), one finds

$$v_1 = iK\partial_K U - \sigma^{(1)} f_1(X) + \text{const } U', \quad (6.120)$$

where $f_1(X)$ is the solution of the equation

$$-K^4 f_1'''' - K^2 f_1'' + DK(Uf_1)' + 3K^2(U^2 f_1)'' = U',$$

$$f_1(X + 2\pi) = f_1(X).$$

Note that $f_1(X)$ is an even function.

At the second order in Q:

$$-K^2 v_2'' - K^4 v_2'''' + DK(Uv_2)' + 3K^2(U^2 v_2)'' =$$
$$(\sigma^{(1)} - iDKU)v_1 + 4iK^4 v_1''' + 2iK^2[(1 - 3U^2)v_1]'$$
$$+\sigma^{(2)} v_0 - 6K^4 v_0'' + K^2(3U^2 - 1)v_0, \quad (6.121)$$

$$v_2(X + 2\pi) = v_2(X).$$

Integrating equation (6.121) over the period and substituting (6.120), one finds

$$\left[\sigma^{(1)}\right]^2 = -\frac{K^2 \int_0^{2\pi} dX U \partial U/\partial K}{\int_0^{2\pi} dX f_1(X)} = -\frac{K^2}{2} \frac{dA(K)/dK}{\langle f_1 \rangle}, \quad (6.122)$$

where $A = \langle U^2 \rangle$.

If $\left[\sigma^{(1)}\right]^2 > 0$, two real roots with opposite signs exist. One of them is positive; therefore, the stationary solution is unstable. If $\left[\sigma^{(1)}\right]^2 < 0$, there are two imaginary roots. In the latter case, the stability with respect to perturbations with small Q depends on the sign of $\sigma^{(2)}$ that can be obtained in a similar way from the equation in the third order of Q.

Thus, one can expect that there are boundaries between the regions of stability and instability of stationary solutions with respect to longwave disturbances that coincide with the extrema of the function $A(K)$.

Note that a similar relation between the non-monotonicity of the dependence between the amplitude and the wavenumber was established for the Kuramoto–Sivashinsky equation [55] and the reaction–diffusion equation [56].

Arbitrary q

For arbitrary values of q, the problem (6.109) is solved numerically [47]. Substituting

$$v(x,q) = \sum_{n=-\infty}^{\infty} \widehat{v}_n(q)e^{inKx}, \quad K = \frac{2\pi}{l} \tag{6.123}$$

into (6.109), one gets the following matrix equation for the eigenvalues and eigenvectors:

$$\sum_{n=-\infty}^{\infty} [M_{mn} - \sigma(q)\delta_{mn}]\widehat{v}_n = 0, \quad -\infty < m < +\infty, \tag{6.124}$$

where

$$M_{mn} = [(mk+q)^2 - (mk+q)^4]\delta_{mn}$$
$$+ iD(mk+q)\widehat{U}_{m-n} - 3(mk+q)^2(\widehat{U^2})_{m-n}; \tag{6.125}$$

here the symbolˆdenotes the Fourier coefficient.

The analysis of the spectrum $\sigma(q)$ leads to the conclusion that for $D > D_0 = \sqrt{2}/3$, there exists a stability interval $K_{left} < K < K_{right}$ for stationary periodic solutions that belong to the main family. The boundaries of the stability interval are presented in Table 6.1. Both boundaries move into the longwave region with the decrease of D and tend to zero when D approaches $D_0 = \sqrt{2}/3$. Recall that D_0 corresponds to the transition between monotonic and oscillatory behavior of the kink tail.

The right boundary of the stability interval always coincides with the global maximum of the function $A(K)$, which corresponds to equation (6.122) with $\langle f_1 \rangle > 0$. Therefore, on the right boundary of the stability interval, a monotonic longwave instability always occurs.

The type of the instability on the left boundary of the stability interval depends on D. For sufficiently large values of D ($D = 5$), as in the case of the Kuramoto–Sivashinsky equation, the destabilization of the stationary solutions for $K < 0.677$ is due to the oscillatory instability with $q \to 0$. For smaller values of D ($D = 2$ and $D = 1$), the dependence of $\mathrm{Re}[\sigma]$ on the quasi-wavenumber q has two maxima near the left stability boundary. For smaller values of D ($D = 2$ and $D = 1$), the dependence of $\mathrm{Re}[\sigma]$ on the quasi-wavenumber q has two maxima near the left stability boundary.

D	K_{left}	K_{right}
∞	0.766 osc.,$q \to 0$	0.838 mon.,$q \to 0$
5	0.677 osc.,$q \to 0$	0.775 mon.,$q \to 0$
2	0.537 osc., $q \to K/2$	0.640 mon.,$q \to 0$
1	0.376 osc.,$q \to K/2$	0.4752 mon.,$q \to 0$
0.8	0.314059 mon.,$q \to 0$	0.4065 mon.,$q \to 0$
0.5	0.111 mon.,$q \to K/2$	0.1767048 mon.,$q \to 0$

Table 6.1 Boundaries of the stability intervals

We have found that the destabilization scenario in this case is different: now the most "dangerous" oscillatory perturbation (the one with the largest real part of σ_1) at the left end of the stability interval corresponds to $q \to K/2$. For $D = 0.8$, a monotonic instability with $q \to 0$ takes place at *both* ends of the interval. Note that the left boundary of the monotonic instability is not connected with the extrema of $A(K)$. As a matter of fact, in the point $K = 0.314059$, the eigenvalue problem (6.116) has one more eigenfunction with $\sigma^{(0)} = 0$, in addition to the solution $v_0 = U'$. This eigenfunction has a nonzero mean value and is related to the bifurcation of another family of asymmetric stationary solutions with a nonzero mean value. For smallest values of D ($D = 0.5$), at the left end of the interval, we have a monotonic instability for $q \to K/2$.

The prediction of the existence of stable stationary solutions matches the results of the direct numerical simulations of equation (6.91) [58] for intermediate values of D, $D_0 < D < D_1$, $D_1 \approx 7$. For $D > D_1$, one have observed irregular spatiotemporal dynamics, qualitatively similar to that of the KS equation, when cell oscillations, splitting and merging occur on the same time scale. The attractor corresponding to the spatiotemporal chaos coexists with stable stationary solutions, but apparently it has a larger attraction basin.

Dynamics of Domain Walls

In the case $0 < D < D_0 = \sqrt{2}/3$, where the kink tails are monotonic, solutions of equation (6.91) exhibit coarsening governed by the interaction between the kinks and their collective motion.

Constant Solutions and Moving Kinks

We shall start the investigation of the kink dynamics governed by equation (6.91) with the consideration of its simplest solutions.

First, let us note that any constant $u = U_0$ is a solution of equation (6.91). In order to study the linear stability of this solution, linearize equation (6.91),

$$u = U_0 + \tilde{u}(x)e^{\sigma_0 t},$$

and find the eigenfunctions and the eigenvalues:

$$\tilde{u}(x) = e^{ikx}, \quad \sigma_0(k) = (1 - 3U_0^2)k^2 - k^4 - ikDU_0. \tag{6.126}$$

Obviously, the constant solutions are linearly stable if

$$|U_0| > 1/\sqrt{3}. \tag{6.127}$$

Let us consider now traveling wave solutions,

$$u = U(X), \quad X = x - vt, \quad v = \text{const}. \tag{6.128}$$

Substituting (6.128) into (6.91) and integrating once with respect to X, one obtains the following ordinary differential equation,

$$U_{XXX} + (U - U^3)_X - vU - \frac{D}{2}U^2 = -J, \quad -\infty < X < \infty, \tag{6.129}$$

where J is the integration constant. *Heteroclinic* solutions ("domain walls" or "kinks") connect two critical points of the dynamical system (6.129),

$$U = U_\pm = -\frac{v}{D} \pm \sqrt{\left(\frac{v}{D}\right)^2 + \frac{2J}{D}}, \tag{6.130}$$

$J > -v^2/2D$. Note that in the case $v \neq 0$, $U_- \neq -U_+$. Depending on the boundary conditions, the heteroclinic solutions can be of two types: *positive kinks*, with

$$U(\pm\infty) = U_\pm, \tag{6.131}$$

and *negative kinks*, or *antikinks*, with

$$U(\pm\infty) = U_\mp. \tag{6.132}$$

Later on, we shall consider only the case with $|U_\pm| > 1/\sqrt{3}$, i.e., when both constant solutions $U = U_\pm$ are stable.

In order to investigate the asymptotics of the heteroclinic solutions at $x \to \pm\infty$, linearize equation (6.129) around the constant solutions, $U = U_\pm$,

$$U(X) = U_\pm + \tilde{U}e^{\kappa X}, \tag{6.133}$$

to obtain the following equation for the eigenvalues κ:

$$\kappa^3 - (3U_{\pm}^2 - 1)\kappa - (v + DU_{\pm}) = 0. \tag{6.134}$$

Depending on the sign of $Q = -[(3U_{\pm}^2 - 1)/3]^3 + [(v + DU_{\pm})/2]^2$, two cases are possible: (i) for $Q \leq 0$ all roots of equation (6.134) are real; (ii) for $Q > 0$ one root is real and two other roots are complex conjugate. In the case (ii), the kinks have oscillatory tails, and one can expect the development of ordered or disordered structures rather than a system of kinks [57, 58]. In the present section, we consider the case (i), when the system evolution is determined by the motion of kinks.

Since $\kappa_1 + \kappa_2 + \kappa_3 = 0$, the roots κ_j, $j = 1, 2, 3$, cannot have the same sign. Taking into account that

$$\kappa_1 \kappa_2 \kappa_3 = v + DU_{\pm} = \pm\sqrt{v^2 + 2DJ}$$

is positive for $U = U_+$ and negative for $U = U_-$, one concludes that in the former case, two eigenvalues are negative and one eigenvalue is positive, while in the latter case, two eigenvalues are positive and one eigenvalue is negative. In other words, for the critical point $U = U_+$ ($U = U_-$), the unstable manifold $W^u(U_+)$ (the stable manifold $W^s(U_-)$) is one-dimensional, while the stable manifold $W^s(U_+)$ (the unstable manifold $W^u(U_-)$) is two-dimensional.

According to its definition, the positive kink corresponds to a trajectory on $M_+ = W^u(U_-) \cap W^s(U_+)$, while the antikink is situated on $M_- = W^u(U_+) \cap W^s(U_-)$; see Figure 6.5. In the former case, M_+ is an intersection of two two-dimensional manifolds in a three-dimensional space, which is generic and exists in a certain region in the parameter space (v, J). Therefore, a two-parameter family of kinks exists for certain intervals of v and J values. For fixed values of U_- and U_+, the velocity of the kink is calculated to be

$$v = -\frac{D}{2}(U_- + U_+).$$

A negative kink (antikink) can exist only if the trajectory $W^u(U_+)$ reaches the vicinity of the point U_- exactly along the trajectory $W^s(U_-)$; see Figure 6.6. Such event has a codimension 2 in the parameter space. Therefore, one can expect that the antikink exists for isolated values of (v, J). Indeed, there exists a solution

$$U = -A \tanh \frac{A}{\sqrt{2}}(X - X_0), \quad A = \sqrt{1 - D/\sqrt{2}}, \quad X_0 = \text{const}, \tag{6.135}$$

for $v = 0$, $J = DA^2/2$. Note that the numerical investigation of equation (6.129) reveals the existence of other isolated heteroclinic ($U_+ \to U_-$) solutions with $v = 0$, $J < DA^2/2$, which correspond to non-monotonic functions $U(X)$. These solutions have never been observed in the simulations of the original problem (6.91), and they are apparently unstable.

Note that for $v = 0$, $U_\pm = \pm A$, the roots of the cubic equation (6.134) are described by simple formulas,

$$\kappa_1 = \mp(A+B)/\sqrt{2}, \quad \kappa_2 = \mp(A-B)/\sqrt{2}, \quad \kappa_3 = \pm A\sqrt{2}, \qquad (6.136)$$

where A is defined by equation (6.135), and

$$B = \sqrt{1 - 3D/\sqrt{2}}. \qquad (6.137)$$

All three roots are real if $D < D_0 = \sqrt{2}/3$.

Antikink-Kink Pairs

Let us consider now the interaction of distant kinks. Since this interaction is expected to decrease exponentially with the distance between the kinks [59], the dynamics is typically determined by the pair of the closest kinks, while the motion of other kinks is negligible compared with the motion of this pair. In the present section, we consider a kink pair described by a *homoclinic solution*, $U = U_h(X)$ ($X = x - vt$, $v = $ const), of the dynamic system (6.129), that satisfies the boundary condition

$$U(-\infty) = U(+\infty) = U_+. \qquad (6.138)$$

Note that if $U = U_h(X)$ is a solution of equation (6.129), then $U = -U_h(-X)$ is also a solution. Therefore, it is sufficient to consider the case $U_+ > 0$ (i.e., the case when the antikink is located to the left from the kink). Numerical solution of equation (6.91) in the form of such antikink–kink pair is shown in Figure 6.7 [60]. The antikink–kink pair moves slowly to the left with a constant speed, as shown in Figure 6.8.

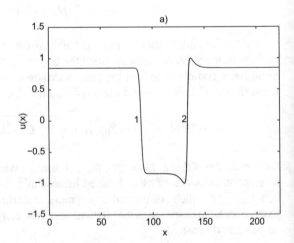

Fig. 6.7 Numerical solution of equation (6.91) in the form of an antikink–kink pair, very slowly moving to the left with a constant speed. $D = 0.4$

The homoclinic trajectory $U_h(X)$ corresponds to the trajectory $W^u(U_+)$ that has to return to the point U_+ along its two-dimensional stable manifold. This is a codimension-1 event; therefore, in the two-dimensional parameter space (v, J), the homoclinic trajectories should exist on a curve (or on a set of curves). Thus, antikink–kink pairs form a *one-parameter* family. Both parameters v and J have to be calculated as functions of the *distance L* between the kink centers.

In order to find these functions in the limit of large L, we use the approach similar to that applied in [61] for the analysis of the interaction between solitons and in [45] for the analysis of the interaction between kinks governed by the Allen–Cahn equation (Section 2.1) and by standard Cahn–Hilliard equation (Section 2.2). We shall construct the solution of the *nonlinear* equation (6.129) as a *linear superposition* of two kink solutions with a certain correction, which is caused by the nonlinearity and has to be calculated. The solvability conditions for that correction determine the parameters v and J (see below).

We expect that at large L, the velocity of a kink pair is small, and for the slowly moving pair containing an antikink, $U(-\infty) = U_+$ will be close to $A = \sqrt{1 - D/\sqrt{2}}$ that corresponds to a sole motionless antikink (6.135). In order to match the condition $U(+\infty) = U_+$, one has to take a kink with the parameter U_+ close to A as well. Thus, one can construct a solution as [45, 61]

$$U(X) = U_1(X) + U_2(X) + A + \tilde{u}(X), \tag{6.139}$$

where $U_1(X)$ is the negative kink described by (6.135) with $X_0 = 0$, and $U_2(X)$ is a representative of the family of kinks corresponding to $v = 0$, $J = DA^2/2$, with the center located in the point $X = L$, and $\tilde{u}(X)$ is a small correction that includes the difference between U_+ for a kink pair and A selected by the motionless negative kink. Unlike the antikink, the kink cannot be presented in an analytical form. Note that the asymptotics of the kink solution for large negative $X - L$ is determined by the eigenvalues (6.136) calculated near the critical point $U = -A$. The left "tail" of the kink is characterized by the smallest positive eigenvalue, $\kappa_2 = (A - B)/\sqrt{2}$, and can be written as

$$U_2(X) + A = u_2(X) \sim a(D) \exp[\kappa_2(X - L)], \tag{6.140}$$

where the constant $a(D)$ can be calculated analytically only in the limit of small D. Using the matched asymptotic expansions, one finds that for $D \ll 1$

$$a(D) \sim -D\sqrt{2}/2, \tag{6.141}$$

(see [59]). In the general case, the constant $a(D)$ is found numerically to be negative (see also [28]).

It is convenient to rewrite equation (6.129) in the form

$$U_{XXX} + (U - U^3)_X - \frac{D}{2}(U^2 - A^2) = q + vU, \quad -\infty < X < \infty, \tag{6.142}$$

where $q = -J + DA^2/2$, which is zero for individual kinks, is expected to be small for a distant kink–antikink pair.

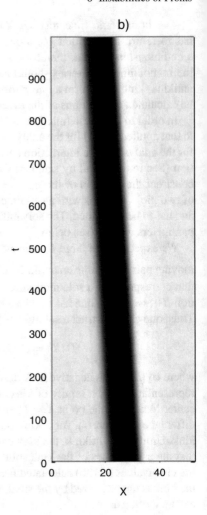

Fig. 6.8 Space–time diagram showing numerical solution of equation (6.91) in the form of a uniformly moving antikink–kink pair (the distance between the kinks is less than that shown in Figure 6.7)

Substitute (6.139) into equation (6.142) and linearize the latter with respect to u. According to the approach of [45, 61], we separately consider the regions around the centers of the left and the right kinks.

Near the left kink ($X = O(1)$), the contribution of the right kink, $U_2(X)$, is exponentially small. We can replace the function $U_2(X) + A$ by its asymptotic limit $u_2(X)$ and linearize equation (6.142) also with respect to u_2. Taking into account that $u_2(X)$ satisfies the equation

$$u_{2XXX} + (1 - 3A^2)u_{2X} - DAu_2 = 0,$$

we obtain the following equation for $\tilde{u}(X)$ valid in the region $X = O(1)$:

$$\tilde{u}_{XXX} + [(1 - 3U_1^2)\tilde{u}]_X - DU_1\tilde{u} = q + vU_1 + [3(U_1^2 - A^2)u_2]_X + D(U_1 + A)u_2.$$
$$(6.143)$$

Recall that the functions $U_1(X)$ and $U_2(X)$ are determined by the formulas (6.135) with $X_0 = 0$, and (6.140), respectively.

The solution of equation (6.143) is bounded if its right-hand side is orthogonal to two bounded solutions of the adjoint problem,

$$u^c_{XXX} + (1 - 3U_1^2)u^c_X + DU_1 u^c = 0, \tag{6.144}$$

that can be found analytically:

$$u^c_-(X) = \frac{\sinh(BX/\sqrt{2})}{\cosh(AX/\sqrt{2})} \tag{6.145}$$

is an odd function, and

$$u^c_+(X) = \frac{\cosh(BX/\sqrt{2})}{\cosh(AX/\sqrt{2})} \tag{6.146}$$

is an even function. Note that the third linearly independent solution of equation (6.144), $u^c_0(X) = 1 + D/3(\sqrt{2} - D)\cosh^2\left(AX/\sqrt{2}\right)$, is unbounded and not relevant. Taking into account that $U_1(X)$ is an odd function, we obtain two equations that determine v and q as functions of L and D:

$$v \int_{-\infty}^{+\infty} dX u^c_- U_1 + 3 \int_{-\infty}^{+\infty} dX u^c_- [(U_1^2 - A^2)u_2]_X + D \int_{-\infty}^{+\infty} dX u^c_-(U_1 + A)u_2 = 0, \tag{6.147}$$

$$q \int_{-\infty}^{+\infty} dX u^c_+ + 3 \int_{-\infty}^{+\infty} dX u^c_+ [(U_1^2 - A^2)u_2]_X + D \int_{-\infty}^{+\infty} dX u^c_+(U_1 + A)u_2 = 0. \tag{6.148}$$

The integrals in (6.147)–(6.148) can be calculated analytically. For example, the explicit formula for the velocity of an antikink–kink pair, $v \equiv v_2(D, L)$, can be written as

$$v_2(D, L) = a(D)C_2(D)e^{-\kappa_2 L}, \tag{6.149}$$

where

$$C_2(D) = \frac{\cos(\beta\pi/2)\{D\sqrt{2} + A^2(3\beta - 1) + \beta[A^2(1 - \beta^2) - D\sqrt{2}]\pi\csc(\beta\pi)\}}{\sqrt{2}\beta\pi}, \tag{6.150}$$

and

$$\beta = \frac{B}{A} = \sqrt{\frac{\sqrt{2} - 3D}{\sqrt{2} - D}}. \tag{6.151}$$

The coefficient $C_2(D)$ is positive for any values of D. In the limit $D \to 0$, $C_2(D) \sim D/2 + O(D^2)$. Using the relation (6.141) and taking into account that $\kappa_2 \sim D/2$, one finds that in the limit $D \ll 1$, $DL \gg 1$,

$$v_2(D, L) \sim -(D^2\sqrt{2}/4)\exp(-DL/2),$$

that coincides with the result of Watson et al. [59]. In the limit $D \to D_0 = \sqrt{2}/3$, $C_2(D) \sim \sqrt{2}/\pi - (D_0 - D)^{1/2}2^{-1/4}\pi + O(D_0 - D)$. As mentioned above, the constant $a(D)$ cannot generally be found analytically for an arbitrary D and has to be found numerically. We have found that $a(D) < 0$ for any $D < D_0$, and therefore $v < 0$, i.e., the antikink–kink pair moves to the left (i.e., the antikink is leading). Similarly, using the transformation $U \to -U, X \to -X$, one can show that the kink–antikink pair moves with the same velocity to the right (hence again the antikink is leading). A space–time diagram showing a moving antikink–kink pair is presented in Figure 6.8.

Near the center of the right (positive) kink, one obtains the following equation:

$$\tilde{u}_{XXX} + [(1 - 3U_2^2)\tilde{u}]_X - DU_2\tilde{u} = q + vU_2 + [3(U_2^2 - A^2)u_1]_X + D(U_2 + A)u_1,$$

$$(6.152)$$

where $U_2(X)$ is the full solution corresponding to the kink, while $u_1(X)$ is the exponential tail of the solution $U_1(X)$ at large X. It is interesting that equation (6.152) provides no additional solvability conditions. The corresponding adjoint problem,

$$u_{XXX}^c + (1 - 3U_2^2)u_X^c + DU_2u^c = 0, \qquad (6.153)$$

has no bounded solutions. Indeed, in the limit $X - L \to +\infty$, the solution can be written as

$$u^c(X) = a_1 e^{k_1 X} + a_2 e^{k_2 X} + a_3 e^{k_3 X},$$

where $k_{1,2,3}$ are the roots of the following characteristic equation,

$$k^3 + (1 - 3A^2)k + DA = 0. \qquad (6.154)$$

Equation (6.154) has two positive roots, k_1 and k_2, and one negative root, k_3. Thus, the bounded solution has to satisfy two conditions, $a_1 = 0$ and $a_2 = 0$. However, if $u^c(X)$ is even with respect to $X - L$, i.e., generally, $u^c(L) \neq 0$, $u_{XX}^c(L) \neq 0$, while $u_X^c(L) = 0$, one has only one fitting parameter, $u_{XX}^c(L)/u^c(L)$ (note that the equation (6.153) is linear) which is not sufficient to satisfy the two conditions. If u_c is odd, $u^c(L) = u_{XX}^c(L) = 0$, $u_X^c \neq 0$, and the solution is unique up to a constant factor; therefore, one has no fitting parameters at all. Thus, generally, equation (6.153) has no bounded solutions.

The abovementioned asymmetry between positive and negative kinks, caused by the differences in the manifold dimensions at infinity, and by the different dimensions of the solutions families, is rather unusual for problems of the interaction of spatially localized objects.

Kink Triplets

Numerical simulations show that kink triplets play an important role in the kink annihilation. In a multikink solution, the motion of a pair of neighboring kinks can be considered isolated until it approaches the kink to the left (to the right) of the antikink–kink (kink–antikink) pair. In both cases, a configuration appears such that the antikink is located between the two kinks. Numerical solution of equation (6.91) in the form of a kink triplet is shown in Figure 6.9 [60].

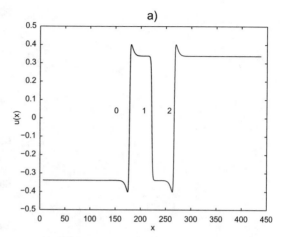

Fig. 6.9 Numerical solution of equation (6.91) in the form of a symmetric kink triplet: 0.2, kinks; 1. antikink. The two kinks a moving with a small, constant speed toward the central, motionless antikink. $D = 0.4$

For the sake of simplicity, later on we shall consider a symmetric system of kinks where the central antikink (kink # 1) is located near the point $x = 0$, while the two adjacent kinks are located at the points $x = \mp L(t)$ (kinks # 0 and # 2). We are interested in finding the instantaneous velocity of the kink # 2, $v_3(L) = dL/dt$. Due to the symmetry of the system, the kink # 0 moves with the velocity $-v_3$, and the kink # 1 does not move at all.

Similarly to (6.139), in the vicinity of the central kink, the solution is represented as an odd function

$$u = U_1 + U_0 + U_2 + \tilde{u} \approx -A\tanh\frac{A}{\sqrt{2}}x - a(D)e^{-\kappa_2(x+L)} + a(D)e^{\kappa_2(x-L)} + u, \quad (6.155)$$

where $\tilde{u}(x)$ describes the distortion of the central kink, decaying at large $|x|$. Since the terms $\tilde{U}(x) = U_0(x) + U_2(x) + \tilde{u}(x)$ are small, one can linearize equation (6.91):

$$\tilde{U}_t + [\tilde{U}_{xx} + (1 - 3U_1^2)\tilde{U}]_{xx} - D(U_1\tilde{U})_x = 0. \quad (6.156)$$

Since the velocity of the kinks is assumed to be exponentially small, one can omit the term with the time derivative in the leading order and to integrate equation (6.156):

$$\tilde{U}_{xxx} + [(1 - 3U_1^2)\tilde{U}]_x - DU_1\tilde{U} = q, \quad (6.157)$$

where the constant q is to be found from the condition of solvability of equation (6.157) in the class of odd functions growing as $\exp(\kappa_2|x|)$ (but not faster) at infinity.

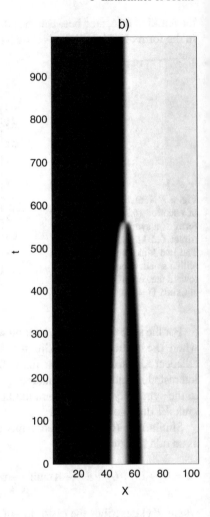

Fig. 6.10 Space–time diagram showing numerical solution of equation (6.91) in the form of a symmetric kink triplet in which the two kinks move toward the central antikink and annihilate (the distance between the kinks is less than that shown in Figure 6.9)

Equation (6.157) with $U_1(x) = -A\tanh(Ax/\sqrt{2})$ is solved analytically. The general odd solution can be presented as the sum of the general odd solution of the homogeneous equation,[1]

$$\tilde{U}_h = C\cosh z[3\cosh(\beta z)(A^2\tanh^2 z - 1)\tanh z + \beta\sinh(\beta z)(1 + 2A^2 - 3A^2\tanh^2 z)],$$
$$(6.158)$$

where $z = Ax/\sqrt{2}$ and C is an arbitrary constant, and the partial solution of the inhomogeneous problem,

$$\tilde{U}_{ih}(z) = qF(z),\qquad\qquad\qquad\qquad (6.159)$$

---

[1] The linearly independent even solutions of the homogeneous problem are $\tilde{U}_1 = \cosh^{-2} z$ and $\tilde{U}_2 = \cosh z[3\sinh(\beta z)(A^2\tanh^2 z - 1)\tanh z + \beta\cosh(\beta z)(1 + 2A^2 - 3A^2\tanh^2 z)]$. Note that only one solution of the linearized problem, $\tilde{U}_1$, is bounded as $z \to \pm\infty$, while the adjoint problem (6.144) has two bounded solutions, (6.145) and (6.146).

where $F(z) = w(z)/\cosh^2 z$ and $w(z)$ can be found in explicit form in terms of hypergeometric and algebraic functions of a new variable, $\xi = \tanh z$. For arbitrary C and Q, the asymptotic expression for the solution $\tilde{U}(X)$ at large X contains two exponentially growing terms,

$$\tilde{U}(X) \sim a_1 \exp(\kappa_1 x) + a_2 \exp(\kappa_2 x),$$

where $\kappa_1 = (A+B)/\sqrt{2}$, $\kappa_2 = (A-B)/\sqrt{2}$. For a solution representing the kink tails, $a_1 \sim e^{-\kappa_1 L}$, $a_2 \sim e^{-\kappa_2 L}$, $a_1 \ll a_2$, and therefore the coefficient a_1 can be neglected in the leading order. This condition determines the relation between C and Q and the final expression for $\tilde{U}(x)$.

For large x, the obtained solution $\tilde{U}(x)$ has the following asymptotics:

$$\tilde{U}(x) \sim qe^{\kappa_2 x} \frac{\sqrt{2}\pi[A^2(3+\beta^2)-5+\beta^2]}{A^5\beta(3-\beta)(9-\beta^2)(1-\beta^2)\cos(\pi\beta/2)}. \tag{6.160}$$

Matching (6.160) with the known asymptotics of the kink tail,

$$\tilde{U}(x) \sim a(D)e^{-\kappa_2 L}e^{\kappa_2 x},$$

one finds

$$q = a(D)e^{-\kappa_2 L}\frac{A^5\beta(3-\beta)(9-\beta^2)(1-\beta^2)\cos(\pi\beta/2)}{\sqrt{2}\pi[A^2(3+\beta^2)-5+\beta^2]}. \tag{6.161}$$

The velocity of the kinks can now be found from the conservation of flux (see [59]). Indeed, equation (6.91) can be written as

$$u_t + Q_x = 0, \tag{6.162}$$

where

$$Q = u_{xxx} + (u - u^3)_x - Du^2/2.$$

For a kink moving with the velocity v, $u = u(X)$, $X = x - vt$, equation (6.162) gives the conservation of the quantity $Q - vu$. Consider the kink #2. Obviously, to the left of the kink center $Q = -DA^2/2 + q$ and $vu = -vA$ in the leading order. The velocity of the kink actually depends on the condition for the flux at infinity. If one assumes that the flux at infinity is $Q = -DA^2/2$ and $vu = vA$ (which corresponds to matching the kink amplitude to the next remote antikink), the conservation law gives

$$-DA^2/2 + q + vA = -DA^2/2 - vA,$$

therefore, $v = -q/(2A)$. Using (6.161), one finds the following expression for the velocity $v \equiv v_3(D,L)$ of the right kink in the triplet (the left kink has the opposite velocity):

$$v_3(D,L) = a(D)C_3(D)e^{-\kappa_2 L}, \tag{6.163}$$

where

$$C_3(D) = \frac{A^4\beta(3-\beta)(9-\beta^2)(1-\beta^2)\cos(\pi\beta/2)}{2\pi\sqrt{2}[5-\beta^2-A^2(3+\beta^2)]}, \tag{6.164}$$

β is defined in (6.151) and $a(D)$ is the same as in equation (6.149) for the velocity of an antikink–kink pair. In the limit $D \to 0$, $C_3(D) \sim D/2 + O(D^2)$, and we reproduce the result of Watson et al. [59]:

$$v_3(D,L) \sim v_2(D,L) \sim -(D^2\sqrt{2}/4)\exp(-DL/2),$$

when $D \ll 1$, $DL \gg 1$. In the limit $D \to D_0$, $C_3(D) \sim 3(D_0-D)^{1/2}2^{-1/4}\pi + O(D_0 - D)$. The sign of equation (6.163) is negative that corresponds to the attraction of the positive kinks to the central negative kink.

The attraction of two kinks to the central antikink leads to their annihilation and to the transformation of the triplet into a new kink. The space–time diagram showing the motion and annihilation of two kinks in a triplet is presented in Figure 6.10. The annihilation stage cannot be studied by means of the asymptotic theory.

In the present section, we have considered two typical problems related to solidification front. The first example is the classical directional solidification. Because of the applied temperature field, the translational invariance is absent, and one could expect a type III instability of the front. However, in the case of a small segregation coefficient, a longwave instability takes place, which can be described by a damped Sivashinsky equation. Another example is the formation of facets in the course of a free propagation of the crystal growth front. In a contradistinction to the potential problem of the anisotropy-induced instability development on a motionless crystal surface, which is governed by a Cahn–Hilliard equation, an additional quadratic nonlinearity, characteristic for propagating fronts of any physical nature, appears. The "hybrid" equation containing both quadratic and cubic nonlinearities has many new features.

References

1. R.A. Fisher, Ann. Eugen. **7**, 355 (1937)
2. A. Kolmogorov, I. Petrovskii, N. Piskunov, Mosc. Univ. Bull. Math. **1**, 1 (1937)
3. Y. Zeldovich, G. Barenblatt, V. Librovich, G. Makhviladze, *Mathematical Theory of Combustion and Explosions* (Plenum Press, New York, 1985)
4. P. Pelce (ed.), *Dynamics of Curved Fronts* (Academic, Boston, 1988)
5. A.I. Volpert, V.A. Volpert, V.A. Volpert, *Traveling Wave Solutions of Parabolic Systems* (American Mathematical Society, Providence, RI, 1994)
6. L.M. Pismen, *Patterns and Interfaces in Dissipative Dynamics* (Springer, Berlin, 2006)
7. G.I. Sivashinsky, Acta Astronaut. **4**, 1177 (1977)
8. B.J. Matkowsky, G.I. Sivashinsky, SIAM J. Appl. Math. **37**, 686 (1979)
9. S.B. Margolis, B.J. Matkowsky, SIAM J. Appl. Math. **45**, 93 (1985)
10. T. Bohr, M.H. Jensen, G. Paladin, A. Vulpiani, *Dynamical Systems Approach to Turbulence* (Cambridge University Press, Cambridge, 1998)
11. I.S. Aranson, L. Kramer, Rev. Mod. Phys. **74**, 99 (2002)
12. E. Knobloch, J. De Luca, Nonlinearity **3**, 975 (1990)

13. B.J. Matkowsky, V.A. Volpert, Physica D **54**, 203 (1992)
14. J. Pelaez, SIAM J. Appl. Math. **47**, 781 (1987)
15. A.A. Golovin, B.J. Matkowsky, A. Bayliss, A.A. Nepomnyashchy, Physica D **129**, 253 (1999)
16. A.A. Golovin, A.A. Nepomnyashchy, B.J. Matkowsky, Physica D **160**, 1 (2001)
17. E.A. Glasman, A.A. Golovin, A.A. Nepomnyashchy, SIAM J. Appl. Math. **65**, 230 (2004)
18. A.A. Golovin, B.J. Matkowsky, A.A. Nepomnyashchy, Physica D **179**, 183 (2003)
19. W.W. Mullins, R.F. Sekerka, J. Appl. Phys. **34**, 323 (1963)
20. W.W. Mullins, R.F. Sekerka, J. Appl. Phys. **35**, 444 (1964)
21. S.H. Davis, *Theory of Solidification* (Cambridge University Press, Cambridge, 1999)
22. D.J. Wollkind, L.A. Segel, Philos. Trans. R. Soc. A **268**, 351 (1970)
23. G.I. Sivashinsky, Physica D **8**, 243 (1983)
24. A. Novick-Cohen, G.I. Sivashinsky, Physica D **20**, 237 (1986)
25. D.J. Bernoff, A.L. Bertozzi, Physica D **85**, 375 (1995)
26. K. Leung, J. Stat. Phys. **61**, 345 (1990)
27. C. Yeung, T. Rogers, A. Hernandes-Machado, D. Jasnow, J. Stat. Phys. **66**, 1071 (1992)
28. C.L. Emmott, A.J. Bray, Phys. Rev. E **54**, 4568 (1996)
29. Y. Saito, M. Uwaha, J. Phys. Soc. Jpn. **65**, 3576 (1996)
30. A.C. Newell, J.A. Whitehead, J. Fluid Mech. **38**, 279303 (1969)
31. R.K. Dodd, J.C. Eilbeck, J.D. Gibbon, H.C. Morris, *Solitons and Nonlinear Wave Equations* (Academic, London, 1982)
32. J. Krug, Adv. Phys. **46**, 139 (1997)
33. A.A. Golovin, *Pattern Formation at Interfaces* (Springer, Wien/New York, 2010), pp. 219–253
34. P.W. Voorhees, S.R. Coriell, G.B. McFadden, R.F. Sekerka, J. Cryst. Growth **67**, 425 (1984)
35. A.R. Umantsev, Sov. Phys. Crystallogr. **30**, 87 (1985)
36. A.R. Umantsev, A.L. Roitburd, Sov. Phys. Solid State **30**, 651 (1988)
37. A.A. Golovin, S.H. Davis, A.A. Nepomnyashchy, Physica D **122**, 202 (1998)
38. G.B. McFadden, S.R. Coriell, R.F. Sekerka, J. Cryst. Growth **91**, 180 (1988)
39. S.R. Coriell, R.F. Sekerka, J. Cryst. Growth **34**, 157 (1976)
40. A.A. Golovin, S.H. Davis, Physica D **116**, 363 (1997)
41. A.R. Umantsev, S.H. Davis, Phys. Rev. A **45**, 7195 (1992)
42. M.L. Frenkel, Physica D **27**, 260 (1987)
43. D.C. Saroka, A.J. Bernoff, Physica D **85**, 348 (1995)
44. J.W. Cahn, J.E. Hilliard, J. Chem. Phys. **28**, 258 (1958)
45. K. Kawasaki, T. Ohta, Physica A **116**, 573 (1982)
46. J. Stewart, N. Goldenfeld, Phys. Rev. A **46**, 6505 (1992)
47. M.A. Zaks, A. Podolny, A.A. Nepomnyashchy, A.A. Golovin, SIAM J. Appl. Math. **66**, 700 (2006)
48. A. Novick-Cohen, L.A. Segel, Physica D **10**, 277 (1984)
49. Y.A. Demekhin, G.Y. Tokarev, V.Y. Shkadov, Physica D **52**, 338 (1991)
50. I.G. Kevrekidis, B. Nicolaenko, J.C. Scovel, SIAM J. Appl. Math. **50**, 760 (1990)
51. K.R. Meyer, Trans. Am. Math. Soc. **149**, 95 (1970)
52. J.J. Gervais, J. Differ. Equ. **75**, 28 (1988)
53. P. Kent, J. Elgin, Nonlinearity **4**, 1045 (1991)
54. P. Kuchment, *Floquet Theory for Partial Differential Equations* (Birkhäuser, Basel/Boston, 1993)
55. A.A. Nepomnyashchy, Fluid Dyn. **9**, 354 (1974)
56. P. Politi, C. Misbah, Phys. Rev. Lett. **92**, 090601 (2004)
57. P. Coullet, C. Elphick, D. Repaux, Phys. Rev. Lett. **58**, 431 (1987)
58. A.A. Golovin, A.A. Nepomnyashchy, S.H. Davis, M.A. Zaks, Phys. Rev. Lett. **86**, 1550 (2001)
59. S.J. Watson, F. Otto, B.Y. Rubinstein, S.H. Davis, Physica D **178**, 127 (2003)
60. A. Podolny, M.A. Zaks, B.Y. Rubinstein, A.A. Golovin, A.A. Nepomnyashchy, Physica D **201**, 291 (2005)
61. K.A. Gorshkov, L.A. Ostrovsky, Physica D **3**, 428 (1981)

Chapter 7
Longwave Modulations of Shortwave Patterns

The description of longwave instabilities would be incomplete without mentioning longwave instabilities of shortwave spatially periodic patterns. The investigation of that kind of instabilities has a long history [1–3], and it is described in the literature in detail [4–9]. Nevertheless, we find it reasonable to present here a description of basic approaches and results in that field, with an emphasis to aspects not discussed in books mentioned above.

In the previous chapters, we considered problems where the longwave instability was related to the existence of a one-parametric family of basic spatially homogeneous solutions. A large-scale spatial modulation of the family parameter (mean temperature, layer thickness, concentration of a surfactant, etc.) created a slowly decaying mode that could be destabilized by a definite physical instability mechanism.

In the present chapter, the "family parameters" which characterize the basic state are the *wavevectors* of the pattern. A large-scale modulation of the set of wavevectors creates *modulational modes* which can be an origin of *modulational instabilities* of patterns. This kind of instabilities is an intrinsic property of patterns, which is crucial for the pattern selection and the development of the spatiotemporal chaos.

We consider longwave modulations of stationary patterns in Section 7.1 and those of wave patterns in Section 7.2.

7.1 Stationary Patterns

7.1.1 Roll Patterns

7.1.1.1 Modulated Patterns

We start with the simplest but widespread kind of patterns, which is usually called *roll pattern* in the context of liquid motions and *stripe pattern* in the context of reaction–diffusion systems.

© Springer Science+Business Media, LLC 2017
S. Shklyaev, A. Nepomnyashchy, *Longwave Instabilities and Patterns in Fluids*,
Advances in Mathematical Fluid Mechanics,
https://doi.org/10.1007/978-1-4939-7590-7_7

As the basic example of a system where the roll pattern is selected, we use the Swift–Hohenberg model [7]

$$\frac{\partial \phi}{\partial t} = \varepsilon^2 \phi - \left(1 + \nabla^2\right)^2 \phi - \phi^3 \tag{7.1}$$

(see Section C.1 of Appendix C). In the Appendix, the selection of rolls among the patterns with a discrete set of basic wavevectors satisfying the condition $|\mathbf{k}_n| = 1$ for any n is considered. However, the set of wavevectors corresponding to unstable modes is actually a ring which contains a continuum of wavevectors with $|k^2 - 1| < \varepsilon$ (see Figure 7.1.) In the present section, we discuss a wider class of solutions corresponding to large-scale modulations of periodic patterns.

Let us consider a roll pattern with the basic wavevector $\mathbf{k} = (1,0)$. As we know, the roll pattern is stable with respect to disturbances with other wavevectors $\mathbf{k}_n$, $|\mathbf{k}_n| = 1$. The analysis done in the previous section uses implicitly the assumption that the difference between the wavevectors $\mathbf{k}_m$ and $\mathbf{k}_n$ is $O(1)$, i.e., the angle between them is $O(1)$. Now we take into account the possibility of large-scale distortion of patterns by means of disturbances with wavevectors close to each other. For that goal, we apply the multiscale analysis. Because the ring of unstable modes, $1 - \varepsilon < k_x^2 + k_y^2 < 1 + \varepsilon$, around the point $k_x = 1$, $k_y = 0$ has the width $O(\varepsilon)$ in k_x-direction and the width $O(\varepsilon^{1/2})$ in k_y-direction (see Figure 7.1), it is natural to assume that the function ϕ depends on the following scaled variables:

$$x_0 = x, \ x_1 = \varepsilon x, \ y_{1/2} = \varepsilon^{1/2} y. \tag{7.2}$$

The Fourier transform of such a function is concentrated around the instability ring. As in the previous section, we use the scaled time variable $T = \varepsilon^2 t$.

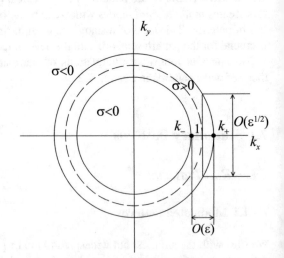

Fig. 7.1 Ring of unstable wavevectors

Following the idea of multiscale expansions (see, e.g., [10] and the previous chapters of this book), we substitute

$$\frac{\partial}{\partial x} = \frac{\partial}{\partial x_0} + \varepsilon \frac{\partial}{\partial x_1}, \quad \frac{\partial}{\partial y} = \varepsilon^{1/2} \frac{\partial}{\partial y_{1/2}}, \quad \frac{\partial}{\partial t} = \varepsilon^2 \frac{\partial}{\partial T}$$

into (7.1) and obtain:

$$\varepsilon^2 \frac{\partial \phi}{\partial T} = \varepsilon^2 \phi - \left[1 + \frac{\partial^2}{\partial x_0^2} + \varepsilon \left(2 \frac{\partial^2}{\partial x_0 \partial x_1} + \frac{\partial^2}{\partial y_{1/2}^2} \right) + \varepsilon^2 \frac{\partial^2}{\partial x_1^2} \right]^2 \phi - \phi^3. \quad (7.3)$$

Now we substitute the solution in the form

$$\phi = \varepsilon \phi_1 + \varepsilon^2 \phi_2 + \varepsilon^3 \phi_3 + \dots, \quad (7.4)$$

and demand boundness with respect to all the variables in each order.

At the order ε, we obtain:

$$- \left(1 + \frac{\partial^2}{\partial x_0^2} \right)^2 \phi_1 = 0. \quad (7.5)$$

The most general bounded solution of this equation,

$$\phi_1 = A(T, x_1, y_{1/2}) e^{ix_0} + A^*(T, x_1, y_{1/2}) e^{-ix_0}, \quad (7.6)$$

describes a large-scale modulation of a roll pattern.

At the order ε^2, we get:

$$- \left(1 + \frac{\partial^2}{\partial x_0^2} \right)^2 \phi_2 = 2 \left(2 \frac{\partial^2}{\partial x_0 \partial x_1} + \frac{\partial^2}{\partial y_{1/2}^2} \right) \left(1 + \frac{\partial^2}{\partial x_0^2} \right) \phi_1. \quad (7.7)$$

Because

$$\left(1 + \frac{\partial^2}{\partial x_0^2} \right) \phi_1 = 0,$$

the right-hand side of (7.7) vanishes, so that the equation is always solvable, and its solution is similar to (7.6):

$$\phi_2 = B(T, x_1, y_{1/2}) e^{ix_0} + B^*(T, x_1, y_{1/2}) e^{-ix_0}.$$

Finally, at the order ε^3, taking into account that

$$\left(1 + \frac{\partial^2}{\partial x_0^2} \right) \phi_1 = \left(1 + \frac{\partial^2}{\partial x_0^2} \right) \phi_2 = 0,$$

we obtain:

$$-\left(1+\frac{\partial^2}{\partial x_0^2}\right)^2 \phi_3 = \frac{\partial \phi_1}{\partial T} - \phi_1 + \phi_3 + \left(2\frac{\partial^2}{\partial x_0 \partial x_1} + \frac{\partial^2}{\partial y_{1/2}^2}\right)^2 \phi_1. \qquad (7.8)$$

The term proportional to $\exp(ix_0)$ in the right-hand side of (7.8) should vanish; otherwise, the solution of (7.8) is not bounded as $x_0 \to \pm\infty$. That gives us the following evolution equation for the envelope function $A(T, x_1, y_{1/2})$, which is called the *Newell–Whitehead–Segel (NWS) equation* [3, 11]:

$$\frac{\partial A}{\partial T} = A - 3|A|^2 A - \left(2i\frac{\partial}{\partial x_1} + \frac{\partial^2}{\partial y_{1/2}^2}\right)^2 A. \qquad (7.9)$$

Rescaling the variables,

$$a = A\sqrt{3}, \ X = x_1/2, \ Y = y_{1/2}/\sqrt{2}, \qquad (7.10)$$

we transform the NWS equation to its standard form:

$$\frac{\partial a}{\partial T} = a - |a|^2 a + \left(\frac{\partial}{\partial X} - \frac{i}{2}\frac{\partial^2}{\partial Y^2}\right)^2 a. \qquad (7.11)$$

Equation (7.11) is not specific for patterns described by the Swift–Hohenberg equations, but is generic for any roll patterns generated by a shortwave monotonic instability in a rotationally invariant system. Therefore, all the results obtained below by means of that equation are generic.

The NWS equation can be written in the form

$$\frac{\partial a}{\partial T} = -\frac{\delta F}{\delta a^*}, \qquad (7.12)$$

where the *Lyapunov functional* is defined as

$$F = \int d\mathbf{r}\left[\frac{1}{2}(|a|^2 - 1)^2 + \left|\left(\frac{\partial}{\partial X} - \frac{i}{2}\frac{\partial^2}{\partial Y^2}\right)a\right|^2\right]. \qquad (7.13)$$

Note that

$$\frac{dF}{dT} = \int d\mathbf{r}\left(\frac{\delta F}{\delta a}\frac{\partial a}{\partial T} + \frac{\delta F}{\delta a^*}\frac{\partial a^*}{\partial T}\right) = -2\int d\mathbf{r}\left|\frac{\partial a}{\partial T}\right|^2 \le 0. \qquad (7.14)$$

Therefore, the system tends to a stationary solution and has no time-periodic or chaotic solutions.

Equation (7.12) has a family of stationary solutions which do not depend on Y:

$$a = a_K(X) = \sqrt{1 - K^2}e^{iK(X - X_0)}, \ -1 < K < 1, \qquad (7.15)$$

where X_0 is an arbitrary constant. For the sake of simplicity, we choose $X_0 = 0$. Taking into account relations (7.10), we find that the corresponding order-parameter field is

$$\phi_1 = Ae^{ix_0} + A^*e^{-ix_0} = 2\sqrt{3}\sqrt{1-K^2}\cos\left(x_0 + \frac{1}{2}Kx_1\right)$$

$$= 2\sqrt{3(1-K^2)}\cos\left[\left(1 + \frac{1}{2}K\varepsilon\right)x\right].$$

Thus, the obtained solutions correspond to roll patterns with the wavenumbers

$$k = 1 + K\varepsilon/2, \quad -1 < K < 1 \tag{7.16}$$

inside the instability interval, generally different from 1.

Let us investigate the stability of roll solutions in the framework of the NWS equation. Linearizing equation (7.11) around the solution (7.15), we obtain the following equation:

$$\frac{d\tilde{a}}{dT} = -(1 - 2K^2)\tilde{a} - (1 - K^2)\tilde{a}^*e^{2iKX} + \left(\frac{\partial}{\partial X} - \frac{i}{2}\frac{\partial^2}{\partial Y^2}\right)^2 \tilde{a}. \tag{7.17}$$

The dependence of one of the coefficients on X can be eliminated by the transformation

$$\tilde{a}(X,T) = b(X,T)e^{iKX};$$

we obtain

$$\frac{db}{dT} = -(1 - 2K^2)b - (1 - K^2)b^* + \left(\frac{\partial}{\partial X} - \frac{i}{2}\frac{\partial^2}{\partial Y^2} + iK\right)^2 b. \tag{7.18}$$

Now we can find the normal modes in the form

$$b = b_1 e^{i(\tilde{K}_X X + \tilde{K}_Y Y) + \sigma T} + b_2 e^{-i(\tilde{K}_X X + \tilde{K}_Y Y) + \sigma^* T} \tag{7.19}$$

(actually the eigenvalues σ are real, because of the property (7.14)). The condition of the existence of nontrivial solutions for the coupled algebraic system for b_1 and b_2^* gives the following expression for two branches of eigenvalues:

$$\sigma_\pm(\tilde{K}_X, \tilde{K}_Y; K) = -1 + K^2 - \tilde{K}_X^2 - \frac{1}{2}K\tilde{K}_Y^2 - \frac{1}{16}\tilde{K}_Y^4$$

$$\pm\sqrt{(1-K^2)^2 + 4\tilde{K}_X^2\left(K + \frac{1}{2}\tilde{K}_Y^2\right)^2}. \tag{7.20}$$

Recall that $\tilde{K}_X, \tilde{K}_Y$ are the components of the wavevector of the disturbance, while K is the parameter of the basic roll solution related to its wavenumber k according to (7.16).

To reveal the instability modes, it is sufficient to consider the branch with the higher value of σ in the limit of longwave disturbances, i.e., for small $\tilde{K}_X, \tilde{K}_Y$:

$$\sigma_+ = -\frac{1-3K^2}{1-K^2}\tilde{K}_X^2 - \frac{1}{2}K\tilde{K}_Y^2 + o(\tilde{K}_X^2, \tilde{K}_Y^2). \tag{7.21}$$

One can see that the roll solutions within the interval $1/3 < K^2 < 1$ are unstable with respect to disturbances with $\tilde{K}_X^2 \neq 0$, $\tilde{K}_Y^2 = 0$, i.e., to *longitudinal modulations*. This kind of instability in nonlinear dissipative systems was discovered by Eckhaus [1], and it is called *Eckhaus instability*.

If the wavenumber of the roll solution satisfies the condition $K < 0$ (i.e., $k < 1$), a *transverse* modulational instability takes place with respect to disturbances with $\tilde{K}_X^2 = 0$, $\tilde{K}_Y^2 \neq 0$, which is called *zigzag instability*.

Prediction of the theory presented above is summarized in Figure 7.2.

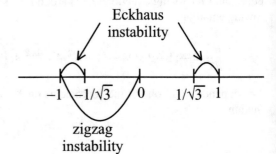

Fig. 7.2 Instabilities of roll patterns

We come to the conclusion that only the roll patterns inside the *stability interval* $0 < K < 1/\sqrt{3}$ are stable. The stability interval is called also *Busse balloon*, for it was first discovered by Busse et al. in the context of the Rayleigh–Bénard convection patterns [2].

7.1.1.2 Nonlinear Phase Diffusion Equation

The longwave nature of two basic instabilities of roll patterns described above shows that the longwave distortions of rolls are of the major interest. Let us take the NWS equation (7.11) as the primary one and consider longwave solutions of that equation:

$$a = R(\xi, \eta, \tau)\exp[i\theta(\xi, \eta, \tau)], \tag{7.22}$$

where

$$\xi = \delta X, \ \eta = \delta^{1/2}Y, \ \tau = \delta^2 T; \ 0 < \delta \ll 1.$$

Needless to say that the small parameter δ has nothing to do with the small parameter ε used in the derivation of the NWS equation, the smallness of ε is due to the smallness of the governing parameter in the SH equation, while δ characterizes the scale of a specific class of solutions of the NWS equation.

Substituting

$$\frac{\partial}{\partial X} = \delta \frac{\partial}{\partial \xi}, \quad \frac{\partial}{\partial Y} = \delta^{1/2} \frac{\partial}{\partial \eta}, \quad \frac{\partial}{\partial T} = \delta^2 \frac{\partial}{\partial \tau}$$

into (7.11) and using the representation (7.22), we obtain:

$$\delta^2 \frac{\partial}{\partial \tau} \left(Re^{i\theta} \right) = (R - R^3)e^{i\theta} + \delta^2 \left(\frac{\partial}{\partial \xi} - \frac{i}{2} \frac{\partial^2}{\partial \eta^2} \right)^2 \left(Re^{i\theta} \right). \tag{7.23}$$

Let us consider solutions in the form

$$R = R_0 + \delta^2 R_2 + \ldots, \quad \theta = \theta_0 + \delta^2 \theta_2 + \ldots. \tag{7.24}$$

After the substitution of (7.24) into (7.23), we find at the leading order:

$$R_0 - R_0^3 = 0.$$

Because we are interested in distorted rolls rather than in the unstable trivial solution, we choose $R_0 = 1$. Note that the leading-order roll amplitude is constant under the action of longwave distortions.

In the next order, a system of coupled equations for R_2 and θ_0 is obtained. After the elimination of R_2, one can obtain the following: *nonlinear phase diffusion equation* [12] (later on, we drop the subscript $_0$):

$$\theta_\tau = \theta_{\xi\xi} - \frac{1}{4}\theta_{\eta\eta\eta\eta} + 2\theta_\eta\theta_{\xi\eta} + \theta_\xi\theta_{\eta\eta} + \frac{3}{2}\theta_\eta^2\theta_{\eta\eta}, \tag{7.25}$$

which governs the longwave phase distortions of roll patterns.

Below we apply the derived equation for consideration of *defects* in roll patterns.

Dislocation

According to relation (7.22), the phase θ is defined modulo 2π in the points where $R \neq 0$ and undefined in the points where $R = 0$. The roll pattern can contain a *point defect* of the following structure. The amplitude $R = 0$ in a certain point, say, in the point $X = Y = 0$. Except this point, the phase is smooth, but going around the point $X = Y = 0$ along a closed circle $X^2 + Y^2 = const$ leads to a phase increment $2n\pi$, where $n = \pm 1$. Such a defect in the roll pattern is called a *dislocation*. We shall apply the nonlinear phase equation for the description of the dislocations in the *far field*, i.e., at a large distance from the center (i.e., for $\xi, \eta = O(1)$).

One can easily show [13] that the stationary equation (7.25) is satisfied by any solution of the *Burgers equation*

$$\theta_\xi = \pm\frac{1}{2}(\theta_{\eta\eta} - \theta_\eta^2). \tag{7.26}$$

A positive dislocation ($n = 1$) is described by the solution of equation (7.26), with the sign + on the right-hand side, which satisfies the boundary conditions on the cut $x = 0, y > 0$:

$$\theta(0^\pm, y) = \pm\pi, \; y > 0. \tag{7.27}$$

By means of the Hopf–Cole transformation

$$\theta = \mp\ln(f), \tag{7.28}$$

equation (7.26) is transformed to

$$f_\xi = \pm\frac{1}{2}f_{\eta\eta}. \tag{7.29}$$

First, we shall find the solution, which corresponds to a positive dislocation, in the region $\xi > 0$ in the form of a *self-similar solution*: $f = f(\zeta)$, $\zeta = \eta/\sqrt{2\xi}$. The corresponding boundary value problem,

$$f'' + 2\zeta f' = 0, \; -\infty < \zeta < \infty; \; f(-\infty) = 1, \; f(\infty) = \exp(-\pi), \tag{7.30}$$

has the following solution:

$$f = \frac{1 + \exp(-\pi)}{2} - \frac{1 - \exp(-\pi)}{2}\operatorname{erf}(\zeta).$$

Solution in the region $\xi > 0$ is calculated in a similar way. Finally, we obtain the following expression for $\theta(\xi, \eta)$:

$$\theta(\xi, \eta) = -\operatorname{sign}(\xi)\ln\left[\frac{1 + \exp(-\pi)}{2} - \frac{1 - \exp(-\pi)}{2}\operatorname{erf}\left(\frac{\eta}{\sqrt{2|\xi|}}\right)\right]. \tag{7.31}$$

The structure of a roll pattern containing a dislocation is shown in Figure 7.3. The solution for a negative dislocation is obtained similarly.

Nonlinear Theory of the Zigzag Instability

In the previous subsection, we have found that a roll pattern with $K < 0$ is subject to a transverse (zigzag) instability with $\tilde{K}_X = 0$, $\tilde{K}_Y \neq 0$. In order to investigate the temporal evolution of a zigzag disturbance on the background of a roll pattern, substitute $\theta = K\xi + \Phi(\eta, \tau)$ into the nonlinear phase equation (7.25). We find:

$$\Phi_\tau = -\frac{1}{4}\Phi_{\eta\eta\eta\eta} + K\Phi_{\eta\eta} + \frac{3}{2}\Phi_\eta^2\Phi_{\eta\eta}. \tag{7.32}$$

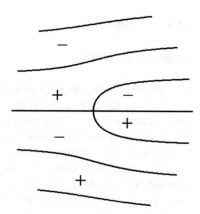

Fig. 7.3 Roll pattern in the
presence of dislocation

For the η-component of the wavevector, $Q = \Phi_\eta$, we obtain a *Cahn–Hilliard equation*,

$$Q_\tau = -\frac{1}{4}Q_{\eta\eta\eta\eta} - (-K)Q_{\eta\eta} + \frac{1}{2}\left(Q^3\right)_{\eta\eta}. \qquad (7.33)$$

As we know, the spatially periodic solutions of equation (7.33) are unstable. The coarsening process leads to the kink solution

$$Q = \sqrt{2(-K)}\tanh\sqrt{2(-K)}\eta,$$

which corresponds to a smooth *domain wall*

$$\theta = K\xi + \ln\cosh\sqrt{2(-K)}\eta \qquad (7.34)$$

between two sets of inclined roll patterns. Isolines of function (7.34) are shown in Figure 7.4.

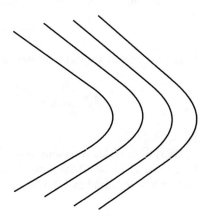

Fig. 7.4 Roll pattern in the
presence of a smooth domain
wall

7.1.1.3 Domain Walls

Amplitude Equations

As explained above, the roll pattern is selected when the off-diagonal elements of
the nonlinear interaction matrix are larger than the diagonal elements, because all
other patterns are unstable. Specifically, it is true for systems governed by the Swift–
Hohenberg equation. The stability arguments cannot determine, however, the direction of the pattern's wavevector: due to the isotropy of the problem, the roll patterns
with different orientations of the wavevector are equally stable. In reality, rolls with
different orientations can develop in different parts of the system forming ordered
domains separated by *domain walls*. In order to describe that situation, we construct
a wider class of solutions to the Swift–Hohenberg equation (7.1),

$$\frac{\partial \phi}{\partial t} = \varepsilon^2 \phi - \left(1 + \nabla^2\right)^2 \phi - \phi^3,$$

than that studied above.

Because we are not going to consider the zigzag instability here, we use a simpler
assumption on the scaling of the solution:

$$\phi = \phi(T, \mathbf{r}_0, \mathbf{r}_1),$$

where $T = \varepsilon^2 t$, $\mathbf{r}_0 = \mathbf{r}$, $\mathbf{r}_1 = \varepsilon \mathbf{r}$. The rescaled Swift–Hohenberg equation reads:

$$\varepsilon^2 \frac{\partial \phi}{\partial T} = \varepsilon^2 \phi - \phi^3 - [(1 + \nabla_0^2) + 2\varepsilon(\nabla_0 \cdot \nabla_1) + \varepsilon^2 \nabla_1^2]^2 \phi. \qquad (7.35)$$

As in (C.2), we assume

$$\phi = \varepsilon \phi_1 + \varepsilon^3 \phi_3 + \dots$$

At the leading order, $O(\varepsilon)$, we find:

$$-(1 + \nabla_0^2)^2 \phi_1 = 0. \qquad (7.36)$$

In order to consider rolls of different orientations, we take

$$\phi_1 = \sum_{n=1}^{N} \left[A_n(T, \mathbf{r}_1) e^{i\mathbf{k}_n \cdot \mathbf{r}_0} + A_n^*(T, \mathbf{r}_1) e^{-i\mathbf{k}_n \cdot \mathbf{r}_0} \right], \qquad (7.37)$$

where $|\mathbf{k}_n| = 1$.

Expansion (7.37) resembles (C.3), but there is an essential difference: now the
functions A_n depend on the slow coordinate $\mathbf{r}_1$. Therefore, we can consider different
roll patterns localized in different spatial regions.

At the order $O(\varepsilon^3)$, we obtain the following generalization of equations (C.5):

$$\frac{\partial A_n}{\partial T} = (1 - 3|A_n|^2 - 6 \sum_{m \neq n} |A_m|^2)A_n + 4(\mathbf{k}_n \cdot \nabla_1)^2 A_n, \; n = 1, \ldots, N. \qquad (7.38)$$

By means of a scale transformation,

$$a_n = A_n \sqrt{3}, \; n = 1, \ldots, N; \mathbf{R} = \mathbf{r}_1/2, \qquad (7.39)$$

the obtained system of coupled NWS equations is transformed to its standard form,

$$\frac{\partial a_n}{\partial T} = (1 - |a_n|^2 - 2 \sum_{m \neq n} |a_m|^2)a_n + (\mathbf{k}_n \cdot \nabla_\mathbf{R})^2 a_n, \; n = 1, \ldots, N. \qquad (7.40)$$

Generally, the system of coupled NWS equations can be written in the form

$$\frac{\partial a_n}{\partial T} = -\frac{\delta F}{\delta a_n^*}, \qquad (7.41)$$

where the Lyapunov functional has the form

$$F = \int d\mathbf{R} L[\{a\}, \{a^*\}, \{\mathbf{k} \cdot \nabla_\mathbf{R} a\}, \{\mathbf{k} \cdot \nabla_\mathbf{R} a^*\}], \qquad (7.42)$$

so that

$$\frac{dF}{dT} \leq 0. \qquad (7.43)$$

Here $\{(\cdot)\}$ means the corresponding set of $(\cdot)_{1,\ldots,N}$. The amplitude equations are:

$$\frac{\partial a_n}{\partial T} = \left(1 - |a_n|^2 - \sum_{m \neq n} \kappa_{mn}|a_m|^2\right)a_n + (\mathbf{k}_n \cdot \nabla_\mathbf{R})^2 a_n, \; n = 1, \ldots, N, \qquad (7.44)$$

where $\kappa_{mn} = \kappa_{nm}$.

Fronts Between Roll Patterns

For the sake of simplicity, let us consider a plane domain wall perpendicular to the axis X, which separates two semi-infinite roll systems with wavevectors $\mathbf{k}_1$ and $\mathbf{k}_2$ [14]. Because the Lyapunov functional densities of both roll patterns are equal, there is no reason for a motion of the domain wall; hence, it is motionless [15]. The problem is governed by the following system of ordinary differential equations:

$$D_1 a_1'' + a_1 - |a_1|^2 a_1 - \kappa |a_2|^2 a_1 = 0, \; D_2 a_2'' + a_2 - |a_2|^2 a_2 - \kappa |a_1|^2 a_2 = 0, \quad (7.45)$$

$$-\infty < X < \infty,$$

where $D_n = (k_{nX})^2$, $'$ means the differentiation with respect to X. In the case of the Swift–Hohenberg equation, $\kappa = 2$. Generally, the roll patterns are selected, if $\kappa > 1$. We assume that at a large distance from the domain wall, there are perfect roll patterns. That leads to boundary conditions:

$$X \to -\infty, \ a_2 \to 0; \ X \to \infty, \ a_1 \to 0. \tag{7.46}$$

Introduce $a_n = R_n \exp(i\theta_n)$, $i = 1, 2$. The imaginary parts of the equations give the relations

$$r_n \theta_n'' + 2r_n' \theta_n' = 0, \ n = 1, 2,$$

or

$$(r_n^2 \theta_n')' = 0, \ n = 1, 2,$$

hence

$$r_n^2 \theta_n' = M_n = const.$$

The boundary conditions (7.46) prescribe $M_1 = M_2 = 0$, so the phases θ_1 and θ_2 are constant. Taking into account the meaning of the phase modulations, we come to the conclusion that a steady domain wall can exist only between patterns with the critical values of wavenumbers $|\mathbf{k}_n|$ ($|\mathbf{k}_n| = 1$ in the case of the SH equation). So, the existence of a domain wall acts as a factor which selects the wavenumber in a much more definite way than the stability arguments presented in Section 7.1.1.1.

The real parts of the equations read (we take into account that $\theta_1' = \theta_2' = 0$):

$$D_1 r_1'' + r_1(1 - r_1^2 - \kappa r_2^2) = 0, \ D_2 r_2'' + r_2(1 - r_2^2 - \kappa r_1^2) = 0, \tag{7.47}$$

$$X \to -\infty: \ r_2 \to 0, \ r_1 \to 1; X \to \infty: \ r_1 \to 0, \ r_2 \to 1. \tag{7.48}$$

The system of equations (7.47) describes the two-dimensional motion of a particle with the Lagrange function (equal to the Lyapunov functional density) $L = K(r_1', r_2') - U(r_1, r_2)$, where

$$K(r_1', r_2') = \frac{1}{2} D_1(r_1')^2 + \frac{1}{2} D_2(r_2')^2$$

(note that the "mass" of the fictitious particle is anisotropic) and

$$U(r_1, r_2) = \frac{1}{2}(r_1^2 + r_2^2) - \frac{1}{4}(r_1^4 + 2\kappa r_1^2 r_2^2 + r_2^4).$$

In the case $q > 1$, when the rolls are stable, the potential $U(r_1, r_2)$ has a minimum in the point $(0, 0)$, two maxima in the points $(1, 0)$ and $(0, 1)$, and a saddle point $(1/\sqrt{1 + \kappa}, 1/\sqrt{1 + \kappa})$. The domain wall solution corresponds to the trajectory that starts in the maximum point $(1, 0)$ at $X \to -\infty$ and tends to another maximum point $(0, 1)$ as $X \to \infty$. An exact solution of (7.47), (7.48) is known for $\kappa = 3$, $D_1 = D_2 = D$:

$$r_1 = \frac{1}{2}\left(1 - \tanh \frac{X}{\sqrt{2D}}\right), \ r_2 = \frac{1}{2}\left(1 + \tanh \frac{X}{\sqrt{2D}}\right) \tag{7.49}$$

(see Figure 7.5).

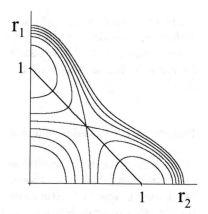

Fig. 7.5 Isolines of the potential $U(r_1, r_2)$ and the domain wall solution

Fronts Between Patterns of Different Types

Fronts between patterns of different types can be considered in the similar way. Assuming that phase modulations of patterns are absent, so that $a_n = r_n$ is real for any $n = 1, \ldots, N$, and the front is flat, we can rewrite the system of equations (7.44) in the form

$$\frac{\partial r_n}{\partial T} = D_n \frac{\partial^2 r_n}{\partial X^2} + \frac{\partial U(r_1, \ldots, r_N)}{\partial r_n}, \quad n = 1, \ldots, N, \tag{7.50}$$

where

$$U(r_1, \ldots, r_N) = \frac{1}{2} \sum_{n=1}^{N} r_n^2 - \frac{1}{4} \sum_{n=1}^{N} r_n^4 - \frac{1}{2} \sum_{n=1}^{N} \sum_{m=1}^{n-1} \kappa_{nm} r_n^2 r_m^2. \tag{7.51}$$

A front propagating with a constant velocity c is described by the solution

$$r_n = r_n(\xi), \quad n = 1, \ldots, N; \quad \xi = X - cT \tag{7.52}$$

of the equation

$$D_n \frac{d^2 r_n}{d\xi^2} + c \frac{dr_n}{d\xi} + \frac{\partial U}{\partial r_n} = 0, \quad n = 1, \ldots, N, \tag{7.53}$$

satisfying the boundary conditions

$$r_n(-\infty) = r_n^-, \quad r_n(\infty) = r_n^+, \tag{7.54}$$

where $\{r_n^-\}$ and $\{r_n^+\}$ correspond to steady patterns on both sides of the front. Multiplying the nth equation of the set (7.53) by $dr_n/d\xi$, integrating with respect to ξ from $-\infty$ to ∞, and taking the sum over n from 1 to N, we find that

$$c = -\frac{U(\infty) - U(-\infty)}{\sum_{n=1}^{N} \int_{-\infty}^{\infty} d\xi (dr_n/d\xi)^2} = \frac{L^+ - L^-}{\sum_{n=1}^{N} \int_{-\infty}^{\infty} d\xi (dr_n/d\xi)^2}, \tag{7.55}$$

where $L^{\pm} = L(r_1^{\pm}, r_2^{\pm}, \ldots, r_N^{\pm})$ are the densities of Lyapunov functionals for the uniform patterns situated at $X \to \pm\infty$.

Note that $c > 0$ if $L^- < L^+$, and $c < 0$ if $L^- > L^+$, i.e., the pattern with the lower Lyapunov functional density ousts the pattern with the higher Lyapunov functional density.

Pattern-Induced Pinning of a Domain Wall

In the framework of the multiscale approach described above, one comes to the conclusion that a domain wall between patterns is proportional to the difference between their Lyapunov functional densities. However, the influence of the underlying periodic pattern leads to some qualitative changes of the domain wall dynamics. Firstly, the motion of the domain wall is an oscillatory process; during one period, one stripe is created or melted [16, 17]. Secondly, because of the pinning effect, there is a finite interval of the parameter value where the domain wall is motionless, i.e., a pattern and a uniform state coexist. Near the threshold of the pattern appearance, that interval is transcendentally small [14, 15], but on a finite distance from the threshold, that effect is significant.

As an example, let us consider the one-dimensional Swift–Hohenberg equation,

$$\phi_t = \left[\gamma - \left(\frac{\partial^2}{\partial x^2} + 1 \right)^2 \right] \phi - \phi^3, \tag{7.56}$$

where $\gamma = O(1)$. The corresponding Lyapunov functional is

$$F\{\phi\} = \int \left\{ -\frac{\gamma}{2}\phi^2 + \frac{1}{4}\phi^4 + \frac{1}{2}\left[\left(\frac{\partial^2}{\partial x^2} + 1 \right) \phi \right]^2 \right\} dx. \tag{7.57}$$

When $0 < \gamma < 1$, periodic patterns exist with wavenumbers k in the interval $1 - \sqrt{\gamma} < k^2 < 1 + \sqrt{\gamma}$, and they are stable in a certain subinterval $k_-(\gamma) < k < k_+(\gamma)$. For $\gamma > 1$, constant nonzero solutions $\phi_{\pm} = \pm\sqrt{\gamma - 1}$ appear. For $\gamma > 3/2$, they are stable with respect to small disturbances; kinks with oscillatory tails connect both stable uniform phases $\phi_{\pm}$ [18].

The value of the Lyapunov functional density for the regular pattern with an optimal wavenumber is lower than that of the uniform state when $\gamma < \gamma_m \approx 6.3$ [16, 18]. Would formula (7.55) be correct at finite γ, the pattern would spread into the uniform state for any $\gamma < \gamma_m$. Actually, that spreading is stopped at much smaller values of γ, $\gamma > \gamma_c \approx 1.7574$. The reason is the self-induced pinning caused by the oscillatory asymptotic perturbation of the uniform state. Similarly, the pinning effect prevents the replacement of a pattern by a uniform state at $\gamma > \gamma_m$.

The stability interval for a finite fragment of patterns sandwiched between semi-infinite regions of a uniform state slightly depends on the length of that fragment [16]. The multistability of pattern fragments of different lengths caused by pinning, which has been revealed for many pattern-forming systems, is known as the "snaking" effect (for a review, see [19]).

Note that a motion of a domain wall can be generated by the noise, which activates the transition from a metastable state to a truly stable, energetically preferred, state [16].

7.1.1.4 Cross–Newell Phase Diffusion Equation

The Newell–Whitehead–Segel equations are valid only near the instability threshold, $\gamma = \varepsilon^2 \ll 1$. However, even far from the instability threshold, the patterns may be subject to large-scale modulations. For their description, another approach can be applied [20].

Derivation of the Cross–Newell Equation

Let us consider the Swift–Hohenberg equation,

$$\phi_t = \gamma\phi - \left(1 + \nabla^2\right)^2 \phi - \phi^3, \tag{7.58}$$

with $\gamma = O(1)$. Equation (7.58) has a class of periodic stationary solutions corresponding to roll patterns:

$$\phi(\mathbf{r}) = \sum_{n=1}^{\infty} A_n \cos(n\mathbf{k} \cdot \mathbf{r}) = f(\theta), \tag{7.59}$$

where $\theta = \mathbf{k} \cdot \mathbf{x}$, $\mathbf{k}$ is a constant wavevector, and $f(\theta)$ is a 2π-periodic function (one can see that only Fourier components with odd n are present in the expansion (7.59) for $f(\theta)$).

As we know, the wavevector $\mathbf{k}$ of the roll pattern is not unique. Therefore, we can imagine a situation when the local wavevector of a roll pattern is a slow function of the coordinate $\mathbf{r}$, and it can slowly change in time:

$$\mathbf{k} = \mathbf{k}(\mathbf{R}, T), \tag{7.60}$$

where $\mathbf{R} = \delta\mathbf{r}$, $\delta \ll 1$; the appropriate time scale is $T = \delta^2 t$. Note that the wavevector for the real function ϕ is defined up to the sign (thus, it is similar to the order parameter of a nematic crystal, the "director"). The wavevector can be considered as a gradient of a certain phase function θ,

$$\mathbf{k} = \nabla\theta.$$

The appropriate scaling of θ is

$$\theta = \frac{1}{\delta}\Theta(\mathbf{R}, T).$$

Let us find the solution of the Swift–Hohenberg equation (7.58), corresponding to a slowly distorted roll pattern with the wavevector field (7.60),

$$\phi = \phi(\theta, \mathbf{R}, T),$$

where ϕ is 2π-periodic in θ. Using the transformation formulas for derivatives

$$\nabla_\mathbf{r}\phi = (\mathbf{k} \cdot \partial_\theta + \delta\nabla_\mathbf{R})\phi,$$
$$\nabla_\mathbf{r}^2\phi = (\mathbf{k} \cdot \partial_\theta + \delta\nabla_\mathbf{R})(\mathbf{k} \cdot \partial_\theta + \delta\nabla_\mathbf{R})\phi$$
$$= \left[k^2\partial_\theta^2 + \delta(2\mathbf{k} \cdot \nabla_\mathbf{R} + \nabla_\mathbf{R} \cdot \mathbf{k})\partial_\theta + \delta^2\nabla_\mathbf{R}^2\right]\phi,$$
$$\partial_t\phi = \partial_t\theta\partial_\theta\phi + \delta^2\partial_T\phi = \delta\partial_T\Theta\partial_\theta\phi + \delta^2\partial_T\phi,$$

we get:

$$\delta\partial_T\Theta\partial_\theta\phi + \delta^2\partial_T\phi = \gamma\phi - \phi^3 - \Big[\left(1 + k^2\partial_\theta^2\right)$$
$$+ \delta\left(2\mathbf{k} \cdot \nabla_\mathbf{R} + \nabla_\mathbf{R} \cdot \mathbf{k}\right)\partial_\theta + \delta^2\nabla_\mathbf{R}^2\Big]^2\phi. \tag{7.61}$$

It is natural to expect that the solution can be expanded in powers of δ as

$$\phi = \phi_0 + \delta\phi_1 + \dots. \tag{7.62}$$

At the order δ^0, we obtain the nonlinear equation,

$$\gamma\phi_0 - \phi_0^3 - \left(1 + k^2\partial_\theta^2\right)^2\phi_0. \tag{7.63}$$

The solution ϕ_0 coincides with the 2π-periodic function $f(\theta)$ described above, which corresponds to an undistorted roll pattern with the wavenumber k.

At the order δ^1, we obtain an inhomogeneous linear equation,

$$\gamma\phi_1 - 3\phi_0^2\phi_1 - \left(1 + k^2\partial_\theta^2\right)^2\phi_1 = \partial_T\Theta\partial_\theta f \tag{7.64}$$

$$+ \left[\left(1 + k^2\partial_\theta^2\right)(2\mathbf{k} \cdot \nabla_\mathbf{R} + \nabla_\mathbf{R} \cdot \mathbf{k}) + (2\mathbf{k} \cdot \nabla_\mathbf{R} + \nabla_\mathbf{R} \cdot \mathbf{k})\left(1 + k^2\partial_\theta^2\right)\right]\partial_\theta f.$$

Equation (7.64) is solvable over the class of 2π-periodic functions only if its right-hand side is orthogonal to the eigenfunction of the homogeneous equation, $\partial_\theta f$. Using the notation

$$\langle g \rangle \equiv \frac{1}{2\pi} \int_0^{2\pi} g \, d\theta,$$

we find the following solvability condition:

$$\tau(k)\partial_T\Theta + \nabla_\mathbf{R} \cdot [B(k)\mathbf{k}] = 0, \tag{7.65}$$

where

$$\tau(k) = \langle(\partial_\theta f)^2\rangle, \tag{7.66}$$

$$B(k) = 2\left[\langle(\partial_\theta f)^2\rangle - k^2\langle(\partial_\theta^2 f)^2\rangle\right]. \tag{7.67}$$

The relation between $\mathbf{k}$ and Θ is $\mathbf{k} = \nabla_{\mathbf{R}}\Theta$; hence,

$$\nabla_{\mathbf{R}} \times \mathbf{k} = 0. \tag{7.68}$$

Equation (7.65) is called the *Cross–Newell equation*. This equation is universal and can be derived for any rotationally isotropic system which produces roll patterns due to a shortwave monotonic instability. Each particular problem is characterized by specific $\tau(k)$ and $B(k)$.

Monoharmonic Approximation

To calculate the functions $\tau(k)$, $B(k)$, one needs the solution of equation (7.63), $f(\theta)$, which cannot be found analytically. However, in a rather wide interval of γ, the higher harmonics A_n in the expansion (7.59) are small with respect to the basic harmonics A_1. Let us truncate the expansion (7.59) using only one term,

$$f \approx A\cos\theta, \tag{7.69}$$

and apply the Galerkin approximation with one Galerkin function, $\cos\theta$. The projection of equation (7.63) on that basic function gives

$$A^2(k^2) = \frac{4}{3}\left[\gamma - \left(1 - k^2\right)^2\right)\right], \ 1 - \sqrt{\gamma} < k^2 < 1 + \sqrt{\gamma}. \tag{7.70}$$

Using the approximation (7.69), (7.70), we find that

$$\langle (\partial_\theta f)^2 \rangle = \langle (\partial_\theta^2 f)^2 \rangle = \frac{A^2}{2},$$

hence

$$\tau(k) = \frac{A^2}{2} = \frac{2}{3}\left[\gamma - \left(1 - k^2\right)^2\right)\right], \tag{7.71}$$

$$B(k) = (1 - k^2)A^2 = \frac{4}{3}(1 - k^2)\left[\gamma - \left(1 - k^2\right)^2\right)\right]. \tag{7.72}$$

The functions $\tau(k)$ and $B(k)$ are defined in the interval $k_L < k < k_R$, where $k_L^2 = 1 - \sqrt{\gamma}$, $k_R^2 = 1 + \sqrt{\gamma}$. The function $\tau(k)$ is positive in the whole interval $k_L < k < k_R$, while the function $B(k)$ is positive in the interval $k_L < k < k_B$ and negative in the interval $k_B < k < k_R$, where $k_B = 1$.

Equation (7.65) can be used for studying the stability of rolls with respect to large-scale modulations far from the threshold point (i.e., for $\gamma = O(1)$) rather than $\gamma \ll 1$ [20]. It turns out that the rolls are subject to a zigzag (transverse) instability if the wavenumber k of the roll pattern has $B(k) > 0$, i.e., in the interval $k_L < k < k_B$. The Eckhaus (longitudinal) modulational instability appears as $d(kB(k))/dk > 0$. The function $kB(k)$ (see formula (7.72)) has a maximum in a certain point $k = k_{EL}$ in the interval (k_L, k_B) and a minimum in a point $k = k_{ER}$ in the interval (k_B, k_R).

Hence, the rolls are subject to the Eckhaus instability if their wavenumber is either in the interval (k_L, k_{EL}) or in the interval (k_{ER}, k_R). Finally, we come to the conclusion that the stability interval ("Busse balloon") is (k_B, k_{ER}), i.e., it is situated between the point k_B where $B(k)$ changes its sign and the point k_{ER} where the function $kB(k)$ has its minimum.

The predictions of the theory are summarized in Figure 7.6.

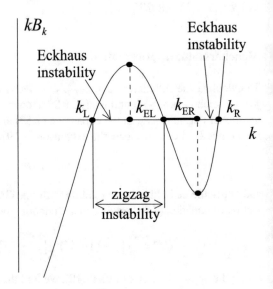

Fig. 7.6 Function kB_k and instabilities of rolls

Disclinations

The Cross–Newell equation can be used for studying a special type of defects in roll patterns, *disclinations* [21, 22]. When going around the core of this type of defects, one observes the pattern vector rotating. As it was noticed above, the local wavevector **k** of a roll pattern is defined up to the sign; hence, **k** is a *director* rather than a vector. Thus, the roll pattern "goes back to itself" after making a full circle around the core if the rotation angle is an integer number n multiplied by π (rather than 2π). Typical examples of disclinations observed in experiments are (i) a focus disclination, generating a target pattern ($n = 2$), (ii) a convex disclination ($n = 1$), (iii) a concave disclination ($n = -1$), and (iv) a saddle disclination ($n = -2$). Corresponding patterns are shown in Figure 7.7.

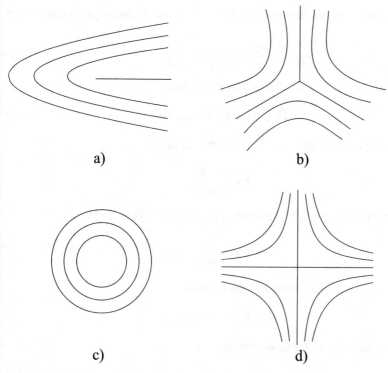

Fig. 7.7 Disclinations of roll patterns: (a) convex disclination, (b) concave disclination, (c) target, and (d) saddle

In the case of a motionless, steady dislocation, the corresponding field $\mathbf{k}(\mathbf{R})$ of the wavevector is governed by the equations

$$\nabla_{\mathbf{R}} \cdot [\mathbf{k}B(k)] = 0, \quad \nabla_{\mathbf{R}} \times \mathbf{k} = 0 \tag{7.73}$$

(except the central point where the wavevector field $\mathbf{k}(\mathbf{R})$ has a singularity and the wavevector is not defined).

Integration of the nonlinear equation (7.73) can be significantly simplified by the Legendre transformation [22]

$$\Theta(X, Y) + \hat{\Theta}(k_X, k_Y) = \mathbf{k} \cdot \mathbf{R}, \tag{7.74}$$

which is equivalent to the inversion of the dependence $\mathbf{k}(\mathbf{R})$,

$$\mathbf{R} = \mathbf{R}(\mathbf{k}),$$

and introduction of a dual function $\hat{\Theta}(\mathbf{k})$,

$$\mathbf{R} = \nabla_{\mathbf{k}}\hat{\Theta}(\mathbf{k}).$$

For $\hat{\Theta}(\mathbf{k})$, one obtains a *linear* equation, which can be written in polar coordinates (k, φ) as

$$\frac{\partial}{\partial k}(kB)\frac{\partial \hat{\Theta}}{\partial k} + \frac{1}{k}\frac{\partial}{\partial k}(kB)\frac{\partial^2 \hat{\Theta}}{\partial^2 \varphi} = 0. \tag{7.75}$$

The latter equation has a set of solutions in the form

$$\hat{\Theta}(k, \varphi) = F_m(k)\cos(m\varphi) \tag{7.76}$$

with an integer m. One can show [22] that the solution

$$\hat{\Theta} \sim \int \frac{dk}{kB(k)}$$

corresponds to a focus disclination with a target pattern, the solution

$$\hat{\Theta} \sim k \int \frac{dk}{k^3 B(k)}\cos\varphi$$

describes a convex disclination, and the solution

$$\hat{\Theta} \sim \ln(k_B - k)\cos 3\varphi$$

gives an asymptotics of a concave disclination.

Non-potential Effects

As shown above, the Swift–Hohenberg model (7.58) is sufficient for the explanation of many features of pattern formation. However, there are several phenomena which need an extension of the Swift–Hohenberg model for their description.

First of all, the Swift–Hohenberg equation (7.58) is a *potential* equation which can be written as

$$\frac{\partial \phi}{\partial t} = -\frac{\delta F(\phi)}{\delta \phi}, \tag{7.77}$$

where the Lyapunov functional

$$F = \int d\mathbf{r}\left[\frac{1}{2}(\nabla^2 \phi)^2 - (\nabla \phi)^2 + \frac{1-\gamma}{2}\phi^2 + \frac{1}{4}\phi^4\right]. \tag{7.78}$$

Hence, the system's dynamics is fully relaxational, i.e., the evolution of the system is characterized by a monotonic decrease of the Lyapunov functional and approaching a final stationary state, which excludes the possibility of oscillatory instabilities, spatiotemporal chaos, etc.

However, dissipative physical systems usually have no Lyapunov functionals. Hence, some features of the pattern formation can be overlooked by using the model (7.58).

In order to eliminate an artificial potential behavior and improve the coincidence of the model predictions with the results of observations and direct numerical simulations, some additional terms can be added to the right-hand side of (7.58), e.g. (see [23]),

$$\partial_t \phi = \left[\gamma - \left(1 + \nabla^2 \right)^2 \right] \phi - a\phi^3 \tag{7.79}$$

$$-b\phi(\nabla\phi)^2 + c\phi^2\nabla^2\phi + d\nabla^2\phi(\nabla\phi)^2 + e(\partial_i\phi)(\partial_j\phi)\partial_i\partial_j\phi.$$

Tuning of the coefficients allows to reproduce details of the stability diagrams obtained numerically in a particular physical problem. Specifically, in addition to the Eckhaus instability (the disturbance wavevector is parallel to that of the roll) and zigzag instability (the disturbance wavevector is orthogonal to that of the roll), one predicts a *skewed-varicose* instability characterized by a disturbance wavevector inclined with respect to the wavevector of the roll.

Mean-Flow Effects

The most remarkable phenomenon that needs an extension of the Swift–Hohenberg model for its explanation is the development of the *spiral-defect chaos* in the Rayleigh–Benard convection [24, 25], which involves rotating spirals, target patterns, dislocations, etc. The origin of the complicated behavior is the creation of a two-dimensional *mean flow*

$$\mathbf{U} = (\partial_y \zeta, -\partial_x \zeta), \tag{7.80}$$

with a vertical vorticity $-\nabla^2\zeta$, in the case when the rolls are curved so that

$$\left(\nabla \left(\nabla^2 \phi \right) \times \nabla\phi \right) \cdot \mathbf{e}_z = \partial_x \nabla^2 \phi \cdot \partial_y \phi - \partial_x \phi \cdot \partial_y \nabla^2 \phi \neq 0$$

[26].

In the presence of a mean flow, the generalized Swift–Hohenberg equation includes the advection of the convective rolls. It can be written, e.g., in the form

$$\left(\partial_t + g_m \mathbf{U} \cdot \nabla \right) \phi = \left[\gamma - \left(1 + \nabla^2 \right)^2 \right] \phi - a\phi^3 + d\nabla^2\phi(\nabla\phi)^2 \tag{7.81}$$

[27, 28]. The generation of the mean flow by the curved rolls is described by a phenomenological equation [27, 29]

$$\left[\tau_\zeta - P \left(\eta\nabla^2 - c^2 \right) \right] \nabla^2 \zeta = \left[\nabla \left(\nabla^2 \phi \right) \times \nabla\phi \right] \cdot \mathbf{e}_z. \tag{7.82}$$

The system (7.80)–(7.82) describes the transient spiral-defect chaos reasonably good [27, 28], though the longtime dynamics may be questionable [29].

7.1.2 Square Patterns

7.1.2.1 Selection of Square Patterns

As the basic model for studying stationary square patterns, let us select the Rayleigh–Bénard convection in a layer with nearly isolated boundaries governed by the system of equations (3.81), (3.82). For the sake of simplicity, take $P \to \infty$; thus, the problem is governed by one amplitude equation,

$$\frac{\partial A}{\partial \tau} = \delta^2 A - (1 + \nabla^2 A)^2 A + \frac{1}{3} \nabla \cdot (|\nabla A|^2 \nabla A) \qquad (7.83)$$

(in this section, we denote the two-dimensional gradient operator as ∇ rather than $\nabla_\perp$). The spatiotemporal dynamics described by equation (7.83) is potential, since that equation can be written in the variational form,

$$\frac{\partial A}{\partial \tau} = -\frac{\delta F}{\delta A}, \qquad (7.84)$$

where the Lyapunov functional F is

$$F = \int d^2 x \left\{ \frac{1}{2} \delta^2 A^2 + \frac{1}{2} [(1 + \nabla^2 A]^2 + \frac{1}{12} |\nabla A|^4 \right\}. \qquad (7.85)$$

As we have seen in Section 3.1.3.2, the interaction of two orthogonal roll systems,

$$A_1 = \left(a_1 e^{ix} + a_1^* e^{-ix} \right) + \left(a_2 e^{iy} + a_2 e^{-iy} \right),$$

is governed by the system of amplitude equations

$$\frac{da_1}{dT} = (1 - |a_1|^2 - \frac{2}{3} |a_2|^2) a_1; \quad \frac{da_2}{dT} = (1 - |a_2|^2 - \frac{2}{3} |a_1|^2) a_2. \qquad (7.86)$$

Because the non-diagonal elements of the matrix of the nonlinear interaction coefficients are smaller than the diagonal elements, the stable stationary state is characterized by a "symbiosis" of two roll systems, i.e., it is a square pattern.

7.1.2.2 Modulated Patterns

Similarly to Section 7.1.1.1, we introduce a set of spatial and temporal variables,

$$x_0 = x, \ y_0 = y, \ x_{1/2} = \varepsilon^{1/2} x, \ y_{1/2} = \varepsilon^{1/2} y,$$

$$x_1 = \varepsilon x, \ y_1 = \varepsilon y, \ T = \varepsilon^2 t;$$

hence,

$$\nabla = \nabla_0 + \varepsilon^{1/2} \nabla_{1/2} + \varepsilon \nabla_1, \ \frac{\partial}{\partial t} = \varepsilon^2 \frac{\partial}{\partial T}.$$

Expanding

$$A = \varepsilon A_1 + \varepsilon^{3/2} A_{3/2} + \varepsilon^2 A_2 + \ldots,$$

we obtain at the leading order:

$$A_1 = a_{11}(x_{1/2}, y_{1/2}, x_1, y_1) e^{ix_0} + a_{21}(x_{1/2}, y_{1/2}, x_1, y_1) e^{iy_0} + c.c.,$$

where $c.c.$ means complex conjugate. At the order $\varepsilon^{3/2}$, we find a similar expression for $A_{3/2}$ with amplitude functions $a_{1,3/2}, a_{2,3/2}$. At the order ε^2, we find that a_{11} does not depend on $x_{1/2}$ and a_{21} does not depend on $y_{1/2}$. At the order $\varepsilon^{5/2}$, we find that $a_{1,3/2}$ does not depend on $x_{1/2}$ and $a_{2,3/2}$ on $y_{1/2}$. Finally, the solvability conditions at the order ε^3 give the following nontrivial equations for the amplitude functions $a_1 \equiv a_{11}, a_2 \equiv a_{21}$:

$$\frac{\partial a_1}{\partial T} = a_1 + 4 \left(\frac{\partial}{\partial x_1} - \frac{i}{2} \frac{\partial^2}{\partial y_{1/2}^2} \right)^2 a_1 - a_1 |a_1|^2 - \frac{2}{3} a_1 |a_2|^2 + 4 \frac{\partial^2 a_{12}}{\partial x_{1/2}^2}, \quad (7.87)$$

$$\frac{\partial a_2}{\partial T} = a_2 + 4 \left(\frac{\partial}{\partial y_1} - \frac{i}{2} \frac{\partial^2}{\partial x_{1/2}^2} \right)^2 a_2 - a_2 |a_2|^2 - \frac{2}{3} a_2 |a_1|^2 + 4 \frac{\partial^2 a_{22}}{\partial y_{1/2}^2}. \quad (7.88)$$

Taking into account that

$$a_1 = a_1(y_{1/2}, x_1, y_1), \quad a_2 = a_2(x_{1/2}, x_1, y_1),$$

we find that the combination

$$c_1 = 4 \frac{\partial^2 a_{12}}{\partial x_{1/2}^2} - \frac{2}{3} a_1 |a_2|^2 \quad (7.89)$$

does not depend on $x_{1/2}$, and

$$c_2 = 4 \frac{\partial^2 a_{22}}{\partial y_{1/2}^2} - \frac{2}{3} a_2 |a_1|^2 \quad (7.90)$$

does not depend on $y_{1/2}$. The functions c_1 and c_2 can be found by averaging of (7.89) and (7.90) over the variables $x_{1/2}$ and $y_{1/2}$, correspondingly:

$$c_1 = -\frac{2}{3} a_1 \langle |a_2|^2 \rangle_{x_{1/2}}, \quad c_2 = -\frac{2}{3} a_2 \langle |a_1|^2 \rangle_{y_{1/2}}. \quad (7.91)$$

Thus, we obtain the following system of amplitude equations for the envelope functions $a_1(y_{1/2}, x_1, y_1), a_2(x_{1/2}, x_1, y_1)$ [30]:

$$\frac{\partial a_1}{\partial T} = a_1 + 4 \left(\frac{\partial}{\partial x_1} - \frac{i}{2} \frac{\partial^2}{\partial y_{1/2}^2} \right)^2 a_1 - a_1 |a_1|^2 - \frac{2}{3} a_1 \langle |a_2|^2 \rangle_{x_{1/2}}, \quad (7.92)$$

$$\frac{\partial a_2}{\partial T} = a_2 + 4\left(\frac{\partial}{\partial y_1} - \frac{i}{2}\frac{\partial^2}{\partial x_{1/2}^2}\right)^2 a_2 - a_2|a_2|^2 - \frac{2}{3}a_2\langle|a_1|^2\rangle_{y_{1/2}}. \tag{7.93}$$

Rescaling the variables (cf. (7.10)), we obtain:

$$\frac{\partial a_1}{\partial T} = \left(\frac{\partial}{\partial x_1} - \frac{i}{2}\frac{\partial^2}{\partial y_{1/2}^2}\right)^2 a_1 + a_1 - a_1|a_1|^2 - \kappa a_1\langle|a_2|^2\rangle_{x_{1/2}}, \tag{7.94}$$

$$\frac{\partial a_2}{\partial T} = \left(\frac{\partial}{\partial y_1} - \frac{i}{2}\frac{\partial^2}{\partial x_{1/2}^2}\right)^2 a_2 + a_2 - a_2|a_2|^2 - \kappa a_2\langle|a_1|^2\rangle_{y_{1/2}}. \tag{7.95}$$

In the case of convection between nearly isolated boundaries (equation (7.83)), $\kappa = 2/3$; for a general system with stable square patterns, $|\kappa| < 1$.

Equations (7.94), (7.95) allow to investigate the modulational instability of a square pattern which consists of rolls with slightly different wavenumbers,

$$a_1^{(0)} = r_1 e^{iK_1 x_1}, \ a_2^{(0)} = r_2 e^{iK_2 y_1},$$

$$r_1^2 = \frac{1 - \kappa - K_1^2 + \kappa K_2^2}{1 - \kappa^2}, \ r_2^2 = \frac{1 - \kappa - K_2^2 + \kappa K_1^2}{1 - \kappa^2},$$

where

$$K_1^2 - \kappa K_2^2 < 1 - \kappa, \ K_2^2 - \kappa K_1^2 < 1 - \kappa.$$

Disregarding dependences of disturbances on $x_{1/2}, y_{1/2}$, one finds that the most unstable mode is the "skew varicose" one with the wavevector inclined with respect to axes x, y, and the stability region of periodic patterns in the plane (K_1, K_2) is described by the inequality [30, 31]

$$1 - \left(1 - \frac{\kappa}{3}\right)(Q_1^2 + Q_2^2) - \left[\frac{\kappa}{3}Q_1^4 - \left(1 - \frac{\kappa^2 a}{3}\right)Q_1^2 Q_2^2 + \frac{\kappa}{3}Q_2^4\right] > 0,$$

where

$$Q_1^2 = \frac{3K_1^2}{1 - \kappa}, \ Q_2^2 = \frac{3K_2^2}{1 - \kappa}.$$

Taking into account the dependences of disturbances on $x_{1/2}, y_{1/2}$, one finds that the pattern is unstable if $K_1 < 0$ or $K_2 < 0$.

7.1.2.3 Dislocations in Square Patterns

For a motionless dislocation in the roll subpattern characterized by the modulation envelope function $a_1(x_1, y_{1/2})$, one finds

$$|a_2|^2 = 1 - \kappa\langle|a_1|^2\rangle = \frac{1}{1 + \kappa},$$

which leads to a closed equation for the amplitude a_1,

$$\left(\frac{\partial}{\partial x_1} - \frac{i}{2}\frac{\partial^2}{\partial y_{1/2}^2}\right)^2 a_1 + \frac{1}{1+K}a_1 - |a_1|^2 a_1 = 0. \tag{7.96}$$

That equation is quite similar to that obtained in the case of a roll pattern, with a changed coefficient. Hence, the phase equation (7.25) and its solution (7.31) are not changed. Also, one can find a solution similar to (7.34), which describes a continuous domain boundary without dislocations (for details, see [30]).

7.1.2.4 Disclinations in Square Patterns

The structure of a disclination in a square pattern can be qualitatively understood on the basis of stationary Cross–Newell equations for a square pattern which are similar to the Cross–Newell equation for a roll pattern considered in Section 7.1.1.4. According to the analysis presented in [32], a pattern formed in a potential system and based on two local wavevectors $\mathbf{k}_1 = \nabla \Theta_1$ and $\mathbf{k}_2 = \nabla \Theta_2$ is described by the stationary Cross–Newell equations

$$\nabla\left[\frac{\partial F}{\partial k_1^2}\mathbf{k}_1\right] + \nabla\left[\frac{\partial F}{\partial(\mathbf{k}_1 \cdot \mathbf{k}_2)}\mathbf{k}_2\right] = 0, \tag{7.97}$$

$$\nabla\left[\frac{\partial F}{\partial k_2^2}\mathbf{k}_2\right] + \nabla\left[\frac{\partial F}{\partial(\mathbf{k}_1 \cdot \mathbf{k}_2)}\mathbf{k}_1\right] = 0, \tag{7.98}$$

where F is the Lyapunov functional of the system.

Let us calculate the stationary solutions of (7.83) for a rectangular pattern (the directions of $\mathbf{n}_i = \mathbf{k}_i/k_i$ are not necessarily orthogonal) at small δ,

$$A = \delta a_1 e^{i(1+\delta K_1)\mathbf{n}_1 \cdot \mathbf{x}} + \delta a_2 e^{i(1+\delta K_2)\mathbf{n}_2 \cdot \mathbf{x}} + \text{c.c.} + \dots \tag{7.99}$$

Substituting (7.99) into (7.85), one finds

$$F = \delta^4 \frac{\kappa_1^2/2 - (2/3)[1 + 2(\mathbf{n}_1 \cdot \mathbf{n}_2)^2]\kappa_1\kappa_2 + \kappa_2^2/2}{1 - (4/9)[1 + 2(\mathbf{n}_1 \cdot \mathbf{n}_2)^2]^2}, \tag{7.100}$$

where $\kappa_1 = 1 - 4K_1^2$ and $\kappa_2 = 1 - 4K_2^2$. Note that due to the definitions of wavevectors, $\mathbf{k}_1$, $\mathbf{k}_2$, $\nabla_\perp \times \mathbf{k}_1 = \nabla_\perp \times \mathbf{k}_2 = 0$.

Since $k_1 = 1 + \delta K_1$, $k_2 = 1 + \delta K_2$, the expressions $B(k_1^2, k_2^2, \mathbf{k}_1 \cdot \mathbf{k}_2) = \partial F/\partial k_1^2$ and $B(k_2^2, k_1^2, \mathbf{k}_1 \cdot \mathbf{k}_2) = \partial F/\partial k_2^2$ are $O(\delta^{-1})$, while $C(k_1^2, k_2^2, \mathbf{k}_1 \cdot \mathbf{k}_2) = \partial F/\partial(\mathbf{k}_1 \cdot \mathbf{k}_2)$ is $O(1)$. Thus, near the threshold (and only there), one can disregard C and write the Cross–Newell equations at the leading order as [33]

$$\nabla[B(k_1^2, k_2^2, \mathbf{k}_1 \cdot \mathbf{k}_2)\mathbf{k}_1] = 0, \tag{7.101}$$

$$\nabla[B(k_2^2, k_1^2, \mathbf{k}_1 \cdot \mathbf{k}_2)\mathbf{k}_2] = 0, \tag{7.102}$$

where (after rescaling)

$$B(k_1^2, k_2^2, \mathbf{k}_1 \cdot \mathbf{k}_2) = 4K_1 \frac{(1 - 4K_1^2) - (2/3)[1 + 2(\mathbf{n}_1 \cdot \mathbf{n}_2)^2](1 - 4K_2^2)}{1 - (4/9)[1 + 2(\mathbf{n}_1 \cdot \mathbf{n}_2)^2]^2}, \qquad (7.103)$$

and $B(k_2^2, k_1^2, \mathbf{k}_1 \cdot \mathbf{k}_2)$ is obtained from (7.103) by replacing 1 by 2 and vice versa, $K_j = \delta^{-1}(k_j - 1)$. For perfect squares, $C = 0$ exactly.

In the case of a disclination, equations (7.101) and (7.102) can be represented as a single equation for a function $\mathbf{k} = \nabla_\perp \Theta$ defined (in the polar coordinates (r, θ)) in the region $0 \le \theta \le 4\pi$, such that $\mathbf{k}_1(\theta) \equiv \mathbf{k}(\theta)$, $\mathbf{k}_2(\theta) \equiv \mathbf{k}(\theta + 2\pi)$, $0 \le \theta \le 2\pi$.

A reasonable approximation to the solution of equations (7.101)–(7.103) can be found if one neglects the dependence of $B(k_1^2, k_2^2, \mathbf{k}_1 \cdot \mathbf{k}_2)$ on all the arguments except the first one, i.e., $B = B(k_1)$ is assumed. In that case, a solution to the Cross–Newell equations for rectangles can be considered as a solution for rolls, but defined in a region $0 \le \theta \le 4\pi$. Following [22], we use the Legendre transformation to obtain a linear equation,

$$k \frac{\partial}{\partial k} \left[kB(k) \frac{\partial \hat{\Theta}}{\partial k} \right] + \frac{\partial}{\partial k}[kB(k)] \frac{\partial \hat{\Theta}}{\partial \phi^2} = 0. \qquad (7.104)$$

An appropriate solution of equation (7.104) is $\hat{\Theta} = F_m(k) \cos m\phi$, where

$$k[kB(k)F_m'(k)]' - m^3 kB(k)F_m(k) = 0. \qquad (7.105)$$

In the case $m = 1$ corresponding to a disclination in a roll pattern, equation (7.105) was solved analytically in [22]. A convex disclination in a square pattern, with the topological charge $1/4$, corresponds to $m = 3$. In that case, an approximate solution of equation (7.105) can be found in the far-field region, $r \gg 1$. Indeed, as follows from [22], the region $r \gg 1$ corresponds to rolls with k close to the value k_0 at the boundary of the zigzag instability, where $B(k_0) = 0$ (in our case $k_0 = 1$, and $K_0 = 0$). Near $k = k_0$,

$$F_m(k) \sim \ln(k_0 - k), \qquad (7.106)$$

and it does not depend on m. The solution (7.106) in the real space is given by $\mathbf{x} = \nabla_{\mathbf{k}} \hat{\Theta}$, which yields

$$r \cos(\theta - \phi) = F_m'(k) \cos m\phi, \qquad (7.107)$$

$$r \sin(\theta - \phi) = -\frac{mF_m(k)}{k} \sin m\phi, \qquad (7.108)$$

where ϕ determines the direction of the vector $\mathbf{k}$. Thus, one gets

$$\theta = \phi + \arctan(\Delta \tan m\phi), \quad \Delta = -\frac{mF_m(k)}{kF_m'(k)}. \qquad (7.109)$$

In the limit (7.106) Δ is small, so that one obtains fast change of θ near $(\pi/m)(2n+1)$ (i.e., sectors with nearly straight rolls) and sectors with $\theta = \phi + n\pi$ (i.e., sectors with targets).

An approximate far-field solution corresponding to a convex disclination with the topological charge $1/4$ is

$$A = \cos(x\cos\phi_1 + y\sin\phi_1) + \cos(x\cos\phi_2 + y\sin\phi_2), \qquad (7.110)$$

where

$$\phi_1 = \{\theta, 0 < \theta < \pi/6; \ \pi/6, \ \pi/6 < \theta < 7\pi/6;$$
$$\theta - \pi, \ 7\pi/6 < \theta < 3\pi/2; \pi/2, \ 3\pi/2 < \theta < 2\pi\}; \qquad (7.111)$$
$$\phi_2 = \{\pi/2, \ 0 < \theta < \pi/2; \ \theta, \ \pi/2 < \theta < 5\pi/6;$$
$$5\pi/6, \ 5\pi/6 < \theta < 11\pi/6; \theta - \pi, \ 11\pi/6 < \theta < 2\pi\}. \qquad (7.112)$$

In [33], a numerical simulation of equation (7.83) has been carried out with initial conditions creating a disclination. The obtained pattern was similar to that described by (7.111) and (7.112) in the far field, but it included also a core with threefold symmetry, which cannot be analyzed in the framework of the longwave approach.

7.1.3 Hexagonal Patterns

7.1.3.1 Modulation of Hexagonal Patterns

Generally, one can expect the selection of hexagonal patterns due to the "symbiotic" mechanism described above in the case, where the nonlinear interaction coefficient M_{nm} in (C.6) is smaller than M_{nn} for wavevectors $\mathbf{k}_m$ and $\mathbf{k}_n$ with the angle $60°$ between them. However, the ubiquity of hexagonal patterns has another explanation, that is, the presence of quadratic nonlinearities in amplitude equations (see Section C.2 in Appendix C).

Let us consider large-scale modulations of hexagonal patterns. Because the procedure of the derivation of the Newell–Whitehead–Segel equations was described in detail for rolls (in Section 7.1.1.1) and for square patterns (in Section 7.1.2.2), we do not repeat it here. One can expect that the system of corresponding amplitude equations will be similar to (7.94), (7.95) but consist of three equations for three amplitudes, a_1, a_2, and a_3, and contain additional quadratic terms, like equations (C.11)–(C.13). It should be noted however that a shorter characteristic scale in the direction orthogonal to $\mathbf{k}_n$ does not appear in the presence of the quadratic term, because of the strong interaction of all the amplitudes through the total phase $\Theta = \theta_1 + \theta_2 + \theta_3$ [35]. It is convenient to rescale the amplitudes in such a way that the coefficient in the quadratic term is equal to 1. Finally, one obtains evolution equations containing only second derivatives in the direction of $\mathbf{k}_n$:

$$\frac{\partial a_j}{\partial T} = (\mathbf{n}_j \cdot \nabla)^2 a_j + \Gamma a_j + a_{j+1}^* a_{j-1}^* - \sum_{k=1}^{3} \kappa_{jk} |a_k|^2 a_j, \qquad (7.113)$$

where the subscripts are defined modulo 3, i.e., a_{j+1} is identified with a_j, $\kappa_{jk} = \kappa_{kj}$. For perfect hexagonal patterns with $|\mathbf{k}_j| = 1$, we can set $\kappa_{jj} = 1$ and $\kappa_{jk} = \kappa$ for $j \neq k$. The amplitude equations in this form first were used in the analysis of a modulated hexagonal pattern by Pomeau [15]. Equation (7.113) can be presented in the form

$$\frac{\partial a_j(\mathbf{x})}{\partial T} = \frac{\delta L}{\delta a_j^*(\mathbf{x})},$$

where

$$L = \int F d\mathbf{x}, \ F = \sum_{j=1}^{3} (\mathbf{n}_j \cdot \nabla) a_j^* (\mathbf{n}_j \cdot \nabla) a_j + U, \qquad (7.114)$$

$$U = -\Gamma \sum_{j=1}^{3} |a_j|^2 - (a_1 a_2 a_3 + a_1^* a_2^* a_3^*) + \frac{1}{2} \sum_{j,k=1}^{3} \kappa_{jk} |a_j|^2 |a_k|^2. \qquad (7.115)$$

In the case of the Boussinesq convection, some more subtle effects have to be taken into account. Though the quadratic term $\alpha(\mathbf{k}_1, \mathbf{k}_2, \mathbf{k}_3) a^* (\mathbf{k}_2) a^* (\mathbf{k}_3)$ in the amplitude equation for $a(\mathbf{k}_1)$ vanishes in the case of equilateral hexagonal patterns, due to the identities

$$\int_V \mathbf{v} \cdot (\mathbf{v} \cdot \nabla) \mathbf{v} d\mathbf{r} = 0, \ \int_V T \mathbf{v} \cdot \nabla T d\mathbf{r} = 0, \qquad (7.116)$$

valid for any velocity field $\mathbf{v}(\mathbf{r})$ and temperature field $T(\mathbf{r})$ satisfying typical conditions on the boundaries of the volume V (rigid or stress-free, isothermal, or insulating boundaries), it reappears in the case where the resonance condition $\mathbf{k}_1 + \mathbf{k}_2 + \mathbf{k}_3 = 0$ is satisfied but the moduli of $\mathbf{k}_j$, $j = 1, 2, 3$, as well as the angle between those vectors, are not equal [36]. Indeed, substituting into (7.116) a solution of the linear problem based on a set of wavenumbers $\mathbf{k}_1$, $\mathbf{k}_2$, and $\mathbf{k}_3$ with arbitrary moduli but satisfying a resonant condition $\mathbf{k}_1 + \mathbf{k}_2 + \mathbf{k}_3 = 0$, we can find that the coupling coefficients $\alpha(\mathbf{k}_1, \mathbf{k}_2, \mathbf{k}_3)$ satisfy the Jacobi identity [37]

$$\alpha(\mathbf{k}_1, \mathbf{k}_2, \mathbf{k}_3) + \alpha(\mathbf{k}_2, \mathbf{k}_3, \mathbf{k}_1) + \alpha(\mathbf{k}_3, \mathbf{k}_1, \mathbf{k}_2) = 0. \qquad (7.117)$$

This identity is a consequence of the fluid energy conservation. Different terms in equation (7.117) are not identical, and therefore, they are not equal to zero. Thus, quadratic terms may be present in amplitude equations for modulated patterns. After rescaling, one obtains the following system of equations:

$$\frac{\partial a_1}{\partial T} = a_1 + (\mathbf{n}_1 \cdot \nabla)^2 a_1 + iK[a_2^*(\tau_3 \cdot \nabla) a_3^* - a_3^*(\tau_2 \cdot \nabla) a_2^*] - (|a_1|^2 + \kappa |a_2|^2 + \kappa |a_3|^2) a_1;$$

$$(7.118)$$

other equations are obtained by a cyclic transposition of subscripts. Here τ_m is a unit vector orthogonal to $\mathbf{n}_m$; we assume that the rotation from $\mathbf{n}_m$ to τ_m is counterclockwise.

The quadratic terms in (7.118), contrary to non-Boussinesq quadratic terms [34, 38], cannot cause a subcritical generation of hexagonal patterns but can support them in the supercritical region and produce an instability of roll patterns. The corresponding analysis can be found in [36, 39]. Contrary to usual NSW equations, equations (7.118) have no Lyapunov functional.

For a non-Boussinesq buoyancy convection and Marangoni convection, non-potential terms in the form

$$iK_0[a_2^*(\mathbf{n}_3 \cdot \nabla)a_3^* + a_3^*(\mathbf{n}_2 \cdot \nabla)a_2^*] \tag{7.119}$$

can be added in the right-hand side of equation (7.118) [40, 41]. Coefficients K_0 and K have been calculated for a number of physical problems [36, 41–44]. The stability analysis in the framework of the amplitude equation containing both types of non-potential terms has been carried out in [45].

7.1.3.2 Fronts

Let us apply amplitude equations (7.113) for the consideration of the dynamics of fronts between different stable patterns, where one of the patterns is the hexagonal one. Assume that the wavenumber of the hexagonal pattern is the critical one, and disregard the phase disturbances. Then we can choose the amplitudes a_j, $j = 1, 2, 3$, as real. Assume also that the front is orthogonal to the direction of $\mathbf{k}_1$, which coincides with the direction of the x-axis. Due to the symmetry, the amplitudes of the other two modes are equal. Thus, we set $a_1 = A$, $a_2 = a_3 = B$, and rewrite equations (7.113) as

$$A_T = A_{XX} + \Gamma A + B^2 - A^3 - 2\kappa AB^2, \quad B_T = \frac{1}{4}B_{XX} + \Gamma B + AB - \kappa A^2 B - (1 + \kappa)B^3. \tag{7.120}$$

We assume $\kappa > 1$. Then there are two multistability regions: for $-1/[4(1 + 2\kappa)] < \Gamma < 0$, the equilibrium state and the hexagonal pattern are stable, and for $1/(\kappa - 1)^2 < \Gamma < (\kappa + 2)/(\kappa - 1)^2$, the roll pattern and the hexagonal pattern are stable. In those regions, the theory of a stable/stable domain wall propagation presented in Section 7.1.1.3 can be applied. The propagation of a stable hexagonal pattern into an *unstable* equilibrium state, which takes place within the interval $0 < \Gamma < 1/(\kappa - 1)^2$, has been explored in [46]. The influence of the non-potential terms in the amplitude equations (see (7.118), (7.119)) was investigated in [47].

7.1.3.3 Defects in Hexagonal Patterns

Penta–Hepta Defect

The experiments [48–50] and numerical simulations [51] show that hexagonal patterns contain often pentagon–heptagon pairs. These defects correspond actually to a bound state of two dislocations on two different roll subsystems forming the hexagonal pattern. Let us use the polar coordinates (r, φ), present the complex amplitudes in the form

$$a_j = R_j(r, \varphi) \exp(i\theta_j(r, \varphi)), \; j = 1, 2, 3,$$

and define topological charges

$$Q_j = \frac{1}{2\pi} \oint_C \nabla\theta_j \cdot d\mathbf{l}, \; j = 1, 2, 3,$$

where C is a contour encircling the defect. In the framework of the large-scale amplitude modulation equations (7.113), the structure in the far field (region where $r \gg 1$) of the defect with topological charges $(Q_1, Q_2, Q_3) = (0, 1, -1)$ can be described analytically as follows [35]:

$$\theta_1 = \frac{1}{2\sqrt{3}}(1 - \cos 2\varphi), \; \theta_{2,3} = \pm\varphi - \frac{1}{2\sqrt{3}}\left[\frac{1}{2} + \cos\left(2\varphi \mp \frac{2\pi}{3}\right)\right]. \quad (7.121)$$

In the presence of non-potential effects (see (7.118), (7.119)), the penta–hepta defects are subject to a spontaneous motion [52, 53]. Moreover, a moving defect can become unstable: two new dislocations are created in the dislocation-free roll system, and each of the two new dislocations forms a penta–hepta defect with a dislocation already present in another roll system. That leads to the transformation of the defect with topological charges $(0, 1, -1)$ into two defects $(1, 0, -1)$ and $(-1, 1, 0)$, which start moving in different directions. Thus, a multiplication of defects takes place [54].

Disclinations

Disclinations in hexagonal patterns have been observed in numerical simulations [33] for the potential Swift–Hohenberg equation with quadratic nonlinearity and for a non-potential damped Kuramoto–Sivashinsky equation,

$$\phi_t = \gamma\phi - (1 + \nabla^2)^2\phi - \frac{1}{2}|\nabla\phi|^2.$$

However, because of the strong coupling of phases in the hexagonal patterns, the Cross–Newell equations have no simple analytical solutions in that case.

7.2 Wave Patterns

Until now, we considered the longwave modulations of stationary patterns which are developed due to a primary monotonic instability. Let us consider now the case when that instability is oscillatory.

As the simplest example, let us consider a one-dimensional two-component *reaction–diffusion system*

$$u_t = D_u u_{xx} + f(u,v), \ v_t = D_v v_{xx} + g(u,v) \tag{7.122}$$

(see, e.g., [55]). A uniform steady state (u_0, v_0) is the solution of the system

$$f(u_0, v_0) = g(u_0, v_0) = 0.$$

Assuming

$$u = u_0 + \tilde{u}e^{ikx+\sigma t}, \ v = v_0 + \tilde{v}e^{ikx+\sigma t}$$

and linearizing equations around the steady state, we obtain the following equation for the growth rate σ:

$$\begin{vmatrix} f_u - D_u k^2 - \sigma & f_v \\ g_u & g_v - D_v k^2 - \sigma \end{vmatrix},$$

or

$$\sigma^2 - (\operatorname{tr} M(k^2))\sigma + \det M(k^2) = 0,$$

where

$$M(k^2) = \begin{pmatrix} f_u - D_u k^2 & f_v \\ g_u & g_v - D_v k^2 \end{pmatrix},$$

hence

$$\sigma = \frac{1}{2}\operatorname{tr} M(k^2) \pm \sqrt{\left(\frac{1}{2}\operatorname{tr} M(k^2)\right)^2 - \det M(k^2)}. \tag{7.123}$$

Assume that

$$\det M(0) > \left(\frac{1}{2}\operatorname{tr} M(0)\right)^2.$$

In that case, the imaginary part of the growth rate is nonzero for sufficiently long-waves. When

$$\operatorname{tr} M(0) = f_u + g_v$$

grows and crosses 0, a *longwave oscillatory instability* is developed in the system.

7.2.1 One-Dimensional Ginzburg–Landau Equation

In a contradistinction to the problems considered in the previous sections, there is no conservation law for system (7.122). Therefore, near the threshold, the generic one-dimensional amplitude equation, which governs a slowly changing in time and space amplitude function $A(t_2, x_1)$, is the complex Ginzburg–Landau equation [56]. In the case of a supercritical Hopf bifurcation, that equation, after rescaling, reads (cf.(7.11)):

$$A_t = A + (1 + i\alpha)A_{xx} - (1 + i\beta)|A|^2 A. \tag{7.124}$$

Recall that this equation governs also the spatiotemporal behavior of the envelope function for patterns created by a shortwave oscillatory instability; see Section 1.3.

Using the notation $A = R \exp(i\theta)$, one can rewrite (7.124) as a system of two real equations

$$R_t = R + \left(R_{xx} - R\theta_x^2\right) - \alpha\left(2R_x\theta_x + R\theta_{xx}\right) - R^3, \tag{7.125}$$

$$R\theta_t = \left(2R_x\theta_x + R\theta_{xx}\right) + \alpha\left(R_{xx} - R\theta_x^2\right) - \beta R^3. \tag{7.126}$$

A vast literature is devoted to the investigation of the dynamics governed by the complex Ginzburg–Landau equation (see review paper [57]) and coupled Ginzburg–Landau equations (see [58] and references therein). Here we discuss only few most basic topics related to longwave modulational instabilities of periodic wave patterns.

7.2.1.1 Periodic Waves

The system of equation (7.125), (7.126) has a one-parametric family of solutions

$$R = R_0(K), \ \theta = Kx - \Omega(K)t, \ |K| < 1, \tag{7.127}$$

where

$$R_0(K) = \sqrt{1 - K^2}, \ \Omega(K) = \beta + (\alpha - \beta)K^2, \tag{7.128}$$

which correspond to spatially periodic solutions $A(x, t)$ of equation (7.124). Similarly to stationary roll solutions studied in Section 5.1.1, these solutions may be stable or unstable, depending on parameters K, α, and β.

7.2.1.2 Stability of Uniform Oscillations

First, let us consider the stability of uniform oscillations, $K = 0$, $R_0(0) = 1$, $\Omega(0) = \beta$. For normal disturbances

$$(\tilde{R}, \tilde{\theta}) \sim e^{\sigma t + i\tilde{K}x},$$

one obtains the following linearized problem:

$$\sigma \tilde{R} = \tilde{R}(1 - \tilde{K}^2) + \alpha \tilde{K}^2 \tilde{\theta} - 3\tilde{R},$$

$$-\beta \tilde{R} + \sigma \tilde{\theta} = -\tilde{K}^2 \tilde{\theta} - 3\beta \tilde{R} - \alpha \tilde{K}^2 \tilde{R};$$

hence,

$$\begin{vmatrix} -2 - \tilde{K}^2 - \sigma & \alpha \tilde{K}^2 \\ -2\beta - \alpha \tilde{K}^2 & -\tilde{K}^2 - \sigma \end{vmatrix} = 0,$$

or

$$\sigma^2 + 2\left(1 + \tilde{K}^2\right)\sigma + 2\tilde{K}^2(1 + \alpha\beta) + (1 + \alpha^2)\tilde{K}^4. \tag{7.129}$$

If $1 + \alpha\beta < 0$, the uniform oscillations are unstable with respect to modulations with the wavenumbers in the interval

$$0 < \tilde{K}^2 < K_m^2 = -\frac{1 + \alpha\beta}{1 + \alpha^2}.$$

This instability is known as the *Benjamin–Feir instability*.

7.2.1.3 Nonlinear Phase Equation

If K_m^2 is small:

$$1 + \alpha\beta = -\varepsilon^2, \ \beta = -\frac{1}{\alpha} - \frac{\varepsilon^2}{\alpha}, \ \varepsilon \ll 1,$$

the nonlinear evolution of the Benjamin–Feir instability can be studied by means of a nonlinear phase equation [59, 60] (cf. Section 7.1.1.2).

The linear theory predicts $\tilde{K} \sim \varepsilon$, $\sigma \sim \varepsilon^4$. Therefore, we assume

$$R = R(X, T), \ \theta = -\beta t + \varepsilon^2 \vartheta(X, T); \ X = \varepsilon x; \ T = \varepsilon^4 t$$

and obtain:

$$\varepsilon^4 R_T = R + \varepsilon^2 \left[(R_{XX} - \varepsilon^4 R \vartheta_X^2) - \alpha \varepsilon^2 (2R_X \vartheta_X + R \vartheta_{XX}) - R^3 \right], \tag{7.130}$$

$$\left(\frac{1}{\alpha} + \frac{\varepsilon^2}{\alpha} \right) R + \varepsilon^6 R \vartheta_T = \varepsilon^2 \left[(2R_X \vartheta_X + R \vartheta_{XX}) \varepsilon^2 + \alpha \left(R_{XX} - \varepsilon^4 R \vartheta_X^2 \right) \right]$$

$$+ R^3 \left(\frac{1}{\alpha} + \frac{\varepsilon^2}{\alpha} \right). \tag{7.131}$$

Let us construct the solution in the form of an asymptotic series in powers of ε^2. We find that the amplitude R is slaved to the phase ϑ_0:

$$r = 1 - \frac{1}{2}\varepsilon^4 \alpha \vartheta_{XX}^{(0)} + \frac{1}{2}\varepsilon^6 \left(-\frac{1}{2}\vartheta_{XXXX}^{(0)} - (\vartheta_X^{(0)})^2 - \alpha \vartheta_{XX}^{(2)} \right) + \dots.$$

The leading-order equation governing the evolution of the phase is the *Kuramoto–Sivashinsky equation*:

$$\vartheta_T^{(0)} = -\vartheta_{XX}^{(0)} - \frac{1}{2}\left(1 + \alpha^2\right)\vartheta_{XXXX}^{(0)} - \left(\alpha + \alpha^{-1}\right)\left(\vartheta_X^{(0)}\right)^2. \tag{7.132}$$

The latter equation is a paradigmatic model for studying the spatiotemporal chaos [61]. It can be written in the form (1.79) for the local wavenumber $q = \theta_X^{(0)}$.

7.2.1.4 Stability of Periodic Waves

A similar analysis can be done for periodic waves (7.127), (7.128) [62]. The longwave asymptotics of the growth rate for the Goldstone mode, which corresponds to spatially periodic shifts of the phase with the period $2\pi/k$, is as follows:

$$\sigma(k) = i\sigma_1 k - \sigma_2 k^2 - i\sigma_3 k^3 + \sigma_4 k^4 + O(k^5), \tag{7.133}$$

where

$$\sigma_1 = -2(\alpha - \beta), \tag{7.134}$$

$$\sigma_2 = 1 + \alpha\beta - \frac{2(1 + \beta^2)K^2}{1 - K^2}, \tag{7.135}$$

$$\sigma_3 = -\frac{2[\alpha(1 - K^2) - 2\beta K^2](1 + \beta^2)K}{(1 - K^2)^2},$$

$$\sigma_4 = -\frac{1 + \beta^2}{2(1 - K^2)^3}[\alpha^2(1 - K^2)^2 - 12\alpha\beta(1 - K^2)K^2 + 4(1 + 5\beta^2)K^4].$$

The corresponding nonlinear phase equation looks as follows [63]:

$$\theta_t = \sigma_1\theta_x + \sigma_2\theta_{xx} + \sigma_3\theta_{xxx} + \sigma_4\theta_{xxxx} + \kappa_1(\theta_x)^2 + \kappa_2(\theta_x^2)_x, \tag{7.136}$$

where

$$\kappa_1 = -(\alpha - \beta), \quad \kappa_2 = -\frac{2K(1 + \beta^2)}{1 - K^2}.$$

In the reference frame moving with the group velocity of waves, $v_g = -\sigma_1$ ($X = x + v_g t$), one obtains the following equation for the local wavenumber $q = \theta_x$:

$$q_t = \sigma_2 q_{XX} + \sigma_3 q_{XXX} + \sigma_4 q_{XXXX} + \kappa_1(q^2)_X + \kappa_2(q^2)_{XX}. \tag{7.137}$$

Thus, one obtains the perturbed Korteweg–de Vries equation (1.77) that was considered in Sections 3.3.1.2 and 5.1.2.1. Recall that this equation includes terms of different orders; hence, it has to be treated in two steps, as explained in Section 5.1.2.1.

7.2.2 2D Complex Ginzburg–Landau Equation: Spiral Wave

In 2D case, the complex Ginzburg–Landau equation reads:

$$A_t = A + (1 + i\alpha)\nabla^2 A - (1 + i\beta)|A|^2 A, \tag{7.138}$$

or $(A = R\exp(i\theta))$

$$R_t = R + \nabla^2 R - R(\nabla\theta)^2 - \alpha(2\nabla R \cdot \nabla\theta + R\nabla^2\theta) - R^3, \tag{7.139}$$

$$R\theta_t = 2\nabla R \cdot \nabla\theta + R\nabla^2\theta + \alpha(\nabla^2 R - R(\nabla\theta)^2) - \beta R^3. \tag{7.140}$$

The 2D complex Ginzburg–Landau equation is characterized by an extremely diverse behavior (see [57]). Here we will discuss only the most remarkable objects typical for this equation, spiral waves.

A spiral wave is a solution of the type

$$R = R(\rho), \ \theta(\rho,\varphi,t) = m\varphi + \psi(\rho) + \Omega t, \tag{7.141}$$

where (ρ,φ) are polar coordinates in the plane (x,y). For the sake of simplicity, later on we shall take $\alpha = 0$ and consider only the waves with $m = 1$. The problem is governed by the following system of equations:

$$R'' + \frac{1}{\rho}R' + R\left[1 - R^2 - (\psi')^2 - \frac{1}{\rho^2}\right] = 0, \tag{7.142}$$

$$q' + \frac{1}{\rho}q + \frac{2R'}{R}q - \beta R^2 = \Omega. \tag{7.143}$$

Here $'$ means differentiation with respect to ρ; $q = \psi'$.

The solution *is not singular* in the point $\rho = 0$, because of the boundary conditions

$$R(0) = 0, \ q(0) = 0. \tag{7.144}$$

At a large distance from the center, the spiral wave becomes indistinguishable from a plane wave with a certain wavenumber k_∞, which is the eigenvalue of the nonlinear problem (7.142), (7.143):

$$R(\infty) = \sqrt{1 - k_\infty^2}, \ q(\infty) = k_\infty. \tag{7.145}$$

The general problem (7.142)–(7.145) can be solved numerically. Here we will present a semi-analytical solution in the limit of small β: $\beta = -\varepsilon$, $|\varepsilon| \ll 1$ [64, 65]. In this limit, one can distinguish between the core region ($\rho = O(1)$) and far-field region ($\rho = O(1/\varepsilon)$), where the asymptotic expansions are different.

7.2.2.1 Inner Expansion

In the region $\rho = O(1)$, we search the solution to the system (7.142)–(7.145) in the form

$$R(\rho) = R_0(\rho) + \varepsilon R_1(\rho) + \dots, \quad q = \varepsilon q_1 + \dots.$$

Also, we assume that $|k_\infty| \ll 1$.

In the leading order, we obtain the following nonlinear problem for $R_0(\rho)$:

$$R_0'' + \frac{1}{\rho} R_0' + \left(1 - \frac{1}{\rho^2} - R_0^2 \right) R_0 = 0; \quad R_0(0) = 0; \quad |R_0(\infty)| < \infty. \tag{7.146}$$

This problem was studied in the context of the vortex core for the Ginzburg–Pitaevskii equation which describes a superfluid flow. It is known that the solution of this problem exists and is unique. The solution can be found only numerically. For large ρ, the asymptotics of the solution is $R_0^2 \sim 1 - 1/\rho^2 + \dots$.

For the local wavenumber q_1, we obtain the linear problem:

$$q_1' + \left(\frac{1}{\rho} + \frac{2R_0'}{R_0} \right) q_1 = 1 - R_0^2, \quad q(0) = 0. \tag{7.147}$$

Its solution can be presented in an explicit way:

$$q_1(\rho) = \frac{1}{\rho R_0^2} \int_0^\rho d\rho' \rho' R_0^2(\rho')[1 - R_0^2(\rho')]. \tag{7.148}$$

At large ρ, the solution (7.148) behaves as

$$q_1(\rho) \sim \frac{1}{\rho}(\ln \rho + C + \dots), \tag{7.149}$$

hence

$$\psi'(\rho) \sim \frac{\varepsilon}{\rho}(\ln \rho + C + \dots). \tag{7.150}$$

A numerical evaluation of the constant C gives $C \approx -0.098$.

7.2.2.2 Outer Expansion

For the construction of the outer expansion, it is better to return to the original system of equations (7.139)–(7.140) and take into account that in the region $\rho \gg 1$, the spatial derivatives of the fields are small. We find that the amplitude field R is slaved to the phase field θ:

$$R^2 \sim 1 - (\nabla \theta)^2.$$

Taking into account that $\Omega = \varepsilon(1 - k_\infty^2)$, we obtain the following nonlinear phase equation:

$$\nabla^2 \theta = \varepsilon \left[k_\infty^2 - (\nabla \theta)^2 \right] = 0. \tag{7.151}$$

This is the *Burgers equation* which can be linearized by means of the *Hopf–Cole transformation*:

$$\theta = -\frac{1}{\varepsilon}\ln F, \tag{7.152}$$

$$\nabla^2 F - \varepsilon^2 k_\infty^2 F = 0. \tag{7.153}$$

Introduce a scaled variable $s \equiv \varepsilon k_\infty \rho$. The spiral-wave solution $\theta = \varphi + \psi(s)$ is transformed to

$$F = e^{-\varepsilon\varphi}H(s),$$

where $H(s)$ satisfies the equation

$$\frac{d^2 H}{ds^2} + \frac{1}{s}\frac{dH}{ds} - \left(1 - \frac{\varepsilon^2}{s^2}\right)H = 0. \tag{7.154}$$

The appropriate solution is the Bessel function of imaginary argument with imaginary index: $H = \text{const } K_{i\varepsilon}(s)$, hence

$$F = \text{const } e^{-\varepsilon\varphi}K_{i\varepsilon}(s), \quad \theta = \varphi - \frac{1}{\varepsilon}\ln K_{i\varepsilon}(\varepsilon k_\infty \rho);$$

$$\psi(\rho) = -\frac{1}{\varepsilon}\ln K_{i\varepsilon}(\varepsilon k_\infty \rho).$$

The derivative $\psi'(\rho)$ should be matched to the expression (7.150) obtained from the inner solution.

7.2.2.3 Matching

Using the asymptotic of the Bessel function for small $\varepsilon k_\infty \rho$,

$$K_{i\varepsilon}(\varepsilon k_\infty \rho) \sim \sin\left(\varepsilon\ln\frac{\varepsilon k_\infty \rho}{2}\right) + \varepsilon\gamma\cos\left(\varepsilon\ln\frac{\varepsilon k_\infty \rho}{2}\right),$$

where $\gamma \approx 0.577$ is the Euler constant, we find:

$$-\frac{1}{\rho}\frac{\cos\left(\varepsilon\ln\frac{\varepsilon k_\infty \rho}{2}\right) - \varepsilon\gamma\sin\left(\varepsilon\ln\frac{\varepsilon k_\infty \rho}{2}\right)}{\sin\left(\varepsilon\ln\frac{\varepsilon k_\infty \rho}{2}\right) + \varepsilon\gamma\cos\left(\varepsilon\ln\frac{\varepsilon k_\infty \rho}{2}\right)} \sim \frac{\varepsilon}{\rho}(\ln\rho + C).$$

The matching can be performed if

$$\varepsilon\ln\frac{\varepsilon k_\infty \rho}{2} = -\frac{\pi}{2} + \delta, |\delta| \ll 1,$$

so that

$$\cos\left(\varepsilon\ln\frac{\varepsilon k_\infty \rho}{2}\right) \sim \delta, \quad \sin\left(\varepsilon\ln\frac{\varepsilon k_\infty \rho}{2}\right) \sim -1,$$

and the asymptotics of the outer solution is given by the expression

$$\psi'(\rho) \sim \frac{1}{\rho}(\delta + \varepsilon\gamma) = \frac{1}{\rho}\left(\frac{\pi}{2} + \varepsilon\ln\frac{\varepsilon k_\infty \rho}{2} + \varepsilon\gamma\right). \tag{7.155}$$

Comparing (7.155) and (7.150), we find the matching condition:

$$\frac{\pi}{2} + \varepsilon\ln\frac{\varepsilon k_\infty}{2} + \varepsilon\gamma = \varepsilon C.$$

Thus, the selected wavenumber k_∞ is determined by the formula

$$k_\infty = \frac{2}{\varepsilon}\exp\left(-\frac{\pi}{2\varepsilon} - \gamma + C\right). \tag{7.156}$$

7.3 Interaction of Longwave and Shortwave Instabilities

As an example of the interaction of longwave and shortwave instability modes, let us consider the development of the convection in a layer with a fixed temperature (rather than heat flux) on the bottom and non-small Biot number on the free surface. Note that effective Biot number is large in the case of liquid evaporation [66]. Also, the analogue of the Biot number in the case of mass transfer (Sherwood number) is not small. In that case, the monotonic neutral curve $M(k)$, which is determined by the following formula [67]:

$$M(k) = \frac{4k(k\cosh k + Bi\sinh k)(\sinh 2k - 2k)(Ga + Ck^2)}{8k^5\cosh k + (\sinh^3 k - k^3\cosh k)(Ga + Ck^2)}, \tag{7.157}$$

has typically two minima. The minimum at $k = 0$,

$$M_l = M(0) = \frac{2}{3}G(1 + Bi), \tag{7.158}$$

corresponds to the longwave instability. The other one, $M_s = M(k_c)$, $k_c \neq 0$, determines the threshold of the shortwave instability.

Near the shortwave instability threshold, $M - M_s = O(\varepsilon^2)$, the longwave amplitude and phase modulation of a shortwave pattern is described by the amplitude function $A(X, \tau)$, $X = \varepsilon x$, $\tau = \varepsilon^2 t$, which satisfies the Ginzburg–Landau equation with real coefficients (because the instability is monotonic). That modulation creates a longwave modulation of pressure, which is a cause of a layer thickness disturbance, $B(X, \tau)$. At the same time, the inhomogeneity of the layer thickness leads to the inhomogeneity of the local Marangoni number, and hence, it influences the growth rate of the shortwave instability.

If $M_l > M_s$, i.e., the surface is stable with respect to longwave deformations at M close to M_s, one obtains, after rescaling, the following coupled system of amplitude equations [68]:

$$\frac{\partial A}{\partial \tau} = A + \frac{\partial^2 A}{\partial X^2} - \lambda |A|^2 A + AB,$$ (7.159)

$$\frac{\partial B}{\partial \tau} = m\frac{\partial^2 B}{\partial X^2} + w\frac{\partial^2 |B|^2}{\partial X^2}.$$ (7.160)

Here $\lambda = 1$, if the shortwave instability is supercritical, and $\lambda = -1$, if it is subcritical. The parameter m has the same sign as $M_l - M_s$, i.e., it is positive, while w can be of either sign.

Stationary solutions of the problem (7.159), (7.160) are determined by the following system of equations:

$$A_{XX} + A - \lambda |A|^2 A + AB = 0,$$ (7.161)

$$mB_{XX} + w(|A|^2)_{XX} = 0.$$ (7.162)

By definition, the averaged deflection of the surface vanishes,

$$\langle B(X) \rangle = 0.$$ (7.163)

The bounded solution $B(X)$ of equation (7.162) satisfying the condition (7.163) is:

$$B(X) = \left(-\frac{w}{m}\right)(|A(X)|^2 - \langle |A|^2 \rangle),$$ (7.164)

which leads to the following closed equation for the amplitude $A(X)$:

$$A_{XX} - \alpha A + \beta |A|^2 A = 0,$$ (7.165)

where

$$\alpha = -1 + \frac{w}{m}\langle |A|^2 \rangle, \quad \beta = -\lambda - \frac{w}{m}.$$ (7.166)

One can see that perfectly periodic stationary patterns with $|A| = const$ do not produce surface deformations; hence, they are not influenced by the deformational mode, but it may be significant for modulated patterns.

In the case $M_l < M_s$, where the longwave instability takes place near the point $M = M_s$, it is necessary to include the stabilizing term with the fourth-order spatial derivative into the amplitude equation. Therefore, one obtains the following rescaled system of equations for A and B [68]:

$$\frac{\partial A}{\partial \tau} = A + \frac{\partial^2 A}{\partial X^2} - \lambda |A|^2 A + AB,$$ (7.167)

$$\frac{\partial B}{\partial \tau} = -|m|\frac{\partial^2 B}{\partial X^2} - s\frac{\partial^4 h}{\partial X^4} + w\frac{\partial^2 |A|^2}{\partial X^2}.$$ (7.168)

The regular shortwave pattern with a flat surface, described by the solution $A_0 = 1$, $B_0 = 0$, can be unstable with respect to disturbances of the kind,

$$\tilde{A} = a_+ e^{ikX+\sigma\tau} + a_- e^{-ikX+\sigma^*\tau},$$

$$\tilde{B} = b_+ e^{ikX+\sigma\tau} + b_- e^{-ikX+\sigma^*\tau}.$$

The monotonic stability curve is given by the relation,

$$m = k^2 s + \frac{2w}{2+k^2}, \tag{7.169}$$

while the boundary of the oscillatory instability is

$$m = k^2 w + 1 + \frac{2}{k^2}. \tag{7.170}$$

The analysis of nonlinear surface waves created by instabilities has been carried out in [68–70]. In some limits, exact solutions have been found. Numerical simulations reveal traveling and standing waves, as well as chaotic regimes.

References

1. W. Eckhaus, *Studies in Non-linear Stability Theory* (Springer, Berlin, 1965)
2. A. Schlüter, D. Lortz, F. Busse, J. Fluid Mech. **23**, 129 (1965)
3. A.C. Newell, J.A. Whitehead, J. Fluid Mech. **38**, 279 (1969)
4. L.M. Pismen, *Patterns and Interfaces in Dissipative Dynamics* (Springer, Berlin/Heidelberg, 2006)
5. R. Hoyle, *Pattern Formation, An Introduction to Methods* (Cambridge University Press, Cambridge, 2006)
6. M. Cross, H. Greenside, *Pattern Formation and Dynamics in Nonequilibrium Systems* (Cambridge University Press, Cambridge, 2009)
7. J.W. Swift, P.C. Hohenberg, Phys. Rev. A **15**, 319 (1977)
8. M.C. Cross, Phys. Fluids **23**, 1727 (1980)
9. J.J. Christensen, A.J. Bray, Phys. Rev. E **58**, 5364 (1998)
10. J. Kevorkian, J.D. Cole, *Multiple Scale and Singular Perturbation Methods* (Springer, New York, 1996)
11. L.A. Segel, J. Fluid Mech. **38**, 203 (1969)
12. E.D. Siggia, A. Zippelius, Phys. Rev. A **24**, 1036 (1981)
13. A.A. Nepomnyashchy, L.M. Pismen, Phys. Lett. A **153**, 427 (1991)
14. B.A. Malomed, A.A. Nepomnyashchy, M.I. Tribelsky, Phys. Rev. A **42** (1990) 7244
15. Y. Pomeau, Physica D **23**, 3 (1986)
16. I.S. Aranson, B.A. Malomed, L.M. Pismen, L.S. Tsimring, Phys. Rev. E **62**, R5 (2000)
17. A. Scheel, Arch. Ration. Mech. Anal. **181**, 505 (2006)
18. K. Ouchi, H. Fujisaka, Phys. Rev. E **54**, 3895 (1996)
19. E. Knobloch, Nonlinearity **21**, T45 (2008)
20. M.C. Cross, A.C. Newell, Physica D **10**, 299 (1984)
21. C. Bowman, A.C. Newell, Rev. Mod. Phys. **70**, 289 (1998)
22. T. Passot, A.C. Newell, Physica D **74**, 301 (1994)
23. H.S. Greenside, M.C. Cross, Phys. Rev. A **31**, 2492 (1985)

24. S.W. Morris, E. Bodenschatz, D. Cannel, G. Ahlers, Phys. Rev. Lett. **71**, 2026 (1993)
25. M. Assenheimer, V. Steinberg, Nature (London) **367**, 345 (1994)
26. L.M. Pismen, Phys. Lett. A **116**, 241 (1986)
27. H.-W. Xi, J.D. Gunton, J. Viñals, Phys. Rev. Lett. **71**, 2030 (1993)
28. M.C. Cross, Y. Tu, Phys. Rev. Lett. **75**, 834 (1995)
29. R. Schmitz, W. Pesch, W. Zimmermann, Phys. Rev. E **65**, 037302 (2002)
30. A. Nepomnyashchy, Int. J. Bifurcation Chaos **4**, 1147 (1994)
31. R.B. Hoyle, Physica D **67**, 198 (1993)
32. R.B. Hoyle, Phys. Rev. E **61**, 2506 (2000)
33. A.A. Golovin, A.A. Nepomnyashchy, Phys. Rev. E **67**, 056202 (2003)
34. F.H. Busse, J. Fluid Mech. **30**, 625 (1967)
35. L.M. Pismen, A.A. Nepomnyashchy, Europhys. Lett. **24**, 461 (1993)
36. E.A. Kuznetsov, A.A. Nepomnyashchy, L.M. Pismen, Phys. Lett. A **205**, 261 (1995)
37. E.A. Kuznetsov, M.D. Spector, J. Appl. Mech. Tech. Phys. **21**(2), 220 (1980)
38. F.H. Busse, Dissertation, Munich (1962). In German
39. A.E. Nuz, A.A. Nepomnyashchy, L.M. Pismen, Physica A **249**, 179 (1998)
40. H.R. Brand, Prog. Theor. Phys. Suppl. **99**, 442 (1989)
41. A.A. Golovin, A.A. Nepomnyashchy, L.M. Pismen, J. Fluid Mech. **341**, 317 (1997)
42. J. Bragard, J. Pontes, M.G. Velarde, Int. J. Bifurcation Chaos **6**, 1665 (1996)
43. J. Bragard, M.G. Velarde, J. Fluid Mech. **368**, 1665 (1998)
44. R. Kuske, P. Milewski, Eur. J. Appl. Math. **10**, 157 (1999)
45. A.E. Nuz, A.A. Nepomnyashchy, A.A. Golovin, A. Hari, L.M. Pismen, Physica D **135**, 233 (2000)
46. L.M. Pismen, A.A. Nepomnyashchy, Europhys. Lett. **27**, 433 (1994)
47. A. Hari, A.A. Nepomnyashchy, Phys. Rev. E **61**, 4835 (2000)
48. P. Cerisier, R. Occelli, C. Perez-Garcia, C. Jamond, J. Phys. (Paris) **48**, 569 (1987)
49. E.L. Koschmieder, *Benard Cells and Taylor Vortices* (Cambridge University Press, Cambridge, 1993)
50. O.V. Afenchenko, A.B. Ezersky, A.V. Nazarovsky, M.G. Velarde, Int. J. Bifurcation Chaos **11**, 1261 (2001)
51. M. Bestehorn, Phys. Rev. E **48**, 3622 (1993)
52. L.S. Tsimring, Phys. Rev. Lett. **74**, 4201 (1995)
53. L.S. Tsimring, Physica D **89**, 368 (1996)
54. P. Colinet, A.A. Nepomnyashchy, J.C. Legros, Europhys. Lett. **57**, 480 (2002)
55. J.D. Murray, *Mathematical Biology* (Springer, Berlin, 1989)
56. Y. Kuramoto, T. Tsuzuki, Prog. Theor. Phys. **55**, 356 (1976)
57. I.S. Aranson, L. Kramer, Rev. Mod. Phys. **74**, 99 (2002)
58. G. Dangelmayr, I. Oprea, J. Nonlinear Sci. **18**, 1 (2008)
59. A.A. Nepomnyashchy, Proc. Perm State Univ. **316**, 105 (1974, in Russian)
60. T. Yamada, Y. Kuramoto, Prog. Theor. Phys. **56**, 681 (1976)
61. T. Bohr, M.H. Jensen, G. Paladin, A. Vulpiani, *Dynamical System Approach to Turbulence* (Cambridge University Press, Cambridge, 1998)
62. J.T. Stuart, R.C. DiPrima, Proc. R. Soc. Lond. Ser. A **372**, 357 (1980)
63. B. Janiaud, A. Pumir, D. Bensimon, V. Croquette, H. Richter, L. Kramer, Physica D **55**, 269 (1992)
64. P.S. Hagan, SIAM J. Appl. Math. **42**, 762 (1982)
65. L.M. Pismen, A.A. Nepomnyashchy, Physica D **54**, 183 (1992)
66. D. Merkt, M. Bestehorn, Physica D **185**, 196 (2003)
67. M. Takashima, J. Phys. Soc. Jpn. **50**, 2745 (1981)
68. A.A. Golovin, A.A. Nepomnyashchy, L.M. Pismen, Phys. Fluids **6**, 34 (1994)
69. A.A. Golovin, A.A. Nepomnyashchy, L.M. Pismen, H. Riecke, Physica D **106**, 131 (1997)
70. P.C. Matthews, S.M. Cox, Nonlinearity **13**, 1293 (2000)

Chapter 8
Control of Longwave Instabilities

In the previous chapters, we have considered longwave instabilities that appear and develop in "a natural way." The applications need, however, *controlling* the instabilities. In some cases, the instabilities should be eliminated, and in other cases, some definite patterns should be selected and controlled.

An efficient way of stabilization/destabilization of a system is an appropriate *prescribed temporal* modulation of system parameters. The paradigmatic example of such a modulation is the stabilization of a pendulum in *an inverted position* by means of vertical vibrations [1]. This kind of control is described in Section 8.1. The influence of a *spatial* parameter modulation on the stability and pattern formation is the subject of Section 8.2. Both approaches can be considered as realizations of some "open-loop" control strategies.

Another way of control is the *feedback control*, where the parameter change is not prescribed from advance but determined by the information provided by measurements. The latter approach can be more efficient and less invasive. Some examples of the feedback control of instabilities are presented in Section 8.3.

8.1 Time-Periodic Action

The behavior of liquids under the action of temporal parameter modulation is the subject of an extensive investigation during the last decades (see [2–4]). The linear stability problem in the case of a time-periodic parameter modulation is described by the system of equations of the kind

$$\sum_j \left(L_{ij} \frac{\partial}{\partial t} + M_{ij} + f_{ij}(t) N_{ij} \right) u_j(\mathbf{x}, t) = 0,$$

© Springer Science+Business Media, LLC 2017
S. Shklyaev, A. Nepomnyashchy, *Longwave Instabilities and Patterns in Fluids*,
Advances in Mathematical Fluid Mechanics,
https://doi.org/10.1007/978-1-4939-7590-7_8

where $u_j(\mathbf{x},t)$ is a multicomponent function characterizing the fields of physical variables, L_{ij}, M_{ij} and N_{ij} are linear operators that may contain differentiation over spatial variables, and f_{ij} are time-periodic functions, $f_{ij}(t+2\pi/\omega) = f_{ij}(t)$. The general solution is constructed as the superposition of Floquet–Bloch functions [5] of the type, $u_j(\mathbf{x},t) = \tilde{u}_j(\mathbf{x},t)\exp(\lambda t)$, where $\tilde{u}_j(\mathbf{x},t+2\pi/\omega) = \tilde{u}_j(\mathbf{x},t)$. The instability takes place when $\mathrm{Re}\lambda$ crosses 0. There are three generic case with $\mathrm{Re}\lambda = 0$ corresponding to different values of the multiplier $m = \exp(2\pi\lambda/\omega)$:
(i) $\mathrm{Im}\lambda = 0$, $m = 1$ (synchronous instability);
(ii) $\mathrm{Im}\lambda = \pm i\omega/2$, $m = -1$ (subharmonic instability);
(iii) $\mathrm{Im}\lambda \neq 0$, $\mathrm{Im}\lambda \neq \pm i\omega/2$, m is a complex number (quasiperiodic or higher resonance instability).

Two main types of parameter variations, namely, vibrations and modulated temperature/heat flux, are considered below. Depending on the relation between the characteristic time scales of a longwave instability and the period of the external action, the parameter modulations fall into three categories: *high-frequency*, *moderate-frequency*, and *low-frequency*. For each of these limits, some specific techniques are used. Note that the influence of vibrations on instabilities in liquids is a subject of a number of books [3, 4]. In this section, we briefly recall the basic works in that field and discuss recent works published after the publication of those books.

8.1.1 High-Frequency Parameter Modulation

When the characteristic time scale of the parameter modulation is high with respect to the characteristic time scale of the flow, the parameter modulation acts through averaged factors, which depend on the amplitude and frequency of the modulation.

8.1.1.1 Influence of High-Frequency Vibrations on the Deformational Marangoni Instability

As the first example of the influence of a parameter modulation on a longwave instability, let us consider the Marangoni convection in a horizontal liquid layer heated from below with the aspect ratio L (see Section 3.2.3). In the absence of vibrations, the rescaled strongly nonlinear longwave evolution equation for the surface deformation is governed by the equation (see (3.228))

$$\frac{\partial h}{\partial \tau} = -\nabla \cdot \left(\frac{MBih^2}{2(1+Bih)^2}\nabla h - \frac{Ga}{3}h^3\nabla h + \frac{C}{3}h^3\nabla\Delta h \right), \qquad (8.1)$$

where $C = \varepsilon^2 Ca = O(1)$, $\tau = \varepsilon^2 t$, $\nabla = \varepsilon^{-1}\nabla_\perp$ (the disjoining pressure, which is relevant only for ultrathin films, and the dependence of the surface temperature on the heat transfer are ignored in that equation). Under the action of high-frequency vertical vibrations with the nondimensional frequency $\Omega = \Omega_* d^2/v$ and amplitude

$B = b/d$ (Ω_* and b are the dimensional frequency and amplitude), the Marangoni instability competes with the Faraday instability. The threshold of the Marangoni instability is not influenced by vibrations [6], while the excitation of the Faraday instability becomes more difficult with the growth of frequency. Therefore, the Marangoni instability can be observed for sufficiently large Ω. The action of vibration provides an additional nonlinear term

$$\nabla \left[\frac{Vi}{12} h^3 \nabla (|\nabla h|^2) \right],$$

in the right-hand side of equation (8.1), where $Vi = B^2 \Omega^2 P / L^2$ [7], which does not break the variational structure of the problem.

Because the characteristic frequency of longwaves is $O(\varepsilon^2)$, where ε is the scale of the wavenumber, the frequency of order of 1 can be considered as a high frequency, and the averaging approach is applicable [8–10]. The corresponding modification of equation (3.228) in the case $\Omega = O(1)$, $B = B_1/\varepsilon$, $B_1 = O(1)$ has been obtained in [11]:

$$\frac{\partial h}{\partial \tau} = \nabla \cdot \left[\frac{h^3}{3} \nabla R - \frac{MBih^2 \nabla h}{2(1 + Bih)^2} - \frac{B_1^2 \Omega^2 P}{2} \mathbf{Q} \right], \tag{8.2}$$

$$R = Gah - C\nabla^2 h - \frac{B_1^2 \Omega^2 P}{2} \nabla (fh\nabla h). \tag{8.3}$$

Here $C = \varepsilon^2 Ca = O(1)$, $\nabla = \varepsilon^{-1} \nabla_\perp$, $\tau = \varepsilon^2 t$. Exact expressions for $\mathbf{Q}$ and f are cumbersome, and they are not presented here (see [11]). In the limit $\Omega \gg 1$, equations obtained in [7] are restored. In the limit $\Omega \ll 1$ (but still $\Omega \gg \varepsilon^2$), one obtains

$$\mathbf{Q} = \frac{2\Omega^2 h^6}{315} [h(\nabla^2 h\nabla h + 9\nabla h \cdot \nabla \nabla h) - 21(\nabla h)^2 \nabla h],$$

$$R = Gah - C\nabla^2 h - \frac{B_1^2 \Omega^4 P}{15} \nabla \cdot (h^5 \nabla h).$$

The analysis confirms the independence of the critical value of the Marangoni number on vibration parameters. Also, even in the case of vibrations, the bifurcation is subcritical, and it leads to the rupture of the film.

Let us mention also the works of [12, 13], where the longwave asymptotics of the neutral curve of the Marangoni instability was studied for a two-layer system.

8.1.1.2 Excitation of Deformational Marangoni Instability by a Heat Flux Modulation

The heat flux modulation has an advantage as a way for controlling the Marangoni instability with respect to vibration, because of the absence of the Faraday instability. Let us consider a horizontal layer with a free surface under the action of a prescribed heat flux on the bottom and the free surface (see Section 3.2.2). When the heat flux is constant ($Bi = 0$), according to (3.260), the critical Marangoni number of the monotonic instability is determined by formula

$$M_m = \frac{48}{1 + 72Ga^{-1}},$$

which describes a smooth transition between the deformational and non-deformational modes [14]. Assume now that the flux on the bottom is a periodic function of time, either (i) with a nonzero mean value, $M(t) = M_0(1 + \delta \cos \Omega t)$, or (ii) with a zero mean value, $M(t) = \Delta M \cos \Omega t$, while the flux on the free surface is equal to that mean value.

In the former case, one obtains the following longwave limit for the growth rate [15]:

$$\sigma = \frac{3k^2}{M_0 + 6} \left[M_0 \left(\frac{Ga}{72} + 1 \right) - \frac{2}{3}Ga + \delta^2 M_0^2 S(\Omega, P) \right], \tag{8.4}$$

where $S(\Omega, P)$ can be negative or positive, depending on ω and P. If $S(\Omega, P) < 0$, the heat modulation suppresses the instability, so that the instability interval of M_0 is bounded from both below and above, and it disappears for sufficiently large δ. If $S(\Omega, P) > 0$, by heating from below ($M_0 > 0$), the instability threshold decreases with M_0. Also, the quiescent state becomes unstable by heating from above ($M_0 < 0$), if M_0 is sufficiently large.

In the case of zero mean flux [15], formula (8.4) is reduced to

$$\sigma = \frac{k^2}{2} \left(-\frac{2}{3}Ga + \Delta M^2 S(\Omega, P) \right), \tag{8.5}$$

hence

$$\Delta M = \pm \sqrt{\frac{2Ga}{3S(\Omega, P)}}. \tag{8.6}$$

The instability takes place only in the intervals of Ω, where $S(\Omega, P) > 0$. For $P = 7$, minimum of $[S(\Omega, P)]^{-1/2}$ is equal to 105.26 at $\Omega = 11.29$. It turns out however that typically the instability threshold for a finite wavelength mode is lower than that of the longwave mode.

A more efficient way of the modulational excitation of the Marangoni convection, when the flux modulation is applied to the free surface, has been considered in [16]. Formula (8.6) is valid, but the expression for $S(\Omega, P)$ is different. For $P = 7$, minimum of $[S(\Omega, P)]^{-1/2}$ is equal to 14.01 at $\Omega = 2.65$, i.e., it is significantly lower than in the case of the bottom flux modulation. In a certain interval of frequency, the longwave instability wins the competition with the subharmonic shortwave instability mode [16, 17].

The nonlinear theory of the instability discussed above has been developed in [18]. Similarly to the case discussed in Section 3.2.2, the critical Marangoni number is a non-monotonic function of the layer thickness. Therefore, a subcritical instability takes place everywhere except the point of the minimum, where the Cahn–Hilliard equation (3.167) is valid. The strongly nonlinear theory gives a pair of amplitude equations, where the equation for the film thickness has the standard form (8.1), while the second equation, which describes the evolution of the temperature averaged over the vertical coordinate and time, is rather cumbersome and not presented here.

8.1.1.3 Excitation of Surfactant-Induced Marangoni Waves by a Heat Flux Modulation

Let us consider now a liquid layer heated from below with an insoluble surfactant adsorbed at the undeformable free surface; the Biot number $Bi = \beta \varepsilon^4$ is small (see Section 4.2.4.6). That system is characterized by the presence of two modes, monotonic and oscillatory ones. Assume that the heat flux at the bottom oscillates; in nondimensional variables, $\partial T / \partial z = -1 + \delta \cos \Omega t$ [19].

The monotonic neutral curve $M(k) = M_m + \varepsilon^2 M_2(K), k = \varepsilon K$, is slightly distorted by the heat flux modulation. The leading-order term (4.392) is unchanged, but the next-order correction is modified,

$$M_2(K) = \frac{60L(4L+N)\beta + 2K^4[8L^2 + 45(4L+N)^2\delta^2 S(\Omega,P)]}{5L^2K^2},$$

where $S(\Omega,P) = A/B$,

$$A = [9\alpha\cosh\alpha + (\alpha^4 - 3\alpha^2 - 9)\sinh\alpha][-\mu\sinh\alpha^* + \mu\cosh\mu(\sinh\alpha^* - \alpha^*) +$$

$$\alpha^*\cosh\alpha^*(\mu - \sinh\mu) + \alpha^*\sinh\mu],$$

$$B = \alpha^{*2}\alpha^5(\alpha^{*2} - \mu^2)\sinh\alpha^*\sinh^2\alpha(\mu\cosh\mu - \sinh\mu),$$

$\alpha = \sqrt{i\Omega}, \mu = \alpha/\sqrt{P}$.

The oscillatory neutral curve is changed in a similar way: M_o is unchanged, while the expression for $M_2(K)$ obtained in [20] gets an addition,

$$\tilde{M}_2(K) = 36K^2(4 + 4L + N)^2\delta^2 S(\Omega,P).$$

8.1.1.4 Longwave Vibrational Instability of Binary Mixtures

Let us consider a channel with rigid boundaries filled by a binary liquid (see Section 4.1). Gershuni et al. [21] applied the high-frequency approach formerly developed for studying the thermovibrational convection under microgravity conditions [22] to the buoyancy convection in a binary liquid. It was found that even in the case of fixed temperature at the boundaries, when only a cellular ($k_c \neq 0$) convection is possible in a pure liquid, there exists a monotonic longwave instability mode, which is of the solutocapillary nature (the leading-order disturbance is the disturbance of concentration). For negative χ, an oscillatory instability is possible by heating from below, and a monotonic instability can appear by heating from above.

In [23], the influence of a vertical vibration on a convective flow of a binary mixture in a vertical channel heated from the side. A longwave instability has been revealed in some intervals of the Soret coefficient values. The case of an inclined layer was considered in [24].

8.1.2 Moderate- and Low-Frequency Parameter Modulation

If the parameter modulation has the same characteristic time scale as the motion in the liquid, then that modulation enters into the amplitude equation directly, rather than through averaged quantities. Below we present some examples.

8.1.2.1 Deformational Marangoni Instability

Let us start with the problem of the monotonic deformational Marangoni instability discussed in Section 8.1.1.1, but this time we assume that the frequency of vibrations is $\Omega = \varepsilon^2 \Omega_2$, $\Omega_2 = O(1)$. To get a finite acceleration, one has to take a large amplitude of vibrations, $B = B_4 \varepsilon^{-4}$, $B_4 = O(1)$. In equation (8.1), one has to replace Ga by $Ga(\tau) = Ga_0 + B_4 \Omega_2^2 \cos \Omega_2 \tau$. Because the modified equation (8.1) is of the zeroth order in t, the influence of the vibration on the temporal evolution of a disturbance $\exp(iKX + \phi(\tau))$ on the background of the base state $h_0 = 1$ is rather trivial [11]:

$$\phi(\tau) = \sigma_0(K)\tau - K^3 \frac{B_4 \Omega_2}{3} \sin \Omega_2 \tau,$$

where $\sigma_0(K)$ is the neutral curve $M = M_0(K)$ corresponding to $Ga = Ga_0$.

However, one has to take into account that for low-frequency vibrations, the standard stability analysis may be insufficient in the presence of even a small noise [25–27]. By slow modulation of the parameter, the system spends some time in the subcritical region of parameter and some time in the supercritical region. During the subcritical time interval, the exponential decay of the disturbance stops as soon as the perturbation amplitude becomes comparable with the noise level, which makes the predictions of the stability analysis irrelevant. The noise is strongly amplified during the supercritical time interval, which can be considered as an "instability" from the practical point of view. Homsy [26] has formulated the *strong global stability* criterion: the system is treated as an "unstable" one if the perturbation grows during a certain interval of time. The application of that criterion gives a lower value of the critical Marangoni number $\tilde{M}_0$ which is obtained from M_0 by the replacement of G_0 with $G_0 - B_4 \Omega_2^2$. In the interval of M between $\tilde{M}_0$ and M_0, the system is sensitive to the noise level. Numerical simulations reveal some nearly periodic oscillations in that parameter region [11]. It is interesting that vibrations prevent the rupture of the film.

In the intermediate region of frequencies, $\Omega = \varepsilon^\beta \Omega_\beta$, $0 < \beta < 2$, there exists a nontrivial distinguished limit [11] characterized by the scales

$$B = B_\beta \varepsilon^{1+3\beta/2}, \; t_\beta = \varepsilon^\beta t, \; \xi = \varepsilon^\alpha x.$$

where $1/2 < \alpha = (2+\beta)/4 < 1$. Corresponding leading-order amplitude equation includes only effects of the surface tension and vibrations:

$$\frac{\partial h}{\partial t_\beta} = \frac{1}{3}\frac{\partial}{\partial \xi}\left[h^3 \frac{\partial}{\partial \xi}\left(B_\beta \Omega_\beta^2 \cos(\Omega_\beta t_\beta)h - C\frac{\partial^2 h}{\partial \xi^2}\right)\right].$$

8.1.2.2 Binary Mixture with a Non-deformable Surface Under the Action of Vibrations

In systems with two slow longwave modes, where an oscillatory instability is possible in the absence of vibrations, the dynamics is much more nontrivial. Let us consider a binary liquid layer with a non-deformable free surface at a low Bi number. Recall (see Section 4.2.1) that an oscillatory instability with the nondimensional frequency $\Omega = \tilde{\Omega}k^2$,

$$\tilde{\Omega}^2 = -\frac{\chi(1+L+L^2)+L^2}{1+\chi}, \tag{8.7}$$

is developed in the region

$$-1 < \chi < \chi_0, \; \chi_0 = -\frac{L^2}{1+L+L^2}$$

at the stability boundary

$$m_0 = \frac{1+L}{1+\chi}, \; m_0 = M_0/48.$$

Assume now that the layer is subject to vertical vibrations with the amplitude b and *low* frequency $\hat{\Omega}$. First, take $Bi = 0$ and choose the nondimensional frequency $\hat{\Omega}d/\kappa$ as q small parameter, $\hat{\omega} \equiv \varepsilon^2$. The leading-order dynamics of a disturbance with the wavenumber $k = \varepsilon K$ is governed by the following equation for the combination of disturbances of temperature T_0 and C_0 [28]:

$$\frac{d^2\rho}{d\tau^2} - \alpha\frac{d\rho}{d\tau} + \phi_\chi K^2 B\frac{d}{d\tau}(\rho\cos\tau) - \rho(\Theta + \Phi B\cos\tau) = 0, \tag{8.8}$$

where

$$\rho = T_0 + \phi C_0, \; \phi = \beta_C \gamma_T/\beta_T \gamma_C, \; \tau = \varepsilon^2,$$
$$\alpha = K^2[m(\chi-1)-L-1], \; \Theta = K^4[m(\chi+L+\chi L)-L], \; \Phi = -K^4[(\phi\chi-1)(L+1)+1],$$

$$B = b\hat{\Omega}^2 d^4\beta_T/320\nu\kappa, \; m = M/48.$$

If $\Theta < 0$, equation (8.8) is similar to the well-known Mathieu equation with damping. According to the Floquet theory, solution of (8.8) can be presented in the form

$$\rho(\tau) = e^{\Lambda \tau} R_0(\tau),$$

where $R_0(\tau)$ is a 2π-periodic function and $\Lambda = \Lambda_r + i\Lambda_i$ is the complex Floquet exponent. The stability boundaries $\mathrm{Re}\Lambda_r = 0$ are of three types: (i) subharmonic instability, $\Lambda_i = 1/2$; (ii) synchronous (harmonic) instability, $\Lambda_i = 0$; (iii) "quasiperiodic" (nonresonant) instability, $0 < |\Lambda_i| < 1/2$ (strictly speaking, only irrational values of $|\Lambda_i|$ create truly quasiperiodic solutions).

The analysis of the parametric excitation of Marangoni oscillations by vibrations has been carried out by expansions at small B and by a direct numerical solution of equation (8.8). For the subharmonic mode, the expansion has the form

$$\rho = \rho_0 + B\rho_1 + \ldots, \quad m = m_0 + Bm_1 + \ldots$$

The mode is excited in the interval of wavenumbers $O(B)$ around the wavenumber $K_{1/2}$ satisfying the resonant relation

$$\omega(K_{1/2}) = K_{1/2}^2 \tilde{\Omega} = \frac{1}{2}$$

(see Figure 8.1). In Figure 8.1,

$$\delta_1 B = 2\tilde{\Omega}\left(\frac{K}{K_{1/2}} - 1\right), \quad \lambda_k = \frac{\lambda}{K_{1/2}}.$$

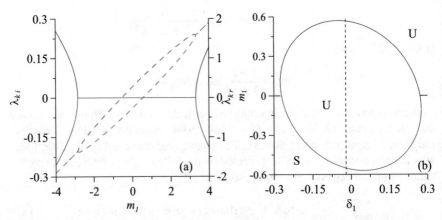

Fig. 8.1 Results on the linear stability for $\gamma(K) \approx 1/2$; $\chi = -0.05$ ($K_* = 1.47$). Panel (a): $\lambda_{kr}(m_1)$ and $\lambda_{ki}(m_1)$ at $\delta_1 = -0.0207$ [vertical dashed line in panel (b)]. Solid and dashed lines correspond to the imaginary and real part of λ_k respectively. Panel (b): $m_{1*}(\delta_1)$ at $\lambda_{kr} = 0$. The capitals "S" and "U" correspond to stability and instability areas

In Figure 8.2, the results obtained by analytical formulas and numerical calculations by the Floquet theory are compared for both synchronous and subharmonic modes.

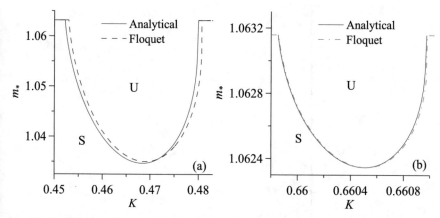

Fig. 8.2 Results on the linear stability for $\gamma(K) \approx 1/2$; $B = 0.05$, $\chi = -0.05$ ($K_* = 1.47$). Panel (a): Subharmonic instability zone (above the curves). Panel (b): Synchronous instability zone (above the curve). The capitals "S" and "U" correspond to stability and instability areas

For a synchronous mode, the excitation takes place around $K_1 = 1/\sqrt{\widetilde{\Omega}}$ in the interval of wavenumbers $O(B^2)$.

Typical pictures of instability domains are shown in Figures 8.3 and 8.4.

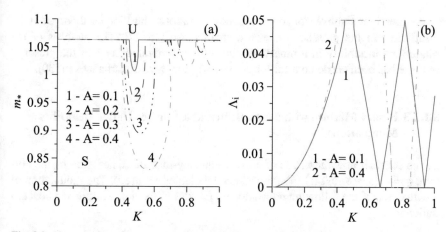

Fig. 8.3 Numerical results at $\chi = -0.05$. Panel (a): Neutral stability curves $m_*(K)$ for different values of the vibration amplitude B. Panel (b): Imaginary part of the Floquet exponent, $\Lambda_i(K)$. Solid and dashed-dotted lines correspond to $B = 0.1$ and $B = 0.4$, respectively. Solid, dashed, and dashed-dotted lines correspond to synchronous, subharmonic, and quasiperiodic modes, respectively

A generalization of the theory for nonzero Bi has been done in [29]. It was shown that the increase of Bi stabilizes the system.

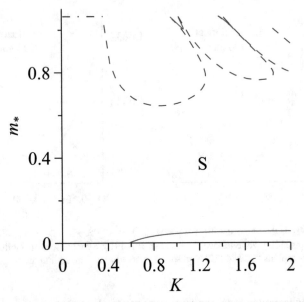

Fig. 8.4 Numerical results at $\chi = -0.05$. Marginal stability curves at $B = 0.6$. Solid, dashed, and dashed-dotted lines correspond to synchronous, subharmonic, and quasiperiodic modes, respectively; the domain of stability is marked with "S"

As mentioned above, for low-frequency vibrations, the Floquet theory may be insufficient in the presence of even a small noise [25–27]. The analysis of the Marangoni instability in a binary mixture filling a vibrated layer in the presence of noise has been carried out for longwave and shortwave disturbances in [30].

8.1.2.3 Binary Mixture with a Non-deformable Surface Under Heat Flux Modulations

Let us consider now the action of a low-frequency-modulated heat flux on the bottom, $q = q_0(1 + B\cos\Omega t)$ (nondimensional variables are used). The modulation of the flux creates the following nonstationary profiles of the temperature and concentration:

$$T_0 = -z + B\frac{\sin\Omega t}{\Omega} + B\cos\Omega t\frac{2 - 6z + 3z^2}{6} + O(\Omega), \qquad (8.9)$$

$$C_0 = \chi z - \chi B\cos\Omega t\frac{2 - 6z + 3z^2}{6} + O(\Omega). \qquad (8.10)$$

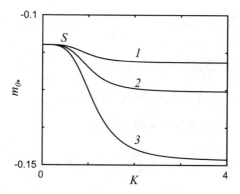

Fig. 8.5 Neutral stability curves $m_{0*}(K)$ for (8.11) and (8.12) for $\chi = -0.1$ and $L = 0.01$. Curves 1, 2, and 3 correspond to $B = 2.5$, 3.75, and 5, respectively. The domain of stability is marked by "S"

For $Bi = 0$ and $\varepsilon = \sqrt{\Omega}$, one obtains the following system of the leading-order equations [31, 32]:

$$\mathcal{L}_1(F,G) = F_\tau - \nabla^2 F + \tilde{m}_0 \nabla^2 h = 0, \tag{8.11}$$

$$\mathcal{L}_2(F,G) = G_\tau - L\nabla^2(G + \chi F) - \chi \tilde{m}_0 \nabla^2 h = 0, \tag{8.12}$$

where $\tilde{m}_0 = m_0\left(1 + \frac{2}{5}B\cos\tau\right)$ is the effective rescaled Marangoni number. Linear equations (8.11) and (8.12) were studied in [31], where the neutral stability curves $m_{0*}(K)$ were obtained. Three modes of longwave perturbations were found: subharmonic ($\Lambda = \pi$ at the stability boundary) with finite K_c for critical perturbations, synchronous ($\Lambda = 0$) with $K_c = 0$, and quasiperiodic ($0 < \Lambda < \pi$), which is not critical. An example of computations for the synchronous mode is shown in Figure 8.5; the synchronous perturbations are critical at negative χ and for heating from above ($m_{0*} < 0$); the critical perturbations correspond to $K = 0$.

At the fourth order in ε, the solvability conditions have the following form:

$$\mathcal{L}_1(Q,R) = -\dot{F} - \frac{m_0}{10}\left(1 + \frac{44}{105}B\cos\tau\right)\nabla^2\left(\nabla^2 h - \frac{1}{2P}h_\tau\right) - \frac{m_0}{21}B\sin\tau\nabla^2 h$$

$$-\left(1 + \frac{2}{5}B\cos\tau\right)\nabla^2\left(\Gamma + \frac{m_0^2}{10P}(\nabla h)^2\right) - \frac{m_0^2}{10P}\left(2 + P + \frac{32 - 6P}{35}B\cos\tau\right)\nabla\cdot(\nabla^2 h\nabla h)$$

$$+ \frac{48m_0^2}{35}\nabla\cdot(|\nabla h|^2\nabla h) - \frac{312}{35P}\mathbf{J}\cdot\nabla F, \tag{8.13}$$

$$\mathcal{L}_2(Q,R) = -\dot{G} + \frac{\chi m_0}{10}\left(1 + \frac{44}{105}B\cos\tau\right)\nabla^2\left(\nabla^2 h - \frac{1}{2P}h_\tau\right)$$

$$+ \frac{\chi m_0}{21}(1 + L^{-1})B\sin\tau\nabla^2 h + \chi\left(1 + \frac{2}{5}b\cos\tau\right)\nabla^2\left(\Gamma + \frac{m_0^2}{10P}(\nabla h)^2\right)$$

$$+\frac{\chi m_0^2}{10P}\left(2+P+S+\frac{32-6(P+S)}{35}b\cos\tau\right)\nabla\cdot(\nabla^2 h\nabla h)$$

$$-\frac{48 m_0^2}{35}\nabla\cdot((\Phi-\nabla h\cdot\nabla F)\nabla h)-\frac{312}{35P}\mathbf{J}\cdot\nabla G, \qquad (8.14)$$

$$\mathbf{J}=\nabla\nabla^{-2}\nabla\cdot(\nabla^2 h\nabla h)-\nabla^2 h\nabla h, \qquad (8.15)$$

where the dots denote derivatives with respect to t_4

$$\Gamma=m_0(Q-R)+m_2 h-\frac{m_0^2}{30}\frac{\chi_{\mathrm{L}}}{L}\left(1-\frac{2}{7}B\cos\tau\right)\nabla^2 h+\frac{3m_0^2}{5}\Phi,$$

$$\Phi=\nabla((1+\chi)F-L^{-1}G)\cdot\nabla h.$$

A number of limiting cases have been studied in [32].

8.1.2.4 System with Two Deformable Boundaries Under the Action of Vibrations or Heating Modulation

Let us consider the two-layer system described by system of equations (3.313). In the case of vibration, the gravity acceleration g in formulas (3.307), (3.308) is replaced with $g(t)=b\Omega^2$, where b and Ω are the amplitude and the frequency of vibration, correspondingly. The simulations carried out in [33] revealed an excitation of subharmonic waves in the interval of Ω around $2\omega_0$, where ω_0 is the typical frequency of Marangoni oscillations. Also, one have found that waves are excited in the interval of Ω around $4\omega_0$, which is a genuine nonlinear effect, not described by the Floquet stability theory.

In [34], the influence of heating modulation on the parametric excitation of waves was studied. The temperature of the gas T_g was constant, while the temperature of the solid substrate changed periodically in time, $T_s(t)=T_s^0+\Theta\sin\Omega t$. Excitation of different planforms of subharmonic modes was observed.

8.2 Spatial Modulation of Parameter

Another possible way of a controlled pattern selection is a spatial modulation of the control parameter. First it was applied by Chen and Whitehead in 1968 [35] in order to create convective roll pattern with a prescribed wavelength. Some examples of spatial modulation of parameters were presented in Section 2.1, where we considered the influence of a spatial parameter inhomogeneities on the dynamics of a system governed by the Allen–Cahn equation and subjected to coarsening. The impact of spatial modulation of parameters on pattern-forming systems can be even more significant.

The action of a resonant spatially periodic forcing on the onset of stationary patterns created by a shortwave instability has been explored rather well [36–44]. The influence of resonant spatial parameter modulation on the shortwave oscillatory instabilities was studied in [45–47].

In the present section, we consider the influence of a *longwave* spatial parameter modulation on the formation of wave patterns in the framework of the CGLE with parametric forcing.

As explained in Chapter 1, the complex Ginzburg–Landau equation can be applied for a description of two types of oscillatory instabilities: the *longwave* instability developing when the neutral curve of the control parameter $R = R(k)$ has its minimum $R = R_c$ at wavenumber $k = 0$ and the *shortwave* instability developing when the neutral curve $R = R(k)$ has its minimum at a finite wavenumber $k = k_c$. The longwave oscillatory instability occurs, for instance, in a reaction–diffusion system when a pure reaction system described by ordinary differential equations (ODE) is subject to Hopf bifurcation at a certain value of the control parameter $R = R_c$. Assuming that the control parameter R is a slow function of space coordinate x,

$$R = R_c + \varepsilon^2 R_2(x_1), \ x_1 = \varepsilon x, \ \varepsilon \ll 1,$$

employing the eigenvectors of the ODE problem at $R = R_c$ with a slowly changing amplitude function $A(x_1, t_2)$, $t_2 = \varepsilon^2 t$, and using the multiscale approach, one can reduce the problem (to the leading order) to a complex Ginzburg–Landau equation

$$\frac{\partial A}{\partial t_2} = \gamma(x_1)A + D\frac{\partial^2 A}{\partial x_1^2} - \kappa |A|^2 A. \tag{8.16}$$

Here $\gamma(x_1) = R_2(x_1)(d\sigma/dR)_{R_c}$ is the spatially modulated growth rate coefficient, whereas complex coefficients D and κ are equal to those in the homogeneous case with $R = R_c$.

The second type of instabilities governed by complex Ginzburg–Landau equation is the shortwave oscillatory instability developing when the minimum of the neutral curve $R = R(k)$ is situated at a finite wavenumber $k = k_c \neq 0$ [48, 49]. For the shortwave instability, the amplitude function A has the meaning of an envelope function of a wave packet with a base wavenumber k_c. If the group velocity of waves at the critical point $R = R_c$, $k = k_c$ is small because of some physical reasons, $v = \varepsilon v_1$, the leading-order evolution of the wave packet takes place on the time scale $t_2 = \varepsilon^2 t$ and is governed by the equation

$$\frac{\partial A}{\partial t_2} + v_1\frac{\partial A}{\partial x_1} = \gamma(x_1)A + D\frac{\partial^2 A}{\partial x_1^2} - \kappa |A|^2 A, \tag{8.17}$$

where $\gamma(x_1) = \gamma_r(x_1) + i\gamma_i(x_1)$ describes the spatially inhomogeneous growth rate and wave frequency, $\kappa = \kappa_r + i\kappa_i$ determines the nonlinear saturation of the instability and nonlinear frequency shift ($\kappa_r > 0$), and $D = D_r + iD_i$ characterizes the diffusion and dispersion of waves ($D_r \geq 0$).

Equation (8.16) can be considered as a particular case of equation (8.17). Therefore in the foregoing, we do not distinguish between these two physically distinct cases. By rescaling and redefining variables, equation (8.17) can be written as

$$\frac{\partial A}{\partial t} + v\frac{\partial A}{\partial x} = [f(x) + ig(x)]A + (1+i\alpha)A_{xx} - (1+i\beta)|A|^2A. \tag{8.18}$$

Here $f(x)$ and $g(x)$ characterize the influence of external spatial inhomogeneity on the growth rate and oscillation frequency, respectively, and coefficients α and β, that describe the dispersion and nonlinear frequency shift, are standard for the Ginzburg–Landau equation.

8.2.1 Localized Gain

Assume that the inhomogeneity of the complex growth rate $\gamma(x)$ has a small spatial scale with respect to that characteristic for the Ginzburg–Landau equation, thus assuming

$$f(x) + ig(x) = -\lambda + \Gamma\delta(x), \tag{8.19}$$

where λ is a real constant (its imaginary part λ_i can be eliminated by the transformation $u \to u\exp(-i\lambda_i t))$[50].

8.2.1.1 Real Amplitude Equation

We will start our analysis with the simplest case of a nonlinear advection–diffusion equation with a *real* variable $u(x,t)$

$$u_t + vu_x = -\lambda u + \Gamma\delta(x)u + u_{xx} + F(u), \tag{8.20}$$

where $F(u)$ is a function satisfying conditions $F(0) = F'(0) = 0$, e.g., $F(u) = -u^2$ or $F(u) = -u^3$. The physical meaning of the variable $u(x,t)$ is a disturbance of a concentration on the background of a certain base state described by the solution $u = 0$. Without loss of generality, below we assume $v \geq 0$.

First, let us consider the evolution of infinitesimal disturbances $\tilde{u}(x)$ on the background of the base state $u = 0$. Taking $\tilde{u}(x) = U(x)\exp(\sigma t)$, we obtain the following equation governing the shape and the growth rate of the disturbance:

$$\sigma U + vU' = [-\lambda + \Gamma\delta(x)]U + U''. \tag{8.21}$$

It is natural to impose the physical condition of the boundness of the solution at the infinity,

$$\lim_{x\to\pm\infty} |U(x)| < \infty. \tag{8.22}$$

The eigenvalue problem (8.21)–(8.22) has two kinds of solutions. The eigenfunctions that do not decay at infinity have the form:

$$U = U_-(x) = ae^{ikx} + be^{v-ik}x, \; x < 0$$

$$U = U_+(x) = e^{ikx}, \; x > 0,$$

where

$$b = \frac{\Gamma}{v - 2ik}, \; a = 1 - b.$$

The corresponding continuous spectrum of eigenvalues

$$\sigma(k) = -\lambda - ivk - k^2 \tag{8.23}$$

is not affected by the localized inhomogeneity. However, a sufficiently strong inhomogeneity can create a *localized* eigenmode decaying at infinity, which has an eigenvalue depending on Γ. Consider solutions in the form

$$U = e^{k_- x}, \; x < 0; \; U = e^{k_+ x}, \; x > 0, \tag{8.24}$$

where $k_\pm$ satisfy the following conditions:

$$\sigma + vk_\pm = -\lambda + k_\pm^2, \tag{8.25}$$

$$\Gamma + d(k_+ - k_-) = 0. \tag{8.26}$$

A localized mode exists if the real parts $\text{Re}(k_-) > 0$, $\text{Re}(k_+) < 0$. From (8.25), one finds that

$$k_+ - k_- = -\sqrt{v^2 + 4d(\sigma + \lambda)}\text{sign}\Gamma, \tag{8.27}$$

hence from (8.26), one concludes that

$$\sqrt{v^2 + 4(\sigma + \lambda)} = |\Gamma|, \tag{8.28}$$

thus

$$\sigma = -\lambda + \frac{\Gamma^2 - v^2}{4}, \tag{8.29}$$

$$k_\pm = \frac{1}{2}(v \mp \Gamma). \tag{8.30}$$

Note that $k_\pm$ are real; hence, a localized mode exists if $k_- > 0$, $k_+ > 0$. Obviously, that is possible in the case $\Gamma > v$. In a contradistinction to the continuous spectrum (8.23), the eigenvalue of a localized mode (8.29) is always real, and it is higher than the real part of any non-localized disturbances.

In the case $\lambda > 0$ (where the background is stable), the localized mode decays in time if $v < \Gamma < \sqrt{v^2 + 4\lambda}$ and grows if $\Gamma > \sqrt{v^2 + 4\lambda}$. In the case $\lambda < 0$ (where the background is unstable), the localized mode grows in the whole region of its existence (faster than any non-localized disturbances).

Thus, in the case of a real amplitude equation, a nonzero group velocity breaks the reflection symmetry of the normal disturbance ($k_+ \neq -k_-$, see (8.30)) and increases the instability threshold, $\Gamma_c = \sqrt{v^2 + 4\lambda}$.

Let us consider now nonlinear localized structures created by the instability.

The Case $v = 0$

In the case $v = 0$, the stationary localized structure created by the monotonic instability of the base state with respect to the localized mode is governed by the equation

$$- \lambda u + \Gamma \delta(x) u + u'' + F(u) = 0 \tag{8.31}$$

with boundary conditions

$$u(\pm\infty) = 0. \tag{8.32}$$

Because of the symmetry of the solution, $u(-x) = u(x)$, it is sufficient to solve the equation

$$- \lambda u + u'' + F(u) = 0 \tag{8.33}$$

in the region $0 < x < \infty$ with boundary conditions

$$\Gamma u(0) + 2u'(0) = 0, \ u(\infty) = 0. \tag{8.34}$$

Recall that $f(0) = f'(0) = 0$. First of all, in the case $\lambda < 0$ (unstable background), equation (8.33) has no nontrivial solutions decaying at the infinity; thus, the boundary value problem (8.33), (8.34) has no nontrivial solutions. Later on, we consider the case $\lambda \geq 0$.

The calculation of the solutions to (8.33), (8.34) is tedious but straightforward. Below we present results of the calculation in the cases $F(u) = -u^2$ and $F(u) = -u^3$.

In the case of a quadratic nonlinearity, $F(u) = -u^2$, the stationary solution in the region $x > 0$ for $\lambda > 0$ is

$$u(x) = \frac{3}{2} \lambda \frac{\Gamma^2 - 4\lambda}{\left[\Gamma \sinh\left(\sqrt{\lambda/2}x \right) + 2\sqrt{\gamma}\cosh\left(\sqrt{\lambda/2}x \right) \right]^2}. \tag{8.35}$$

In the region $x < 0$, $u(x) = u(-x)$. At the critical value of Γ, $\Gamma_c = 2\sqrt{\lambda}$, a two-sided bifurcation takes place. The nontrivial solution in $u(x)$ is positive and stable for $\Gamma > \Gamma_c$, and it is negative and unstable for $\Gamma < \Gamma_c$.

It is interesting that a localized solution exists also in the case of a neutrally stable background, $\lambda = 0$. In that case, the solution in the region $x > 0$ is

$$u(x) = \frac{6}{(x + 4/\Gamma)^2},$$

thus the decay at infinity is algebraic rather than exponential.

In the case of a cubic nonlinearity, the solution is

$$u(x) = \pm \frac{\sqrt{2\lambda(\Gamma^2 - 4\lambda)}}{\Gamma\sinh(\sqrt{\lambda}x) + 2\sqrt{\lambda}\cosh(\sqrt{\lambda}x)}, \quad x > 0. \tag{8.36}$$

A supercritical pitchfork bifurcation takes place. Note that the solution (8.36) belongs to the family of localized solutions obtained in [51]. For $\lambda = 0$, the localized solution is

$$u(x) = \frac{\sqrt{2}}{|x| + 2/\Gamma}$$

(see [52]).

The Case $v \neq 0$

In the case $v \neq 0$, the stationary localized structure created by the localized mode is governed by the boundary value problem

$$-\lambda u - vu' + \Gamma\delta(x)u + u'' + F(u) = 0 \tag{8.37}$$

with boundary conditions

$$u(\pm\infty) = 0. \tag{8.38}$$

No solutions are possible in the case $\lambda < 0$, because in that case, the point $u = 0$ is a repeller in the framework of equation (8.37) (recall that $v > 0$). In the case $\lambda > 0$, the solution is possible but its analytical computation is a formidable task. Here we present only the result of the bifurcation analysis in the case $F(u) = -u^3$.
  In the vicinity of the critical value $\Gamma_c = \sqrt{v^2 + 4\lambda}$,

$$\Gamma = \Gamma_c + \varepsilon^2\gamma, \quad \varepsilon^2 \ll 1,$$

one can construct an asymptotic expansion in the form

$$u(x) = \varepsilon u_1(x) + \varepsilon^3 u_3(x) + \dots,$$

which describes the bifurcation of a nontrivial localized solution of equation (8.37) from the trivial solution $u(x) = 0$. The leading-order equation is linear, and its solution can be found analytically:

$$u_1(x) = Ae^{k_- x}, \ x < 0; \ u(x) = Ae^{k_+ x}, \ x > 0, \tag{8.39}$$

where

$$k_\pm = \frac{v \mp \sqrt{v^2 + 4\lambda}}{2}. \tag{8.40}$$

The solvability condition of the equation for $u_3(x)$ obtained in order $O(\varepsilon^3)$ determines the amplitude A:

$$\left(\frac{1}{4k_- - v} - \frac{1}{4k_+ - v}\right) A^2 = \gamma. \tag{8.41}$$

Because

$$4k_- - v = v + 2\sqrt{v^2 + 4\lambda} > 0, \ 4k_+ - v = v - 2\sqrt{v^2 + 4\lambda} < 0,$$

the solution exists only for $\gamma \geq 0$, i.e., there is a supercritical pitchfork bifurcation of two branches of localized solutions (with positive and negative A) from the trivial solution $u = 0$. In the case $v = 0$,

$$A^2 = 2\sqrt{\lambda}\gamma, \tag{8.42}$$

which is compatible with (8.36).

8.2.1.2 Complex Ginzburg–Landau Equation

The results of the previous section can be easily extended to the case of the Ginzburg–Landau equation with the real coefficients,

$$u_t + v u_x = -\lambda u + \Gamma \delta(x) u + u_{xx} - |u|^2 u. \tag{8.43}$$

In the present section, we consider the full complex Ginzburg–Landau equation,

$$u_t + v u_x = [-\lambda + (\Gamma_r + i\Gamma_i)\delta(x)]u + (1 + i\alpha)u_{xx} - (1 + i\beta)|u|^2 u. \tag{8.44}$$

Linear Stability Theory

Let us linearize equation (8.44) and search the solution in the form $u(x,t) = U(x)\exp(\sigma t)$. We find:

$$\sigma U + v U' = [-\lambda + (\Gamma_r + i\Gamma_i)\delta(x)]U + (1 + i\alpha)U''. \tag{8.45}$$

In the case $v = 0$, this problem has been solved in [51]. Using ansatz (8.24) for localized solutions of that equation, we find the following expression for the growth rate of disturbances:

$$\sigma = -\lambda + \frac{(\Gamma_r + i\Gamma_i)^2 - v^2}{4(1 + i\alpha)}, \tag{8.46}$$

(cf. (8.29)). The threshold value of $\Gamma_{r,c}$ is determined by the relation

$$\mathrm{Re}(\sigma) = \frac{(\Gamma_{r,c}^2 - \Gamma_i^2 - v^2) + 2\alpha \Gamma_{r,c}\Gamma_i}{4(1 + \alpha^2)} - \lambda = 0. \tag{8.47}$$

Note that the obtained expression resembles that of the *absolute* instability threshold in the absence of the inhomogeneity,

$$\lambda = -\frac{v^2}{4(1+\alpha^2)}.$$

The instability boundary is a hyperbola in the plane (Γ_c, Γ_i) with the asymptotes

$$\Gamma_i/\Gamma_r = \alpha \pm \sqrt{\alpha^2 + 1}.$$

The explicit formula for the threshold value is

$$\Gamma_{r,c} = -\alpha\Gamma_i + \sqrt{(4\lambda + \Gamma_i^2)(1+\alpha^2) + v^2}. \tag{8.48}$$

The values of $k_\pm$ in the expression (8.24) are:

$$k_\pm = \frac{v \mp \sqrt{v^2 + 4(1+i\alpha)\lambda}}{2(1+i\alpha)}. \tag{8.49}$$

Note that conditions of the boundness of the eigenfunctions on the infinity, $\mathrm{Re}(k_-) > 0$, $\mathrm{Re}(k_+) < 0$, are satisfied. Indeed,

$$\mathrm{Re}(k_\pm) = \frac{v \mp \Gamma_r \mp \Gamma_i \alpha}{2(1+\alpha^2)}. \tag{8.50}$$

Substituting the critical value of Γ_r, (8.48), we find that at $\Gamma_r = \Gamma_{r,c}$,

$$\mathrm{Re}(k_\pm) = v \mp \sqrt{v^2 + (1+\alpha^2)(\Gamma_i^2 + 4\lambda)},$$

hence the boundness conditions are satisfied. Obviously, they are satisfied for $\Gamma_r > \Gamma_{r,c}$, because $\mathrm{Re}(k_+)$ $(\mathrm{Re}(k_-))$ decreases (grows) with the growth of Γ_r.

In the limit $\alpha \gg 1$, the instability condition becomes $\Gamma_r\Gamma_i > 2\lambda\alpha$ (see [52]). Note that it does not depend on v. However, the velocity v influences the critical value of frequency,

$$\omega_0 = -Im\sigma = \frac{\Gamma_r^2 - \Gamma_i^2 - v^2}{4\alpha} = \frac{4\alpha^2\lambda^2 - v^2\Gamma_i^2 - \Gamma_i^4}{4\alpha\Gamma_i^2}. \tag{8.51}$$

Thus, in the case of a complex amplitude equation, the nonzero group velocity breaks the reflection symmetry of solutions ($k_+ \neq -k_-$; see equation (8.49)), typically enhances the instability threshold (see equation (8.48)), and influences the frequency of the oscillatory instability (see equation (8.51)).

Nonlinear Localized Structures

Let us consider the bifurcation of solutions in the framework of equation (8.44). In the vicinity of the threshold value of Γ_r,

$$\Gamma_r = \Gamma_{r,c} + \varepsilon^2\gamma, \quad \varepsilon^2 \ll 1,$$

a nontrivial solution of (8.44) can be constructed in the form

$$u = \varepsilon u_1(x,t_0,t_2) + \varepsilon^3 u_3(x,t_0,t_2) + \dots,$$

where $t_0 = t$, $t_2 = \varepsilon^2 t$. At the leading-order $O(\varepsilon)$, the linear problem considered in the previous section is reproduced, hence

$$u_1(x,t_0,t_2) = U(x)e^{-i(\omega_0 t_0 + \omega_2 t_2)}, \tag{8.52}$$

where $U(x)$ is defined by (8.24). At the order $O(\varepsilon^3)$, we obtain the following equation:

$$u_{3,t_0} + vu_{3,x} + [\lambda - (\Gamma_{r,c} + i\Gamma_i)\delta(x)]u_3 - (1+i\alpha)u_{3,xx}$$
$$= -u_{1,t_2} + \gamma\delta(x)u_1 - (1+i\beta)|u_1|^2 u_1. \tag{8.53}$$

Assuming

$$u_3(x,t_0,t_2) = U_3^{\pm} e^{-i(\omega_0 t_0 + \omega_2 t_2)} \tag{8.54}$$

(the upper and lower signs correspond to regions $x > 0$ and $x < 0$), we obtain

$$vU_{3,x}^{\pm} + (\lambda - i\omega_0)U_3^{\pm} - (1+i\alpha)U_{3,xx}^{\pm}$$

$$= i\omega_2 Ae^{k_{\pm}x} - (1+i\beta)|A|^2 Ae^{\tilde{k}_{\pm}x},$$

where $\tilde{k}_{\pm} = 3Rek_{\pm} + iImk_{\pm}$, and hence

$$U_{3,x}^{\pm} = A_3^{\pm} e^{k_{\pm}x} + \frac{i\omega_2 A}{v - 2k_{\pm}(1+i\alpha)} xe^{k_{\pm}x} - \frac{(1+i\beta)|A|^2 A}{\lambda - i\omega_0 + v\tilde{k}_{\pm} - (1+i\alpha)v\tilde{k}_{\pm}^2} e^{\tilde{k}_{\pm}x},$$

where $A_3^{\pm}$ are constants. The condition of the continuity of U_3 in the point $x = 0$ and the condition

$$-(1+i\alpha)(U_{3,x}^+ - U_{3,x}^-)|_{x=0} - (\Gamma_{r,c} + i\Gamma_i)U_3(0) = \gamma,$$

which is obtained by the integration of (8.53) in the interval $-\delta \leq x \leq \delta$ and by taking the limit $\delta \to 0$, give a system of two algebraic linear equations for $A_3^{\pm}$ with a zero determinant. The solvability condition of this system gives the following relation:

$$-(1+i\beta)|A|^2 \left[\frac{1}{v - 2q_+(1+i\alpha)} - \frac{1}{v - 2q_-(1+i\alpha)} \right] + \frac{\gamma + i\omega_2}{1 + i\alpha} = 0, \tag{8.55}$$

where $q_{\pm} = k_{\pm} + Rek_{\pm}$. Multiplying relation (8.55) by $1 + i\alpha$ and taking the real part, we find the bifurcation equation,

$$|A|^2 Re\left\{ (1+i\alpha)(1+i\beta) \left[\frac{1}{v - 2q_+(1+i\alpha)} - \frac{1}{v - 2q_-(1+i\alpha)} \right] \right\} = \gamma, \tag{8.56}$$

which is the generalization of equation (8.41) in the case of the complex amplitude function.

8.2.2 Smooth Gain Modulation

In the present subsection, we consider smooth longwave modulations of the linear growth rate with the large spatial scale compared to characteristic length scale of complex Ginzburg–Landau equation, i.e., $f = f(kx)$, $g = g(kx)$, where $k \ll 1$ [53, 54]. With $A(x,t) = R(x,t)\exp[i\Theta(x,t)]$, we obtain the following system of equations:

$$R_t + vR_x = (f - R^2 - \Theta_x^2)R - \alpha(2R_x\Theta_x + R\Theta_{xx}) + R_{xx}, \quad (8.57)$$
$$R(\Theta_t + v\Theta_x) = (g - \alpha\Theta_x^2 - \beta R^2)R + 2R_x\Theta_x + R\Theta_{xx} + \alpha R_{xx}.$$

We consider spatially modulated and temporally monochromatic waves with

$$R = R(x), \quad \Theta = \theta(x) - \omega t, \quad (8.58)$$

where ω is constant.

Defining $Q = \theta_x$ (local wavenumber), assuming $R = R(kx)$, $Q = Q(kx)$, and introducing rescaled spatial variable $X = kx$, we find that the monochromatic wave is governed by the following system of equations [55]:

$$(f - R^2 - Q^2)R - k[vR_X + \alpha(2R_XQ + RQ_X)] + k^2 R_{XX} = 0, \quad (8.59)$$
$$(g + \omega - vQ - \alpha Q^2 - \beta R^2)R + k(2R_XQ + RQ_X) + \alpha k^2 R_{XX} = 0.$$

In the case of a non-modulated growth rate parameter ($f = 1$, $g = 0$), there exists a family of monochromatic spatially periodic waves determined by relations (cf. (7.128))

$$R^2 = 1 - Q^2, \quad \omega = \beta + vQ + (\alpha - \beta)Q^2. \quad (8.60)$$

Recall also that the group velocity of a longwave disturbance propagating *on the background of a periodic wave* is proportional to $\alpha - \beta$ (see (7.134)).

In the presence of a parameter modulation, the solution for equation (8.269) can be found as a power series in terms of small k:

$$R = R_0 + kR_1 + \ldots, \quad Q = Q_0 + kQ_1 + \ldots. \quad (8.61)$$

Note that due to the singular nature of equation (8.59), expansion (8.61) is actually valid for an outer solution). We will not consider solutions containing internal layers.

At the zeroth order in k, we obtain a system of algebraic equations

$$(f - R_0^2 - Q_0^2)R_0 = 0, \quad (g + \omega - vQ_0 - \alpha Q_0^2 - \beta R_0^2)R_0 = 0.$$

For nontrivial solutions with $R_0 \neq 0$, the wave amplitude R_0 is slaved to the local wavenumber Q_0 through the relation $R_0^2 = f - Q_0^2$, whereas the wavenumber satisfies the quadratic equation

$$(\alpha - \beta)Q_0^2 + vQ_0 - \omega + \beta f - g = 0. \quad (8.62)$$

Obviously, the case $\alpha = \beta$, when the group velocity of the disturbances on the background of a periodic wave vanishes, is a resonant one; thus, that case has to be considered separately.

8.2.2.1 The Case $\alpha \neq \beta$

Monochromatic Waves

In the nonresonant case with $\alpha \neq \beta$, equation (8.62) has real solutions

$$Q_0^\pm = \frac{-v \pm \sqrt{D(X)}}{2(\alpha - \beta)}, \tag{8.63}$$

if the value of

$$D(X) = v^2 + 4(\alpha - \beta)(\omega - \beta f + g) \tag{8.64}$$

is nonnegative everywhere in the region. For given functions $f(X)$ and $g(X)$, the conditions $D(x) \geq 0$ and $R_0^2 = f - Q_0^2 > 0$ impose restrictions on possible values of frequency ω and thus determine the region of existence of the monochromatic wave.

In the case of $\alpha - \beta > 0$, the value of D grows with ω; therefore, the condition $D(x) \geq 0$ is satisfied for $\omega \geq \omega_*$, where $\omega_* = \beta f - g - v^2/4(\alpha - \beta)$. At the point $\omega = \omega_*$, both branches of solution in equation (8.63) merge at $Q_0 = Q_* = -v/2(\alpha - \beta)$. If $f > v^2/4(\alpha - \beta)^2$, the branches $Q_0^\pm$ satisfy the condition $Q_0^2 < f$ in the interval of frequencies $\omega_* \leq \omega < \omega_\pm$, where $\omega_\pm = \alpha f - g \pm v\sqrt{f}$. If $0 < f < v^2/4(\alpha - \beta)^2$, only one of the branches $Q_0^\pm$ can satisfy that condition. In particular, if $v > 0$, then the branch Q_0^+ exists as $\omega_- < \omega < \omega_+$. For $v < 0$ the branch Q_0^- exists as $\omega_+ < \omega < \omega_-$.

In case of $\alpha - \beta < 0$, the solutions exist in the region $\omega \leq \omega_*$. Similarly, for $f > v^2/4(\alpha - \beta)^2$, the branches $Q_0^\pm$ exist in the intervals $\omega_\mp < \omega \leq \omega_*$, while for $0 < f < v^2/4(\alpha - \beta)^2$, the branch Q_0^+ exists in the interval $\omega_- < \omega < \omega_+$. If $v < 0$, and the branch Q_0^- exists in the interval $\omega_+ < \omega < \omega_-$, if $v > 0$.

Specifically, in the case $v = 0$, $g(X) = 0$, we find that [53]

$$Q_0^2 = \frac{\omega - \beta f}{\alpha - \beta}, \quad R_0^2 = -\frac{\omega - \alpha f}{\alpha - \beta}. \tag{8.65}$$

The admissible values of ω are inside an interval $\omega_- < \omega < \omega_+$. On one border of this interval, $Q_0 \to 0$, and on the other border, $R_0 \to 0$. For each admissible value of ω, there are two solutions with different signs of Q_0. The interval disappears when the conditions $Q_0^2 > 0$ and $R_0^2 > 0$ cannot be satisfied for some X.

The first-order corrections R_1 and Q_1 are determined by the system of equations

$$2R_0^2 R_1 + 2Q_0 R_0 Q_1 = -\alpha(2R_{0X} Q_0 + R_0 Q_{0X}),$$

$$2\beta R_0^2 R_1 + 2\alpha Q_0 R_0 Q_1 = 2R_{0X} Q_0 + R_0 Q_{0X},$$

hence

$$R_1 = -\frac{1+\alpha^2}{2(\alpha-\beta)R_0^2}(2R_{0X}Q_0 + R_0Q_{0X}),$$ (8.66)

$$Q_1 = \frac{1+\alpha\beta}{2(\alpha-\beta)Q_0R_0}(2R_{0X}Q_0 + R_0Q_{0X}).$$ (8.67)

Note that for a spatially periodic modulation $(f(X), g(X))$ functions $R_0(X)$, $Q_0(X)$ are also periodic, but the solution

$$A(x,t) \sim R_0(kx)\exp\left[i\int_0^x Q_0(kz)dz - i\omega t\right]$$ (8.68)

is quasiperiodic.

Stability of Monochromatic Waves

For the sake of simplicity, we describe the stability of monochromatic waves only in the case where $v = 0$, $g(x) = 0$. Recall that in the case of a non-modulated growth rate parameter $(f = 1, g = 0)$, the monochromatic wave (8.60) is stable within the "Busse balloon"

$$Q^2 < Q_m^2 = \frac{1+\alpha\beta}{3+\alpha\beta+2\beta^2}$$ (8.69)

(see (7.135)).

If f is modulated, linearizing equations

$$R_t = (f - R^2 - Q^2)R - k\alpha(2R_XQ + RQ_X) + k^2R_{XX},$$ (8.70)
$$Q_t = -k(\alpha Q^2 + \beta R^2)_X + k^2(2R_XK/R + K_X)_X + k^3\alpha(R_{XX}/R)_X,$$

around the monochromatic solution $(R(X), Q(X))$, we obtain the following system of equations for the disturbances $(\tilde{R}(X), \tilde{Q}(X))\exp(\sigma t)$:

$$\sigma\tilde{R} = (f - 3R^2 - Q^2)\tilde{R} - 2QR\tilde{Q} - k\alpha(2R_X\tilde{Q} + 2\tilde{R}_XQ + \tilde{R}Q_X + R\tilde{Q}_X) + k^2\tilde{R}_{XX},$$ (8.71)
$$\sigma\tilde{Q} = -k(2\alpha Q\tilde{Q} + 2\beta R\tilde{R})_X + k^2(2\tilde{R}_XQ/R + 2R_X\tilde{Q}/R - 2R_XQ\tilde{R}/R^2 + \tilde{Q}_X)_X + \alpha k^3(\tilde{R}_{XX}/R - R_{XX}\tilde{R}/R^2)_X.$$

Now we apply expansions in powers of k for $Q, R, \tilde{Q}, \tilde{R}$, and σ.
At the zeroth order,

$$\sigma_0\tilde{R}_0 = -2R_0^2\tilde{R}_0 - 2Q_0R_0\tilde{Q}_0, \quad \sigma_0\tilde{Q}_0 = 0.$$

For non-decaying disturbances,

$$\sigma_0 = 0, \tilde{R}_0 = -Q_0\tilde{Q}_0/R_0.$$ (8.72)

At the first order, we find:

$$2R_0^2 \tilde{R}_1 + 2Q_0 R_0 \tilde{Q}_1 = -\sigma_1 \tilde{R}_0 - (2Q_0 Q_1 + 6R_0 R_1)\tilde{R}_0$$

$$- 2(Q_0 R_1 + Q_1 R_0)\tilde{Q}_0 - \alpha(2R_{0X}\tilde{Q}_0 + 2\tilde{R}_{0X}Q_0 + \tilde{R}_0 Q_{0X} + R_0 \tilde{Q}_{0X}), \qquad (8.73)$$

$$\sigma_1 \tilde{Q}_0 = -(2\alpha Q_0 \tilde{Q}_0 + 2\beta R_0 \tilde{R}_0)_X. \qquad (8.74)$$

Substituting relation (8.72) into equation (8.74), we find that

$$\tilde{Q}_0(X) = \frac{1}{Q_0(X)} \exp\left[-\frac{\sigma_1}{2(\alpha - \beta)} \int_0^X \frac{d\xi}{Q_0(\xi)}\right] \qquad (8.75)$$

up to an arbitrary coefficient. As $Q_0(\xi)$ is real and does not change its sign, the integral $\int_0^X d\xi / Q_0(\xi)$ is real and nonzero. On the other hand, the function (8.75) is bounded at $X \to \pm\infty$ only if σ_1 is purely imaginary, $\sigma_1 = -i\omega_1$.

Thus, we obtain a one-parametric family of perturbations,

$$\tilde{Q}_0(X) = \frac{1}{Q_0(X)} \exp(i\Phi_0(X)), \ \tilde{R}_0(X) = -\frac{1}{R_0(X)} \exp(i\Phi_0(X)), \qquad (8.76)$$

where

$$\Phi_0(X) = \frac{\omega_1}{2(\alpha - \beta)} \int_0^X \frac{d\xi}{Q_0(\xi)}. \qquad (8.77)$$

As $\mathrm{Re}[\sigma_1] = 0$, we cannot determine whether the perturbations grow or decay. An explicit expression for $\tilde{R}_1$ can be found from algebraic equation (8.73); however, it is cumbersome, and it is not presented here.

The stability criterion is obtained in the second order from the equation for $\tilde{Q}_1$ which can be written in the following form:

$$i\omega_1 \tilde{Q}_1 + 2(\alpha - \beta)(Q_0 \tilde{Q}_1)_X = \left(-\frac{\sigma_2}{Q_0} + F\right) \exp(i\Phi_0), \qquad (8.78)$$

where F is a (cumbersome) expression containing the functions that have already been found. Substituting

$$\tilde{Q}_1 = \frac{\Psi}{Q_0} \exp(i\Phi_0),$$

we obtain the following equation for Ψ:

$$2(\alpha - \beta)\Psi_X = -\frac{\sigma_2}{Q_0} + F.$$

If the external modulation function $f(X)$ is periodic with the period L, then the solution is bounded at the infinity and periodic with the period L under the condition

$$\int_0^L \left(-\frac{\sigma_2}{Q_0(\xi)} + F(\xi)\right) d\xi = 0.$$

Therefore,

$$\sigma_2 = \frac{\langle F \rangle}{\langle Q_0^{-1} \rangle},$$

where

$$\langle u \rangle \equiv \frac{1}{L} \int_0^L u(\xi) d\xi.$$

In the case of a nonperiodic external forcing, the variables are averaged over the entire region $-\infty < X < \infty$.

Substituting explicit expression of F and collecting all the terms that are not derivatives of periodic functions and hence contribute to the integral, we find:

$$\sigma_2 = \frac{\omega_1^2}{4(\alpha - \beta)^2} \cdot \frac{-(1 + \alpha\beta)\langle Q_0^{-3} \rangle + 2(1 + \beta^2)\langle Q_0^{-1} R_0^{-2} \rangle}{\langle Q_0^{-1} \rangle}. \qquad (8.79)$$

Thus, the monochromatic wave is stable if for this wave

$$\frac{\langle Q_0^{-1} R_0^{-2} \rangle}{\langle Q_0^{-3} \rangle} < \frac{1 + \alpha\beta}{1 + \beta^2} \qquad (8.80)$$

and it is unstable otherwise. The functions $Q_0(X)$ and $R_0(X)$ are determined by formulas (8.65).

In the absence of the modulation ($f = 1$), the Eckhaus instability criterion (8.69) is recovered from equation (8.80). On the border of the interval of admissible values of ω, where $R_0 \to 0$, the left-hand side of equation (8.80) diverges; hence, the monochromatic wave is unstable. On another border, where $Q_0 \to 0$, the left-hand side of equation (8.80) vanishes; hence, the monochromatic wave is stable for $1 + \alpha\beta > 0$. The latter inequality coincides with the condition of the absence of a Benjamin–Feir instability for non-modulated waves. Thus, if this inequality is satisfied, the interval of existence of a monochromatic wave always contains a subinterval where that wave is stable.

8.2.2.2 The Case $\alpha = \beta$

Let us describe briefly the solutions in the *resonant case*, $\alpha = \beta$, where the group velocity of phase disturbances vanishes (see (7.134)) and the effect of the forcing is especially strong. A temporally monochromatic solution in the resonant case can be found for a weak modulation, $f = 1 + kF$. Expanding $\omega(k) = \omega_0 + \omega_1 k + \ldots$, we find that

$$R_0^2 = 1 - Q_0^2, \quad \omega_0 = \alpha. \qquad (8.81)$$

For the local wavenumber Q_0, we obtain the following nonlinear equation:

$$(1 + \alpha^2) \frac{1 - 3Q_0^2}{1 - Q_0^2} Q_{0y} = \alpha F(ky) - \omega_1. \qquad (8.82)$$

equation (8.82) has a bounded solution if $\omega_1 = \alpha \langle F \rangle$, where $\langle F \rangle$ is the spatial average of F. The left-hand side of this equation is typical for phase disturbances of CGLE. The origin of term $\alpha F(ky)$ on the left-hand side is originated from the local nonlinear frequency shift, produced by the parameter modulation. Integrating equation (8.82), we find an expression for $Q_0(y)$ an implicit form:

$$H(Q_0(x)) = H(Q_0(0)) + \frac{\alpha}{1+\alpha^2} \int_0^x [F(kz) - \langle F \rangle] dz, \qquad (8.83)$$

where

$$H(Q_0) \equiv 3Q_0 - \frac{1}{2} \ln \frac{1+Q_0}{1-Q_0}. \qquad (8.84)$$

If the wavevector Q_0 is in the region of the Eckhaus stability, $Q_0^2 < 1/3$, and the value of $H(Q_0(y))$ in (8.83) belongs to the interval $(-H_m, H_m)$, with $H_m = H(1/\sqrt{3})$, then the solution for equation (8.83) exists and is uniquely defined.

8.3 Feedback Control

The approaches mentioned above can provide a *passive* suppression of instability and can be considered as "open-loop" control strategies. Recently, another approach based on the *feedback (online) control* has been suggested. The feedback control of the collective behavior of nonlinear systems can be much more efficient, because in many cases, it is not invasive and nearly vanishes when the desired state is achieved. However, in the case of a *large-aspect ratio* system, that is a difficult task, because the system has many degrees of freedom. Let us briefly describe some most relevant publications related to controlling flow instabilities (for comprehensive reviews of the subject, see [56, 57]).

The use of the feedback control for the stabilization of large-aspect ratio systems was demonstrated first in the case of the Rayleigh–Bénard convection, both theoretically [58, 59] and in experiments [60, 61]. Still, there are only few works on the feedback of the interfacial convection in large-aspect ratio containers.

Benz et al. [62] suppressed the development of hydrothermal waves in a thin layer of silicone oil subject to a horizontal temperature gradient. A nonintrusive sensing of free-surface temperature perturbation has been carried out by an infrared camera. The signal from the camera has been converted into a video signal and used for the real-time determination of the hydrothermal-wave phase velocity. The control was achieved by heating the free surface with a sheet of infrared radiation along lines parallel to the crests of the hydrothermal waves corresponding to low-temperature disturbances.

Grigoriev [63] considered the problem of the stabilization of instabilities which are developed by the evaporation of thin liquid films. He derived the nonlinear evolution equation for the film surface deformation in the presence of a volumetric absorption of the electromagnetic radiation and an absorption at the bottom interface. Several feedback control strategies for the suppression of instabilities in pure liquids and binary solutions have been suggested and analyzed.

The feedback control has been successfully applied for the suppression of the fingering contact-line instability in a thermocapillary-driven fluid film. In [64], a slip model was applied for the description of the spreading of a liquid film under the action of the thermocapillary effect and suggested a control algorithm which makes the dynamics asymptotically stable without increasing the transient amplification of disturbances characteristic for the uncontrolled contact line. The proposed control algorithm has been verified experimentally in [65]. The shape of the contact line has been monitored by computer-controlled digitalization and processing of video images from high-resolution CCD cameras. The control has been achieved by suitable spatial and temporal perturbation of the light intensity. Let us mention also the application of the feedback control to the suppression of the rivulet formation and film rupture in a locally heated falling film [66].

Below, we describe the application of a feedback control to a number of systems governed by longwave amplitude equations.

8.3.1 Feedback Control of a Supercritical Instability: Kuramoto–Sivashinsky Equation

We select this equation as a paradigmatic example of supercritical instability creating complex patterns, which is not plagued with the danger of blow-up. In that case, the feedback control approach can be applied for a total suppression of the instability by means of a linear control.

First, let us consider the *linearized* Kuramoto–Sivashinsky equation

$$\frac{\partial a}{\partial t} + \frac{\partial^4 a}{\partial x^4} + \frac{\partial^2 a}{\partial x^2} = 0 \tag{8.85}$$

satisfying the periodicity condition

$$a(x - L, t) = a(x + L, t). \tag{8.86}$$

By rescaling

$$X = x(\pi/L), \quad T = t(\pi/L)^2, \tag{8.87}$$

we rewrite problem (8.85), (8.86) in the form

$$\frac{\partial a}{\partial T} + v\frac{\partial^4 a}{\partial X^4} + \frac{\partial^2 a}{\partial X^2} = 0, \tag{8.88}$$

$$a(X - \pi, t) = a(X + \pi, t),$$

where $v = (\pi/L)^2$. Using the Fourier components,

$$\hat{a}_n(t) = \int_{-\pi}^{\pi} a(X, t) \exp(-inX)\frac{dX}{2\pi},$$

we can present (8.88) in the form of an infinite system of ordinary differential equations,

$$\frac{d\hat{a}_n}{dt} = \sigma_n \hat{a}_n, \ n = 0, \pm 1, \pm 2, \dots, \tag{8.89}$$

where $\sigma_n = n^2 - \nu n^4$. Below we consider the subsystem with $n > 0$, because $\hat{a}_0(t) = \hat{a}_0(0)$, $\hat{a}_{-n}(t) = \hat{a}_n^*(t)$. Note that $\sigma_n > 0$ only for a finite number of Fourier harmonics, $n < (1/\nu)^{1/2}$.

The goal of the *control* is the modification of system (8.89) in order to make the solution $\hat{a}_n = 0$, $n = 1, 2, \dots$ exponentially stable. Let us introduce a control of unstable modes by means of l *actuators*,

$$\frac{d\hat{a}_n}{dt} = \sigma_n \hat{a}_n + \sum_{i=1}^{l} b_i^n u_i(t), \ n = 1, 2, \dots, l. \tag{8.90}$$

The *linear control* is determined by the relation

$$u_i(t) = \sum_{j=1}^{l} K_{ij} \hat{a}_j, \ i = 1, \dots, l, \tag{8.91}$$

where K_{ij} and b_i are chosen in such a way that the eigenvalues of the matrix

$$M_{ij} = \sigma_i \delta_{ij} + \sum_{k=1}^{l} b_k^i K_{kj} \tag{8.92}$$

have negative real parts. That gives the exponential stability of solution $a = 0$.

The same idea can be applied to the nonlinear Kuramoto–Sivashinsky equation. Christofides [67, 68] has proved that the solution $a = 0$ of the controlled nonlinear Kuramoto–Sivashinsky equation,

$$\frac{\partial a}{\partial t} + \nu \frac{\partial^4 a}{\partial X^4} + \frac{\partial^2 a}{\partial X^2} + \frac{\partial a^2}{\partial X} = \sum_{n=1}^{l} \exp(inX) \sum_{i=1}^{l} b_i^n u_i(t), \tag{8.93}$$

with $u_i(t)$ determined by (8.91), is exponentially stable under the same conditions on the eigenvalues of the matrix M_{ij} as in the linear case.

A direct measurement of Fourier components $\hat{a}_j$ needed for construction of the control function $u_i(t)$ may be inconvenient. A more flexible approach, with the control based on measurements of quantities

$$y_j(t) = \int_{-\pi}^{\pi} c_j(X) a(X, t) dX, \ j = 1, \dots, N, \tag{8.94}$$

where $c_j(X)$ is a function characterizing the jth sensor, can be applied for controlling the state $a = 0$. The linear control can be constructed as

$$u_i(t) = \sum_{j=1}^{N} K_{ij} y_j(t), \ i = 1, \dots, l, \tag{8.95}$$

with an appropriate choice of the matrix K_{ij}.

For instance, function $c_j(X)$ can be chosen as a characteristic function of the jth interval inside $[-\pi, \pi]$, and the term

$$-\mu \sum_{j=1}^{N} y_j(t) c_j(X)$$

can be added in the right-hand side of the controlled equation [69, 70].

Another approach has been suggested by Christofides [67, 68]. The linear control $u_i(t)$ is determined by formulas similar to (8.91),

$$u_i(t) = \sum_{j=1}^{l} K_{ij} \tilde{a}_j, \ i = 1,\dots,l. \tag{8.96}$$

However, instead of a direct measurement of Fourier components, one *computes* $\tilde{a}_j(t)$ by solving equations

$$\frac{d\tilde{a}_i}{dt} =) = \sum_{j=1}^{l} M_{ij} \tilde{a}_j + \sum_{j=1}^{l} L_{ij}(y_j - \tilde{y}_j), \ i = 1,\dots,l, \tag{8.97}$$

where $y_j(t)$ are the results of measurements, while $\tilde{y}_j(t)$ are calculated as

$$\tilde{y}_j(t) = \int_{-\pi}^{\pi} c_j^*(X) \sum_{n=1}^{l} [\tilde{a}_n(t) \exp(inX) + \tilde{a}_n^*(t) \exp(-inX)] \, dX, \tag{8.98}$$

and L_{ij} is the appropriate matrix. Under conditions formulated in [67, 68], this *output feedback control* provides an exponential stability of the solution $a = 0$.

A generalization of the approach described above, which allows to stabilize different solutions (not necessarily the trivial one), was suggested in [71]. It was applied to the Kawahara equation (1.78) in [72] and to equations governing falling liquid film flows in [73].

8.3.2 Feedback Control of Subcritical Instabilities

In the case of a subcritical instability, the suppression of a linear instability may be insufficient.

Among the subcritical instabilities, one can mention the generation of hexagonal Marangoni cells in a heated liquid layer due to Pearson's instability (with weakly deformed surface) and non-saturable Sterling–Scriven's deformational instability which leads to the film rupture.

Several works are devoted to a theoretical analysis of the feasibility of a feedback control for the suppression of Rayleigh–Bénard–Marangoni instability in a fluid layer heated from below. Bau [74] applied the linear proportional control,

$$\theta(x,y,0,t) = -K\theta(x,y,1,t), \tag{8.99}$$

where θ is the deviation of the fluid's temperature from its conductive value. By an appropriate choice of the controller gain, K, both Pearson's instability mode (non-deforming surface) and Sterling-Scriven's instability mode (deforming surface) are suppressed. However, the control strategy (8.99) turns out to be non-efficient for longwave disturbances. At sufficiently large values of K, an oscillatory instability is generated. The linear proportional feedback control has been used for suppression of Marangoni instability in variable viscosity fluids [75], in a rotating fluid layer [76], and in a layer with internal heat source [77].

More general feedback control laws,

$$\theta(x,y,0,t) = -K_p\theta(x,y,1,t) - K_q\theta^2(x,y,1,t) - K_c\theta^3(x,y,1,t) \qquad (8.100)$$

and

$$\partial\theta(x,y,0,t)/\partial z = -K_p\theta(x,y,1,t) - K_q\theta^2(x,y,1,t) - K_c\theta^3(x,y,1,t), \qquad (8.101)$$

have been used in [78] for the nonlinear stabilization of the Pearson–Nield instability in the presence of both thermocapillarity and buoyancy (but in the absence of the surface deformation). Besides the linear stabilization of the quiescent state, the non-linear control altered a wide range of weakly nonlinear properties of the subcritical hexagonal convection.

A suppression of the longwave deformational Marangoni mode has been achieved in [79] where a nonlinear feedback control strategy,

$$\theta(x,y,0,t) = K_1h + K_2h^2 + K_3h^3, \qquad (8.102)$$

has been applied; here h is the deformation of the free surface. The appropriate values of the gain constants, K_1, K_2, and K_3, have been determined by means of a longwave weakly nonlinear analysis. The nonlinear feedback control is efficient for both suppressing the linear instability and eliminating nonlinear subcritical instability.

Another efficient approach is the *global* feedback control, which allows to deal with a finite number of the most important parameters of the system [56]. Its application to subcritically unstable systems is considered below.

8.3.2.1 Sivashinsky Equation

The subcritical instability is characteristic for systems governed by the Sivashinsky equation (1.81). A number of examples are presented in other chapters of the book (see Sections 3.2.2, 6.2.1).

Formulation of the Problem

As the basic example, we select the feedback control of the solidification front instability discussed in Section 6.2.1. As we have seen, that instability, which is governed by the damped Sivashinsky equation (6.67), is subcritical, i.e., the finite-amplitude disturbances grow in the subcritical region. The nonlinear instability leads to an unbounded growth of disturbances. Our goal is the change *of the instability type* rather than a shift of the instability threshold. It is quite natural to control the process by the pulling speed V and the applied temperature gradient G. The problem is governed by system (6.61)–(6.64), but expression $1 + h_t$ in (6.61) and (6.63) is replaced by $V(t) + h_t$, and parameter W in (6.62) is replaced by $WG(t)$.

Amplitude Equation

Let us repeat the derivation of the amplitude equation (6.67) assuming

$$V(\tau) = 1 + \varepsilon v(\tau), \; G(\tau) = 1 + \varepsilon g(\tau). \tag{8.103}$$

We obtain:

$$h_\tau^{(0)} + \beta \Delta_\perp^2 h^{(0)} + [s + v(\tau) - g(\tau)]\Delta_\perp h^{(0)} + \kappa h^{(0)} - \frac{1}{2}\Delta_\perp (h^{(0)})^2 + v_\tau = 0. \tag{8.104}$$

We define

$$h(\mathbf{X}, \tau) = h^{(0)}(\mathbf{X}, \tau) + f(\tau), \; f(\tau) = \exp(-\kappa\tau) \int_0^\tau v_s(s)\exp(\kappa s)ds, \tag{8.105}$$

and make the coordinate transformation $\mathbf{X} = \beta^{1/2}\mathbf{X}'$, $\tau = \beta^{1/2}T$ to rewrite equation (6.67) in the form

$$h_T + \Delta_\perp^2 h + [s + U(T)]\Delta_\perp h + \alpha h - \frac{1}{2}\Delta_\perp (h^2) = 0, \tag{8.106}$$

where

$$U(T) = v(T) - g(T) + f(T). \tag{8.107}$$

Equation (8.106) can be written as

$$h_T + \alpha h + \nabla_\perp \cdot \mathbf{Q} = 0, \; \mathbf{Q} = \nabla_\perp \{\Delta_\perp h + [s + U(T)]h - \frac{1}{2}h^2\}. \tag{8.108}$$

Consider equation (8.108) in a finite domain D with an area A and the boundary ∂D, and define

$$\langle h \rangle_T = A^{-1} \int_D h(\mathbf{X}, T)d\mathbf{X}. \tag{8.109}$$

For periodic boundary conditions, as well as for the boundary condition $Q_n = 0$ on ∂D, $\langle h(T) \rangle$ satisfies the equation

$$\langle h \rangle_T + \alpha \langle h \rangle = 0. \tag{8.110}$$

Therefore, $\langle h \rangle \to 0$ as $T \to \infty$ for any $\alpha > 0$. The same result can be obtained in an infinite region, with $\langle h \rangle$ defined as the limit of the expression (8.109) where the integration domain D is expanded, for periodic boundary conditions or if $\mathbf{Q}$ vanishes at infinity.

Control Function

The function $U(T)$ in equation (8.106) is determined by the slowly varying pulling speed and temperature gradient (see equation (8.107)) that can be adjusted with a feedback to the solidification front perturbations. Thus, the function $U(T)$ is the control function that can be taken as a function of the interface shape $h(\mathbf{X}, T)$. The objective of the feedback control is the elimination of the subcritical instability leading to the blow-up characterized by the divergence of the minimum value of $h(\mathbf{X}, T)$ (formation of deep cells). For that purpose, we choose

$$U(T) = P \min_{\mathbf{X}} h(\mathbf{X}, T), \qquad (8.111)$$

where P is a positive coefficient, the *control parameter*. The effect of such control will be the increase of the effective viscosity when the negative peaks of $h(X, T)$ appear. Note that the evolution of infinitesimal disturbances of the planar front, $h = 0$, described by the linearized problem, is not affected by the chosen control. As without any control, the planar front is linearly stable for $\alpha > 1/4$, while for $0 < \alpha < 1/4$, $s = 1$, it is unstable with respect to disturbances with the wavenumbers in the interval

$$k_-^2 < k^2 < k_+^2, \; k_\pm^2 = \frac{1 \pm \sqrt{1 - 4\alpha}}{2}. \qquad (8.112)$$

One-Dimensional Front

Consider the 1+1 case of a two-dimensional crystal and melt with a one-dimensional interface described by the 1D version of equation (8.106) with control functional (8.11

$$h_T + h_{XXXX} + [s + P \min_X h(X, T)]h_{XX} + \alpha h - \frac{1}{2}(h^2)_{XX} = 0, \qquad (8.113)$$

and investigate the influence of the feedback control parameter, P, on one-dimensional stationary solutions of equation (8.113), both in the subcritical ($s = -1$) and supercritical ($s = 1$) cases. The disappearance of stationary solutions in the subcritical region and the appearance of stable stationary solutions in the supercritical region can be considered as the suppression of the subcritical instability and its transformation into a supercritical instability.

Complete Solute Rejection, $\alpha = 0$

We start with the consideration of an important, analytically tractable case of a complete solute rejection, corresponding to a zero segregation coefficient, $\alpha = 0$. In this case, one-dimensional stationary deformations of the solidification front in the presence of the feedback control are described by the following equation:

$$h_{XXXX} + S_p h_{XX} - \frac{1}{2}(h^2)_{XX} = 0, \; S_p = s + P \min_X h(X). \qquad (8.114)$$

Equation (8.114) is solved in the infinite region, under the conditions

$$|h(\pm\infty)| < \infty, \quad \langle h \rangle = 0.$$

There is a one-parameter family of solutions of equation (8.114) satisfying these conditions (see Section 3.2.2.6):

$$h(q,X) = \frac{3S_p}{F(q)}\left[1 - \frac{E(q)}{K(q)} - q^2\mathrm{sn}^2\left(\frac{X-X_0}{\Delta};q\right)\right], \quad F(q) = 2 - q^2 - 3\frac{E(q)}{K(q)};$$

$$\Delta = 2\sqrt{-\frac{F(q)}{S_p}}; \quad X_0 = \mathrm{const}, \tag{8.115}$$

where q is the modulus of the elliptic function [80], $0 < q^2 < 1$; $E(q)$ and $K(q)$ are the complete elliptic integrals of the first and second kind, respectively. The solution (8.115) exists for $S_p < 0$, if $F(q) > 0$, $q_* < q \le 1$, and for $S_p > 0$, if $F(q) > 0$, $0 < q < q_*$, where $q_* \approx 0.9804$. For $q < 1$ solution (8.114) is periodic, with the period

$$L = 4K(q)\sqrt{-\frac{F(q)}{S_p}}. \tag{8.116}$$

For $q = 1$ it corresponds to a localized solution,

$$h(1,X) = 3S_p\cosh^{-2}\frac{\sqrt{-S_p}(X-X_0)}{2}. \tag{8.117}$$

The minimum value of $h(q,X)$,

$$h_{min} = \min_X h(q,X) = 3S_p\frac{1 - E(q)/K(q)}{F(q)} < 0 \tag{8.118}$$

does not depend on the sign of S_p.

The case of the absence of control ($P = 0$) has been considered in Section 3.2.2.6 in another physical context. In the subcritical case ($W > W_c$, $s = -1$) there is a family of periodic solutions with $q_* < q < 1$. For $q \to q_*$, the period of the solution tends to zero and $h_{min} \to -\infty$. In the opposite limit, $q \to 1$, the period tends to infinity, which corresponds to the localized solution (8.117). Thus, the spatially periodic solutions in the subcritical region exist for any values of the spatial period L. All of them are unstable [81]. The stable manifold of such a solution separates the periodic initial conditions corresponding to decaying finite-amplitude disturbances, and the periodic initial conditions which lead eventually to the blow-up. In the supercritical case ($W < W_c$, $s = 1$), there are periodic solutions with $0 < q < q_*$. Also, $h_{min} \to 0$ for $q \to 0$, while the spatial period defined by equation (8.116) tends to $L_* = 2\pi$ corresponding to the marginal linear mode. All supercritical periodic solutions have $L < L_*$ that correspond to the region where the planar solidification front, $h = 0$, is stable. Therefore, these periodic solutions are unstable. There are no stationary solutions with wavelengths in the linear instability region, $L > L_*$. The linearly unstable mode grows without saturation until the blow-up occurs in a finite time.

Let us consider now stationary solutions of equation (8.114) in the presence of the global feedback control ($P > 0$). From (8.118), (8.115), one finds

$$S_p = s \left[1 - 3P \frac{1 - E(q)/K(q)}{F(q)} \right]^{-1}. \tag{8.119}$$

For solutions with $S_p > 0$, $F(q) < 0$, $0 < q < q_*$, the signs of S_p and s coincide and, therefore, these solutions exist in the supercritical region, $s = 1$. The spatial period of the supercritical solutions in the presence of control is

$$L = 4K(q) \sqrt{\frac{3E(q)}{K(q)} - 2 + q^2 + 3P \left[1 - \frac{E(q)}{K(q)} \right]}. \tag{8.120}$$

For $q > 0$ the period increase with P, and for $q = q_*$ it is finite.

For solutions with $q_* < q < 1$, $F(q) < 0$, the signs of S_p and s do not always coincide. For any q, there exists the critical value $P_c(q)$,

$$P_c(q) = \frac{1}{h_{min}(q)} = \frac{F(q)}{3[1 - E(q)/K(q)]}, \tag{8.121}$$

such that for $P < P_c(q)$ the solution with the fixed value of q exists in the subcritical region, $s < 0$, while for $P > P_c(q)$ it disappears in the subcritical region but reappears in the supercritical region, $s > 0$.

The minimum of the function $|h_{min}(q)|$ is achieved for $q \to 1$, and it corresponds to the solution (8.117). Since

$$\lim_{q \to 1} |h_{min}(q)| = 3$$

one finds

$$\max_q P_c(q) = P_c(1) = \frac{1}{3}. \tag{8.122}$$

For the localized solution (8.117),

$$S_p = -\frac{s}{1 - 3P}.$$

In the subcritical region ($s = -1$), this solution exists for $P < 1/3$, and in the supercritical region ($s = 1$), it exists for $P > 1/3$.

Thus, there is a critical value of the control parameter,

$$P_c = \frac{1}{3}, \tag{8.123}$$

such that for $0 < P \leq 1/3$ the stationary solutions exist both in the subcritical and in the supercritical region, while for $P > 1/3$, no stationary solutions exist in the subcritical region, but in the supercritical region, there are periodic solutions with the period that is given by equation (8.120) for all $0 < q < 1$. One can show that the supercritical stationary solutions exist for any $L > L_*$. Thus, for $P > 1/3$, for any wavelength in the instability interval of the planar front, there exists a supercritical stationary solution, i.e., the growth of any unstable periodic disturbance saturates.

Note that the disappearance of the subcritical stationary solutions is determined by the behavior of "large" localized solutions in the limit $P \to 1/3^-$,

$$h(X) = -\frac{3}{1-3P}\cosh^{-2}\frac{X-X_0}{2\sqrt{1-3P}}. \tag{8.124}$$

This also is the leading-order stationary solution of equation (8.113) with $\alpha > 0$, $s = -1$:

$$h_{XXXX} + [-1 + P\min_X h(X)]h_{XX} - \frac{1}{2}(h^2)_{XX} + \alpha h = 0. \tag{8.125}$$

Therefore, the value $P = 1/3$ is the critical value of the control parameter at which unstable subcritical stationary solutions disappear for any α.

Region Near the Instability Threshold

Near the instability threshold, $\alpha = 1/4 - \delta^2$, one obtains the globally controlled Ginzburg–Landau equation (cf. (6.72))

$$A_{T_2} = A + 2A_{X_1X_1} + \frac{2}{9}|A|^2A - p(\max_{X_1}|A|)A. \tag{8.126}$$

We consider that case in more detail in Section 8.3.2.2. Here we note that the periodic solutions of that equation,

$$A_\pm = r_\pm(Q)e^{i(QX_1+\phi)}, \; r_\pm(Q) = \frac{1}{4}\left[9p \pm 3\sqrt{9p^2 - 8(1-2Q^2)}\right],$$

$(Q < \sqrt{2}/2, \; p > (2\sqrt{2}/3)\sqrt{1-3Q^2})$, are unstable. However, equation (8.126) has *localized solutions* in the form,

$$A_\pm^s(X_1) = a_\pm e^{i\phi}\cosh^{-1}[b_\pm(X_1 - \bar{X}_1)],$$

where $\bar{X}_1$ and ϕ are arbitrary, and

$$a_\pm = \frac{9p \pm 3\sqrt{9p^2 - 4}}{2}, \; b_\pm = \frac{a_\pm}{3\sqrt{2}}; \; p > \frac{2}{3}.$$

Numerical simulations of (8.126) show that for sufficiently small initial conditions, the solution tends to A_-^s. Initial conditions with large amplitude produce blow-up.

Intermediate Values of α

For arbitrary values of α, the analysis can be carried out numerically [82]. Periodic solutions turn out to be unstable with respect to a modulational instability. For sufficiently large L, the system evolves to stationary localized solutions. For relatively small α, the tails of the localized solution are monotonic. For α sufficiently close to 1/4, the tails of the localized solution are oscillatory, and its shape is close to the envelope localized solution. The behavior of the tails is determined by the sign of the expression

$$d(P,\alpha) = \frac{1}{4}[P|h_{min}(P,\alpha)| - 1]^2 - \alpha,$$

where $h_{min}(P,\alpha) = \min_X h(X)$ for the localized solution. For $d > 0$, the tails are monotonic, while for $d < 0$ the tails are oscillatory.

Two-Dimensional Solidification Front

In the two-dimensional case, it is difficult to obtain analytical solutions of equation (8.106) with control (8.111). Still, when $1/4 - \alpha$ is small, the interval of the unstable wavenumbers is narrow, and one can expect the development of slowly modulated patterns. The generic two-dimensional pattern is the hexagonal one that occurs via a transcritical bifurcation [83]. In this case, the method of amplitude equations is not fully justified except for the case when the quadratic nonlinear resonant interaction coefficient is small. Nevertheless, the method of amplitude equations usually gives quite reasonable predictions (see, e.g., [84]).

Define $\alpha = 1/4 - \gamma\delta^2$, where γ can be of either sign. The first-order solution is considered to be a superposition of three modulated one-dimensional patterns that form together a hexagonal pattern:

$$h^{(1)} = \sum_{m=1}^{3} \{A_m \exp[ik_c(\mathbf{n}_m \cdot \mathbf{X}_0)] + A_m^* \exp[-ik_c(\mathbf{n}_m \cdot \mathbf{X}_0)]\}, \qquad (8.127)$$

where $k_c = \sqrt{2}/2$, and the unit vectors $\mathbf{n}_m$, $m = 1,2,3$ can be chosen as follows:

$$\mathbf{n}_1 = (1,0); \mathbf{n}_2 = (-1/2, \sqrt{3}/2); \mathbf{n}_3 = (-1/2, -\sqrt{3}/2).$$

Introduce two time scales, $T_1 = \delta T$ and $T_2 = \delta^2 T$, and write together the terms of order δ and δ^2 in the following three amplitude equations:

$$A_{1,T} = -\frac{\delta}{2}A_2^*A_3^* + \delta^2 \left[2(\mathbf{n} \cdot \nabla_\perp)^2 A_1 + \gamma A_1 + i\sqrt{2}(\mathbf{n} \cdot \nabla_\perp)(A_2^*A_3^*) + \right.$$

$$\left. \left(\frac{2}{9}|A_1|^2 + \frac{3}{4}|A_2|^2 + \frac{3}{4}|A_3|^2\right) A_1 - p\max_{\mathbf{X}_1}(|A_1| + |A_2| + |A_3|)A_1 \right], \qquad (8.128)$$

where the other two equations are obtained by cyclic permutations of the subscripts of A_m, $\mathbf{n}_m$, $m = 1,2,3$.

The numerical simulation of the amplitude equations (8.128) [82] reveals the localization of each amplitude in the presence of control ($p > 0$). With the growth of p, the size of the localized pattern grows.

For arbitrary α, direct numerical simulations of the controlled 2D Sivashinsky equations (8.106), (8.111) for $s = 1$, $P > 1/3$ have been carried out [82]. For relatively small α, the stationary solution is localized and axially symmetric. For larger values of α, the axial symmetry is broken; the solution is now symmetric only with respect to rotation by an angle multiple to $\pi/6$. The increase of α leads to the formation of a hexagonal pattern in the central part of the localized structure. When α approaches 1/4, the spatial size of the localized state grows, so that for a fixed size of the computational domain, one observes modulated patterns rather than localized structures. The increase of the control parameter P yields the same effect.

8.3.2.2 Subcritical Ginzburg–Landau Equation with Real Coefficients

In the present section, we will consider in more detail the problem described by equation (8.126).

For further analysis, let us rescale equation (8.126) and rewrite it in the standard form,

$$A_t = A + A_{xx} + |A|^2 A - p(\max_x |A|)A. \tag{8.129}$$

Stability of Periodic Patterns

First, consider non-modulated patterns with the spatial period $2\pi/k_c$. In that case, the amplitude A is independent of x and is described by the dynamical system

$$A_t = A + |A|^2 A - p|A|A. \tag{8.130}$$

Besides the trivial solution $A = 0$, equation (8.130) has two nontrivial equilibrium states, $A_{\pm} = r_{\pm} \exp(i\phi)$, where ϕ is arbitrary, and

$$r_{\pm} = \frac{p \pm \sqrt{p^2 - 4}}{2}, \quad p > 2. \tag{8.131}$$

In the framework of equation (8.130), the solutions $A = 0$ and $A = A_+$ are unstable, while the solution $A = A_-$ is stable. However, in the framework of equation (8.129), when the spatially nonuniform perturbations are allowed, the solution $A = A_-$, too, turns out to be unstable with respect to spatial modulations. Indeed, consider the perturbed solution of equation (8.129) in the form $A(x,t) = A_- + \tilde{A}(x,t)$, where $\tilde{A}(x,t) = \tilde{A}_0(t) + \tilde{A}_K(t)\exp(iKx) + \tilde{A}_{-K}(T)\exp(-iKx)$, and the phase of A_- can be taken as zero since it does not affect the solution stability. Linearize equation (8.129) to obtain the following system of linear equations for $\tilde{A}_0$, $\tilde{A}_K$, and $\tilde{A}_{-K}$:

$$\frac{d\tilde{A}_0}{dT} = \left(1 + 2r_-^2 - pr_-\right)\tilde{A}_0 + r_-^2\tilde{A}_0^* - pr_- \max_x(\tilde{A}(x,t)), \tag{8.132}$$

$$\frac{d\tilde{A}_K}{dt} = \left(1 - K^2 + 2r_-^2 - pr_-\right)\tilde{A}_K + r_-^2\tilde{A}_{-K}^*, \tag{8.133}$$

$$\frac{d\tilde{A}_{-K}}{dt} = \left(1 - K^2 + r_-^2 - pr_-\right)\tilde{A}_{-K} + r_-^2\tilde{A}_K^*. \tag{8.134}$$

Note that $\tilde{A}_0$ does not appear in equations (8.133) and (8.134); therefore, the system of these two equations can be considered separately. One can see that there exists a normal mode $\tilde{A}_K = \tilde{A}_{-K}^* \sim \exp(\sigma t)$ with the growth rate

$$\sigma = 2r_-^2 - K^2,$$

positive for $K^2 < 2r^2$. Thus, the non-modulated periodic structure is unstable with respect to amplitude modulations.

The modulational instability of other periodic solutions of equation (8.129),

$$A_\pm = r_\pm(Q)e^{i(Qx+\phi)}, \; r_\pm(Q) = \frac{p}{2} \pm \frac{1}{2}\sqrt{p^2 - 4(1 - Q^2)]}, \tag{8.135}$$

can be investigated in a similar way. Solution r_- of (8.135) is unstable with respect to perturbations $\tilde{A}(x,t) = [\tilde{A}_0(t) + \tilde{A}_K(t)\exp(iKx) + \tilde{A}_{-K}\exp(-iKx)]\exp(iQx)$ for

$$K^2 < 2r_-^2(Q) + 4Q^2. \tag{8.136}$$

Stationary Localized Solutions

Besides the constant and periodic solutions described above, equation (8.129) has localized solutions. We will construct such solutions both in the subcritical region and the supercritical region, i.e., we consider equation

$$A_t = sA + A_{xx} + |A|^2A - p(\max_x |A|)A, \tag{8.137}$$

where $s = \pm 1$.

Equation (8.137) has a stationary localized solution

$$A(x) = A_0(x) = R(x - x_0)e^{i\Theta}, \tag{8.138}$$

where x_0 and Θ are arbitrary constants,

$$R(y) = \sqrt{2}k(q)\cosh^{-1}[k(q)y], \; -\infty < y < \infty, \tag{8.139}$$

and $k(q)$ is a positive root of the quadratic equation

$$k^2 - 2kq + s = 0, \; q = p/\sqrt{2}. \tag{8.140}$$

In the subcritical region, $s = -1$, there exists only one solution branch

$$k(q) = q + \sqrt{q^2 + 1}, \; -\infty < q < \infty. \tag{8.141}$$

Specifically, for $q = 0$, i.e., in the absence of control, the localized solution has the form

$$R(y) = \sqrt{2}\cosh^{-1} y, \quad -\infty < y < \infty. \tag{8.142}$$

In the supercritical region, $s = 1$, there are two branches of solutions,

$$k(q) = q \pm \sqrt{q^2 - 1}, \quad q \geq 1. \tag{8.143}$$

Note that for any localized solution, the effective linear growth rate

$$\sigma_0 = s - K[A_0] = s - 2qk(q) = -[k(q)]^2 < 0 \tag{8.144}$$

in the whole region of the existence of localized solutions.

Formulation of the Stability Problem

Obviously, the stability of a localized solution does not depend on x_0 and Θ. Below, we fix $x_0 = \Theta = 0$; thus, $A_0(x) = R(x)$ is real. In order to investigate the stability, we consider the evolution of a disturbance on the background of a stationary solution. Linearizing equation (8.137) around the localized solution (8.138),

$$A(x,t) = A_0(x) + \tilde{A}(x,t), \tag{8.145}$$

we find:

$$\frac{\partial \tilde{A}(x,t)}{\partial T} = \frac{\partial^2 \tilde{A}(x,t)}{\partial x^2} + [s - 2qk(q) + 2A_0^2(x)]\tilde{A}(x,t) + \tag{8.146}$$

$$A_0^2(x)\tilde{A}^*(x,t) - 2qk(q)\mathrm{Re}\tilde{A}(0,t).$$

It is assumed that $|\tilde{A}(x,t)|$ is bounded as $x \to \pm\infty$. Define $\tilde{A}(x,t) = \tilde{A}_r(x,t) + i\tilde{A}_i(x,t)$, where $\tilde{A}_r$ and $\tilde{A}_i$ are real functions. The problems for $\tilde{A}_r$ and $\tilde{A}_i$ are not coupled:

$$\frac{\partial \tilde{A}_r(x,t)}{\partial T} = \frac{\partial^2 \tilde{A}_r(x,t)}{\partial x^2} + [s - 2qk(q) + 3A_0^2(x)]\tilde{A}_r(x,t) - 2qk(q)\tilde{A}_r(0,t); \tag{8.147}$$

$$\frac{\partial \tilde{A}_i(x,t)}{\partial t} = \frac{\partial^2 \tilde{A}_i(x,t)}{\partial x^2} + [s - 2qk(q) + A_0^2(x)]\tilde{A}_i(x,t); \tag{8.148}$$

recall that

$$A_0(x) = \sqrt{2}k(q)\cosh^{-1}[k(q)x].$$

Later on, we rescale the coordinate, $z \equiv k(q)x$, and consider normal modes,

$$\tilde{A}_r(x,t) = u(z)e^{\sigma t}, \quad \tilde{A}_i(x,t) = v(z)e^{\sigma t}.$$

Taking into account relation (8.140), we obtain an equation which is equally valid in the subcritical and supercritical cases:

$$k^2 u'' + \left(-k^2 - \sigma + \frac{6k^2}{\cosh^2 z} \right) u = 2kq \frac{u(0)}{\cosh z}; \; |u| < \infty, \; z \to \pm\infty; \qquad (8.149)$$

$$k^2 v'' + \left(-k^2 - \sigma + \frac{2k^2}{\cosh^2 z} \right) v = 0; \; |v| < \infty, \; z \to \pm\infty, \qquad (8.150)$$

where $'$ means the differentiation with respect to z.

Problem (8.149) describes amplitude disturbances of the localized solution, while (8.150) describes its phase disturbances.

Phase Disturbances

Let us start with problem (8.150). Rewriting this problem as

$$-v'' + (1 - 2\cosh^{-2} z)v = -(\sigma/k^2)v; \; |v| < \infty, \; z \to \pm\infty, \qquad (8.151)$$

we obtain a well-known eigenvalue problem for the Schrödinger equation which is exactly solvable (see, e.g., [85] or [86]). The continuum spectrum of the problem is $-(\sigma/k^2) > 1$; hence, it does not produce an instability. The only discrete eigenvalue is $\sigma = 0$; the corresponding eigenfunction is

$$v(z) = \cosh^{-1} z,$$

which corresponds to an infinitesimal change of Θ in (8.138).

Amplitude Disturbances

In the present subsection, we analyze the nonlocal eigenvalue problem (8.149).

In the case $q = 0$, the localized solution exists only in subcritical region, $s = -1$, and is described by (8.139) with $k = 1$. The eigenvalue problem (8.149) can be presented as

$$-u'' + (1 - 6\cosh^{-2} z)u = -\sigma u; \; |u| < \infty, \; z \to \pm\infty. \qquad (8.152)$$

Again, the continuum spectrum of the problem is located at $\sigma < -1$ and does not produce any instability. The discrete spectrum includes two eigenvalues [85, 86]:

$$\sigma = 0, \; u = \sinh z \cosh^{-2} z;$$

$$\sigma = 3, \; u = \cosh^{-2} z.$$

The first mode corresponds to a translation of the localized solution (an infinitesimal change of x_0 in (8.138)). The second mode is the origin of the instability of an uncontrolled subcritical localized solution.

Below we consider equation (8.149) in the general case, $q \neq 0$. Obviously, $\sigma < -k^2 < 0$ for any disturbances which do not decay as $z \to \infty$. Hence, for the stability analysis, it is sufficient to consider localized solutions with $\text{Re}\,\sigma > -k^2$. Due to the symmetry of equation (8.149), any eigenfunctions can be presented as either even or odd functions of z. For odd eigenfunctions, $u(0) = 0$, we return to the uncontrolled case discussed above. Therefore, later on we consider only even solutions of equation (8.149).

Let us fix the norm of the eigenfunction $u(z)$ by the condition

$$u(0) = 1$$

and present the eigenvalue problem in the form

$$u'' + \left(\frac{6}{\cosh^2 z} - r^2 \right) u = m \frac{1}{\cosh z}, \quad -\infty < z < \infty; \tag{8.153}$$

$$|u| \to 0, \ z \to \pm\infty; \tag{8.154}$$

$$u(0) = 1, \tag{8.155}$$

where

$$r^2 = \frac{\sigma + k^2}{k^2}, \quad m = \frac{s + k^2}{k^2}.$$

According to (8.141), in the subcritical region , $s = -1$,

$$m = \frac{2q}{q + \sqrt{q^2 + 1}}.$$

Hence, $0 < m < 1$ for a stabilizing control ($q > 0$) and $-\infty < m < 0$ for a destabilizing control ($q < 0$). In the supercritical region, $s = 1$,

$$m = \frac{2q}{q \pm \sqrt{q^2 - 1}},$$

where $q \geq 1$ (see (8.143)). One can see that $1 < m \leq 2$ for the upper branch and $2 \leq m < \infty$ for the lower branch. Thus, the stability of localized solutions in all the cases mentioned above can be studied by means of equation (8.153).

The eigenvalue σ is above the continuous spectrum, if $\text{Re}(r^2) > 0$. The instability corresponds to $\text{Re}(r^2) > 1$.

The general solution of equation (8.153) can be presented as

$$u(z) = u_0(z) + u_p(z),$$

where u_0 is the general solution of the homogeneous equation and u_p is a particular solution of the inhomogeneous equation. Because we are interested only in even solutions of the problem, it is sufficient to consider the region $0 \leq z < \infty$.

The general solution of the homogeneous equation is

$$u_0(z) = C_+^h \, P_2^r(\tanh z) + C_-^h \, P_2^{-r}(\tanh z), \tag{8.156}$$

where $P_n^m(x)$ denotes the associated Legendre polynomial. The particular solution of the inhomogeneous equation can be found using the method of constant variation, which leads to the expression

$$u_0(z) = C_+^i(z)\, P_2^r(\tanh z) + C_-^i(z)\, P_2^{-r}(\tanh z), \tag{8.157}$$

where

$$C_{\pm}^i(z) = \pm \frac{\pi m \exp[(1-r^2)\tau/(m-1)]}{2\sin \pi r} \int \frac{P_2^{\mp r}(\tanh z)}{\cosh z}\, dz =$$
$$\pm \frac{\pi m \exp[(1-r^2)\tau/(m-1)]}{2\sin \pi r} \int \frac{P_2^{\mp r}(y)}{\sqrt{1-y^2}}\, dy . \tag{8.158}$$

The computation of the integral in (8.158) gives the following expression:

$$C_{\pm}^i = \pm \frac{\pi m \exp[(1-r^2)\tau]/(m-1)}{2\Gamma(3\pm r)\sin \pi r} \left[3\left(I_3^{(\mp r+3)/2}(w) - 2I_3^{(\mp r+1)/2}(w) + I_3^{(\mp r-1)/2}(w) \right) \right.$$
$$\left. \pm\, 3p\left(I_2^{(\mp r+1)/2}(w) - I_2^{(\mp r-1)/2}(w) \right) + (r^2-1)I_1^{(\mp r-1)/2}(w) \right], \tag{8.159}$$

where $w = (1+y)/(1-y) = (1+\tanh z)/(1-\tanh z)$,

$$I_n^{\alpha}(w) = \int \frac{w^{\alpha}}{(w+1)^n}\, dw = \frac{w^{1+\alpha}}{1+\alpha}\, {}_2F_1(n, 1+\alpha; 2+\alpha; -w). \tag{8.160}$$

Finally, the general solution of the problem reads

$$u(z) = (C_+^h + C_+^i)P_2^r(\tanh z) + (C_-^h + C_-^i)P_2^{-r}(\tanh z). \tag{8.161}$$

Condition (8.154) leads to the following values of the coefficients:

$$C_+^h = -\frac{\pi m \exp[(1-r^2)\tau/(m-1)]}{2\Gamma(3+r)\sin \pi r} \frac{\pi(1-r^2)}{2} \sec \frac{\pi r}{2}, \quad C_-^h = 0. \tag{8.162}$$

Finally, substituting equations (8.161), (8.159), and (8.162) into condition (8.155) and using the properties of the Γ-function and hypergeometric functions [87], we arrive to the following relation:

$$\frac{m}{2r(4-r^2)} \left\{ 3r + \frac{r^2-1}{2} \left[\psi\left(\frac{r+1}{4}\right) - \psi\left(\frac{r+3}{4}\right) \right] \right\} = 1. \tag{8.163}$$

where $\psi(x) = \Gamma'(x)/\Gamma(x)$ is the logarithmic derivative of Γ-function. Equation (8.163) describes the dependence of the growth rate on the control parameter for both subcritical and supercritical regions in an implicit form.

For monotonic disturbances (real r), the dependence $m(r)$ can be found explicitly. One can see that for any $m < 2$, there are two values of r, one of which is larger than 1. That means that pulse solutions in the subcritical region and the upper branch of pulse solutions in the supercritical region are unstable.

In the region $2 < m < m_* = 2.02193$, both values of r are real and less than 1. For $m > m_*$ the two eigenvalues of r are complex conjugate with $Re(r^2) < 1$. For large m, the leading-order terms of the asymptotic expansion for $r(m)$ can be presented in the form

$$r = \frac{\pi(m-4)}{4} e^{-\pi\sqrt{m-5}/2} + i\sqrt{m-5}; \tag{8.164}$$

hence, the stability condition $Re(r^2) < 1$ is not violated.

Thus, the lower branch of pulse solutions in the supercritical region is stable in the whole region of its existence $m > 2$, i.e., $p > \sqrt{2}$. The stability of pulse solutions under the global control was observed in numerical simulations [82, 88].

Instability of Localized Solutions Under the Action of Delayed Control

Usually, in systems with feedback control, there is a delay between the measurement of the system parameters by sensors and the application of control action by actuators. In some systems, this delay is small and can be neglected. The analysis described above is valid for this case. In the present section, we consider the general case when a delay in feedback control is present. Thus, we consider a more general nonlinear control,

$$K[A] = -p\max_x |A(x, t - \tau)|, \; p > 0, \tag{8.165}$$

where $\tau = \text{const}$ is the control delay. Obviously, the stationary solutions of equation (8.137) are not affected by the delay. However, the control delay may change the stability properties of solutions and create new dynamic regimes.

Linear Stability Theory

In the presence of the delay, equation (8.163) becomes

$$\frac{m\exp[(1-r^2)\tau/(m-1)]}{2r(4-r^2)} \left\{ 3r + \frac{r^2-1}{2}\left[\psi\left(\frac{r+1}{4}\right) - \psi\left(\frac{r+3}{4}\right)\right]\right\} = 1. \tag{8.166}$$

Obviously, the monotonic stability boundary, $r^2 = 1$, is not changed by the delay. Hence, the boundary between monotonically stable and monotonically unstable solutions, $m = 2$, is unchanged; hence, the upper-branch pulses are unstable. However, the delay can produce an oscillatory instability of supercritical pulse [88]. The end point of the oscillatory instability boundary corresponds to $p = \sqrt{2}$ and $\tau = (1 - \ln 2)/3 \approx 0.1$. For $p \gg 1$, $\tau \to \pi/2$.

The Variational Approach

The nonlinear development of the oscillatory instability can be studied either by means of a direct simulation of equation (8.129) [88] or by means of a variational approach [89, 90]. The latter approach has been widely used for solving problems of nonlinear optics governed primarily by partial differential equations [91], including nonconservative problems [92, 93]. The equations that are used for the treatment of complex dissipative systems are written in the following form:

$$F[u] = Q, \tag{8.167}$$

where $F[u]$ is a conservative part of (8.167), that is, there exists a Lagrangian $L[u, u^*]$ such that

$$\left(\frac{\delta L}{\delta u}\right)^* = \frac{\delta L}{\delta u^*} = F[u], \tag{8.168}$$

and Q is a nonconservative part of (8.167). If one utilizes an ansatz of the form

$$u = u(b_1(t), b_2(t), \dots, b_N(t),), \tag{8.169}$$

then the variational technique which is used in [92] can be represented by the following system of equations:

$$\frac{d}{dt}\frac{\partial <L>}{\partial (b_j)_t} - \frac{\partial <L>}{\partial b_j} = 2\mathrm{Re}\int_{-\infty}^{\infty} Q\frac{\partial u^*}{\partial b_j}dx,$$

$$j = 1, \dots, N, \tag{8.170}$$

where

$$<L> = \int_{-\infty}^{\infty} L[u(b_1(t), b_2(t), \dots, b_N(t), x), u^*(b_1(t), b_2(t), \dots, b_N(T), x)] dx.$$

$$\tag{8.171}$$

We apply the variational method to the equation

$$A_t = A + A_{xx} + |A|^2 A + K[A]A, \tag{8.172}$$

which is valid in the supercritical region, with a control functional of the form (8.165). Equation (8.172) is presented in the form

$$iA_t = i\left((1 + K[A])A + A_{xx} + |A|^2 A\right). \tag{8.173}$$

The right-hand side of (8.173) is the nonconservative part, Q. The Lagrangian of the left-hand side of (8.173) is as follows:

$$L = \frac{i}{2}\left(AA_t^* - A^*A_t\right). \tag{8.174}$$

For studying the dynamics of a localized solution, we use the ansatz compatible with the exact localized solution (8.138), (8.139),

$$A(x, t) = \frac{C(t)}{\cosh[\kappa(t)(x - x_0)]}e^{i\Theta} \tag{8.175}$$

with $C(t)$, $\kappa(t)$ playing the role of $b_1(t)$, $b_2(t)$. The control functional can be written in the form $K(A) = -pC(t - \tau)$. The obtained evolution equations for $C(t)$, $\kappa(t)$ are as follows:

$$C_t(t) = \frac{1}{9}C(t)\left[9 - 9pC(t - \tau) + 8C^2(t) - 7\kappa^2(T)\right],$$

$$\kappa_t(t) = \frac{4}{9}\kappa(t)\left[C^2(t) - 2\kappa^2(t)\right]. \tag{8.176}$$

The stationary nontrivial solutions of (8.176) are given by formulas

$$C = p \pm \sqrt{p^2 - 2},$$

$$\kappa = \frac{\sqrt{2}}{2}C, \tag{8.177}$$

which reproduce the exact solution (8.139), (8.141).

We linearize system (8.176) around solution (8.177). The linear growth rate of the perturbed solution, λ, is a solution of the following characteristic equation:

$$\lambda^2 - \lambda\left(\frac{16}{9}\kappa^2 - \sqrt{2}p\kappa e^{-\lambda\tau}\right) + \frac{16}{9}\kappa^3(\sqrt{2}pe^{-\lambda\tau} - 2\kappa) = 0, \tag{8.178}$$

with κ defined by the relation

$$\kappa(p) = \frac{\sqrt{2}p}{2} \pm \sqrt{\frac{p^2}{2} - 1}. \tag{8.179}$$

Setting $\lambda = 0$ in (8.178), one can obtain the monotonic instability boundary $\kappa = p/\sqrt{2}$ which is equivalent to a condition $p = \sqrt{2}$. Note that for $p < \sqrt{2}$ the stationary solutions do not exist. The condition $p = \sqrt{2}$ corresponds to a saddle-node bifurcation point where two stationary solutions, a stable one and an unstable one, are born. As mentioned above, the stable branch corresponds to the sign "-" in equation (8.177).

The oscillatory instability boundary is obtained by substituting $\lambda = iw$ in (8.178). The following system of equations is obtained:

$$-w^2 + w\sqrt{2}p\kappa\sin(w\tau) + \frac{16}{9}\kappa^3[\sqrt{2}p\cos(w\tau) - 2\kappa] = 0, \tag{8.180}$$

$$w\left[\frac{16}{9}\kappa^2 - \sqrt{2}p\kappa\cos(w\tau)\right] + \frac{16\sqrt{2}}{9}p\kappa^3\sin(w\tau) = 0, \tag{8.181}$$

with κ defined by (8.179). First, we can find asymptotes for $p \to \sqrt{2}$ and for $p \to \infty$. For $p \to \sqrt{2}$, it holds $\kappa \to 1$ and $w \to 0$; therefore, $\sin(w\tau) \sim w\tau$ and $\cos(w\tau) \sim 1$. Then from (8.181), it follows $\tau = 0.0625$. Recall that the exact value of a delay parameter τ for $p = \sqrt{2}$, obtained in [88], is $\tau = (1 - \ln(2))/3 \approx 0.107$.

For $p \to \infty$, it holds $\kappa = p/\sqrt{2} - \sqrt{p^2/2 - 1} \sim 1/(\sqrt{2}p)$, and thus in the leading order, we obtain

$$-w^2 + w\sin(w\tau) = 0, \tag{8.182}$$

$$w\cos(w\tau) = 0. \tag{8.183}$$

Assuming $w \neq 0$, we obtain from (8.183) that $w\tau = \pi/2$. Then from (8.182), it follows $w = 1$, and therefore $\tau = \pi/2$ for $p \to \infty$, which is the same as in [88].

Next, we can derive an analytic expression for the dependence of a delay parameter τ on a control parameter p. From (8.180) and (8.181), one can express $\sin(w\tau)$ and $\cos(w\tau)$ in terms of w, p and $\kappa(p)$. Using the trigonometry identity $\sin^2(u) + \cos^2(u) = 1$, one obtains the following equation for w:

$$w^4 - 2w^2\left(p^2\kappa^2 - \frac{416}{81}\kappa^4\right) + \frac{512}{81}\kappa^6(2\kappa^2 - p^2) = 0. \tag{8.184}$$

The solution of (8.184) that matches the asymptotes found above is as follows:

$$w = p\kappa\left[1 - \frac{416\kappa^2}{81p^2} + \left(1 - \frac{320\kappa^2}{81p^2} + \frac{90112\kappa^4}{6561p^4}\right)^{1/2}\right]^{1/2}. \tag{8.185}$$

Now we have

$$\tau = \frac{1}{w}\arcsin\left[\frac{\sqrt{2}w(w^2 + 32\kappa^4/9) - 256\sqrt{2}\kappa^4 w/81}{2p\kappa(w^2 + 256\kappa^4/81)}\right]. \tag{8.186}$$

The region of stability of the localized solutions is shown in Figure 8.6. The comparison of the obtained oscillatory instability boundary with the exact results from [88] shows that the distance between the two boundaries is rather small and negligible for large values of p. Therefore, we conclude that the low-dimensional model gives a good approximation to a complete system especially with the increase of a control parameter.

The low-dimensional model (8.176) allows finding the type of the Hopf bifurcation on the oscillatory instability boundary in a simpler way than the exact approach. For this purpose, a small disturbance is added to the critical value of a delay parameter, i.e., we consider $\tau(\varepsilon) = \tau_0 + \varepsilon^2\tau_2 + \dots$. It has been shown that the type of the bifurcation is changed at $p_* \approx 2.08$: the Hopf bifurcation is subcritical (the small solution exists for $\tau_2 < 0$), as $p < p_*$, and supercritical ($\tau_2 > 0$), as $p > p_*$ (see Figure 8.7). The value of p_*, found numerically in [88], is $p_* \approx 2.022$.

The nonlinear simulations of (8.176) confirm the conclusions of the bifurcation theory (see Figure 8.8). For $p < p_*$, $p_* \approx 2.08$, the Hopf bifurcation is subcritical and the solution blows up for $\tau > \tau_0(p)$. For $p > p_*$ the bifurcation is supercritical and the solution exhibits stable oscillatory regime for $\tau > \tau_0(p)$. With an increase of τ for sufficiently large values of p, the oscillatory regime is characterized by strong variations of the amplitude $C(t)$. During a long time interval, which is large as compared with the delay parameter τ, the amplitude $C(t)$ is small and grows in an exponential way. Then it reaches its maximum value and decreases down to its minimum value during a relatively short period of time, comparable with τ. The

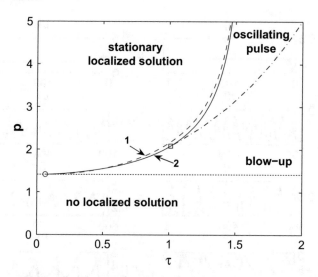

Fig. 8.6 Exact (line 1) and approximate (line 2) stability boundaries for the stationary solution and the blow-up boundary (dash-dotted line) for the oscillating solution. The circle corresponds to $p = \sqrt{2}$ and $\tau = 0.0625$. The square at $p \approx 2.0828$, $\tau \approx 1.01116$ corresponds to the merging point of the approximate stability and the blow-up boundaries

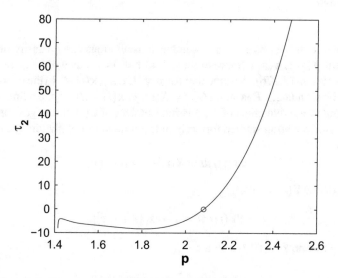

Fig. 8.7 Bifurcation parameter τ_2 for different values of p

oscillations resemble a limit cycle close to a homoclinic bifurcation, though there is no homoclinic bifurcation in the present system. The further increase of a delay parameter leads to a blow-up of the solution (see Figure 8.6).

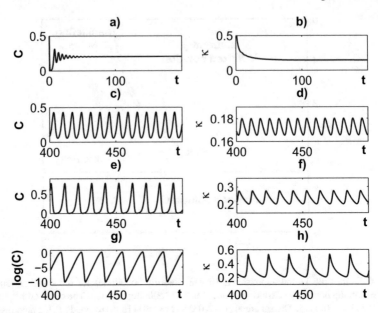

Fig. 8.8 Regimes for $C(t)$ and $\kappa(t)$ for $p = 5$ and: a), b) $\tau = 1.3$; c), d) $\tau = 1.6$; e), f) $\tau = 1.9$; g), h) $\tau = 1.99657$.

The nature of the observed delay-induced oscillations can be easily understood in the limit $p \gg 1$, $\tau \gg 1$ (more exactly, $1 \ll \tau \ll \ln(p)$, as will be shown below). Consider system (8.176). Assume that for $p \gg 1$, $C(t), \kappa(t) \ll 1$ (the conditions for that will be seen later). Rescaling $C(t) = X(t)/p$, $\kappa(t) = Z(t)/p$, we find that at the leading order, the dynamics of $X(t)$ is independent of $Z(t)$ and is governed by the delay logistic equation studied formerly in the context of population dynamics (see [94–96])

$$dX(t)/dt = X(t)[1 - X(t - \tau)], \tag{8.187}$$

or, if we define $Y(t) = \ln[X(t)]$,

$$dY(t)/dt = 1 - \exp[Y(t - \tau)]. \tag{8.188}$$

Rescaling $t = \tau s$, $Y = \tau y$, we obtain

$$dy(s)/ds = 1 - \exp[\tau y(s - 1)]. \tag{8.189}$$

Assume now that $\tau \gg 1$. In that case, we can construct the solution at the leading order analytically.

First, let us consider the time interval where $y(s - 1) < 0$. In that case, $y = s + const$ up to exponentially small terms. Shifting the origin of s in such a way that

$$y(s) = s, \tag{8.190}$$

we find that this epoch finishes as $s = 1$.

Consider now the time interval $1 < s < 2$. In that region, we can use the solution (8.190) for $y(s-1)$; hence,

$$y'(s) = 1 - \exp[\tau(s-1)]. \tag{8.191}$$

Except the small region $s = 1 + O(1/\tau)$, the second term prevails, hence at the leading order

$$y(s) = -\exp[\tau(s-1)]/\tau + O(1). \tag{8.192}$$

Thus, $y(s)$ strongly decreases and reaches

$$y(2) = -\exp(\tau)/\tau + O(1) \tag{8.193}$$

at $s = 2$.

For $s > 2$, $y(s-1)$ is again negative, except a small region $s = 2 + O(1/\tau)$. We get again $dy/ds = 1$ and

$$y(s) = y(2) + s - 2 = s - \exp(\tau)/\tau + O(1). \tag{8.194}$$

Now we can see that the duration of the epoch of the linear growth is $\sim \exp(\tau)/\tau$. Then $y(s-1)$ changes its sign again.

So, we have an alternation of two kinds of epochs: (i) linear growth from $y \sim -\exp(\tau)/\tau$ until $y = 1$ during the time interval $\Delta s_I \sim \exp(\tau)/\tau$ and (ii) exponential change from $y = 1$ until $y \sim -\exp(\tau)/\tau$ during the time interval $\Delta s_{II} \sim 1 + o(1/\tau)$. Returning to t and $X(t)$, we see that $X(t)$ grows exponentially from $\exp[-\exp(\tau)]$ to $\exp(\tau)$ during the time interval $\Delta t_I \sim \exp(\tau)$ and then decreases superexponentially from $\exp(\tau)$ to $\exp[-\exp(\tau)]$ during the time interval $\Delta t_{II} \sim \tau$.

In the general case, the oscillations of the amplitude $C(t)$ are influenced by the term nonlinear in $C(t)$ and coupled with oscillations of the parameter $\kappa(t)$. However, the nature of the oscillations is unchanged.

The origin of the blow-up can be analyzed with the help of the Poincaré mappings. For fixed values of p and τ, we have measured the successive maximal values C_n of $C(t)$ and analyzed the dependence $f(C_n) = C_{n+1}$. The obtained mappings have the parabolic shape typical for the tangent (saddle-node) bifurcation (see Figure 8.9(a), (b)), for any values of the control parameters. No growth of the oscillations period is observed by approaching the bifurcation point. Thus, the disappearance of the time-periodic solution is caused by the saddle-node rather than homoclinic bifurcation.

Let us compare the predictions of the low-dimensional model with the results of a direct numerical simulation of the original equation (8.172) for $p = 5$ and for $\tau < \tau_b$ and $\tau > \tau_b$, where τ_b corresponds to the blow-up boundary. In this case, the values of C_n correspond to the maximal values of $|A(t)|$ during the oscillations of the pulse. As for the low-dimensional model, the period of oscillations does not change significantly by crossing the blow-up threshold. The plots of $f(C_n) = C_{n+1}$ and $f(C_n) = C_n$ are shown in Figure 8.9(c), (d), and they have no qualitative distinctions from the low-dimensional case.

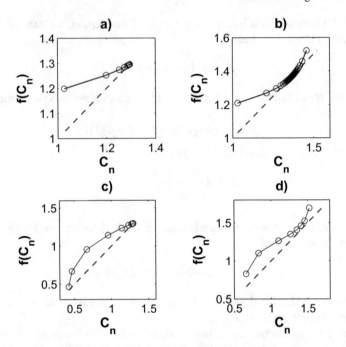

Fig. 8.9 Poincare mappings for $p = 5$. For variational model (8.176) with $\tau_b = 1.99658$: a) $\tau = 1.996$; b) $\tau = 1.9966$. For direct numerical simulation of equation (8.172) with $\tau_{blow-up=1.951}$: c) $\tau = 1.95$; d) $\tau = 1.952$. Solid line: $f(C_n) = C_{n+1}$. Dashed line: $f(C_n) = C_n$

8.3.2.3 Interacting Longwave and Shortwave Modes

Let us discuss now the case where the liquid layer is subjected to a shortwave instability, described by a complex amplitude function $A(X, \tau)$, but the modulations of
the shortwave patterns create a longwave surface deformation characterized by a real
amplitude function $B(X, \tau)$ (see Section 7.3). We will consider the case where the
shortwave instability is subcritical, i.e., $\lambda = -1$ and $1 - w/m > 0$ in (7.159), (7.160);
hence, the bifurcating solutions are unstable. Similar to the problem considered in
Section 8.3.2.2, our goal is to suppress the blow-up of solutions and transform the
subcritical instability into a supercritical one by means of a global feedback control,

$$A_{tau} = (1 - p \max_X |A|)A + A_{XX} + |A|^2 A + AB, \qquad (8.195)$$

$$B_{tau} = mB_{XX} + w|A|^2_{XX}.$$

Here p is the parameter characterizing the control strength. This way of control
can be easily implemented by changing the global parameters of the system (see
Section 8.3.2.2).

Stationary Patterns

Because the structures obtained in numerics [97] are characterized by a constant value of the phase, we fix that value equal to zero and arrive to the following system of equations for real variables A and B:

$$A_{XX} + (1 - p \max_X A)A + A^3 + AB = 0, \tag{8.196}$$

$$mB_{XX} + w(A^2)_{XX} = 0. \tag{8.197}$$

Only sign-preserving (positive) solutions corresponding to structures observed in numerical simulations are discussed below. Recall that

$$\langle B(X) \rangle = 0, \tag{8.198}$$

(cf. equation (7.163)).

The bounded solution $B(X)$ of equation (8.197) satisfying the condition (8.198) is:

$$B(X) = \left(-\frac{w}{m} \right) (A^2(X) - \langle A^2 \rangle), \tag{8.199}$$

which leads to the following equation for the amplitude $A(X)$:

$$A_{XX} - \alpha A + \beta A^3 = 0, \tag{8.200}$$

where

$$\alpha = -1 + p \max_X |A| + \frac{w}{m} \langle A^2 \rangle, \quad \beta = 1 - \frac{w}{m}. \tag{8.201}$$

Integration of (8.271) gives the relation

$$\frac{A_X^2}{2} - \frac{\alpha A^2}{2} + \frac{\beta A^4}{4} = E = \text{const.} \tag{8.202}$$

The positive solutions exist when $\alpha > 0$, $\beta > 0$, $-\alpha^2/4\beta < E < 0$, and they can be presented in terms of Jacobi elliptic functions,

$$A(X) = A_m \mathrm{dn} \left[A_m \sqrt{\frac{\beta}{2}} (X - X_0); s \right], \tag{8.203}$$

where A_m is connected with the parameter α by the relation $\alpha = (2 - s^2)\beta A_m^2/2$, and s is the modulus of the elliptic function. With $\langle A^2 \rangle = A_m^2 E(s)/K(s)$ (where $K(s)$ and $E(s)$ are the complete elliptic integrals of the first and the second kind, respectively) equation (8.201) gives the following relation that determines the amplitude A_m as a function of s:

$$\left[\frac{1}{2} \left(1 - \frac{w}{m} \right) (2 - s^2) + \frac{w}{2m} \frac{E(s)}{K(s)} \right] A_m^2 - pA_m + 1 = 0. \tag{8.204}$$

The period of the solution (8.203) is

$$L = \frac{2\sqrt{2}K(s)}{A_m\sqrt{1 - w/m}}.$$ (8.205)

In the limit $s \to 1$ ($E \to 0$), the periodic solution tends to a a localized solution such that both $A(X)$ and $B(X)$ tend to zero at infinity. In this case,

$$B(X) = -\frac{w}{m}A^2,$$ (8.206)

and the function $A(X)$ satisfies the equation

$$A_{XX} + (1 - p\max_X A)A + \left(1 - \frac{w}{m}\right)A^3 = 0.$$ (8.207)

Two nonzero localized solutions of (8.207), $(A_{\pm}(X), B_{\pm}(X))$ exist in the following interval of the coupling parameter w:

$$w_{min} = m(2 - p^2)/2 \le w < m,$$ (8.208)

and are given by

$$A_{\pm}(X) = a_{\pm}\cosh^{-1}[k_{\pm}(X - X_0)], \quad B_{\pm}(x) = -\frac{w}{m}a^2\cosh^{-2}[k_{\pm}(x - x_0)],$$ (8.209)

where

$$a_{\pm} = \frac{p \pm \sqrt{p^2 - 2(1 - w/m)}}{1 - w/m}, \quad k_{\pm} = \frac{1}{\sqrt{2}}\sqrt{1 - \frac{w}{m}a_{\pm}}.$$ (8.210)

Stability of Localized Pulses

Let us check the stability of solutions (8.209), (8.210) near the bifurcation point

$$p = p_* = \sqrt{2\beta}, \quad \beta = 1 - \frac{w}{m}$$ (8.211)

where two branches of stationary solutions are born due to a saddle-node bifurcation.

Let us parametrize the family of solutions (8.209), (8.210) using parameter k. Take $X_0 = 0$ (obviously, the stability does not depend on X_0). Both branches of solutions can be written as

$$A = a\cosh^{-1}(kX), \quad B = -\frac{w}{m}A^2,$$ (8.212)

where

$$a = \sqrt{\frac{2}{\beta}}k, \quad p = \sqrt{\frac{\beta}{2}}\frac{k^2 + 1}{k}.$$ (8.213)

Solutions with $k < 1$ correspond to the lower branch, while solutions with $k > 1$ correspond to the upper branch.

The time evolution of real disturbances $(\tilde{A}^r, \tilde{B})$ on the background of (8.212) is governed by the following system of functional–differential equations:

$$\tilde{A}^r_\tau = \tilde{A}^r_{XX} + (-k^2 + 3A^2 + B)\tilde{A}^r + A\tilde{B} - pA\tilde{A}^r(0, \tau), \qquad (8.214)$$

$$\tilde{B}_\tau = m\tilde{B}_{XX} + 2w(A\tilde{A}^r)_{XX}. \qquad (8.215)$$

Defining $z = kX$ and assuming $\tilde{A}^r = u(z)\exp(\sigma t)$, $\tilde{B} = v(z)\exp(\sigma t)$, we obtain the following system:

$$u_{zz}(z) + \left[-1 - \frac{\sigma}{k^2} + \frac{2(2+\beta)}{\beta \cosh^2 z}\right] u(z) + \frac{\sqrt{2}}{k\sqrt{\beta}} \frac{v(z)}{\cosh z} = \left(1 + \frac{1}{k^2}\right) \frac{u(0)}{\cosh z}, \quad (8.216)$$

$$v_{zz}(z) + \frac{2w}{m}\sqrt{\frac{2}{\beta}} k \left[\frac{u(z)}{\cosh z}\right]_{zz} - \frac{\sigma}{mk^2} v(z) = 0. \qquad (8.217)$$

Due to the symmetry of the problem, both components of the eigenfunction $(u(z), v(z))$ are either even or odd. For odd eigenfunctions, $u(0) = 0$, while for even ones, generally $u(0) \neq 0$. In the present section, we consider even disturbances and normalize the solution of the problem (8.216), (8.217) by the condition

$$u(0) = 1. \qquad (8.218)$$

For $k = 1$, i.e., in the point of the saddle-node bifurcation leading to the creation of two branches of solutions for $p > \sqrt{2\beta}$, the problem (8.216), (8.217) has the following exact solution:

$$\sigma = 0, \; u = u_0 = \frac{1}{\cosh z} - z\sinh z\cosh^2 z, \; v = v_0 = -\frac{2w}{m}\sqrt{\frac{2}{\beta}} \frac{u_0}{\cosh z}. \quad (8.219)$$

Near that point, we apply the expansion

$$\frac{1}{k^2} = 1 + \varepsilon, \; \sigma = \varepsilon\sigma_1 + \dots, \; u = u_0 + \varepsilon u_1 + \dots, \; v = v_0 + \varepsilon v_1 + \dots.$$

Note that $k = 1 - \varepsilon/2 + \dots$; hence, $\varepsilon > 0$ ($\varepsilon < 0$) corresponds to the lower (upper) branch of solutions.

At the first order in ε, the following system is obtained:

$$(u_1)_{zz} + \left[-1 + \frac{2(2+\beta)}{\beta \cosh^2 z}\right] u_1 + \sqrt{\frac{2}{\beta}} \frac{v_1}{\cosh z} = \frac{1}{\cosh z} + \sigma_1 u_0 - \frac{1}{2}\sqrt{\frac{2}{\beta}} \frac{v_0}{\cosh z},$$

$$(8.220)$$

$$(v_1)_{zz} + \frac{2w}{m}\sqrt{\frac{2}{\beta}} \left(\frac{u_1}{\cosh z}\right)_{zz} = \frac{\sigma_1}{m} v_0 + \frac{w}{m}\sqrt{\frac{2}{\beta}} \left(\frac{u_0}{\cosh z}\right)_{zz}. \qquad (8.221)$$

Eliminating v_1, we find:

$$(u_1)_{zz} + \left(-1 + \frac{6}{\cosh^2 z}\right) u_1 = \frac{1}{\cosh z} + \sigma_1 \left[u_0 + \frac{2}{\beta} \frac{w}{m^2} \frac{z \sinh z}{\cosh^2 z}\right]; \qquad (8.222)$$

$$|u_1| < \infty, \; z \to \pm\infty.$$

Problem (8.222) is solvable when its right-hand side is orthogonal to the solution of the homogeneous problem, which is u_0; hence,

$$\sigma_1 = -\frac{\int_{-\infty}^{\infty} u_0 \cosh^{-1}(z) \, dz}{\int_{-\infty}^{\infty} u_0 [u_0 + (2wz \sinh z)/(\beta m^2 \cosh^2 z)] \, dz}. \qquad (8.223)$$

Calculating the integral, we find:

$$\sigma_1 = -\frac{18}{(12 + \pi^2) - 2(\pi^2 - 6)w/(\beta m^2)}. \qquad (8.224)$$

One can see that if

$$w < w_* = \frac{m^2}{m + 2(\pi^2 - 6)/(\pi^2 + 12)}, \qquad (8.225)$$

then $\sigma_1 < 0$. That means that in the region (8.225), the upper branch (that with $\varepsilon < 0$) is unstable, while the lower branch (that with $\varepsilon > 0$) is stable with respect to *even* real disturbances. In the contradistinction to the case where the zero mode is absent, for $w > w_*$ the lower branch becomes unstable, and the upper branch becomes stable with respect to that kind of disturbances. However, in the next section, it will be shown that the stationary solutions become unstable with respect to *odd* disturbances leading to the development of traveling waves at the value $w = w_0 = m^2/(m+2) < w_*$.

Traveling Waves

The mechanism of a traveling wave propagation is similar to that known for an activator–inhibitor system [94], where the component A plays the role of an activator and the quantity $-B$ corresponds to the concentration of an inhibitor. The coupling between the components is different from that used in the standard FitzHugh–Nagumo (FHN) model. Nevertheless, the distribution of the inhibitor, $-B$, in a moving pulse is similar to that in an FHN model (see [98]): the component B is positive in the front part of the pulse and negative in the back part of the pulse. Therefore, the growth of the component A is enhanced in the front part and suppressed in the back part, which leads to the effective motion of the pulse as the whole.

In the present section, we find the transition threshold and analyze the bifurcation of the traveling wave solutions moving either to the right,

$$A(X, \tau) = A_r(X_r), \; B(X, \tau) = B_r(X_r), \; X_r = X - c\tau, \; c > 0, \qquad (8.226)$$

or a traveling wave moving to the left,

$$A(x, \tau) = A_l(X_l), \; B(X, \tau) = B_l(X_l), \; X_l = X + c\tau, \; c > 0. \tag{8.227}$$

Later on, we consider the wave (8.226) and omit the index r.

Transition's Threshold

The traveling wave solution satisfies the system of equations

$$A_{XX} + cA_X + A(1 - p\max_X |A|) + A^3 + AB = 0, \tag{8.228}$$

$$mB_{XX} + cB_X + w(A^2)_{XX} = 0, \tag{8.229}$$

with periodicity conditions

$$A(X + L) = A(X), \; B(X + L) = B(X), \tag{8.230}$$

and the condition of zero mean value of B:

$$\langle B \rangle = \frac{1}{L} \int_{-L/2}^{L/2} B(X)dX = 0. \tag{8.231}$$

Let us define the operator

$$\partial_X^{-1} B = \int_0^X B(y)dy - \langle \int_0^X B(y)dy \rangle, \tag{8.232}$$

and rewrite equation (3.86) in the form

$$B = -\frac{w}{m}(A^2 - \langle A^2 \rangle) - \frac{c}{m}\partial_X^{-1}B. \tag{8.233}$$

Substituting (8.233) into (8.228), we obtain:

$$A_{XX} + A(1 - p\max_X A + \frac{w}{m}\langle A^2 \rangle) + (1 - \frac{w}{m})A^3 = c(-A_X + \frac{1}{m}A\partial_X^{-1}B). \tag{8.234}$$

Near the bifurcation point, we expect the following expansions:

$$A = A_0 + \varepsilon A_1 + \varepsilon^2 A_2 + \dots, \; B = B_0 + \varepsilon B_1 + \varepsilon^2 B^2 + \dots,$$

$$c = \varepsilon c_1 + \dots, \; w = w_0 + \varepsilon w_2 + \dots. \tag{8.235}$$

The zeroth order terms A_0 and B_0 correspond to the stationary solutions (8.203), (8.199). Let us choose the constant $x_0 = 0$ so that these functions are even. Then the correction A_1 is an odd function satisfying the following equation:

$$(A_1)_{XX} + A_1 \left(1 - p \max_X A_0 + \frac{w_0}{m} \langle A_0^2 \rangle \right) + 3\left(1 - \frac{w_0}{m}\right) A_0^2 A_1 = c_1 [-(A_0)_X + \frac{1}{m} A_0 \partial_X^{-1} B_0],$$

$$(8.236)$$

$$A_1(X+L) = A_1(X), \tag{8.237}$$

(note that the addition of an odd function A_1 does not change the maximum value of $A_0 + \varepsilon A_1$ in the order $O(\varepsilon)$). The self-adjoint homogeneous problem corresponding to the linear operator on the left-hand side of (8.236) has a solution $A_1^h(X) = (A_0)_X(X)$. The solvability condition for the inhomogeneous problem (8.236), (8.237) is the condition of the orthogonality of the right-hand side to that solution,

$$- \langle [(A_0)_X]^2 \rangle + \frac{1}{m} \langle (A_0)_X A_0 \partial_X^{-1} B_0 \rangle = 0. \tag{8.238}$$

Taking into account that $(A_0)_X A_0 = -(m/2w_0)(B_0)_X$ and integrating the second integral in (8.238) by parts, we find:

$$- \langle [(A_0)_X]^2 \rangle + \frac{1}{2w_0} \langle B_0^2 \rangle = 0. \tag{8.239}$$

Substituting

$$B_0 = -\frac{w_0}{m}(A_0^2 - \langle A_0^2 \rangle)$$

(according to (8.199)), we obtain the solvability condition in the form

$$- \langle [(A_0)_X]^2 \rangle + \frac{w_0}{2m^2}[\langle A_0^4 \rangle - (\langle A_0^2 \rangle)^2] = 0. \tag{8.240}$$

Substituting the solution (8.203) for A_0, we arrive at the following expression for the threshold of the transition from the stationary to traveling wave solutions:

$$w_0 = \frac{m^2}{m + f(s)}, \tag{8.241}$$

where

$$f(s) = \frac{-(1 - s^2) + 2(2 - s^2)E(s)/K(s) - 3[E(s)/K(s)]^2}{-2(1 - s^2) + (2 - s^2)E(s)/K(s)}. \tag{8.242}$$

In the limit of the localized solution ($s \to 1$), one finds

$$w_0 = \frac{m^2}{m + 2}, \tag{8.243}$$

while in the limit of small-amplitude waves ($s \to 0$),

$$w_0 = \frac{m^2}{m + 1}. \tag{8.244}$$

Note that the instability threshold does not depend on the control parameter p but only on the zero-mode damping coefficient m.

Bifurcation Analysis

The bifurcation of traveling wave solutions reveals a surprising dependence of the bifurcation parameter $[dc^2(w)/dw]|_{w=w_0}$ on the period L. To explain that unusual phenomenon, we perform the nonlinear analysis of the bifurcation in the limit $L \gg 1$. In that case, the elliptic function (8.203) can be approximated by the localized solution (8.209).

For small c, iterating the relation (8.233), we find

$$B = -\frac{w}{m}\left[1 - \frac{c}{m}\partial_X^{-1} + \left(\frac{c}{m}\right)^2 \partial_X^{-2} - \left(\frac{c}{m}\right)^3 \partial_X^{-3} + O(c^4)\right](A^2 - \langle A^2 \rangle). \quad (8.245)$$

Substituting (8.245) into (8.228), we obtain a closed integrodifferential equation for $A(X)$:

$$A_{XX} + cA_X + A(1 - p\max_X A) + A^3 -$$

$$\frac{w}{m}A\left[1 - \frac{c}{m}\partial_X^{-1} + \left(\frac{c}{m}\right)^2 \partial_X^{-2} - \left(\frac{c}{m}\right)^3 \partial_X^{-3} + O(c^4)\right](A^2 - \langle A^2 \rangle). \quad (8.246)$$

We search the solution in the interval $-L/2 \leq X \leq L/2$.

Substitute expansion (8.235) into (8.246) and equate the terms of the same order in ε. At the zeroth order, we obtain the equation which is identical to that obtained above for the stationary solution:

$$(A_0)_{XX} + A_0(1 - p\max_X A_0) + (A_0)^3 - \frac{w_0}{m}A_0(A_0^2 - \langle A_0^2 \rangle) = 0. \quad (8.247)$$

In the limit of large L, the solution $A_0(X)$ can be approximated by the localized solution (we choose the origin is such a way that $A_0(X)$ is an even function):

$$A_0(X) = a\cosh^{-1} kX, \quad (8.248)$$

where

$$a = \left[p - \sqrt{p^2 - 2\left(1 - \frac{w_0}{m}\right)}\right]\left(1 - \frac{w_0}{m}\right)^{-1},$$

$$k^2 = \frac{1}{2}\left(1 - \frac{w_0}{m}\right)a^2.$$

At the first order, as shown in the previous subsection, the solution is governed by the equation

$$\hat{L}A_1 = (A_1)_{XX} + A_1(1 - pa) + 3\left(1 - \frac{w_0}{m}\right)A_0^2 A_1 =$$

$$c_1\left[-(A_0)_X - \frac{w_0}{m^2}A_0\partial_X^{-1}(A_0^2 - \langle A_0^2 \rangle)\right]. \quad (8.249)$$

The solvability condition for (8.249) is the orthogonality of the right-hand side to $(A_0)_X$ (the integral is calculated on the interval $(-L/2, L/2)$). In the limit of large L,

$$\partial_X^{-1}(A_0^2 - \langle A_0^2 \rangle) = \frac{a^2}{k}\left(\tanh(kX) - \frac{2X}{L}\right). \tag{8.250}$$

Because the function $(A_0)_X$ is exponentially small in the region $|X| \gg 1$, the behavior of all the functions is relevant only in the region $X = O(1)$. Finally, we obtain the threshold value (8.243). In the bifurcation point, the right-hand side of (8.249) is $O(1/L)$; hence, the solution $A_1 = C(A_0)_X + O(1/L)$, which corresponds to the possibility of an arbitrary translation of the solution, due to the symmetry of the original problem. Here and below, we shall select the solution with $C = 0$; hence, $A_1(X) = O(1/L)$ (note that

$$B_1(X) = \frac{c_1 w_0}{m^2}\partial_X^{-1}(A_0^2 - \langle A_0^2 \rangle) = \frac{c_1 w_0 a^2}{m^2 k}\left(\tanh(kX) - \frac{2X}{L}\right)$$

is not small).

At the second order, the equation is as follows:

$$\hat{L}A_2 = pA_0(X)\max_X(A_2) + \frac{w_2}{m}A_0(A_0^2 - \langle A_0^2 \rangle) +$$

$$\frac{w_0 c_1^2}{m^3}A_0\partial_X^{-2}(A_0^2 - \langle A_0^2 \rangle) - \frac{2w_0 A_0}{m}\langle A_0 A_2 \rangle. \tag{8.251}$$

The function on the right-hand side is even; hence, the solvability condition is always satisfied. The solution can be presented in the form

$$A_2 = w_2 A_2^w + c_1^2 A_2^c,$$

where A_2^w and A_2^c are solutions of the following equations:

$$\hat{L}A_2^w = pA_0(X)\max_X(A_2^w) + \frac{1}{m}A_0(A_0^2 - \langle A_0^2 \rangle) - \frac{2w_0 A_0}{m}\langle A_0 A_2^w \rangle; \tag{8.252}$$

$$\hat{L}A_2^c = pA_0(X)\max_X(A_2^c) + \frac{w_0}{m^3}A_0\partial_X^{-2}(A_0^2 - \langle A_0^2 \rangle) - \frac{2w_0 A_0}{m}\langle A_0 A_2^c \rangle. \tag{8.253}$$

Note that

$$\partial_X^{-2}(A_0^2 - \langle A_0^2 \rangle) = \frac{a^2}{k}\left(\frac{1}{k}\ln\cosh(kX) - \frac{x^2}{L} - \frac{L}{6}\right). \tag{8.254}$$

Thus, in the region $X = O(1)$, the corresponding term is $O(L)$. As noted above, the bifurcation equation is determined by the values of the functions in the region $X = O(1)$. In this region, the solutions in the leading order of L are as follows:

$$A_2^w = \frac{a^3}{2m(k^2-1)}\cosh^{-1}kX - \frac{a^3(1+k^2)}{4mk(k^2-1)}X\sinh kX\cosh^{-2}kX + O(1/L);$$
$$(8.255)$$

$$A_2^c = -\frac{w_0 a^3 L}{6m^3 k(k^2-1)}(\cosh^{-1}kX - kX\sinh kX\cosh^{-2}kX) + O(1). \qquad (8.256)$$

At the third order, we obtain the following equation:

$$\hat{L}A_3 = -c_1(A_2)_X - \frac{w_2 c_1}{m^2}A_0\partial_X^{-1}(A_0^2 - \langle A_0^2\rangle) - \frac{w_0 c_1}{m^2}A_2\partial_X^{-1}(A_0^2 - \langle A_0^2\rangle) - \quad (8.257)$$

$$\frac{2w_0 c_1}{m^2}A_0\partial_X^{-1}(A_0 A_2 - \langle A_0 A_2\rangle) - \frac{w_0 c_1^3}{m^4}A_0\partial_X^{-3}(A_0^2 - \langle A_0^2\rangle).$$

Substituting (8.248), (8.255), and (8.256) into the solvability condition of (8.257), we obtain the bifurcation equation

$$c_1(w_2 I_w + c_1^2 I_c) = 0, \qquad (8.258)$$

where

$$I_w = \langle A_{0X}A_{2X}^w\rangle + \frac{1}{m^2}\langle A_{0X}A_0\partial_X^{-1}(A_0^2 - \langle A_0^2\rangle)\rangle$$

$$+ \frac{w_0}{m^2}\langle A_{0X}A_2^w\partial_X^{-1}(A_0^2 - \langle A_0^2\rangle)\rangle$$

$$+ \frac{2w_0}{m^2}\langle A_{0X}A_0\partial_X^{-1}(A_0 A_2^w - \langle A_0 A_2^w\rangle)\rangle,$$

$$I_c = \langle A_{0X}A_{2X}^c\rangle + \frac{w_0}{m^2}\langle A_{0X}A_2^c\partial_X^{-1}(A_0^2 - \langle A_0^2\rangle)\rangle$$

$$+ \frac{2w_0}{m^2}\langle A_{0X}A_0\partial_X^{-1}(A_0 A_2^c - \langle A_0 A_2^c\rangle)\rangle$$

$$+ \frac{w_0}{m^4}\langle A_{0X}A_0\partial_X^{-3}(A_0^2 - \langle A_0^2\rangle)\rangle.$$

Using integration by parts, one finds that

$$I_w = \langle A_2^w\left[-A_{0XX} + \frac{w_0}{m^2}A_{0X}\partial_X^{-1}(A_0^2 - \langle A_0^2\rangle) - \frac{w_0}{m^2}A_0^3\right]\rangle - \frac{1}{m^2}\langle A_0^4\rangle + O(1/L^2).$$

Evaluation of this expression, which is done with the help of the relation

$$A_{0XX} + \frac{w_0}{m^2}A_{0X}\partial_X^{-1}(A_0^2 - \langle A_0^2\rangle) + \frac{w_0}{m^2}A_0^3 = O(1/L^2)$$

(for $X = O(1)$), gives

$$I_w = -\frac{a^4}{3w_0 kL} + O(1/L^2).$$

Similarly, I_c can be transformed to the following expression:

$$I_c = \frac{2w_0}{m^2}\langle A_2^c A_{0X}A_0\partial_X^{-1}(A_0^2 - \langle A_0^2\rangle)\rangle + \frac{w_0 a^2 L}{12m^4 k}\langle A_0^2\rangle + O(1/L).$$

Finally, we obtain

$$I_c = \frac{w_0 a^4}{6m^4 k^2} + O(1/L).$$

Using the obtained values of the coefficients I_w and I_c, we find that at the threshold point,

$$\frac{dc^2(w)}{dw} = \frac{c_1^2}{w_2} = -\frac{I_w}{I_c} = \frac{2k(m+2)^2}{L}. \tag{8.259}$$

Note that this quantity tends to zero as $L \to \infty$.

8.3.2.4 Subcritical Ginzburg–Landau Equation with Complex Coefficients

Let us discuss now the influence of the global feedback control on patterns created by an oscillatory instability and governed by the subcritical Ginzburg–Landau equation with complex coefficients,

$$A_t = \mu A + (b_1 + ib_2)A_{xx} - (c_1 + ic_2)|A|^2 A, \tag{8.260}$$

for the complex amplitude $A(x,t)$ of an unstable mode. A subcritical instability takes place of $c_1 < 0$. The physical examples of subcritical oscillatory instabilities can be found in [99–101]. One could expect that in the case of a subcritical instability, all the solutions blow up in finite time for $\mu > 0$. Nevertheless, in the case of a focusing nonlinearity ($c_2/b_2 < 0$), stable bounded solutions have been found analytically and observed in numerical simulations in a certain parameter region [102–104].

Exact Pulse Solutions

In the present section, we will discuss only spatially localized, pulse-like, structures. The description of spatially periodic waves generated by subcritical oscillatory instability under the action of feedback control can be found in [105, 106].

By rescaling, the subcritical complex Ginzburg–Landau equation (CGLE) under a feedback control can be rewritten as

$$A_t = A + (1+ib)A_{xx} - (-1+ic)|A|^2 A + K(A)A. \tag{8.261}$$

Feedback control is imposed by adding a term $K(A)A$ to the right-hand side of the equation (8.261), with a control functional $K(A)$ of the form

$$K(A) = -p \max_x |A|. \tag{8.262}$$

Solutions can be presented in the form

$$A(x,t) = R(x,t)e^{i\theta(x,t)}, \tag{8.263}$$

where $R(x,t)$ and $\theta(x,t)$ are real functions, and thus the control functional can be written as $K(A) = -p\max_x R$. Substituting (8.263) into the equation (8.261) and denoting $\max_x R \equiv R_{max}$, we obtain the following system of two real equations:

$$R_t = R_{xx} - R\theta_x^2 - b(2\theta_x R_x + R\theta_{xx}) + R^3 + (1 - pR_{max})R,$$
$$R\theta_t = b(R_{xx} - R\theta_x^2) + (2\theta_x R_x + R\theta_{xx}) - cR^3. \tag{8.264}$$

Note that the parameter $\mu = 1 - pR_{max}$ is the effective linear growth rate parameter. The system of equations (8.264) has the following exact pulse solutions:

$$R(x,t) = \frac{C}{\cosh \kappa x}, \qquad \theta(x,t) = \gamma \ln \cosh \kappa x - \Omega t, \tag{8.265}$$

where

$$C = \frac{p}{2(1-\alpha)} \left[1 \pm \sqrt{1 - 4(1-\alpha)/p^2} \right] > 0,$$

$$\alpha = \frac{1}{12(1+b^2)} \left[\sqrt{9(bc-1)^2 + 8(b+c)^2} - 3(bc-1) + 4b(b+c) \right],$$

$$\gamma = \frac{1}{b+c} \left[6\alpha(1+b^2) + 3(bc-1) - 2b(b+c) \right],$$

$$\kappa^2 = C^2 \frac{1}{3\gamma} \frac{b+c}{1+b^2},$$

$$\Omega = b\kappa^2 - \gamma\kappa^2 + cC^2. \tag{8.266}$$

For $\alpha < 1$, two solutions for C exist for $p > 2\sqrt{1-\alpha}$, and the linear growth rate $\mu = 1 - pC < 0$. Therefore, the solutions can be stable in this region of parameters or, in other notation, for $\{-b - 3\sqrt{b^2+1} < c < -b + 3\sqrt{b^2+1}\}$.

For $\alpha > 1$, i.e., for $c > -b + 3\sqrt{b^2+1}$ or for $c < -b - 3\sqrt{b^2+1}$, there is only one solution for $C > 0$, and there holds $\mu = 1 - pC > 0$. Because μ is the effective linear growth rate, the solution (8.265) is obviously unstable with respect to disturbances that do not decay as $x \to \pm\infty$.

In the latter case, it is interesting to investigate the nonlinear dynamics produced by the instability of pulse solutions. In [107], the problem was studied by means of a direct numerical simulation of (8.261). A remarkable observation done in [107] is that in many cases, the dynamic regime is determined by an indirect interaction of pulses through the applied feedback control. In a certain region of parameters, a coexistence of pulses leads to the development of multi-pulse regimes. In another region of parameters, the pulses compete; therefore, only one pulse "rules" for some time, until it gets "overthrown" by another pulse. Below we provide an explanation of those phenomena.

Variational Approach

We apply the variational method to equation (8.261), with a control functional of the form (8.262). Equation (8.261) is presented in the form

$$iA_t + bA_{xx} - c|A|^2 A = i\left((1 - K(A))A + A_{xx} + |A|^2 A\right) \qquad (8.267)$$

(cf. (8.173)). The right-hand side of (8.267) is the nonconservative part, Q. The Lagrangian of the left-hand side of (8.267) is as follows:

$$L = \frac{i}{2}(AA_t^* - A^*A_t) + b|A_x|^2 + \frac{c}{2}|A|^4. \qquad (8.268)$$

For studying the dynamics of a one-pulse solution, we use the ansatz compatible with the exact pulse solution (8.265),

$$R(x,t) = \frac{C(t)}{\cosh[\kappa(t)x]}, \qquad \theta(x,t) = \gamma(t)\ln\cosh[\kappa(t)x] + \phi(t). \qquad (8.269)$$

The evolution equations for $C(t)$, $\kappa(t)$, $\gamma(t)$, and $\phi(t)$ are as follows:

$$C_t = \frac{1}{9}C\left[9 - 9pC + 8C^2 - (\gamma^2 + 6b\gamma + 7)\kappa^2\right],$$

$$\kappa_t = \frac{4}{9}\kappa\left[C^2 - (3b\gamma + 2 - \gamma^2)\kappa^2\right],$$

$$\gamma_t = \frac{2}{3}C^2(c - \gamma) + \frac{2}{3}\kappa^2(2b - \gamma)(1 + \gamma^2),$$

$$\phi_t = -\frac{1}{9}C^2\left[6\ln 2\,(\gamma - c) - 4(\gamma - 3c)\right]$$

$$+ \frac{1}{9}\kappa^2\left[(6\ln 2 - 4)(2b - \gamma)(1 + \gamma^2) - 7b - b\gamma^2 + 6\gamma\right]. \qquad (8.270)$$

According to definitions, $C(t)$ and $\kappa(t)$ are nonnegative.

For the investigation of interaction of two distant pulse-like solutions, we use the ansatz $A = A_1 + A_2$, $A_j = R_j \exp[i\theta_j]$, where

$$R_j(x,t) = \frac{C_j(t)}{\cosh[\kappa_j(t)(x - x_j)]}, \quad \theta_j(x,t) = \gamma_j(t)\ln\cosh[\kappa_j(t)(x - x_j)] + \phi_j(t), \quad j = 1,2$$

$$(8.271)$$

for each of the solitary wave, which leads to the following variational model (the overlap of solitary waves is disregarded):

$$\dot{C}_j = \frac{1}{9}C_j\left[9 - 9p\max\{C_1, C_2\} + 8C_j^2 - (\gamma_j^2 + 6b\gamma_j + 7)\kappa_j^2\right],$$

$$\dot{\kappa}_j = \frac{4}{9}\kappa_j\left[C_j^2 - (3b\gamma_j + 2 - \gamma_j^2)\kappa_j^2\right],$$

$$\dot{\gamma}_j = \frac{2}{3} C_j^2 (c - \gamma_j) + \frac{2}{3} \kappa_j^2 (2b - \gamma_j)(1 + \gamma_j^2), \quad j = 1, 2. \tag{8.272}$$

Equations for $\dot{\phi}_j$ do not influence the dynamics and are not written here.

Stationary Solutions

In the framework of (8.270), the exact solution (8.266) is reproduced. When the model (8.272) is used, we obtain four kinds of nontrivial stationary solutions $u_j = (C_j, \kappa_j, \gamma_j)$, $j = 1, 2$: (*i*) $C_1 \neq 0, C_2 = 0$; this leads to a solution $u_1 = (C, \kappa, \gamma)$, $u_2 = (0, 0, \gamma_2)$, where C, κ, γ are given by (8.266) and γ_2 is arbitrary; (*ii*) $C_1 = 0, C_2 \neq 0$; this leads to a solution $u_1 = (0, 0, \gamma_1)$, $u_2 = (C, \kappa, \gamma)$, where C, κ, γ are given by (8.266) and γ_1 is arbitrary; (*iii*) $u_1 = u_2 = (C, \kappa, \gamma)$, where C, κ, γ are given by (8.266); (*iv*) on the line $\alpha = 1$ (or $pC = 1$), there exist two more families of stationary solutions; the first family is $u_1 = (C, \kappa, \gamma)$, $u_2 = (C_2, \kappa_2, \gamma)$, where C_2 is an arbitrary number satisfying the condition $0 < C_2 < C$ and $\kappa_2 = C_2 / \sqrt{1 + b\gamma}$; the second one is $u_1 = (C_1, \kappa_1, \gamma)$, $u_2 = (C, \kappa, \gamma)$, $0 < C_1 < C$ and $\kappa_1 = C_1 // \sqrt{1 + b\gamma}$. We shall call solutions of the types (*i*) and (*ii*) single-pulse solutions, solutions of the type (*iii*) two-pulse solutions, and solutions of the type (*iv*) mixed-mode solutions. Below we consider the stability of stationary solutions.

Stability of the Pulse to the Disturbances of Its Shape

In order to investigate the internal stability of the pulse solution, we linearized the system (8.270) around the solution (8.266). The linear growth rate of the perturbed solution, λ, is a solution of the following characteristic equation:

$$\lambda^3 + a_1 \lambda^2 + a_2 \lambda + a_3 = 0, \tag{8.273}$$

where

$$a_1 = \frac{1}{9} \left[C(9p - 16C) + 2\kappa^2 (2\gamma^2 + 9b\gamma + 17) \right],$$

$$a_2 = \frac{2}{81} \kappa^2 \left[C(9p - 16C)(2\gamma^2 + 9b\gamma + 17) + 4\kappa^2 (64 \right.$$
$$\left. + 18b^2 + 11\gamma^2 + 36b^2\gamma^2 + 69b\gamma + \gamma^4 - 3b\gamma^3) \right],$$

$$a_3 = \frac{16}{9} \kappa^4 (1 + b^2)(2 + \gamma^2)(2 - pC). \tag{8.274}$$

The monotonic instability boundary $\lambda = 0$ is equivalent to a condition $a_3 = 0$, that is, to a condition $pC = 2$. Solving the equation $pC = 2$ leads to $p^2 = 4(1 - \alpha)$, which corresponds to the line of merging of solutions $C_{\pm}$ (see (8.266)). Thus, the boundary $\lambda = 0$ corresponds to a transition from one branch of solution to another. For the upper branch of solution $C = C_+$, it holds $pC > 2$ and $a_3 < 0$; hence, the product of the eigenvalues is positive and therefore at least one of the eigenvalues λ must be positive; thus, the upper branch of C is unstable. For the lower branch of solution $C = C_-$, it holds $pC < 2$ and $a_3 > 0$. That branch of solutions can be either stable (for sufficiently large p) or unstable.

Stability of the Pulse to the Appearance of Another Pulse

The stability of the solution with regard to the appearance of the second soliton can be studied in the framework of the model (8.272). We linearize the system (8.272) around the solution (8.266) for A_1 and around a solution $(0,0,\gamma)$ for A_2. The linear growth rate of the perturbed solution, λ, is a solution of the following characteristic equation:

$$\lambda^2(1 - pC - \lambda)(\lambda^3 + a_1\lambda^2 + a_2\lambda + a_3) = 0, \tag{8.275}$$

where a_1, a_2, a_3 are given by (8.274). The boundary $\lambda = 0$ is equivalent to conditions $a_3 = 0$ or $1 - pC = 0$, that is, to conditions $pC = 2$ or $pC = 1$. The latter condition is equivalent to conditions $c = -b \pm 3\sqrt{1 + b^2}$. Therefore, for $1 - pC > 0$, it holds $\lambda = 1 - pC > 0$ and the single-pulse solution is unstable. For $1 - pC < 0$, it holds $\lambda = 1 - pC < 0$ and the single-pulse solution is stable for $C = C_-$.

Stability of a Two-Pulse Solution

The monotonic stability of a two-pulse solution can be studied by linearizing the system (8.272) around the solution (8.266) for both A_1 and A_2 and assuming a certain relation between the perturbations $\tilde{C}_1$, $\tilde{C}_2$, e.g., $\tilde{C}_1 > \tilde{C}_2$. Then the characteristic equation is as follows:

$$(\lambda^3 + a_1\lambda^2 + a_2\lambda + a_3)(\lambda^3 + \hat{a}_1\lambda^2 + \hat{a}_2\lambda + \hat{a}_3) = 0, \tag{8.276}$$

where a_1, a_2, a_3 are given by (8.274) and

$$\hat{a}_1 = \frac{2}{9}\kappa^2 \left[1 - 15b\gamma + 10\gamma^2\right],$$

$$\hat{a}_2 = \frac{8}{81}\kappa^4 \left[4(\gamma^2 - 3b\gamma - 2)(2\gamma^2 + 9b\gamma + 17) + (64 \right.$$
$$\left. + 18b^2 + 11\gamma^2 + 36b^2\gamma^2 + 69b\gamma + \gamma^4 - 3b\gamma^3)\right],$$

$$\hat{a}_3 = \frac{32}{9}\kappa^4(1 + b^2)(2 + \gamma^2)(1 - pC). \tag{8.277}$$

The boundary $\lambda = 0$ is equivalent to conditions $a_3 = 0$ or $\hat{a}_3 = 0$, that is, to conditions $pC = 2$ or $pC = 1$. The latter condition is equivalent to conditions $c = -b \pm 3\sqrt{1 + b^2}$. For $1 - pC < 0$, it holds $\hat{a}_3 < 0$; therefore, at least one of the eigenvalues λ is positive. For $1 - pC > 0$, it holds $\hat{a}_3 > 0$ and hence $\lambda < 0$. Thus the two-pulse solution is unstable for $1 - pC < 0$ and stable for $1 - pC > 0$ and $C = C_-$, and this is opposite to the stability of the single-pulse solution.

Stability of Mixed-Mode Solutions

For $\alpha = 1$, it holds $pC = 1$; therefore, the stationary value of C reduces to $C = 1/p$. The condition $\alpha = 1$ is equivalent to conditions $c = -b \pm 3\sqrt{1+b^2}$. For the upper branch, $c = -b + 3\sqrt{1+b^2}$, the stationary value of γ reduces to $\gamma = b + \sqrt{1+b^2}$, and for the lower branch, $c = -b - 3\sqrt{1+b^2}$; it follows $\gamma = b - \sqrt{1+b^2}$. It follows also that the stationary value of κ satisfies a condition $\kappa^2 = C^2/(1+b\gamma)$.

For the investigation of stability of mixed-mode solutions, we linearized the system (8.272) around the solution (C, κ, γ) for A_1 and around (C_2, κ_2, γ), for A_2, with C, κ, γ described above and $\kappa_2^2 = C_2^2/(1+b\gamma)$. The obtained characteristic equation is as follows:

$$\lambda \left(\lambda^2 + \tilde{a}_1\lambda + \tilde{a}_2\right)\left(\lambda^3 + a_1\lambda^2 + a_2\lambda + a_3\right) = 0, \qquad (8.278)$$

where

$$\tilde{a}_1 = \frac{1}{9}\kappa_2^2(5\gamma^2 + 17),$$

$$\tilde{a}_2 = -\frac{32}{9}\kappa_2^4(1+b^2)b\gamma, \qquad (8.279)$$

and a_1, a_2, a_3 are given by (8.274). For $pC = 1$, it holds $a_1 > 0$ and $a_3 > 0$. Hence, the stability boundary $\lambda = 0$ is equivalent to a condition $b = 0$.

In the case $\alpha = 1$, the condition $a_2 > 0$ is fulfilled for $p^2 > 32\text{sign}(\gamma)b\sqrt{1+b^2}/(13\gamma^2 + 25)$. For $c = -b + 3\sqrt{1+b^2}$, it holds $\gamma > 0$ and therefore $a_2 > 0$ for all values of p if $b < 0$. For $c = -b - 3\sqrt{1+b^2}$, it holds $\gamma < 0$ and therefore $a_2 > 0$ for all values of p if $b > 0$. Next, the coefficient $\tilde{a}_2$ is positive for $b < 0$ on the upper branch and for $b > 0$ on the lower branch. Thus, the mixed-mode solution is stable for $b < 0$ on the upper branch of $\alpha = 1$ and for $b > 0$ on the lower branch and unstable otherwise.

Nonlinear Dynamics Produced by Monotonic Instabilities of a Pulse Solution

Below we describe the results of numerical simulations based on the variational model (8.272).

In accordance with the prediction of the linear theory, in the region I, $\{-b - 3\sqrt{b^2+1} < c < -b + 3\sqrt{b^2+1}\}$, the only stable state corresponds to a single-pulse solution. This observation coincides also with that obtained in [107] by means of the direct numerical simulations of the original equation (8.261). An example of the transient evolution of the dynamical system (8.272) in the case $b = -5$, $c = 10$ is shown in Figure 8.10(a). The pulse with the higher amplitude survives and tends to its stationary shape, while the pulse with the lower amplitude decays.

In the regions II, $\{c < -b - 3\sqrt{b^2+1}\} \cup \{c > -b + 3\sqrt{b^2+1}\}$, the single-soliton solution is unstable with respect to the development of the second soliton. However, we observe two qualitatively different two-soliton regimes. For positive c, near the

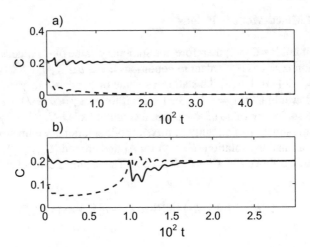

Fig. 8.10 Regimes for points: a) $(-5, 10)$, b) $(-3, 18)$, and $p = 5$. Solid line, $C_1(t)$; dashed line, $C_2(t)$

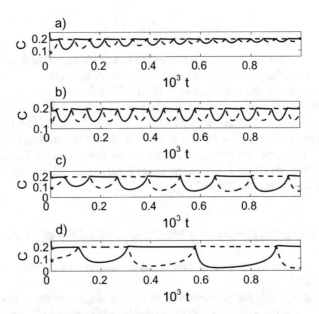

Fig. 8.11 Evolution of regimes for $p = 5$ and points: a) $(2, 12)$, b) $(2.22, 12)$, c) $(3, 12)$, d) $(4, 12)$. Solid line, $C_1(t)$; dashed line, $C_2(t)$

left boundary of the region II, a stationary two-pulse state is observed; see Figures 8.10(b) and 8.11(a). Near the right boundary of the region II, the system tends to an oscillatory regime with alternating pulses (see Figure 8.11(c) and (d)). The boundary between two regimes is the oscillatory instability boundary of a two-pulse solution (see Figure 8.11(b)), which depends on p and can be obtained numerically.

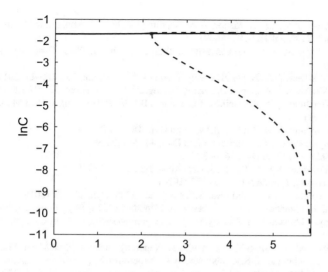

Fig. 8.12 Stationary and oscillatory regimes for $c = 12$ and $p = 5$: solid line, stationary value of C; dashed lines, maximum and minimum values of C for an oscillating solution

Numerical simulations (Figure 8.12) show the supercritical character of the Hopf bifurcation on the oscillatory instability boundary of the two-pulse solution.

The results of numerical simulations of the nonlinear dynamics of solitary waves are described in more detail in [90, 108].

References

1. P.L. Kapitza, Sov. Phys. JETP **21**, 588 (1951)
2. S. Fauve, *Dynamics of Nonlinear and Disordered Systems*, ed. by G. Martínez-Mekler, T.H. Seligman (World Scientific, Singapore, 1995)
3. G.Z. Gershuni, D.V. Lyubimov, *Thermal Vibrational Convection* (Wiley, New York, 1998)
4. D.V. Lyubimov, T.P. Lyubimova, A.A. Cherepanov, *Dynamics of Interfaces in Vibration Fields* (Fizmatlit, Moscow, 2004, in Russian)
5. W. Magnus, S. Winkler, *Hill's Equation* (Dover, New York, 2004)
6. R.V. Birikh, V.A. Briskman, A.A. Cherepanov, M.G. Velarde, J. Colloid Interface Sci. **238** 16 (2001)
7. U. Thiele, J.M. Vega, E. Knobloch, J. Fluid Mech. **546**, 61 (2006)
8. S. Shklyaev, M. Khenner, A.A. Alabuzhev, Phys. Rev. E **77**, 036320 (2008)
9. S. Shklyaev, A.A. Alabuzhev, M. Khenner, Phys. Rev. E **79**, 051603 (2009)
10. E.S. Benilov, M. Chugunova, Phys. Rev. E **81**, 036302 (2010)
11. S. Shklyaev, A.A. Alabuzhev, M. Khenner, Phys. Rev. E **92**, 013019 (2015)
12. S.M. Zen'kovskaya, V.A. Novosiadlyi, J. Appl. Math. Mech. **73**, 271 (2009)
13. S.M. Zen'kovskaya, V.A. Novosiadlyi, J. Eng. Math. **69**, 277 (2011)
14. P.L. Garcia-Ybarra, J.L. Castillo, M.G. Velarde, Phys. Fluids **30**, 2655 (1987)
15. A.C. Or, R.E. Kelly, J. Fluid Mech. **456**, 161 (2002)
16. B.L. Smorodin, A.B. Mikishev, A.A. Nepomnyashchy, B.I. Myznikova, Phys. Fluids **21**, 062102 (2009)

17. B.L. Smorodin, A.B. Mikishev, A.A. Nepomnyashchy, B.I. Myznikova, Microgravity Sci. Technol. **21**(Suppl 1), S193 (2009)
18. A.B. Mikishev, A.A. Nepomnyashchy, B.L. Smorodin, J. Phys. Conf. Ser. **216**, 012004 (2010)
19. A.B. Mikishev, A.A. Nepomnyashchy, J. Adhes. Sci. Technol. **25** 1411–23 (2011)
20. A.B. Mikishev, A.A. Nepomnyashchy, Microgravity Sci. Technol. **22**, 415 (2010)
21. G.Z. Gershuni, A.K. Kolesnikov, J.C. Legros, B.I. Myznikova, Int. J. Heat Mass Transf. **42**, 547 (1999)
22. G.Z. Gershuni, E.M. Zhukhovitsky, Fluid Dyn. **16**, 498 (1981)
23. B.I. Myznikova, B.L. Smorodin, Fluid Dyn. **44**, 240 (2009)
24. S.V. Shklyaev, Fluid Dyn. **36**, 682 (2001)
25. R.G. Finucane, R.E. Kelly, Int. J. Heat Mass Transf. **19**, 71 (1976)
26. G.M. Homsy, J. Fluid Mech. **62**, 387 (1974)
27. D.V. Lyubimov, T.P. Lyubimova, B.S. Maryshev, Fluid Dyn. **45**, 859 (2010)
28. I.S. Fayzrakhmanova, S.V. Shklyaev, A.A. Nepomnyashchy, Phys. Fluids **22**, 104101 (2010)
29. I.S. Fayzrakhmanova, S.V. Shklyaev, A.A. Nepomnyashchy, Eur. Phys. J. Spec. Top. **192**, 95 (2011)
30. I.S. Fayzrakhmanova, S.V. Shklyaev, A.A. Nepomnyashchy, J. Fluid Mech. **714**, 190 (2013)
31. I.S. Fayzrakhmanova, S.V. Shklyaev, A.A. Nepomnyashchy, *Without Bounds: A Scientific Canvas of Nonlinearity and Complex Dynamics* (Springer, Berlin, 2013), p. 133
32. I.S. Fayzrakhmanova, S.V. Shklyaev, A.A. Nepomnyashchy, Fluid Dyn. Res. **46**, 041411 (2014)
33. A.A. Nepomnyashchy, I.B. Simanovskii, J. Fluid Mech. **726**, 476 (2013)
34. A.A. Nepomnyashchy, I.B. Simanovskii, J. Fluid Mech. **771**, 159 (2015)
35. M.N. Chen, J.A. Whitehead, J. Fluid Mech. **31**, 1 (1968)
36. L.P. Vozovoi, A.A. Nepomnyashchy, *Hydrodynamics*, pt. 7, ed. by E.M. Zhukhovitskii (Perm State Pedagogical Institute, Perm, 1974, in Russian)
37. L.P. Vozovoi, A.A. Nepomnyashchy, J. Appl. Math. Mech. **43**, 1080 (1979)
38. P. Coullet, C. Elphick, D. Repaux, Phys. Rev. Lett. **58**, 431 (1987)
39. L.M. Pismen, Phys. Rev. Lett. **59**, 2740 (1987)
40. R. Manor, A. Hagberg, E. Meron, Europhys. Lett. **83**, 10005 (2008)
41. G. Freund, W. Pesch, W. Zimmermann, J. Fluid Mech. **673**, 318 (2011)
42. Y. Mau, L. Haim, A. Hagberg, E. Meron, Phys. Rev. E **88**, 032917 (2013)
43. S. Weiss, G. Seiden, E. Bodenschatz, J. Fluid Mech. **756**, 293 (2014)
44. L. Haim, Y. Mau, E. Meron, Phys. Rev. E **90**, 022904 (2014)
45. A.A. Nepomnyashchy, J. Appl. Math. Mech. **52**, 677 (1988)
46. C. Utzny, W. Zimmermann, M. Bär, Europhys. Lett. **57**, 113 (2002)
47. M. Hammele, W. Zimmermann, Phys. Rev. E **73**, 066211 (2006)
48. K. Stewartson, J.T. Stuart, J. Fluid Mech. **48**, 529 (1971)
49. A.C. Newell, Lect. Appl. Math. **15**, 157 (1974)
50. A.A. Nepomnyashchy, *Spontaneous Symmetry Breaking, Self-Trapping and Josephson Oscillations* (Springer, Heidelberg, 2013), p. 309
51. C.-K. Lam, B.A. Malomed, K.W. Chow, P.K.A. Wai, Eur. Phys. J. Spec. Top. **173**, 233 (2009)
52. C.H. Tsang, B.A. Malomed, C.K. Lam, K.W. Chow, Eur. Phys. J. D **59**, 81 (2010)
53. S.I. Abarzhi, O. Desjardins, A. Nepomnyashchy, H. Pitsch, Phys. Rev. E **75**, 046208 (2007)
54. A. Nepomnyashchy, S.I. Abarzhi, Phys. Rev. E **81**, 037202 (2010)
55. B.A. Malomed, Phys. Rev. E **47**, R2257 (1993)
56. A.S. Mikhailov, K. Showalter, Phys. Rep. **425**, 79 (2006)
57. E. Schöll, H.G. Schuster (ed.), *Handbook of Chaos Control*, 2nd edn. (Wiley, Weinheim, 2008)
58. J. Tang, H.H. Bau, Phys. Rev. Lett. **70**, 1795 (1993)
59. J. Tang, H.H. Bau, Proc. R. Soc. Lond. A **447**, 587 (1994)
60. L.E. Howle, Phys. Fluids **9**, 1861 (1997)
61. J. Tang, H.H. Bau, J. Fluid Mech. **363**, 153 (1998)
62. S. Benz, P. Hintz, R.J. Riley, G.P. Neitzel, J. Fluid Mech. **359**, 165 (1998)

63. R.O. Grigoriev, Phys. Fluids **14**, 1895 (2002)
64. R.O. Grigoriev, Phys. Fluids **15**, 1363 (2003)
65. N. Garnier, R.O. Grigoriev, M.F. Schatz, Phys. Rev. Lett. **91**, 054501 (2003)
66. A.M. Frank, Phys. Fluids **18**, 078106 (2006)
67. P.D. Christofides, Feedback control of the Kuramoto-Sivashinsky equation, in *Proceedings of the 37th IEEE Conference on Decision and Control* (IEEE, Piscataway, 1998), pp. 4646–4651
68. A. Armaou, P.D. Christofides, Physica D **137**, 49 (2000)
69. A. Azouani, E.S. Titi, Evol. Equ. Control Theory **3**, 579 (2014)
70. E. Lunasin, E.S. Titi, arXiv: 1506.03709v2 [math.AP] (2017)
71. S.N. Gomes, M. Pradas, S. Kalliadasis, D.T. Papageorgiou, G.A. Pavliotic, Phys. Rev. E **92**, 022912 (2015)
72. S.N. Gomes, D.T. Papageorgiou, G.A. Pavliotic, IMA J. Appl. Math. **82**, 158 (2017)
73. A.B. Thompson, S.N. Gomes, G.A. Pavliotic, D.T. Papageorgiou, Phys. Fluids **28**, 012107 (2016)
74. H.H. Bau, Int. J. Heat Mass Transf. **42**, 1327 (1999)
75. I. Hashim, S.A. Kechil, Fluid Dyn. Res. **41**, 045504 (2009)
76. Z. Siri, Z. Mustafa, I. Hashim, Int. J. Heat Mass Transf. **52**, 5770 (2009)
77. I. Hashim, H. Othman, S.A. Kechil, Int. Commun. Heat Mass Transfer **36**, 161 (2009)
78. A.C. Or, R.E. Kelly, J. Fluid Mech. **440**, 27 (2001)
79. A.C. Or, R.E. Kelly, L. Cortelezzi, J.L. Speyer, J. Fluid Mech. **387**, 321 (1999)
80. I.S. Gradshteyn, I.M. Ryzhik, *Tables of Integrals, Series and Products*, 4th edn. (Academic, London, 1965)
81. I.B. Simanovskii, A.A. Nepomnyashchy, *Convective Instabilities in Systems with Interface* (Gordon and Breach, Amsterdam, 1993)
82. A.A. Nepomnyashchy, A.A. Golovin, V. Gubareva, V. Panfilov, Physica D **199**, 61 (2004)
83. F.H. Busse, J. Fluid Mech. **30**, 625 (1967)
84. A.A. Golovin, A.A. Nepomnyashchy, L.M. Pismen, J. Fluid Mech. **341**, 317 (1997)
85. L.D. Landau, E.M. Lifshitz, *Quantum Mechanics: Nonrelativistic Theory* (Pergamon, Oxford, 1965)
86. A.D. Polyanin, V.F. Zaitsev, *Handbook of Exact Solutions for Ordinary Differential Equations* (Chapman and Hall/CRC, Boca Raton, 2003)
87. A.P. Prudnikov, Y.A. Brychkov, O.I. Marichev, *Integral and Series, Vol. 3: More Special Functions, Gordon and Breach Science Publishers* (Gordon and Breach Science Publishers, New York, 1990)
88. B.Y. Rubinstein, A.A. Nepomnyashchy, A.A. Golovin, Phys. Rev. E **75**, 046213 (2007)
89. Y. Kanevsky, A.A. Nepomnyashchy, Phys. Rev. E **76**, 066305 (2007)
90. Y. Kanevsky, A.A. Nepomnyashchy, Physica D **239**, 87 (2010)
91. B.A. Malomed, Prog. Opt. **43**, 71 (2002)
92. S.C. Cerda, S.B. Cavalcanti, J.M. Hickmann, Eur. Phys. J. D **1**, 313 (1998)
93. V. Skarka, N.B. Aleksić, Phys. Rev. Lett. **96**, 013903 (2006)
94. J.D. Murray, *Mathematical Biology* (Springer, Berlin, 1989)
95. I. Györi, G. Ladas, *Oscillation Theory of Delay Differential Equations* (Clarendon Press, Oxford, 1991)
96. L.N. Erbe, Q. Kong, B.G. Zhang, *Oscillation Theory for Functional Differential Equation* (Marcel Dekker, New York, 1995)
97. A.A. Golovin, Y. Kanevsky, A.A. Nepomnyashchy, Phys. Rev. E **79**, 046218 (2009)
98. G. Röder, G. Bordyugov, H. Engel, M. Falcke, Phys. Rev. E **75**, 036202 (2007)
99. W. Schöpf, W. Zimmermann, Phys. Rev. E **47**, 1739 (1993)
100. J.D. Moores, Opt. Commun. **96**, 65 (1993)
101. S.H. Davis, *Theory of Solidification* (Cambridge University Press, Cambridge, 2001)
102. L.M. Hocking, K. Stewartson, Proc. R. Soc. A **326**, 289 (1972)
103. C.S. Bretherton, E.A. Spiegel, Phys. Lett. A **96**, 152 (1983)
104. W. Schöpf, L. Kramer, Phys. Rev. Lett. **66**, 2316 (1991)
105. Y. Kanevsky, A.A. Nepomnyashchy, Phys. Lett. A **372**, 7156 (2008)
106. Y. Kanevsky, A.A. Nepomnyashchy, Physica D **240**, 1036 (2011)
107. A.A. Golovin, A.A. Nepomnyashchy, Phys. Rev. E **73**, 046212 (2006)
108. Y. Kanevsky, A.A. Nepomnyashchy, Math. Model. Nat. Phenom. **7**(2), 83 (2012)

Chapter 9
Outlook

In conclusion, we present an outlook of some important directions of the investigation that are still on the stage of development.

9.1 Influence of Lateral Boundaries

If the layer is bounded by rigid lateral boundaries, the longwave partial differential equations governing the evolution of interfaces should be supplemented by corresponding boundary conditions. The formulation of those conditions in the framework of the longwave approach is a highly nontrivial problem, because it has to include additional physical phenomena. Among them are the slip of the contact line [1–3], hysteresis of the contact angle [4, 5], as well as dissipative effects beyond the longwave approximation [6–9].

In the most of publications, the analysis of pattern formation is done for an infinite region or for artificial, spatially periodic, boundary conditions that may look reasonable in the case of a shortwave instability, if the imposed spatial period is large with respect to characteristic scale of the spontaneously generated patterns. It is known, however, that even in the case of stationary patterns, the influence of distant lateral boundaries is nontrivial and unexpectedly strong: on a long time scale, the lateral boundaries significantly diminish the interval of selected wavenumbers [10, 11]. In the case of waves, the influence of lateral boundaries is expected to be much stronger. In the case of an *absolute* instability, where disturbances grow in each point, one can expect that the presence of boundaries does not destroy the patterns but acts as an additional factor of the wavenumber and planform selection. In the case of a *convective* instability, where the wave disturbances grow while moving with their group velocity but do not grow locally, the formation of pattern is possible only due to a sufficiently strong reflection of waves on the lateral boundaries, which creates a global instability, or under the action of noise. For shortwave oscillatory

© Springer Science+Business Media, LLC 2017

S. Shklyaev, A. Nepomnyashchy, *Longwave Instabilities and Patterns in Fluids*,

Advances in Mathematical Fluid Mechanics,

https://doi.org/10.1007/978-1-4939-7590-7_9

instabilities, it is known that lateral walls can completely change the patterns, e.g., stationary patterns can be created instead of oscillatory ones [12]. Experiments have revealed a number of nontrivial phenomena characterizing the reflection of waves generated by instabilities [13]. In the case of the waves governed by the coupled Ginzburg–Landau equations, the influence of boundaries has been studied in [8].

In the case of a longwave instability, the scale of the patterns observed near the threshold is determined just by the lateral size of the system. Therefore, the lateral boundary conditions cannot be ignored. For longwave instabilities, boundary conditions on lateral boundaries have never been systematically derived, because the main assumptions of the longwave approach are typically broken near the boundaries. In some particular cases, the problem was tackled in [6, 7, 9, 14, 15].

9.2 Evaporative Convection

A liquid in a layer with a free surface is typically subject to evaporation. Due to the latent heat, the evaporation leads to a reduction of the temperature at the liquid surface, which can be a source of buoyancy and surface tension driven instabilities [16–20]. The evaporation is also significant in the case of instabilities caused by an external heating. It leads to a wavenumber dependence of the effective Biot number [21]. While in the case of a non-deformable surface the evaporation provides a mechanism of negative feedback that suppresses stationary Marangoni instability [16], the situation is different in the case of a deformable interface. In a wave trough, the interface is closer to the hot bottom, and therefore it is warmer. That leads to an enhanced evaporation; hence the trough becomes deeper. Thus, evaporation can strengthen deformational Marangoni and van der Waals instabilities. Moreover, it creates a deformational instability even in the absence of other destabilizing factors [22]. One more instability mechanism is caused by vapor recoil. The enhanced evaporation in a warm interface depression creates an enhanced normal stress due to vapor recoil that amplifies the depression [23–25]. Vapor recoil can create a deformational instability also through inertial effects [23].

In the case of a binary solution, evaporating effects can strongly influence the pattern formation [26, 27]. Evaporation-induced longwave oscillatory Marangoni instabilities have been predicted for solutions [28] and in the presence of insoluble surfactants [29].

The difficulties of the theoretical analysis of the evaporative convection are twofold. First, the evaporation at the interface between gas and liquid phases involves a number of additional physical effects. Second, the problem is intrinsically non-stationary, because of the temporal change of the layer thickness. Until recently, the "freezing" approach was used for the stability analysis, which is not always justified. For instance, in the case of the evaporation of the solution, the transients in the gas phase [30] and the process of diffusion can be as slow as that of evaporation, and the evolution of disturbances is essentially non-exponential [31]. A comprehensive pattern formation theory for evaporative longwave instabilities is still in progress.

9.3 Longwave Instabilities in Viscoelastic Liquids and Biofluids

The viscoelasticity, which introduces additional temporal scales, can affect significantly the development of thin layer instabilities. An important example of a system where such instabilities can develop is the mucus film in human lungs. That is a two-layer system, where the bottom liquid is Newtonian and the upper liquid is viscoelastic [32]. The pulmonary surfactants play the crucial role in the prevention of mucus film instabilities, which can be a source of lung disorders [33, 34].

While the shortwave instabilities in non-Newtonian fluids are rather well studied, the investigation of longwave oscillatory instabilities has been started very recently [35, 36]. In the literature, there exist some attempts to derive a closed lubrication approximation model for viscoelastic films (see, e.g., [37]). However, they are based on the linearization of the rheological relation, which ignores the convective terms. In the case of a sufficiently large relaxation time, these terms become an essential part of a self-consistent description [36, 38]. The problem becomes strongly nonlinear, and its dimension cannot be reduced.

The development of self-consistent models for the description of viscoelastic thin-film flows and surfactant-induced instabilities, specifically, for medical applications, is a subject of research [39–44].

9.4 Isothermic Vibrational Instabilities in Large Aspect Ratio Containers

The nonlinear theory of Faraday waves in the presence of longwave surface deformations is a quite nontrivial field for application of the longwave asymptotic approach. Despite the presence of a number of small parameters that may be used for the simplification of the problem (separation of the vertical scale and the horizontal scale, high or low dimensionless vibration frequency, low Reynolds number, etc.), a correct multiscale description of the relevant physical phenomena (generation of a streaming flow, longwave surface deformation, spatial modulation of waves) turned out to be a difficult task. A critical analysis of the existing approaches using the longwave approximation for isothermic and non-isothermic vibrating films has been done recently in [45]. A self-consistent analysis of the linear problem has been carried out by Mancebo and Vega [46]; in [47] the longwave theory of the damping of the Rayleigh-Taylor instability by vibrations has been developed. The self-consistent nonlinear theory has been developed in [48] in the limit of small viscosity and in [49] for finite viscosity. Some generalizations of the nonlinear theory are presented in [50]. An extension of the longwave approach based on a coupled system of evolution equations for the surface deformation and the flow rate has been applied for studying Faraday patterns in [51–54].

Another relevant problem is the layer instability under horizontal vibrations [14, 55–57]. In the case of a two-layer system, horizontal vibrations generate a "frozen wave" [58–62]. Dynamics of the interface between *miscible* liquids subjected to horizontal vibrations has been studied recently experimentally and numerically [63, 64].

9.5 Reaction-Diffusion-Convection Systems

Autocatalytic chemical reactions, the most familiar of which is the Belousov–Zhabotinsky reaction [65, 66], can generate a variety of wave patterns: fronts, solitary waves, target-like patterns, spiral waves, etc. [67]. The inhomogeneities of density and surface concentration generated by chemical patterns give rise to buoyancy and Marangoni convection. Coupling of reaction-diffusion and convection effects creates a plethora of dynamic behaviors [68–72]. The Marangoni convection induced by chemical wave propagation was studied in [73–79]. The development of the buoyancy-driven convection by reaction front propagation was studied in [80–85].

References

1. S.H. Davis, J. Fluid Mech. **98**, 225 (1980)
2. L.M. Hocking, J. Fluid Mech. **179**, 253 (1987)
3. J. Miles, J. Fluid Mech. **222**, 197 (1991)
4. G.W. Young, S.H. Davis, J. Fluid Mech. **174**, 327 (1987)
5. L.M. Hocking, J. Fluid Mech. **179**, 267 (1987)
6. A.K. Sen, S.H. Davis, J. Fluid Mech. **121**, 163 (1982)
7. G.W. Young, S.H. Davis, Q. Appl. Math. **42**, 403 (1985)
8. C. Martel, J.M. Vega, Nonlinearity **9**, 1129 (1996)
9. P.F. Mendez, T.W. Eagar, J. Appl. Mech. **80**, 011009 (2013)
10. M.C. Cross, P.G. Daniels, P.C. Hohenberg, E.D. Siggia, Phys. Rev. Lett. **45**, 898 (1980)
11. S. Zaleski, Lect. Notes Phys. **210**, 84 (1984)
12. J. Priede, G. Gerbeth, Phys. Rev. E **56**, 4187 (1997)
13. H. Linde, X.-L. Chu, M.G. Velarde, W. Waldhelm, Phys. Fluids A **5**, 3162 (1993)
14. F. Varas, J.M. Vega, J. Fluid Mech. **579**, 271 (2007)
15. A.A. Nepomnyashchy, I.B. Simanovskii, J. Fluid Mech. **771**, 159 (2015)
16. D. Merkt, M. Bestehorn, Physica D **185**, 196 (2003)
17. H. Mancini, D. Maza, Europhys. Lett. **66**, 812 (2004)
18. B. Haut, P. Colinet, J. Colloid Interface Sci. **285**, 296 (2005)
19. J. Margerit, M. Dondlinger, P.C. Dauby, J. Colloid Interface Sci. **290**, 220 (2005)
20. B. Scheid, J. Margerit, C.S. Iorio, L. Joannes, M. Heraud, P. Queeckers, P.C. Dauby, P. Colinet, Exp. Fluids **52**, 1107 (2012)
21. H. Machrafi, A. Rednikov, P. Colinet, P.C. Dauby, Phys. Rev. E **91**, 053018 (2015)
22. R.O. Grigoriev, Phys. Fluids **14**, 1895 (2002)
23. H.J. Palmer, J. Fluid Mech. **75**, 487 (1976)
24. A. Prosperetti, M.S. Plesset, Phys. Fluids **27**, 1590 (1984)
25. J.P. Burelbach, S.G. Bankoff, S.H. Davis, J. Fluid Mech. **195**, 463 (1988)
26. S.G. Yiantsios, B.G. Higgins, Phys. Fluids **22**, 022102 (2010)
27. J. Zhang, A. Oron, R.P. Behringer, Phys. Fluids **23**, 072102 (2011)

28. S.K. Serpetsi, S.G. Yantsios, Phys. Fluids **24**, 122104 (2012)
29. A.B. Mikishev, A.A. Nepomnyashchy, Phys. Fluids **25**, 054109 (2013)
30. H. Machrafi, A. Rednikov, P. Colinet, P.C. Dauby, Phys. Fluids **25**, 084106 (2013)
31. A. Nepomnyashchy, A. Golovin, A. Tikhomirova, V. Volpert, *Proceedings 10th Jubilee National Congress on Theoretical and Applied Mechanics, Varna 13–16 Sept 2005*, vol. 521, ed. by M. Drinov (Academic Publishing House, Sofia, 2005)
32. J.H. Widdicombe, S.J. Bastacky, D.X.-Y. Wu, C.Y. Lee, Eur. Respir. J. **10**, 2892 (1997)
33. E. Putman, W. Liese, W.F. Voorhout, L. van Bree, L.M.G. van Golde, H.P. Haagsman, Toxicol. Appl. Pharmacol. **142**, 288 (1997)
34. A. Akella, S.B. Deshpande, Ind. J. Exp. Biol. **51**, 5 (2013)
35. D. Comissiong, R.A. Kraenkel, M.A. Manna, Proc. R. Soc. A **465**, 109 (2009)
36. D. Halpern, H. Fujioka, J.B. Grotberg, Phys. Fluids **22**, 011901 (2010)
37. M. Rauscher, A. Münch, B. Wagner, R. Blossey, Eur. Phys. J. E **17**, 373 (2005)
38. Y.L. Zhang, O.K. Matar, R.V. Craster, J. Non-Newtonian Fluid Mech. **105**, 53 (2002)
39. C.F. Tai, S. Bian, D. Halpern, Y. Zheng, M. Filoche, J.B. Grotberg, J. Fluid Mech. **677**, 483 (2011)
40. H. Fujioka, D. Halpern, D.P. Gayer III, J. Biomech. **46**, 319 (2013)
41. Z.Q. Zhou, J. Peng, Y.J. Zhang, W.L. Zhuge, Non-Newtonian Fluid Mech. **204**, 94 (2014)
42. R. Levy, D.B. Hill, M.G. Forest, J.B. Grotberg, Integr. Comp. Biol. **54**, 985 (2014)
43. M. Benzaquen, T. Salez, E. Raphael, Europhys. Lett. **106**, 36003 (2014)
44. E. Hermans, M.S. Bhamla, P. Kao, G.G. Fuller, J. Vermant, Soft Matter **11**, 8048 (2015)
45. S. Shklyaev, A.A. Alabuzhev, M. Khenner, Phys. Rev. E **92**, 013019 (2015)
46. F.J. Mancebo, J.M. Vega, J. Fluid Mech. **467**, 307 (2002)
47. V. Lapuerta, F.J. Mancebo, J.M. Vega, Phys. Rev. E **64**, 016318 (2001)
48. F.J. Mancebo, J.M. Vega, Physica D **197**, 346 (2004)
49. F.J. Mancebo, J.M. Vega, J. Fluid Mech. **560**, 369 (2006)
50. S. Shklyaev, M. Khenner, A.A. Alabuzhev, Phys. Rev. E **77**, 036320 (2008)
51. M. Bestehorn, Phys. Fluids **25**, 114106 (2013)
52. M. Bestehorn, Q. Han, A. Oron, Phys. Rev. E **88**, 023025 (2013)
53. M. Bestehorn, A. Pototsky, Phys. Rev. Fluids **1**, 063905 (2016)
54. S. Richter, M. Bestehorn, Eur. Phys. J. Spec. Top. **226**, 1253 (2017)
55. C.-S. Yih, J. Fluid Mech. **31**, 737 (1968)
56. S. Shklyaev, A.A. Alabuzhev, M. Khenner, Phys. Rev. E **79**, 051603 (2009)
57. E.S. Benilov, M. Chugunova, Phys. Rev. E **81**, 036302 (2010)
58. G.H. Wolf, Phys. Rev. Lett. **24**, 444 (1970)
59. D.V. Lyubimov, A. Cherepanov, Fluid Dyn. **21**, 849 (1987)
60. M.V. Khenner, D.V. Lyubimov, T.S. Belozerova, B. Roux, Eur. J. Mech. B Fluids **18**, 1085 (1999)
61. A.A. Ivanova, V.G. Kozlov, P. Evesque, Fluid Dyn. **36**, 362 (2001)
62. G. Gandikota, D. Chatain, S. Amiroudine, T. Lyubimova, D. Beysens, Phys. Rev. E **89**, 012309 (2014)
63. Y. Gaponenko, M. Torregrosa, V. Yasnou, A. Mialdun, V. Shevtsova, J. Fluid Mech. **784**, 342 (2015)
64. V. Shevtsova, Y. Gaponenko, V. Yasnou, A. Mialdun, A. Nepomnyashchy, Langmuir **31**, 5550 (2015)
65. B.P. Belousov, *Sbornik Referatov po Radiatsionnoy Meditsine*, vol. 145 (Medgiz, Moscow, 1958, in Russian)
66. A.M. Zhabotinsky, Proc. Acad. Sci. USSR **157**, 392 (1964)
67. E. Meron, Phys. Rep. **218**, 1 (1992)
68. J. D'Hernoncourt, A. Zebib, A. De Wit, Chaos **17**, 013109 (2007)
69. C. Almarcha, P.M.J. Trevelyan, P. Grosfils, A. De Wit, Phys. Rev. Lett. **104**, 044501 (2010)
70. H. Miike, K. Miura, A. Nomura, T. Sakurai, Physica D **239**, 808 (2010)
71. L. Sebestikova, M.J.B. Hauser, Phys. Rev. E **85**, 036303 (2012)
72. F. Rossi, M.A. Budroni, N. Marchettini, J. Carballido-Landeira, Chaos **22**, 037109 (2012)
73. K. Matthiessen, S.C. Müller, Phys. Rev. E **52**, 492 (1995)

74. B.S. Martincigh, M.J.B. Hauser, R.H. Simoyi, Phys. Rev. E **52**, 6146 (1995)
75. K. Matthiessen, H. Wilke, S.C. Müller, Phys. Rev. E **53**, 6085 (1996)
76. B.S. Martincigh, C.R. Chinake, T. Howes, R.H. Simoyi, Phys. Rev. E **55**, 7299 (1997)
77. L.M. Pismen, Phys. Rev. Lett. **78**, 382 (1997)
78. L. Rongy, A. De Wit, Phys. Rev. E **77**, 046310 (2008)
79. L.M. Pismen, Math. Model. Nat. Phenom. **6**, 48 (2011)
80. M.R. Carey, S.W. Morris, P. Kolodner, Phys. Rev. E **53**, 6012 (1996)
81. M. Böckmann, S.C. Müller, Phys. Rev. Lett. **85**, 2506 (2000)
82. A. De Wit, Phys. Rev. Lett. **87**, 054502 (2001)
83. Y. Shi, K. Eckert, Chem. Eng. Sci. **61**, 5523 (2006)
84. L. Rongy, P.M.J. Trevelyan, A. De Wit, Phys. Rev. Lett. **101**, 084503 (2008)
85. L. Rongy, P.M.J. Trevelyan, A. De Wit, Chem. Eng. Sci. **65**, 2382 (2010)

Appendix A
Solvability Conditions for an Inhomogeneous Linear Boundary Value Problem

Let us consider a *homogeneous* linear equation of the nth order,

$$L(u) = p_0(x)u^{(n)} + p_1(x)u^{(n-1)} + \ldots p_{n-1}(x)u' + p_n(x)u = 0, \ a < x < b, \quad (A.1)$$

where all functions $p_i(x)$ are continuous, with *homogeneous* linear boundary conditions

$$U_i(u) = \alpha_i^1 u(a) + \alpha_i^2 u'(a) + \ldots + \alpha_i^n u^{(n-1)}(a) +$$
$$\alpha_i^{n+1} u(b) + \alpha_i^{n+2} u'(b) + \ldots + \alpha_i^{2n} u^{(n-1)}(b) = 0, \quad (A.2)$$

where all the constant $2n$-component vectors

$$(\alpha_i^1, \ldots, \alpha_i^{2n}), \ i = 1, \ldots, n, \quad (A.3)$$

are linearly independent.

Assume that the boundary value problem (A.1), (A.2) has k linearly independent solutions. For instance, for the *eigenvalue problem*

$$L(u) = [K(x)u']' - G(x; \lambda)u = 0, \ a < x < b, \quad (A.4)$$

$$U_1(u) = \alpha u(a) + \beta u'(a) = 0, \ \alpha^2 + \beta^2 \neq 0; \quad (A.5)$$

$$U_2(u) = \gamma u(b) + \delta u'(b) = 0, \ \gamma^2 + \delta^2 \neq 0, \quad (A.6)$$

k is either 0 (if λ is not an eigenvalue) or 1 (if λ is an eigenvalue).

Define operator

$$\bar{L}(v) = (-1)^n (p_0(x)v)^{(n)} + (-1)^{n-1}(p_1(x)v)^{(n-1)} + \ldots + (-1)(p_{n-1}(x)v)' + p_n(x)v = 0, \quad (A.7)$$

which is called the *adjoint operator*. Note that in the case (A.4), $\bar{L}(v) = L(v)$. Also, define the *bilinear concominant*

$$P(u,v) = u[p_{n-1}v - (p_{n-2}v)' + \ldots + (-1)^{n-1}(p_0v)^{(n-1)}] +$$

© Springer Science+Business Media, LLC 2017
S. Shklyaev, A. Nepomnyashchy, *Longwave Instabilities and Patterns in Fluids*,
Advances in Mathematical Fluid Mechanics,
https://doi.org/10.1007/978-1-4939-7590-7

$$u'[p_{n-2}v - (p_{n-3}v)' + \ldots + (-1)^{n-2}(p_0v)^{(n-2)}] + \ldots + u^{(n-1)}p_0v. \tag{A.8}$$

One can show that

$$vL(u) - u\bar{L}(v) = \frac{d}{dx}P(u,v) \tag{A.9}$$

(the *Lagrange's identity*), hence

$$\int_a^b [vL(u) - u\bar{L}(v)]dx = P(u,v)|_a^b. \tag{A.10}$$

Let us define now a basis in the space of $2n$-dimensional vectors $(\alpha_i^1, \ldots, \alpha_i^{2n})$, $i = 1, \ldots, 2n$, in such a way that the first n vectors (A.3) correspond to given boundary conditions, while the vectors with $i = n+1, \ldots, 2n$ are arbitrary linearly independent vectors. Define also $U_i(u)$, $i = n+1, \ldots, 2n$ according to formulas (A.2). For instance, in the case (A.5), (A.6), if $\alpha \neq 0$, $\gamma \neq 0$, one can choose $U_3(u) = u'(a)$, $U_4(u) = u'(b)$. The expression in the right-hand side of (A.10) can be presented as

$$P(u,v)|_a^b = \sum_{i,j=1}^{2n} U_j D_{ji} v_i, \tag{A.11}$$

where $\{D_{ji}\}$ is a certain matrix,

$$v_1 = v(a), \ldots, v_n = v^{(n-1)}(a), \; v_{n+1} = v(b), \ldots, v_{2n} = v^{(n-1)}(b). \tag{A.12}$$

Define

$$V_{(2n+1)-j} = \sum_{i=1}^{2n} D_{ji} v_i, \; j = 1, \ldots, 2n. \tag{A.13}$$

Then formula (A.10) can be written as

$$\int_a^b [vL(u) - u\bar{L}(v)]dx = U_1(u)V_{2n}(v) + U_2(u)V_{2n-1}(v) + \ldots + U_{2n}(u)V_1(v) \tag{A.14}$$

(the *Green's formula*).

Define the boundary value problem

$$\bar{L}(v) = 0, \; a < x < b; \; V_1(v) = V_2(v) = \ldots = V_n(v) = 0, \tag{A.15}$$

which is called the *adjoint boundary value problem*. In the case of problems (A.4)–(A.6), the adjoint boundary value problem is identical to the primary boundary value problem, up to the numeration of the boundary conditions:

$$L(v) = (Kv')' - Gv = 0; \; V_1(v) = \gamma v(b) + \delta v'(b) = 0, \; V_2(v) = \alpha v(a) + \beta v'(b) = 0.$$

In that case the problem is *self-adjoint*.

According to the *Fredholm theorem*, the number of the linearly independent solutions of (A.15) is equal to k, i.e., the number of the linearly independent solutions of (A.1), (A.2).

Let us consider now the inhomogeneous problem,

$$L(u) = f(x), \ a < x < b; \ U_i(u) = C_i, \ i = 1, \ldots, n. \tag{A.16}$$

If $k \neq 0$, using the Green's formula (A.14), we find the necessary conditions of the solution existence: the relations

$$\int_a^b vf(x)dx = C_1 V_{2n}(v) + C_2)V_{2n-1}(v) + \ldots + C_n V_{n+1}(v) \tag{A.17}$$

should be satisfied for each of k linearly independent solutions of the adjoint boundary value problem (A.15). The *Fredholm theorem* tells us that these *solvability conditions* are not only necessary but also sufficient for the existence of the solutions to problem (A.16). The solutions form a k-parameter family, because any solution of the homogeneous problem (A.1), (A.2) can be added.

If $k = 0$, the solution to (A.16) exists for any $f(x)$, and it is unique.

The two possibilities described above form the *Fredholm alternative*.

For more details, see [1].

Reference

1. E.L. Ince, *Ordinary Differential Equations* (Dover, Mineola, 1956)

Appendix B
Types of Bifurcations

In the present appendix, we give simple examples of the most typical local bifurcations. For a systematic study of that topic, the book of Guckenheimer and Holmes [1] is recommended.

B.1 Monotonic Instability

Consider a dynamical system governed by the equation

$$\frac{dx}{dt} = f(x,\mu),$$

(B.1)

where $x(t)$ is an n-dimensional vector function and μ is a parameter. For the sake of simplicity, we shall consider the case $n = 1$ in the present section.

Stationary solutions $x = x(\mu)$ of equation (B.1) are determined implicitly by equation

$$f(x,\mu) = 0.$$

(B.2)

Assume that a certain solution $x = x_0$ of (B.1) is known for $\mu = \mu_0$. We are going to investigate the dependence $x(\mu)$ for μ near that point. Without loss of generality, we assume below that $\mu = 0$, $x_0 = 0$, i.e.,

$$f(0,0) = 0.$$

(B.3)

Expand $f(x,\mu)$ into Taylor series near the point $(0,0)$:

$$f(x,\mu) = f_x x + f_\mu \mu + \frac{1}{2} f_{xx} x^2 + f_{x\mu} x\mu + \frac{1}{2} f_{\mu\mu} + \ldots;$$

(B.4)

here the subscript denotes a partial derivative, and all the derivatives are taken in the point $(0,0)$.

© Springer Science+Business Media, LLC 2017
S. Shklyaev, A. Nepomnyashchy, *Longwave Instabilities and Patterns in Fluids*,
Advances in Mathematical Fluid Mechanics,
https://doi.org/10.1007/978-1-4939-7590-7

Let us consider the linear stability of solution in the point $\mu = 0$. Linearizing equation (B.1) and taking $x(t) = z\exp(\lambda t)$, we find $\lambda = f_x$. Thus the solution is linearly stable ($\lambda < 0$) if $f_x < 0$ and unstable if $f_x > 0$. In both cases, the solution $x(\mu)$ is unique near the *regular point* $\mu = 0$:

$$x(\mu) = -\mu\frac{f_\mu}{f_x} + O(\mu^2).$$

If $f_x = 0$ at $\mu = 0$, then $\lambda = 0$, i.e., $\mu = 0$ is the border of the monotonic instability. That is the *bifurcation point*: in its vicinity, the uniqueness of solution is broken.

Below we consider main types of bifurcations.

B.1.1 Saddle-Node Bifurcation

If $f_{xx} \neq 0$ and $f_\mu \neq 0$, the leading terms in equation (B.2) are:

$$f_\mu\mu + \frac{1}{2}f_{xx}x^2 + \ldots = 0.$$

Thus,

$$x_\pm = \pm\sqrt{-2\frac{f_\mu}{f_{xx}}\mu} + O(\mu), \tag{B.5}$$

i.e., there are two solutions in the region of μ where the expression under the root sign is positive. At $\mu = 0$ both solutions merge, and they annihilate when μ changes its sign.

Linearizing the corresponding dynamical equation,

$$\frac{dx}{dt} = f_\mu\mu + \frac{1}{2}f_{xx}x^2 \tag{B.6}$$

around the stationary solutions (B.5),

$$x = x_\pm + z\exp(\lambda_\pm t),$$

we find

$$\lambda_\pm(\mu) = f_{xx}x_\pm. \tag{B.7}$$

Thus, one of the solutions (B.5) is stable, and another one is unstable. In multidimensional case ($n > 1$), the corresponding stationary points are node and saddle; therefore that type of bifurcations is called *saddle-node bifurcation*. That is the most generic type of bifurcations with $\lambda = 0$.

B.1.2 Transcritical Bifurcation

Assume that because of a certain symmetry of the problem

$$f(0,\mu) = 0 \tag{B.8}$$

for all μ, i.e., solution $x(\mu) = 0$ exists for all μ. In that case all the derivatives of $f(x,\mu)$ with respect to μ vanish at $x = 0$:

$$f_\mu = f_{\mu\mu} = \ldots = 0.$$

Then the lowest terms in (B.1) are:

$$\frac{dx}{dt} = f_{x\mu}x\mu + \frac{1}{2}f_{xx}x^2 + \ldots \tag{B.9}$$

Assume that $f_{x\mu} \neq 0$. Then there are two solutions (except the case $\mu = 0$, where both solutions merge):

$$x_1 = 0; \; x_2 = -\frac{2f_{x\mu}}{f_{xx}}\mu + \ldots . \tag{B.10}$$

The linearized problem determines the eigenvalue,

$$\lambda = f_{x\mu}\mu + f_{xx}x. \tag{B.11}$$

For the first solution,

$$\lambda_1 = f_{x\mu}\mu,$$

and for the second solution,

$$\lambda_2 = -f_{x\mu}\mu.$$

Thus, if $f_{x\mu}\mu < 0$, solution $x_1(\mu)$ is stable and solution $x_2(\mu)$ is unstable. By crossing the value $\mu = 0$, where the eigenvalue is equal to zero for each solution, the stability of each solution changes: for $f_{x\mu}\mu > 0$, solution $x_1(\mu)$ is unstable and solution $x_2(\mu)$ is stable. This is the *transcritical bifurcation*, which is a generic bifurcation in the case (B.8).

An example of a transcritical bifurcation is the onset of convection in a non-Boussinesq fluid. Indeed, the base solution corresponding to a quiescent state exists for any values of the Rayleigh number, but it becomes unstable with its growth. For a non-Boussinesq fluid, the amplitude equation contains a quadratic term, which leads to a generic transcritical bifurcation. In the case of a Boussinesq fluid, because of a specific feature of nonlinearity, only cubic nonlinear terms appear in the amplitude equation. In that case, a *pitchfork bifurcation* takes place, which is described below.

B.1.3 Pitchfork Bifurcation

Assume that $f(-x,\mu) = -f(x,\mu)$; thus all derivatives of an even order vanish at $x = 0$. Then the leading-order dynamic equation is

$$\frac{dx}{dt} = f_{x\mu}x\mu + \frac{1}{6}f_{xxx}x^3 + \dots \tag{B.12}$$

In that case, in addition to solution

$$x_1(\mu) = 0, \tag{B.13}$$

more two solutions,

$$x_{2,3} = \pm\sqrt{-\frac{6f_{x\mu}}{f_{xxx}}} + o(\sqrt{\mu}), \tag{B.14}$$

appear in the region of μ where the expression under the root sign is positive. Solutions (B.13) and (B.14) form a "pitchfork" with three teeth.

The eigenvalue for a disturbance around solution y_m, $m = 1,2,3$, is determined by the formula

$$\lambda_m = f_{x\mu}\mu + \frac{1}{2}f_{xxx}y_m^2. \tag{B.15}$$

We find that

$$\lambda_1 = f_{x\mu}\mu$$

and

$$\lambda_{2,3} = -2f_{x\mu}\mu.$$

Thus, solution $x_1 = 0$ is stable if $f_{x\mu}\mu \leq 0$ and unstable otherwise. As to solutions $x_{2,3}$, there are two qualitatively different cases.

I. If $f_{xxx} < 0$, then solutions (B.14) exist in the region $f_{x\mu}\mu > 0$, where solution $x_1 = 0$ is unstable. This is the *supercritical* pitchfork bifurcation. In that case, both solutions (B.14) are stable.

II. If $f_{xxx} > 0$, then solutions (B.14) exist in the region $f_{x\mu}\mu < 0$, where solution $x_1 = 0$ is stable, while solutions (B.14) are unstable. This is the *subcritical* pitchfork bifurcation. In the region $f_{x\mu}\mu > 0$, the only stationary solution is $x_1 = 0$, which is unstable. In the framework of the two-term equation (B.12), any solution with initial condition $x(0) \neq 0$ tends to infinity during a finite time. Actually, higher-order terms have to be incorporated into the analysis.

B.2 Oscillatory Instability

In the case where the linear instability border of a stationary solution corresponds to a pair of imaginary eigenvalues, $\lambda = \pm i\omega$, a periodic solution is born in addition to the stationary solution. This is the *Hopf bifurcation*(see [2]). Near the threshold, this type of bifurcation can be modeled by the following equation for the complex amplitude $z(t) = x(t) + iy(t)$:

$$\frac{dz}{dt} = \lambda z - \kappa |z|^2 z, \tag{B.16}$$

where $\lambda = \mu - i\omega_0$, $\kappa = \kappa_r + i\kappa_i$ are complex numbers. By rescaling, one can fix $\omega_0 = 1$, $\kappa_r = \pm 1$. Actually, (B.16) is a *normal form* for that kind of bifurcations, i.e., an autonomous system of ordinary differential equations can be transformed to that form near the threshold of the oscillatory instability by a certain transformation of variables.

Equation (B.16) has a solution $z = 0$ for any values of μ. It is stable as $\mu < 0$ and unstable as $\mu > 0$. If $\kappa_r > 0$, a stable *limit cycle* $(x(t), y(t))$ determined by formula

$$z(t) = re^{-i\omega(t-t_0)}, \ r = \sqrt{\frac{\mu}{\kappa_r}}, \ \omega = \omega_0 - \frac{\kappa_i}{\kappa_r}\mu \tag{B.17}$$

exists in the region $\mu > 0$ (*supercritical* Hopf bifurcation). If $\kappa_r < 0$, solution (B.17) exists in the region $\mu < 0$, where it is unstable (*subcritical* Hopf bifurcation). The description of bifurcations in the case of higher-order degeneracies can be found in [3].

References

1. J. Guckenheimer, Ph. Holmes, *Nonlinear Oscillations, Dynamical Systems, and Bifurcations of Vector Fields* (Springer, New York, 1983)
2. J.E. Marsden, M. McCracken, *The Hopf Bifurcation and Its Applications* (Springer, New York, 1976)
3. V.I. Arnold, *Geometrical Methods in the Theory of Ordinary Differential Equations* (Springer, New York, 1988)

Appendix C
Stationary Pattern Selection

In the present appendix, we consider the stationary pattern selection. The material of this appendix is not new, but it is given here for the reader's convenience. We do not consider the pattern formation in the case where the instability takes place for two significantly different values of the wavevector, and that circumstance leads to formation of *superlattices* [1].

In the main text of the book, many examples of *rotationally invariant* problems have been presented, where the growth rate σ of a disturbance with the wavevector $\mathbf{K} = (K_x, K_y)$ depends only on its absolute value K, $\sigma = \sigma(K)$. In that case, the linear stability theory predicts the growth of disturbances with arbitrary directions of wavevectors. One could expect that the generation of waves with different orientations would produce a spatially disordered state (*weak turbulence* [2] or *turbulent crystal* [3]). Actually, the strong nonlinear interaction between disturbances leads typically to selection of spatially ordered patterns.

C.1 Amplitude Equations with Cubic Nonlinearity

Let us start the discussion of pattern selection with the model equation,

$$\frac{\partial \phi}{\partial t} = \varepsilon^2 \phi - \left(1 + \nabla^2\right)^2 \phi - \phi^3, \tag{C.1}$$

suggested by Swift and Hohenberg [4] as a model for the description of convective pattern in a fluid layer situated between *perfectly conducting* rigid boundaries and heated from below. Here ϕ is the order parameter proportional to the vertical fluid velocity, and ε^2 is proportional to the difference between the actual temperature drop across the layer and its critical value corresponding to the instability threshold. The linear operator reproduces qualitatively the typical dependence of the growth

© Springer Science+Business Media, LLC 2017
S. Shklyaev, A. Nepomnyashchy, *Longwave Instabilities and Patterns in Fluids*,
Advances in Mathematical Fluid Mechanics,
https://doi.org/10.1007/978-1-4939-7590-7

rate on the wavenumber near the threshold of a shortwave monotonic instability (see Figure 1.1), but the nonlinear interaction between disturbances created by the nonlinear term is much simpler than the actual one in the Rayleigh–Bénard problem: the dependence of the nonlinear interaction coefficient on the wavevectors of disturbances is completely ignored. For the description of the actual nonlinear interaction in the convection problem, see [5–7]. Because equation (C.1) is drastically simpler that the original problem, it serves as a benchmark for the application of different tools of the pattern formation theory. The numerical simulations show that the pattern formation in the framework of equation (C.1) creates locally one-dimensional (roll) patterns [8].

For small ε, that result can be obtained by means of the analysis similar to that done in Section 3.1.3.2 in the case of poorly conducting boundaries, where square patterns are selected. The bounded solutions of equation (C.1) are constructed in the form

$$\phi(\mathbf{x}, T) = \varepsilon(\phi_1(\mathbf{x}, T) + \varepsilon^2 \phi_3(\mathbf{x}, T) + \ldots), \tag{C.2}$$

where $T = \varepsilon^2 t$. Substituting (C.2) into (C.1), we find at the leading order:

$$\phi_1(\mathbf{x}, T) = \sum_{n=1}^{N} \left[A_n(T) e^{i\mathbf{k}_n \cdot \mathbf{x}} + A_n^*(T) e^{-i\mathbf{k}_n \cdot \mathbf{x}} \right], \tag{C.3}$$

where $|\mathbf{k}_n| = 1$.

At the order $O(\varepsilon^3)$, the following equation is obtained:

$$-(1 + \nabla^2)^2 \phi_3 = \frac{\partial \phi_1}{\partial T} - \phi_1 + \phi_1^3. \tag{C.4}$$

Substituting (C.3) into (C.4), we find that the right-hand side of the equation contains nonlinear terms in the form $A_n^2 A_n^* \exp(i\mathbf{k}_n \cdot \mathbf{x})$ ("self-interaction" terms) and those in the form $A_n A_m A_m^* \exp(i\mathbf{k}_n \cdot \mathbf{x})$ ("cross-interaction" terms). The corresponding wavevector arrangements are shown in Figure C.1. Equation (C.4) has bounded solution only if the sum of all the terms, linear and nonlinear, containing $\exp(i\mathbf{k}_n \cdot \mathbf{x})$ vanishes for each n. That leads to the following set of *amplitude equations*

$$\frac{dA_n}{dT} = (1 - 3|A_n|^2 - 6 \sum_{m \neq n} |A_m|^2) A_n. \tag{C.5}$$

Taking $A_n = R_n \exp(i\theta_n)$, $n = 1, \ldots, N$, we obtain *the species competition* problem,

$$\frac{dI_n}{dT} = \frac{1}{2}(1 - M_{nn} I_n - M_{nm} \sum_{m \neq n} I_m) I_n, \quad n = 1, \ldots, N, \tag{C.6}$$

where $I_n = R_n^2$.

Though system (C.6) has been derived for a particular equation, (C.1), that system is rather generic. It should be noted however that while the coefficients M_{nn} are independent on the direction of vector $\mathbf{k}_n$ for any rotationally invariant problems, the independence of the coefficients M_{nm} on the angle θ_{nm} between the wavevectors $\mathbf{k}_n$ and $\mathbf{k}_m$ is a specific feature of the Swift–Hohenberg model. Typically, M_{nm} is an even function of θ_{nm}, i.e., $M_{nm} = M_{mn}$. Note that the last equality can be violated

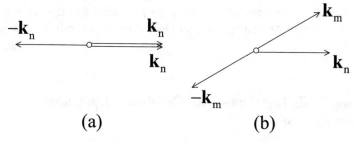

Fig. C.1 Wavevectors corresponding to (a) self-interaction (M_{nn}) and (b) cross-interaction (M_{nm}, $n \neq m$)

if the problem includes pseudovector variables, e.g., in the presence of rotation or magnetic field.

Let us consider the case $N = 2$, which corresponds to a competition of two rolls with some wavevectors $\mathbf{k}_1$ and $\mathbf{k}_2$, in more detail:

$$\frac{dI_1}{dT} = \frac{1}{2}(1 - M_{11}I_1 - M_{12}I_2)I_1, \quad \frac{dI_2}{dT} = \frac{1}{2}(1 - M_{12}I_1 - M_{11}I_2)I_2.$$

The phase portrait of that system in plane (I_1, I_2) is shown in Figure C.2.

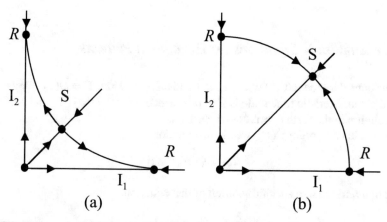

Fig. C.2 Pattern selection between rolls and squares: (a) $M_{12} > M_{11}$; (b) $M_{12} < M_{11}$

In the case of the Swift–Hohenberg equation, $M_{nm} > M_{nn} > 0$ for $n \neq m$, which corresponds to a competitive nonlinear interaction (see Figure C.2(a)). In the case $N = 2$, only solutions $I_1 = 1/M_{11}$, $I_2 = 0$ and $I_1 = 0$, $I_2 = 1/M_{11}$, which correspond to rolls of different orientation, are stable, while all other stationary solutions are unstable. That is correct for any number N of competing "species" [2].

If $M_{12} < M_{11}$, a "symbiotic" square pattern is selected by the system dynamics (see Figure C.2(b)). An example of that kind of nonlinear interaction is given in Section 3.1.3.2.

C.2 Amplitude Equations with Quadratic and Cubic Nonlinearities

The Swift–Hohenberg model (C.1) corresponds to the Boussinesq approximation for the description of convection [9]: the dependence of thermophysical parameters of the fluids on temperature is disregarded. If that dependence is taken into account, the quadratic nonlinearity appears in the amplitude equations [10]. We shall use the following phenomenological modification of the SH equation for the non-Boussinesq convection:

$$\frac{\partial \phi}{\partial t} = \gamma\phi - \left(1 + \nabla^2\right)^2 \phi + \alpha\phi^2 - \phi^3. \tag{C.7}$$

The coefficient α characterizing the non-Boussinesq properties of the fluid can have any sign. Also, we shall consider the system both above the linear instability threshold (for $\gamma > 0$) and below that threshold (for $\gamma < 0$).

C.2.1 Amplitude Equations for Hexagonal Patterns

Let us consider model (C.7) with $|\alpha| \ll 1$ and take $\gamma = \Gamma\alpha^2$, $\Gamma = O(1)$. We assume that the characteristic time scale is $T = \alpha^2 t$, hence $\partial/\partial t = \alpha^2\partial/\partial T$, and construct the solution in the form $\phi = \alpha\phi_1 + \alpha^3\phi_3 + \dots$.

At the leading order $O(\alpha)$, we obtain equation

$$-(1 + \nabla^2)^2\phi_1 = 0.$$

We are interested in a specific solution of that equation,

$$\phi_1(\mathbf{r}, T) = \sum_{n=1}^{3} \left[a_n(T)e^{i\mathbf{k}_n\cdot\mathbf{r}} + a_n^*(T)e^{-i\mathbf{k}_n\cdot\mathbf{r}}\right], \tag{C.8}$$

where the angle between the unit vectors $\mathbf{k}_n$, $n = 1, 2, 3$, is $120°$, so that

$$\mathbf{k}_1 + \mathbf{k}_2 + \mathbf{k}_3 = 0 \tag{C.9}$$

(see Figure C.3).

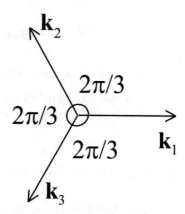

Fig. C.3 Basic wavevectors
of hexagonal patterns

At the order $O(\alpha^3)$, we obtain the following equation:

$$- (1 + \nabla^2)^2 \phi_3 = \frac{\partial \phi_1}{\partial T} - \Gamma \phi_1 - \phi_1^2 + \phi_1^3. \tag{C.10}$$

Because of the relations

$$(-\mathbf{k}_2) + (-\mathbf{k}_3) = \mathbf{k}_1, \ (-\mathbf{k}_3) + (-\mathbf{k}_1) = \mathbf{k}_2, \ (-\mathbf{k}_1) + (-\mathbf{k}_2) = \mathbf{k}_3,$$

the quadratic nonlinear term in the right-hand side of (C.10) generates additional quadratic terms in the solvability conditions. For instance, the condition of vanishing for the Fourier component of the right-hand side with the wavevector $\mathbf{k}_1$ leads to the following amplitude equation:

$$\frac{da_1}{dT} = \Gamma a_1 + 2a_2^* a_3^* - 3|a_1|^2 a_1 - 6(|a_2|^2 + |a_3|^2) a_1. \tag{C.11}$$

Two other amplitude equations are obtained from (C.11) by the cyclic permutation of the subscripts 1, 2, and 3:

$$\frac{da_2}{dT} = \Gamma a_2 + 2a_3^* a_1^* - 3|a_2|^2 a_2 - 6(|a_3|^2 + |a_1|^2) a_2, \tag{C.12}$$

$$\frac{da_3}{dT} = \Gamma a_3 + 2a_1^* a_2^* - 3|a_3|^2 a_3 - 6(|a_1|^2 + |a_2|^2) a_3. \tag{C.13}$$

Systems (C.11)–(C.13) can be written as

$$\frac{da_n}{dT} = -\frac{dU}{da_n^*}, \ n = 1, 2, 3, \tag{C.14}$$

where the Lyapunov function

$$U(a_1,a_1^*,a_2,a_2^*,a_3,a_3^*) = \sum_{n=1}^{3} \left(-\Gamma|a_n|^2 + \frac{3}{2}|a_n|^4 \right) \qquad (C.15)$$

$$-2(a_1a_2a_3 + a_1^*a_2^*a_3^*) + 3(|a_1|^2|a_2|^2 + |a_1|^2|a_3|^2 + |a_2|^2|a_3|^2)$$

(a_n and a_n^* are considered as independent variables). Therefore, $dU/dT \le 0$, and $dU/dT = 0$ only for stationary solutions.

C.2.2 Stationary Solutions

Let us consider the stationary solutions of the system of amplitude equations (C.11)–(C.13) and their stability.

C.2.2.1 Quiescent State

The solution $a_1 = a_2 = a_3 = 0$ corresponds to the quiescent state (no convection). The linearized equations for disturbances are:

$$\frac{d\tilde{a}_1}{dT} = \Gamma\tilde{a}_1, \quad \frac{d\tilde{a}_2}{dT} = \Gamma\tilde{a}_2, \quad \frac{d\tilde{a}_3}{dT} = \Gamma\tilde{a}_3. \qquad (C.16)$$

For normal modes,

$$\tilde{a}_1,\tilde{a}_2,\tilde{a}_3 \sim e^{\sigma T},$$

the eigenvalue $\sigma = \Gamma$, hence the quiescent state is stable as $\Gamma < 0$ and unstable as $\Gamma > 0$.

C.2.2.2 Rolls

Consider the solution $a_1 = \sqrt{\Gamma/3}\exp i\theta_1$, $a_2 = a_3 = 0$. This solution exists only for $\Gamma > 0$. Linearizing systems (C.11)–(C.13) around this solution, we find that the system is splitted into two systems: a separate equation for $\tilde{a}_1$ and a coupled system of equations for $\tilde{a}_2$ and $\tilde{a}_3$. The equation for $\tilde{a}_1$ is

$$\frac{d\tilde{a}_1}{dT} = \Gamma\tilde{a}_1 - 6|a_1|^2\tilde{a}_1 - 3a_1^2\tilde{a}_1^*. \qquad (C.17)$$

Let us substitute the expression for a_1 and define $\tilde{a}_1 = b_1 \exp i\theta_1$. We get

$$\frac{db_1}{dT} = -\Gamma(b_1 + b_1^*). \qquad (C.18)$$

Thus, the real (amplitude) disturbance decays with the rate $\sigma = -2\Gamma$, while the imaginary (phase) disturbance is neutral: $\sigma = 0$. The neutral disturbance corresponds to an infinitesimal change of the phase θ_1, i.e., a spatial shift of the roll system as a whole. The system of equations for $\tilde{a}_2$ and $\tilde{a}_3$ reads:

$$\frac{d\tilde{a}_2}{dT} = \Gamma\tilde{a}_2 + 2a_1^*\tilde{a}_3^* - 6|a_1|^2\tilde{a}_2, \tag{C.19}$$

$$\frac{d\tilde{a}_3}{dT} = \Gamma\tilde{a}_3 + 2a_1^*\tilde{a}_2^* - 6|a_1|^2\tilde{a}_3. \tag{C.20}$$

Taking the complex conjugate of (C.20) and assuming

$$\tilde{a}_2, \tilde{a}_3^* \sim e^{\sigma T},$$

we find that the obtained algebraic system has a nontrivial solution if

$$(\Gamma - 6|a_1|^2 - \sigma)^2 - 4|a_1|^2 = 0,$$

hence

$$\sigma = -\Gamma \pm 2\sqrt{\Gamma/3}. \tag{C.21}$$

Thus, the rolls are unstable in the interval $0 < \Gamma < \Gamma_1$ and become stable as $\Gamma > \Gamma_1$, where $\Gamma_1 = 4/3$.

Obviously, the same result is obtained for two other roll solutions, (i) $a_2 = \sqrt{\Gamma/3}\exp i\theta_2$, $a_3 = a_1 = 0$ and (ii) $A_3 = \sqrt{\Gamma/3}\exp i\theta_3$, $a_1 = a_2 = 0$.

So, in the interval $0 < \Gamma < \Gamma_1$, neither quiescent state nor roll patterns are stable. One has to investigate other critical points of the Lyapunov function (C.15).

C.2.2.3 Hexagons

Let us assume now that all the amplitudes a_n, $n = 1, 2, 3$ are not equal to zero, and present them in the form $a_n(T) = R_n(T)\exp[i\theta_n(T)]$. Equation (C.11) gives rise to the following equations for the real functions:

$$\frac{dR_1}{dT} = (\Gamma - 3R_1^2 - 6R_2^2 - 6R_3^2)R_1 + 2R_2R_3\cos(\theta_1 + \theta_2 + \theta_3), \tag{C.22}$$

$$R_1\frac{d\theta_1}{dT} = -2R_2R_3\sin(\theta_1 + \theta_2 + \theta_3). \tag{C.23}$$

Four more equations are obtained by the cyclic permutation of the subscripts 1, 2, and 3.

Note that due to the quadratic terms in the amplitude equations, the phases θ_n, $n = 1, 2, 3$ are not constant. Adding the equations for θ_n and denoting $\Theta = \theta_1 + \theta_2 + \theta_3$, we obtain the following equation for the time evolution of Θ which describes the phase synchronization of the roll subsystems:

$$\frac{d\Theta}{dT} = -Q\sin\Theta, \tag{C.24}$$

where

$$Q = \frac{2R_2R_3}{R_1} + \frac{2R_3R_1}{R_2} + \frac{2R_1R_2}{R_3} > 0. \tag{C.25}$$

There are two different stationary solutions of equation (C.24), $\Theta = 0$ and $\Theta = \pi$ (adding $2\pi n$ with integer n to Θ does not change the planform of ϕ_1). Linearizing the equation for a disturbance $\tilde{\Theta}$, we find that

$$\frac{d\tilde{\Theta}}{dT} = -Q\cos\Theta \cdot \tilde{\Theta}. \tag{C.26}$$

Hence, the invariant manifold $\Theta = 0$ is attracting, while the manifold $\Theta = \pi$ is repelling. Below, we shall consider the dynamics on the manifold $\Theta = 0$:

$$\frac{dR_1}{dT} = (\Gamma - 3R_1^2 - 6R_2^2 - 6R_3^2)R_1 + 2R_2R_3; \tag{C.27}$$

two other equations are obtained by the cyclic permutation of the subscripts 1, 2, and 3.

There is a stationary solution which corresponds to a hexagonal pattern: $R_1 = R_2 = R_3 \equiv R$, where R satisfies the equation

$$15R^2 - 2R - \Gamma = 0.$$

Because $R > 0$ by definition, we find that there are two branches of solutions:

$$R_+ = \frac{1+\sqrt{1+15\Gamma}}{15}, \, \Gamma \geq \Gamma_2 = -\frac{1}{15} \tag{C.28}$$

and

$$R_- = \frac{1-\sqrt{1+15\Gamma}}{15}, \, \Gamma_2 \leq \Gamma < 0. \tag{C.29}$$

At the point $\Gamma = \Gamma_2$, both branches merge, $R_+ = R_- = 1/15$.

Let us consider now the stability of the hexagons on the manifold $\Theta = 0$. Linearizing (C.27) and other two dynamic equations, we obtain the following system for the evolution of disturbances:

$$\frac{d\tilde{R}_1}{dT} = c_1\tilde{R}_1 + c_2(\tilde{R}_2 + \tilde{R}_3), \, \frac{d\tilde{R}_2}{dT} = c_1\tilde{R}_2 + c_2(\tilde{R}_3 + \tilde{R}_1), \, \frac{d\tilde{R}_3}{dT} = c_1\tilde{R}_3 + c_2(\tilde{R}_1 + \tilde{R}_2), \tag{C.30}$$

where

$$c_1 = -2R - 6R^2, \, c_2 = 2R - 12R^2. \tag{C.31}$$

For the normal modes, $R_n \sim \exp(\sigma T)$, we obtain the relation:

$$\sigma^3 - 3c_1\sigma^2 + 3(c_1^2 - c_2^2)\sigma - (c_1 - c_2)^2(c_1 + 2c_2) = 0. \tag{C.32}$$

According to the Descartes rule, all the roots of (C.32) are negative, so that the hexagons are stable, if the following conditions are satisfied: (i) $-3c_1 > 0$; (ii) $3(c_1^2 - c_2^2) > 0$; and (iii)$-(c_1 - c_2)^2(c_1 + 2c_2) > 0$.

Condition (i) is satisfied for any $R > 0$. Substituting (C.31) into condition (ii), we find $R < 2/3$; the latter inequality is satisfied for the upper branch (C.28), if $\Gamma < \Gamma_3 = 16/3$, as well as for the whole lower branch (C.29). The condition (iii) gives $R > 1/15$; hence the lower branch is unstable. Finally, we obtain that the upper branch is stable in the interval $\Gamma_2 < \Gamma < \Gamma_3$, where $\Gamma_2 = -1/15, \Gamma_3 = 16/3$.

Because $\Gamma_2 < 0$, both the quiescent state and the upper branch of hexagons are stable, i.e., provide a local minimum of the Lyapunov function U, in the interval $\Gamma_2 < \Gamma < 0$. Thus, the transition between the quiescent state and the hexagonal pattern in the presence of a cubic term in the Lyapunov function is similar to a *first-order phase transition* which takes place in the presence of a cubic term in the free energy Ginzburg–Landau functional [11], in a contradistinction to the transition between the quiescent state and the roll pattern in the absence of a cubic term in the Lyapunov function, which is similar to a *second-order phase transition*. The lower branch of hexagons corresponds to a saddle point. Its stable manifold separates the attraction basins of two stable nodes corresponding to the quiescent state and to the upper branch of hexagons.

Recall that the roll pattern becomes stable as $\Gamma > \Gamma_1 = 4/3$. Hence, in the interval $\Gamma_1 < \Gamma < \Gamma_3$, the Lyapunov function has four local minima, three of them correspond to three types of roll patterns, and one of them corresponds to hexagons. The basins of attractions between them are separated by stable manifolds of some more saddle-point stationary solutions, corresponding to rectangles (e.g., $R_1 = R_2 \neq 0, R_3 = 0$) and "skewed hexagons" (e.g., $R_1 = R_2 \neq R_3 \neq 0$). Finding the latter solutions is suggested to readers as an exercise.

The theory presented above describes patterns with $|\mathbf{k}_n| = 1, n = 1, 2, 3$. However, the resonant condition (C.9) can be satisfied also in the case, when the values of $|\mathbf{k}_n|$ are slightly different from $120°$. Such *nonequilateral* hexagonal patterns have been considered in [12].

C.2.3 Rules of Pattern Selection

Let us consider now a general pattern-forming system with the set of basic wavevectors $\{\mathbf{k}_i\}$, $|\mathbf{k}_i| = 1, i = 1, \ldots, N$, and arbitrary coefficients in cubic terms. By rescaling, we rewrite the system of amplitude equations in the following way:

$$\frac{da_i}{dT} = \sigma a_i + a_l^* a_m^* - |a_i|^2 a_i - \sum_{j=1}^{N} T(\theta_{ij})|a_j|^2 a_i, \quad i = 1 \ldots, N, \qquad (C.33)$$

where the nonlinear interaction coefficient T depends on the angle θ_{ij} between the wavevectors $\mathbf{k}_i$ and $\mathbf{k}_j$, and the quadratic term is present only for resonant triplets $\{\mathbf{k}_i, \mathbf{k}_l, \mathbf{k}_m\}$ with $\mathbf{k}_i + \mathbf{k}_l + \mathbf{k}_m = 0$. The competition between three basic patterns, namely, rolls, squares, and hexagons, is determined by the values of coefficients $T_1 = T(\pi/6)$, $T_2 = T(\pi/3)$, and $T_3 = T(\pi/2)$. The pattern selection in the framework of system (C.33) has been studied in many works (see, e.g., [13, 14]). Here we shortly summarize the main results on the pattern selection in the case where all the patterns are supercritical.

1. We already know (see Section C.1) that squares are stable versus rolls if $T_3 < 1$ and unstable otherwise.

2. If rolls are selected versus squares ($T_3 > 1$) and $T_2 > 1$, then the stability region of hexagons,

$$-\frac{1}{4(1+2T_2)} < \sigma < \frac{2+T_2}{(T_2-1)^2}, \tag{C.34}$$

overlaps with the stability region of rolls (see the previous subsection),

$$\sigma > \frac{1}{(T_2-1)^2}. \tag{C.35}$$

The corresponding bifurcation diagram is shown in Figure C.4. If $T_2 < 1$, hexagons are never unstable with respect to the transition to rolls.

3. If squares are selected versus rolls ($T_3 < 1$), then the competition between hexagons and squares depends on the values of the combinations $Q_1 = 1 - 2T_1 + 2T_2 - T_3$ and $Q_2 = 1 - T_1 - T_2 + T_3$.

If $Q_1 > 0$, hexagons become unstable with respect to squares as

$$\sigma > \frac{2T_1 + T_3}{Q_3^2}.$$

Otherwise, hexagons are never transformed into squares directly. They become unstable to rolls for sufficiently large σ (see (C.34)), and then the rolls are transformed into squares.

If $Q_2 > 0$, squares are always unstable with respect to hexagons. If $Q_2 < 0$, squares are stable as

$$\sigma > \frac{1 + T_3}{Q_2^2}.$$

C.3 Quasiperiodic Patterns

The description of the stationary pattern selection would be incomplete without mentioning quasiperiodic patterns with the number of selected modes $N > 3$.

In the case of a rotationally invariant problem, where the growth rate $\sigma(k)$ depends only on the wavenumber but does not depend on the direction of the wavevector, one can formally construct a solution of the leading-order problem in the form (C.3) with an arbitrary set of basic vectors $\{\mathbf{k}_n\}$, $n = 1, \ldots, N$. Patterns with $N > 3$ are typically quasiperiodic in space rather than periodic. If there are no triplets $\{\mathbf{k}_n, \mathbf{k}_l, \mathbf{k}_m\}$ of wavevectors satisfying the condition $\mathbf{k}_n + \mathbf{k}_l + \mathbf{k}_m = 0$, the evolution

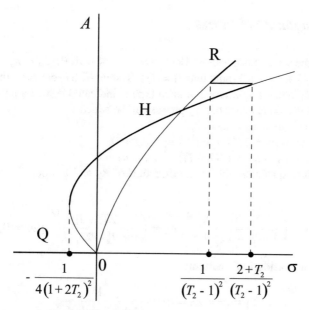

Fig. C.4 Bifurcation diagram for hexagonal patterns (H) vs roll patterns (R) and quiescent state (Q)

equations for amplitudes $A_i(T)$ in the supercritical region are governed by a generalization of system (C.5),

$$\frac{dA_n}{dT} = (1 - M_{nn}|A_n|^2 - \sum_{m\neq n} M_{nm}|A_m|^2)A_n, \quad n = 1,\ldots,N \qquad (C.36)$$

with a certain nonlinear interaction matrix M_{nm}. Later on, we assume that $M_{nn} > 0$, i.e., the roll pattern is supercritical. By rescaling, system (C.36) can be rewritten as

$$\frac{da_n}{dT} = \left[1 - |a_n|^2 - \sum_{m\neq n} T(\theta_{nm})|a_m|^2\right]a_i, \qquad (C.37)$$

where θ_{nm} is the angle between the wavevectors $\mathbf{k}_n$ and $\mathbf{k}_m$, $T(\theta_{nm}) = M_{nm}/M_{nn}$. Below we assume that relations $T(\theta) = T(-\theta) = T(\pi - \theta)$ are satisfied. Note that those relations may be violated in the case of a rotating layer. In the framework of system (C.37) (i.e., in the absence of quadratic terms), the phases of amplitudes are constant and arbitrary.

In the presence of resonant triplets, amplitude equations (C.33) have to be used.

The quasiperiodic structures are much less widespread than rolls, squares, and hexagons, because of much more restrictive stability conditions. Below we present several examples [15].

C.3.1 Octagonal Structures

Let us take the set of basic vectors $\{\mathbf{k}_n\}$, $n = 1,\ldots,4$ with $\theta_{n,n+1} = \pi/4$ and denote $T_n = T(n\pi/4)$, $n = 1,2,3$ (note that $T_1 = T_3$). System (C.37) can have the following patterns (only one representative of each type is indicated below; a permutation of subscripts is possible, and arbitrary phases can be added):

(i) rolls: $a_1 = 1$, $a_2 = a_3 = a_4 = 0$;
(ii) squares: $a_1 = a_3 = 1/(1+T_2)^{1/2}$, $a_2 = a_4 = 0$;
(iii) rectangles: $a_1 = a_2 = 1/(1+T_1)^{1/2}$, $a_3 = a_4 = 0$;
(iv) a structure quasiperiodic in the direction of $\mathbf{k}_2$ and periodic in the direction of $\mathbf{k}_4$:

$$a_1 = a_3 = \left(\frac{1-T_1}{1+T_2-2T_1^2}\right)^{1/2}, \quad a_2 = \left(\frac{1-2T_1+T_2}{1+T_2-2T_1^2}\right)^{1/2}, \quad a_4 = 0; \qquad (C.38)$$

(v) octagonal quasiperiodic pattern:

$$a_1 = a_2 = a_3 = a_4 = \frac{1}{(1+2T_1+T_2)^{1/2}}. \qquad (C.39)$$

Each pattern exists in the supercritical region if the amplitudes are real.

The necessary stability conditions *in the framework of the set of basic vectors* ("internal stability") of the abovementioned patterns are as follows:

(i) rolls: $T_1 > 1$, $T_2 > 1$;
(ii) squares: $T_1 > (1+T_2)/2$, $T_2 < 1$;
(iii) rectangles: $T_1 < 1$, $T_2 < 1$;
(iv) solution (C.38): always unstable;
(v) octagonal quasiperiodic pattern: $T_1 < (1+T_2)/2$, $T_2 < 1$.

If the stability of the octagonal pattern is considered in the framework of the system (C.33), i.e., *resonant triplets are included*, and additional stability condition has to be added [16]:

$$\sigma^{1/2} \geq \frac{2(1+2T_1+T_2)^{1/2}}{\tilde{T} - (1+2T_1+T_2)},$$

where

$$\tilde{T} = T(\pi/12) + T(\pi/6) + T(\pi/3) + T(5\pi/12).$$

If disturbances with *arbitrary directions of wavevectors* are taken into account, the octagonal pattern is stable if *for any* θ

$$\hat{T} \geq 1 + 2T_1 + T_2,$$

where

$$\hat{T} = T(\theta) + T(\pi/4 - \theta) + T(\pi/2 - \theta) + T(3\pi/4 - \theta).$$

The influence of higher-order nonlinearities on the pattern stability has been considered in [15].

C.3.2 Decagonal Structures

For the set of basic vectors $\{\mathbf{k}_n\}$, $n = 1,\ldots,5$ with $\theta_{n,n+1} = \pi/5$, $T_n = T(n\pi/5)$, $n = 1,\ldots,4$ ($T_1 = T_4$, $T_2 = T_3$), one obtains the following stationary solutions:
(i) rolls: $a_1 = 1$, $a_2 = a_3 = a_4 = a_5 = 0$;
(ii) rectangles I: $a_1 = a_2 = 1/(1+T_1)^{1/2}$, $a_3 = a_4 = a_5 = 0$;
(iii) rectangles II: $a_1 = a_3 = 1/(1+T_2)^{1/2}$, $a_2 = a_4 = a_5 = 0$;
(iv) asymmetric structure I: $a_1 = a_3 = [(1-T_1)/Z_1]^{1/2}$, $a_2 = [(1+T_2-2T_1)/Z_1]^{1/2}$, $a_4 = a_5 = 0$, $Z_1 = 1+T_2-2T_1^2$;
(v) asymmetric structure II: $a_1 = a_5 = [(1-T_2)/Z_2]^{1/2}$, $a_3 = [(1+T_1-2T_2)/Z_2]^{1/2}$, $a_2 = a_4 = 0$, $Z_2 = 1+T_1-2T_2^2$;
(vi) asymmetric structure III: $a_1 = a_4 = [(1-T_2)/Z_3]^{1/2}$, $a_2 = a_3 = [(1-T_1)/Z_3]^{1/2}$, $a_5 = 0$, $Z_3 = 1+T_1+T_2-T_1^2-T_1T_2-T_2^2$;
(vii) decagonal quasiperiodic structure: $a_1 = a_2 = a_3 = a_4 = a_5 = 1/(1+2T_1+2T_2)^{1/2}$;
Criteria of the internal stability for the abovementioned patterns: (i) rolls: $T_1 > 1$, $T_2 > 1$;
(ii) rectangles I: $T_1 < 1$, $T_2 > 1$;
(iii) rectangles II: $T_1 > 1$, $T_2 < 1$;
(iv) asymmetric structure I: $1 + \omega_1 T_1 - \omega_2 T_2 < 0$, $T_2 < 1$, where

$$\omega_1 = 2\cos\frac{2\pi}{5} = \frac{\sqrt{5}-1}{2}, \quad \omega_2 = 2\cos\frac{\pi}{5} = \frac{\sqrt{5}+1}{2};$$

(v) asymmetric structure II: $1 + \omega_1 T_2 - \omega_2 T_1 < 0$, $T_1 < 1$;
(vi) asymmetric structure III: always unstable;
(vii) decagonal quasiperiodic structure: $1 + \omega_1 T_1 - \omega_2 T_2 > 0$, $1 + \omega_1 T_2 - \omega_2 T_1 > 0$.
For more details, see [15].

C.3.3 Dodecagonal Structures

Let us consider now the case of basic vectors $\{\mathbf{k}_n\}$, $n = 1,\ldots,6$ with $\theta_{n,n+1} = \pi/6$, and denote $T_n = T(n\pi/6)$, $n = 1,\ldots,5$ in the framework of system (C.33). Define $\mathbf{k}_{n+6} = -\mathbf{k}_n$, $a_{n+6} = \bar{a}_n$, $i = 1,\ldots,6$. Let us present the complex amplitudes as $a_n = A_n \exp(i\phi_n)$ for $n = 1,\ldots,12$, so that $A_{i+6} = A_i$, $\phi_{n+6} = \phi_n + \pi$. There are two resonant triplets:

$$\mathbf{k}_1 + \mathbf{k}_5 + \mathbf{k}_9 = 0, \quad \mathbf{k}_2 + \mathbf{k}_6 + \mathbf{k}_{10} = 0.$$

System (C.33) can be rewritten as

$$\frac{dA_n}{dT} = (\sigma - A_n^2 - \sum_{m \neq n} T_{|m-n|}A_m^2)A_n + A_{n+4}A_{n+8}\cos\Phi_n, \qquad (C.40)$$

$$A_n\frac{d\phi_n}{dT} = -A_{n+4}A_{n+8}\sin\Phi_n, \qquad (C.41)$$

where $\Phi_n = \phi_n + \phi_{n+4} + \phi_{n+8}$, $\phi_n = \phi_{n-12}$ for $n > 12$. Obviously, stationary states correspond to $\sin \Phi_n = 0$.

The set of stationary patterns can be analyzed similarly to previous sections. Here we discuss only the dodecagonal quasiperiodic structures

$$A_1 = A_2 = \ldots = A_6 = A = \frac{1 \pm (1 + 4\sigma Q_0)^{1/2}}{2Q_0},$$

where

$$Q_0 = 1 + 2T_1 + 2T_2 + T_3. \tag{C.42}$$

For additional information, see [15].

The stability analysis shows that as in the case of hexagons, the lower branch is always unstable. For the upper branch, the stability is determined by *amplitude* disturbances. The corresponding six eigenvalues are:

$$\sigma_0 = A - 2Q_0A^2, \ \sigma_1 = \sigma_5 = -2A - 2Q_1A^2, \ \sigma_2 = \sigma_4 = -2A - 2Q_2A^2, \ \sigma_3 = A - 2Q_3A^2,$$

where Q_0 is defined by equation (C.42),

$$Q_1 = 1 + T_1 - T_2 - T_3, \ Q_2 = 1 + T_3 - T_1 - T_2, \ Q_3 = 1 - 2T_1 + 2T_2 - T_3.$$

One can show that the dodecagonal quasistationary pattern can be stable only if $Q_0 > 0$, $Q_3 > 0$; the stability conditions are

$$A > \frac{1}{2Q_0}, \ A > \frac{1}{2Q_3}. \tag{C.43}$$

If Q_1 and/or Q_2 is negative, additional stability conditions appear:

$$A < -\frac{1}{Q_1} \text{ and/or } A < -\frac{1}{Q_2}. \tag{C.44}$$

Conditions (C.43), (C.44) are compatible, if

$$Q_1 + 2Q_0 > 0, \ Q_2 + 2Q_0 > 0, \ Q_1 + 2Q_3 > 0, \ Q_2 + 2Q_3 > 0.$$

Note that the stability of a dodecagonal structure was first considered by Busse [17] who showed that this structure is unstable in the case of a non-Boussinesq Rayleigh convection, in a contradistinction to a hexagonal structure.

References

1. S.L. Judd, M. Silber, Physica D **136**, 45 (2000)
2. V.E. Zakharov, V.S. L'vov, G. Falkovich, *Kolmogorov Spectra of Turbulence* (Springer, Berlin, 1992)
3. A.C. Newell, Y. Pomeau, J. Phys. A **26**, L429 (1993)

4. J.W. Swift, P.C. Hohenberg, Phys. Rev. A **15**, 319 (1977)
5. A. Schlüter, D. Lortz, F. Busse, J. Fluid Mech. **23**, 129 (1965)
6. A.C. Newell, J.A. Whitehead, J. Fluid Mech. **38**, 279 (1969)
7. M.C. Cross, Phys. Fluids **23**, 1727 (1980)
8. J.J. Christensen, A.J. Bray, Phys. Rev. E **58**, 5364 (1998)
9. G.Z. Gershuni, E.M. Zhukhovitsky, *Convective Stability of Incompressible Fluid* (Keter, Jerusalem, 1976)
10. F.H. Busse, J. Fluid Mech. **30**, 625 (1967)
11. L.D. Landau, E.M. Lifshitz, *Statistical Physics* (Pergamon Press, Oxford, 1980)
12. B.A. Malomed, A.A. Nepomnyashchy, A.E. Nuz, Physica D **70**, 357 (1994)
13. E.A. Kuznetsov, M.D. Spektor, J. Appl. Mech. Techn. Phys. **21**, 220 (1980)
14. B.A. Malomed, M.I. Tribelsky, Sov. Phys. JETP **65**, 305 (1987)
15. B.A. Malomed, A.A. Nepomnyashchy, M.I. Tribelsky, Sov. Phys. JETP **69**, 388 (1989)
16. B.A. Malomed, A.A. Nepomnyashchy, M.I. Tribelsky, Sov. Phys. Techn. Phys. **13**, 487 (1987).
17. F.H. Busse, Das Stabilitätsverhalten der Zellularkonvektion bei endlicher Amplitude. Ph.D. Thesis, Munich (1962)

Appendix D
Regular Wave Patterns

In this appendix we present the main results on the planform selection of wave patterns following seminal papers [1] and [2] on the stability of perfect periodic patterns on square (Section D.1) and hexagonal (Section D.2) lattices, respectively. Those results are used in Chapter 4.

D.1 Limit Cycles and Their Stability for Wave Patterns with Square Symmetry

D.1.1 The Set of Amplitude Equations and Limit Cycles

Let us represent the amplitude function h (which can be any scalar perturbation field, for instance, the temperature) as the superposition of two pairs of counter-propagating traveling waves:

$$h = \left(A_1(t_{42})e^{i\mathbf{k}_1\cdot\mathbf{r}} + A_2(t_{42})e^{-i\mathbf{k}_1\cdot\mathbf{r}} + B_1(t_{42})e^{i\mathbf{k}_2\cdot\mathbf{r}} + B_2(t_{42})e^{-i\mathbf{k}_2\cdot\mathbf{r}} \right) e^{-i\gamma_{0i}t_{40}} + c.c.,$$
(D.1)

where $\mathbf{k}_j = \mathbf{k}_j^{(S)}$, as they are introduced in (4.100); see also Figure 4.2.

The complex amplitudes depend on the slow time t_{42}; the evolution in this time is governed by the following set of amplitude equations:

$$\dot{A}_1 = \left(\gamma - N_0 a_1^2 - N_\pi a_2^2 - N_{\pi/2}\left(b_1^2 + b_2^2 \right) \right) A_1 - K_{\pi/2} B_1 B_2 A_2^*,$$
(D.2)

$$\dot{A}_2 = \left(\gamma - N_0 a_2^2 - N_\pi a_1^2 - N_{\pi/2}\left(b_1^2 + b_2^2 \right) \right) A_2 - K_{\pi/2} B_1 B_2 A_1^*,$$
(D.3)

and the similar pair of equations produced from (D.2) and (D.3) by a permutation of A and B. Hereafter $a_j = |A_j|$ and $b_j = |B_j|$ ($j = 1, 2$); a dot denotes the derivative with respect to t_{42}.

© Springer Science+Business Media, LLC 2017
S. Shklyaev, A. Nepomnyashchy, *Longwave Instabilities and Patterns in Fluids*,
Advances in Mathematical Fluid Mechanics,
https://doi.org/10.1007/978-1-4939-7590-7

This equation has five types of solutions corresponding to various patterns (limit cycles).

(i) Traveling rolls (TRs):

$$a_1 = \sqrt{\frac{\gamma_r}{N_{0r}}}, \quad A_2 = B_1 = B_2 = 0; \tag{D.4}$$

(ii) Standing rolls (SRs):

$$A_1 = A_2, \quad a_1 = \sqrt{\frac{\gamma_r}{(N_0 + N_\pi)_r}}, \quad B_1 = B_2 = 0; \tag{D.5}$$

(iii) Traveling squares (TSs):

$$A_1 = B_1 \neq 0, \quad a_1 = \sqrt{\frac{\gamma_r}{(N_0 + N_{\pi/2})_r}}, \quad A_2 = B_2 = 0; \tag{D.6}$$

(iv) Standing squares (StSs):

$$A_1 = A_2 = B_1 = B_2, \quad a_1 = \sqrt{\frac{\gamma_r}{(N_0 + N_\pi + 2N_{\pi/2} + K_{\pi/2})_r}}; \tag{D.7}$$

(v) Alternating rolls (ARs):

$$A_1 = A_2 = iB_1 = iB_2, \quad a_1 = \sqrt{\frac{\gamma_r}{(N_0 + N_\pi + 2N_{\pi/2} - K_{\pi/2})_r}}. \tag{D.8}$$

The snapshots corresponding to these five solutions are shown in Figures D.1–D.5. These figures demonstrate the distribution of h in the $x - -y$ plane (layer plane). It is clear that all these solutions, except for StSs, have twin patterns, which can be produced from the presented ones by permutations of subscripts $(1 \leftrightarrow 2)$ and/or amplitudes $(A \leftrightarrow B)$ and their combination.

Fig. D.1 The pattern corresponding to TRs; this pattern moves along the x-axis (horizontal one)

Swift [3] found also a pattern with the broken symmetry, so-called standing cross rolls (SCRs), for which

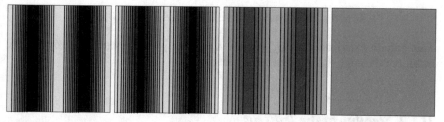

Fig. D.2 Snapshots for SRs during a quarter of the time period; in the next quarter of the period, the snapshots are repeated in the reverse order and with inversion of h

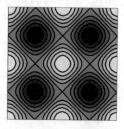

Fig. D.3 The pattern corresponding to TSs; this pattern moves along the diagonal of the square

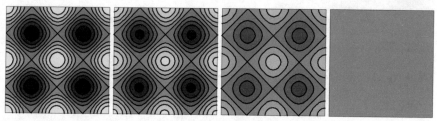

Fig. D.4 Snapshots for StSs during a quarter of the time period; in the next quarter of the period, the snapshots are repeated in the reverse order and with inversion of h

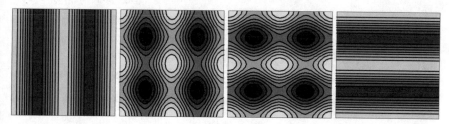

Fig. D.5 Snapshots for ARs during a quarter of the time period; in the next quarter of the period, the snapshots are repeated in the reverse order and with inversion of h

$$A_1 = A_2, \ B_1 = B_2, \ a_1 \neq b_1, \ \Phi = \arg(A_1) - \arg(B_1) \neq 0.$$

This pattern exists only if $N_{0,\pi/2,\pi}$ and $K_{\pi/2}$ satisfy certain inequalities. Near the stability threshold, SCRs are always unstable.

D.1.2 Stability of the Patterns

Here we present only the results of stability analysis for the abovementioned patterns. A reader interested in details of that analysis is referred to the original paper [1].

The conditions listed below are the *stability conditions*; therefore, a pattern is stable only if all the conditions are met.

(i) TRs

$$N_{0r} > 0, \ (N_\pi - N_0)_r > 0, \ (N_{\pi/2} - N_0)_r > 0;$$

(ii) SRs

$$(N_0 + N_\pi)_r > 0, \ (N_0 - N_\pi)_r > 0, \ f_r > 0, \ |f|^2 - |K_{\pi/2}|^2 > 0;$$

(iii) TSs

$$(N_0 + N_{\pi/2})_r > 0, \ (N_0 - N_{\pi/2})_r > 0, \ (N_\pi - N_0 + K_{\pi/2})_r > 0, \ (N_\pi - N_0 - K_{\pi/2})_r > 0.$$

(iv) StSs

$$(N_0 + N_\pi + 2N_{\pi/2} + K_{\pi/2})_r > 0, \ -(3K_{\pi/2} + f)_r > 0, \ |K_{\pi/2}|^2 + \mathrm{Re}(fK_{\pi/2}^*) > 0,$$
$$(N_0 - N_\pi - K_{\pi/2})_r > 0;$$

(v) ARs

$$(N_0 + N_\pi + 2N_{\pi/2} - K_{\pi/2})_r > 0, \ (3K_{\pi/2} - f)_r > 0, \tag{D.9}$$
$$(N_0 - N_\pi + K_{\pi/2})_r > 0, \tag{D.10}$$
$$|K_{\pi/2}|^2 - \mathrm{Re}(fK_{\pi/2}^*) > 0; \tag{D.11}$$

Here $f = 2N_{\pi/2} - (N_0 + N_\pi)$.

For all the patterns, the first condition corresponds to the supercritical bifurcation. In other words, the solutions listed in Section D.1.1 exist above the stability threshold, $\gamma_r > 0$. Recall that SCRs are always unstable within (D.2) and (D.3).

D.2 Limit Cycles and Their Stability for Wave Patterns with Hexagonal Symmetry

D.2.1 The Set of Amplitude Equations

For the hexagonal lattice, the amplitude function h is presented as the superposition of three pairs of counter-propagating waves:

$$h = \left(A_1(t_{42})e^{i\mathbf{k}_1\cdot\mathbf{r}} + A_2(t_{42})e^{-i\mathbf{k}_1\cdot\mathbf{r}} + B_1(t_{42})e^{i\mathbf{k}_2\cdot\mathbf{r}} + B_2(t_{42})e^{-i\mathbf{k}_2\cdot\mathbf{r}}\right.$$
$$\left. + C_1(t_{42})e^{i\mathbf{k}_3\cdot\mathbf{r}} + C_2(t_{42})e^{-i\mathbf{k}_3\cdot\mathbf{r}}\right) e^{-i\gamma_{0i}t_{40}} + c.c., \quad (D.12)$$

where $\mathbf{k}_j = \mathbf{k}_j^{(H)}$; see (4.100) and Figure 4.2.

The evolution of the complex amplitudes in slow time t_{42} is governed by equations:

$$\dot{A}_1 = \left(\gamma - N_0 a_1^2 - N_\pi a_2^2 - N_{2\pi/3} S_1 - N_{\pi/3} S_2\right) A_1$$
$$- K_{\pi/3}\left(B_1 B_2 + C_1 C_2\right) A_2^*, \quad (D.13)$$

$$\dot{A}_2 = \left(\gamma - N_0 a_2^2 - N_\pi a_2^2 - N_{\pi/3} S_1 - N_{2\pi/3} S_2\right) A_2$$
$$- K_{\pi/3}\left(B_1 B_2 + C_1 C_2\right) A_1^*, \quad (D.14)$$

where $a_j = |A_j|$ and $S_j = b_j^2 + c_j^2$ ($j = 1, 2$). Two remaining pairs of equations are produced from (D.13) and (D.14) by a cyclic permutation of A, B, and C.

D.2.2 Limit Cycles

The limit cycles within the set of amplitude equations (D.13) and (D.14) were studied by Roberts et al. [2]; see also a brief survey of the results in [4]. Hereafter we adopt the classification of the patterns from the latter paper [4] with the only exception for alternating rolls (rectangular), which were termed as wavy rolls 1 in the cited papers. Again, below we present a single representative for a family of equivalent solutions. Following the cited papers, we consider only symmetric patterns, for which all the nonzero amplitudes have the same absolute value. For instance, a rectangular analogue of SRCs is not discussed.

The six limit cycles listed in [2] belong to the rhombic (rectangular in the real space) lattice. They can be readily produced from the patterns listed in Section D.1.2. The 1D patterns, (i) TRs and (ii) SRs, are given by (D.4) and (D.5), respectively; the snapshots for these patterns are depicted in Figures D.1 and D.2, respectively.

Square patterns, TRs, StSs, and ARs, are transformed into rectangular ones as follows. There are two types of traveling patterns: (iii) traveling rectangles 1

(TRas1) and (iv) traveling rectangles 2 (TRas2), which are produced from the solution given by (D.6) by the replacement of $N_{\pi/2}$ with either $N_{2\pi/3}$ or $N_{\pi/3}$, respectively. (v) Standing rectangles (SRa) and (vi) alternating rolls (rectangular) (ARs-R) are given by (D.7) and (D.8), respectively, with the replacement of $2N_{\pi/2}$ with $N_{\pi/3} + N_{2\pi/3}$. The evolutions of h for these four patterns are presented in Figures D.6–D.9.

As we stated above, ARs-R coincide with wavy rolls 1 (WR1) ($A_1 = A_2 = B_1 = -B_2$) in classification of Ref. [4]. Indeed, it is obvious that the former pattern is transformed into ARs-R by the shift of the origin along the x-axis: $x \to x + \pi/k$.

Fig. D.6 Snapshot for TRas1; this pattern moves along the vector $\mathbf{k}_1 + \mathbf{k}_2$ (along the longest side of the rectangle)

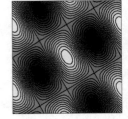

Fig. D.7 Snapshot for TRas2; this pattern moves along the vector $\mathbf{k}_1 - \mathbf{k}_3$ (along the shortest side of the rectangle)

Fig. D.8 Snapshots for SRas during a quarter of the time period; in the next quarter of the period, the snapshots are repeated in the reverse order and with inversion of h

The remaining five patterns are genuine hexagonal patterns:
(vii) Oscillating triangles (OTs)

Fig. D.9 Snapshots for AR-Rs during a quarter of the time period; in the next quarter of the period, the snapshots are repeated in the reverse order and with inversion of h

$$a_1 = b_1 = c_1 = \sqrt{\frac{\gamma_r}{(N_0 + 2N_{2\pi/3})_r}}, \quad a_2 = b_2 = c_2 = 0; \qquad (D.15)$$

(viii) Standing hexagons (SHs)

$$A_j = B_j = C_j = A, \quad |A| = \sqrt{\frac{\gamma_r}{(N_0 + N_\pi + 2(N_{\pi/3} + N_{2\pi/3} + K_{\pi/3}))_r}} \qquad (D.16)$$

(ix) Standing regular triangles (SRTs)

$$A_1 = -A_2 = B_1 = -B_2 = C_1 = -C_2 = A,$$
$$|A| = \sqrt{\frac{\gamma_r}{(N_0 + N_\pi + 2(N_{\pi/3} + N_{2\pi/3} + K_{\pi/3}))_r}} \qquad (D.17)$$

(x) Twisted rectangles (TwRs)

$$A_j = B_j e^{2i\pi/3} = C_j e^{4i\pi/3} = A, \quad |A| = \sqrt{\frac{\gamma_r}{(N_0 + N_\pi + 2(N_{\pi/3} + N_{2\pi/3}) - K_{\pi/3})_r}} \qquad (D.18)$$

(xi) Wavy rolls 2 (WRs2)

$$A_1 = -A_2 = B_1 e^{2i\pi/3} = -B_2 e^{2i\pi/3} = C_1 e^{4i\pi/3} = -C_2 e^{4i\pi/3} = A,$$
$$|A| = \sqrt{\frac{\gamma_r}{(N_0 + N_\pi + 2(N_{\pi/3} + N_{2\pi/3}) - K_{\pi/3})_r}} \qquad (D.19)$$

In fact, within amplitude equations (D.13) and (D.14), the pairs SHs & SRTs and TwRs & WRs2 are representatives of a wider class of solutions: as shown in [2], the amplitude equations are invariant under more general transformation, $(A_2, B_2, C_2) \to e^{i\nu}(A_2, B_2, C_2)$ for any real ν. (It is clear that, say, SRTs can be produced from SHs for $\nu = \pi$.) This degeneracy is removed by accounting for quintic nonlinear terms.

The snapshots for h corresponding to patterns (viii)–(xi) are shown in Figures D.10–D.14.

Fig. D.10 Snapshots for OTs during a quarter of the time period; in the next quarter of the period, the snapshots are repeated in the reverse order and with inversion of both X-axis (reflection of the pattern with respect to the vertical midline) and h

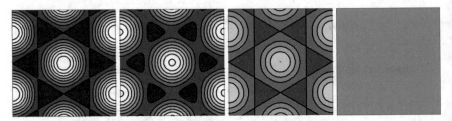

Fig. D.11 Snapshots for SHs during a quarter of the time period; in the next quarter of the period, the snapshots are repeated in the reverse order and with inversion of h

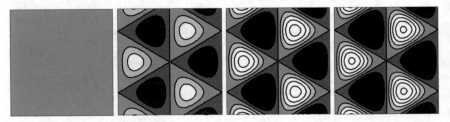

Fig. D.12 Snapshots for SRTs during a quarter of the time period; in the next quarter of the period, the snapshots are repeated in the reverse order and with inversion of h

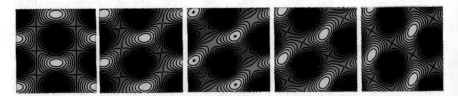

Fig. D.13 Snapshots for TwRs during a one sixth of the time period

Fig. D.14 Snapshots for WRs2 in a one sixth of the time period

D.2.3 Stability of the Limit Cycles

Similarly to the analysis of patterns on a square lattice, we only present the corresponding stability conditions, referring to the original paper [2] for the detail. The corresponding solution is stable, only if all the conditions listed below are satisfied.

(i) TRs

$$N_{0r} > 0, \ (N_\pi - N_0)_r > 0, \ (N_{2\pi/3} - N_0)_r > 0, \ (N_{\pi/3} - N_0)_r > 0,;$$

(ii) SRs

$$(N_0 + N_\pi)_r > 0, \ (N_0 - N_\pi)_r > 0, \ F_r > 0, \ |F|^2 - |K_{\pi/3}|^2 > 0;$$

(iii) TRas1

$$(N_0 + N_{2\pi/3})_r > 0, \ (N_0 - N_{2\pi/3})_r > 0, \ (N_{2\pi/3} - N_0)_r > 0,$$
$$(N_\pi - N_0 + N_{\pi/3} - N_{2\pi/3} + K_{\pi/3})_r > 0, \quad \text{(D.20)}$$
$$(N_\pi - N_0 + N_{\pi/3} - N_{2\pi/3} - K_{\pi/3})_r > 0, \ (2N_{\pi/3} - N_0 - N_{2\pi/3})_r > 0.$$

(iv) TRas2

$$(N_0 + N_{\pi/3})_r > 0, \ (N_0 - N_{\pi/3})_r > 0, \ (N_{2\pi/3} - N_0)_r > 0,$$
$$(N_\pi - N_0 - N_{\pi/3} + N_{2\pi/3} + K_{\pi/3})_r > 0, \quad \text{(D.21)}$$
$$(N_\pi - N_0 - N_{\pi/3} + N_{2\pi/3} - K_{\pi/3})_r > 0, \ (2N_{\pi/3} - N_0 - N_{2\pi/3})_r > 0.$$

(v) SRas

$$(N_0 + N_\pi + N_{\pi/3} + N_{2\pi/3} + K_{\pi/3})_r > 0, \ (N_0 - N_\pi + N_{\pi/3} - N_{2\pi/3} - K_{\pi/3})_r > 0,$$
$$(N_0 - N_\pi - N_{\pi/3} + N_{2\pi/3} - K_{\pi/3})_r > 0, \ (F - K_{\pi/3})_r > 0, \ -(3K_{\pi/3} + f)_r > 0,$$
$$D_r > 0, \ 4|K_{\pi/3}|^2 - |F - K_{\pi/3}|^2 > 0;$$

(vi) ARs-R

$$(N_0 + N_\pi + N_{\pi/3} + N_{2\pi/3} - K_{\pi/3})_r > 0, \ (N_0 - N_\pi + N_{\pi/3} - N_{2\pi/3} + K_{\pi/3})_r > 0,$$
$$(N_0 - N_\pi - N_{\pi/3} + N_{2\pi/3} + K_{\pi/3})_r > 0, \ (F + K_{\pi/3})_r > 0, \ (3K_{\pi/3} - f)_r > 0,$$
$$|K_{\pi/3}|^2 - \text{Re}(F K_{\pi/3}^*) > 0;$$

(vii) OTs

$$(N_0 + 2N_{2\pi/3})_r > 0, \ (N_0 - N_{2\pi/3})_r > 0, \ (N_\pi - N_0 + 2(N_{\pi/3} - N_{2\pi/3} + K_{\pi/3}))_r > 0,$$
$$(N_\pi - N_0 + 2(N_{\pi/3} - N_{2\pi/3} - K_{\pi/3}))_r > 0;$$

(viii), (ix) SHs, SRTs

$$(N_0 + N_\pi + 2(N_{\pi/3} + N_{2\pi/3} + K_{\pi/3}))_r > 0, \ (N_0 - N_\pi + 2(N_{2\pi/3} - N_{\pi/3} - K_{\pi/3}))_r > 0,$$

$$(N_0 - N_\pi + N_{\pi/3} - N_{2\pi/3} - 2K_{\pi/3})_r > 0, \ (N_0 + N_\pi - N_{\pi/3} - N_{2\pi/3} - 4K_{\pi/3})_r > 0, \ D_r > 0;$$

(x), (xi) TwRs, WR2s

$$(N_0 + N_\pi + 2(N_{\pi/3} + N_{2\pi/3}) - K_{\pi/3})_r > 0, \ (N_0 - N_\pi + 2(N_{2\pi/3} - N_{\pi/3}) + K_{\pi/3})_r > 0,$$
$$(N_0 - N_\pi - N_{2\pi/3} + N_{\pi/3} + K_{\pi/3})_r > 0, \ \mathrm{Re}(T \pm \sqrt{T^2 + 6D}) > 0;$$

Here $F = N_{\pi/3} + N_{2\pi/3} - N_0 - N_\pi$, $D = |K_{\pi/3}|^2 + F K_{\pi/3}^*$, and $T = 3K_{\pi/3}^* - (K_{\pi/3} + F)_r$.

In particular, it is clear that TRas1 are always unstable.

References

1. M. Silber, E. Knobloch, Nonlinearity **4**, 1063 (1991)
2. M. Roberts, J.W. Swift, D. H. Wagner, Contemp. Math. **56**, 283 (1986)
3. J.W. Swift, Nonlinearity **1**, 333 (1988)
4. T. Clune, E. Knobloch, Physica D **74**, 151 (1994)

Appendix E
Resonant Perturbations

In this appendix we consider the dynamics of resonant perturbations within the set of nonlocal amplitude equations, which are solvability conditions for the linear nonhomogeneous problems (4.275) and (4.276), which can be written in terms of dynamical system (4.278). To that end we disturb the base solution h_0, which possesses a certain symmetry (e.g., it belongs to a lattice), adding the perturbation breaking that symmetry. Therefore, the analysis is similar to finding the so-called Busse balloons within the local partial differential amplitude equations, where the Floquet–Bloch formalism can be applied.

In Section E.1 the resonant perturbations are studied for arbitrary h_0; six classes of resonant perturbations are found. On the base of that analysis, in Section E.2 the stability of ARs-R is considered. One can readily extend this analysis to any pattern on a rhombic lattice. That study is reduced to a square lattice, i.e., the particular case of the rhombic lattice with the right angle, in Section E.3. It is shown that several stability conditions for ARs cannot be obtained just by setting $\theta = \pi/2$ in the results of Section E.2.

E.1 General Analysis of Resonances

In this section we analyze *resonant* perturbations for (4.278) while perturbing the base solution h_0 given by (4.77) by a small perturbation

$$h' = C_1(t_4)e^{i(\mathbf{P}_1 \cdot \mathbf{R} - \Omega p_1^2 \tau)} + C_2(t_4)e^{i(\mathbf{P}_2 \cdot \mathbf{R} - \Omega p_2^2 \tau)}. \tag{E.1}$$

Equation (E.1) represents the simplest ansatz for a *resonant* perturbation, a superposition of two interacting waves. A single wave, which does not excite any other Fourier harmonics via the interaction with the base solution, is termed as *simple* perturbation, and it is analyzed in Section 4.2.1.4. In fact, the simple perturbation can be treated as a particular case of resonant perturbations; see Class IVa below.

© Springer Science+Business Media, LLC 2017
S. Shklyaev, A. Nepomnyashchy, *Longwave Instabilities and Patterns in Fluids*,
Advances in Mathematical Fluid Mechanics,
https://doi.org/10.1007/978-1-4939-7590-7

More complex resonant disturbances that involve more than two interacting waves can be decomposed into resonant pairs of form (E.1). All such intersections of the resonant classes presented below should be considered separately. An example of such intersection is presented below in Section E.2 for h' given by (E.16); another example is given by (E.65). Therefore, the analysis below is general, irrespective of the symmetry of the base solution and even of a particular problem described by (4.278).

Substituting $h = h_0 + h'$ into (4.278) and linearizing the equations with respect to the perturbation amplitudes C_1 and C_2, one can find six classes of perturbations. Two of them (*Classes I* and *II*) stem from the quadratic interactions, and the remaining four (*Classes III–VI*) are generated by the cubic interactions. For the quadratic resonance, two perturbation waves interact via a base wave with a specific wavevector $\mathbf{n}$. Here $\mathbf{n}$ is any of the wavevectors $\{\mathbf{k}\}$, which appears in the base solution, (4.77). For the cubic resonance, the interaction involves two base wavevectors $\mathbf{n}$ and $\mathbf{m}$ (each pair of $\mathbf{n}$ and $\mathbf{m}$ in the base solution has to be considered separately) and one of the perturbations giving rise to another perturbation wave and vice versa. (It is clear that if the resonant conditions are valid for $\mathbf{n}$, $\mathbf{m}$, $\mathbf{p}_1$ producing the wave with the wavevector $\mathbf{p}_2$, the triplet $\mathbf{n}$, $\mathbf{m}$, $\mathbf{p}_2$ is also resonant giving rise to the wave with $\mathbf{p}_1$.)

The resonant classes mentioned above produce the following terms in the amplitude equations:

Class I

$$\dot{C}_1 \sim -N^{(1)}_{\mathbf{n}\mathbf{p}_2}A_{\mathbf{n}}C_2, \quad \dot{C}_2 \sim -N^{(2)}_{\mathbf{p}_1\mathbf{n}}A^*_{\mathbf{n}}C_1. \tag{E.2}$$

Here and below we present only resonant interactions between two waves omitting other terms corresponding to *simple* perturbations; see *Class IVa* below.

The resonant condition for this class is

$$\mathbf{n} \cdot \mathbf{p}_2 = 0. \tag{E.3}$$

The manifold determined by the perturbation wavevector $\mathbf{p}_2$ is a straight line, perpendicular to the wavevector $\mathbf{n}$, whereas $\mathbf{p}_1 = \mathbf{p}_2 + \mathbf{n}$.

Class II

$$\dot{C}_1 \sim -N^{(2)}_{\mathbf{n}\mathbf{p}_2}A_{\mathbf{n}}C^*_2, \quad \dot{C}_2 \sim -N^{(2)}_{\mathbf{n}\mathbf{p}_1}A_{\mathbf{n}}C^*_1. \tag{E.4}$$

The resonant condition for this class is

$$\left(\mathbf{p}_2 - \frac{\mathbf{n}}{2}\right)^2 = \frac{n^2}{4}. \tag{E.5}$$

Thus, $\mathbf{p}_2$ describes the circle of radius $n/2$ with the center at $\mathbf{n}/2$; the second wavevector given by $\mathbf{p}_1 = \mathbf{n} - \mathbf{p}_2$ ends at the diametrically opposite point of the same circle.

Class III

$$\dot{C}_1 \sim -K^{(1)}_{\mathbf{n}\mathbf{m}\mathbf{p}_2}A_{\mathbf{n}}A_{\mathbf{m}}C_2, \dot{C}_2 \sim -K^{(3)}_{\mathbf{p}_1\mathbf{n}\mathbf{m}}A^*_{\mathbf{n}}A^*_{\mathbf{m}}C_1. \tag{E.6}$$

The resonant condition is

$$(\mathbf{n}+\mathbf{m})\cdot\mathbf{p}_2 + \mathbf{n}\cdot\mathbf{m} = 0. \tag{E.7}$$

Class IV

$$\dot{C}_1 \sim -K^{(2)}_{\mathbf{np}_2\mathbf{m}} A_{\mathbf{n}} A_{\mathbf{m}}^* C_2, \dot{C}_2 \sim -K^{(2)}_{\mathbf{p}_1\mathbf{mn}} A_{\mathbf{n}}^* A_{\mathbf{m}} C_1. \tag{E.8}$$

The resonant condition is

$$(\mathbf{n}-\mathbf{m})\cdot(\mathbf{p}_2-\mathbf{m}) = 0. \tag{E.9}$$

The manifolds determined by the wavevector $\mathbf{p}_2$ for *Classes III* and *IV* are the straight lines; the second resonant wavevector is $\mathbf{p}_1 = \mathbf{n}\pm\mathbf{m}+\mathbf{p}_2$, respectively.

It is also important to emphasize that for $\mathbf{n}=\mathbf{m}$ and, hence, $\mathbf{p}_1 = \mathbf{p}_2$, (E.9) is satisfied for any $\mathbf{p}_1$ and $\mathbf{n}$, resulting in

Class IVa

$$\dot{C}_1 \sim -K_{\mathbf{np}_1} |A_{\mathbf{n}}|^2 C_1. \tag{E.10}$$

This is the *simple* interaction, which provides a nonlinear self-interaction for any $\mathbf{p}_1$; $K_{\mathbf{np}_1}$ is given by

$$K_{\mathbf{k}_1\mathbf{k}_2} = K^{(2)}_{\mathbf{k}_1\mathbf{k}_2\mathbf{k}_1} = (2\tilde{\alpha}+\tilde{\beta})(\mathbf{k}_1\cdot\mathbf{k}_2)^2 + \tilde{\beta}k_1^2 k_2^2$$
$$-\tilde{J}_c(k_1^2 - k_2^2)\frac{k_1^2 k_2^2 - (\mathbf{k}_1\cdot\mathbf{k}_2)^2}{(\mathbf{k}_1-\mathbf{k}_2)^2}. \tag{E.11}$$

Recall that this type of nonlinear interaction contributes to the evolution of all perturbations listed in this section.

Class V

$$\dot{C}_1 \sim -K^{(2)}_{\mathbf{nmp}_2} A_{\mathbf{n}} A_{\mathbf{m}} C_2^*, \dot{C}_2 \sim -K^{(2)}_{\mathbf{nmp}_1} A_{\mathbf{n}} A_{\mathbf{m}} C_1^*. \tag{E.12}$$

The resonant condition is

$$\left(\mathbf{p}_2 - \frac{\mathbf{n}+\mathbf{m}}{2}\right)^2 = \frac{(\mathbf{m}-\mathbf{n})^2}{4}. \tag{E.13}$$

Class VI

$$\dot{C}_1 \sim -K^{(3)}_{\mathbf{np}_2\mathbf{m}} A_{\mathbf{n}} A_{\mathbf{m}}^* C_2^*, \dot{C}_2 \sim -K^{(3)}_{\mathbf{np}_1\mathbf{m}} A_{\mathbf{n}} A_{\mathbf{m}}^* C_1^*. \tag{E.14}$$

The resonant condition is

$$\left(\mathbf{p}_2 - \frac{\mathbf{n}-\mathbf{m}}{2}\right)^2 = \frac{(\mathbf{m}+\mathbf{n})^2}{4} - m^2. \tag{E.15}$$

The manifolds determined by the wavevector $\mathbf{p}_2$ for *Classes V* and *VI* are circles. The second wavevector is given by $\mathbf{p}_1 = \mathbf{n} \pm \mathbf{m} - \mathbf{p}_2$, respectively; therefore (similarly to *Class II*), $\mathbf{p}_1$ and $\mathbf{p}_2$ are directed to the antipodal point of the same circle. The *resonant* perturbations in *Class VI* are possible if for the pair $\{\mathbf{n}, \mathbf{m}\}$, the condition $|\mathbf{n} + \mathbf{m}| \geq 2m$ holds true.

E.2 Resonant Perturbations for ARs-R

In this section we apply the general theory of resonant perturbations (see Section E.1) to the stability of ARs-R. Seven resonant families listed below are found. In fact, this analysis can be easily modified for any pattern on a rhombic lattice as explained below.

(i) The perturbation of h has the form

$$h' = B_1 e^{i(\mathbf{q}_1 \cdot \mathbf{R} - \Omega_1 t_4)} + \left(B_2 e^{i\mathbf{q}_2 \cdot \mathbf{R}} + B_3 e^{i\mathbf{q}_3 \cdot \mathbf{R}}\right) e^{-i\Omega_2 t_4},$$

$$\mathbf{q}_1 = K(0,q), \quad \mathbf{q}_{2,3} = K(\pm 1, q), \quad \Omega_j = \tilde{\Omega}|\mathbf{q}_j|^2 \ (j = 1,2,3). \tag{E.16}$$

With an arbitrary q, the evolution of the amplitudes is governed by the dynamical system

$$\dot{B}_1 = \lambda_{0,q} B_1 - U_{12} a e^{i\omega t_4}(B_2 + B_3), \tag{E.17}$$

$$\dot{B}_2 = \lambda_{1,q} B_2 - U_{21} a e^{-i\omega t_4} B_1 - U_{23} a^2 B_3, \tag{E.18}$$

$$\dot{B}_3 = \lambda_{-1,q} B_3 - U_{21} a e^{-i\omega t_4} B_2 - U_{23} a^2 B_2. \tag{E.19}$$

Here and below $\lambda_{pq} = \gamma_{pq} - K_{pq} a^2$ for any p and q with K_{pq} given by (4.291) and

$$U_{12} = N^{(2)}_{\mathbf{q}_2, \mathbf{K}_1} = N^{(2)}_{\mathbf{q}_3, -\mathbf{K}_1} = \mu_1 q^2, \quad U_{21} = N^{(1)}_{\mathbf{K}_1, \mathbf{q}_1} = N^{(1)}_{-\mathbf{K}_1, \mathbf{q}_1} = \mu_2 q^2, \tag{E.20}$$

$$U_{23} = K^{(2)}_{-\mathbf{K}_1, \mathbf{q}_2, \mathbf{K}_1} = K^{(2)}_{\mathbf{K}_1, \mathbf{q}_3, -\mathbf{K}_1} = 2\alpha + \beta \left(2 - q^2\right) - J_c q^2. \tag{E.21}$$

Recall that $\mu_{1,2} = \tilde{\mu}_{1,2} K^4$, $\alpha = \tilde{\alpha} K^4$, $\beta = \tilde{\beta} K^4$, $J_c = \tilde{J}_c K^4$; the definitions of $N^{(1,2)}_{\mathbf{k}_1, \mathbf{k}_2}$ and $K^{(2)}_{\mathbf{k}_1 \mathbf{k}_2 \mathbf{k}_3}$ and the coefficients bearing tildes are given by (4.280)–(2.173).

The *resonant* family (i) is an intersection of two classes listed in Section E.1: *Class I* with $\mathbf{n} = \mathbf{K}_1$ (and $\mathbf{n} = -\mathbf{K}_1$), $\mathbf{p}_1 = \mathbf{q}_2$, and $\mathbf{p}_2 = \mathbf{q}_1$, which provides interaction of waves with the amplitudes B_2 and B_1 (the similar interaction with $\mathbf{p}_1 = \mathbf{q}_3$ also couples the waves with the amplitudes B_3 and B_1), and *Class IV* with $\mathbf{n} = -\mathbf{m} = \mathbf{K}_1$ and $\mathbf{p}_{1,2} = \mathbf{q}_{2,3}$ (interaction between B_2 and B_3).

There also exists another family, where the interaction is provided by the waves with wavevectors $\pm \mathbf{K}_2$; the stability conditions for this family are clearly the same. Below we do not discuss similar families anymore.

Conventional replacement of the variables ($\tilde{B}_1 = B_1 e^{-i\omega t_4}$, $\tilde{B}_{2,3} = B_{2,3}$) converts (E.17)–(E.19) into the equations with constant coefficients. This allows us to introduce the growth rate λ solving the following cubic equation:

$$(\lambda - \lambda_{0q} - i\omega)(\lambda - \lambda_{1q})(\lambda - \lambda_{-1,q}) - U_{23}^2 a^4 (\lambda - \lambda_{0q} - i\omega)$$
$$-U_{12}U_{21}a^2(2\lambda - \lambda_{1q} - \lambda_{-1,q} - 2a^2 U_{23}) = 0. \tag{E.22}$$

The set of equations (E.17)–(E.19) remains valid for SSs and SRs; for TSs and TRs, $B_3 = 0$ and the third equation is not needed. Of course, for any pattern the self-interaction coefficients K_{pq} can be different from those given by (4.291); they are determined as

$$K_{pq} = \sum_{\mathbf{K}} K_{\mathbf{K},-\mathbf{K},\mathbf{p}}^{(2)}$$

where $\mathbf{K}$ is the wavevector of each base wave and $\mathbf{p}$ is the wavevector of perturbations. Additionally the amplitudes a and frequencies ω for the base solution have to be chosen for the corresponding pattern. Below we do not discuss these changes in K_{pq}, a, ω anymore.

For the remaining resonant families, the perturbation is given by (E.1); below we only specify the wavevectors $\mathbf{p}_{1,2} = K(p_{1,2}, q_{1,2})$ of interacting waves and the amplitude equations for the perturbations:

(ii) $\mathbf{p}_1 = K\left(\frac{3}{2}, q\right)$, $\mathbf{p}_2 = K\left(-\frac{1}{2}, q\right)$ for any real q.
This is a particular case of *Class III* with $\mathbf{n} = \mathbf{m} = \mathbf{K}_1$.
The perturbation amplitudes evolve according to the equations:

$$\dot{C}_1 = \lambda_{\frac{3}{2},q} C_1 - U_{12} a^2 e^{-2i\omega t_4} C_2, \tag{E.23}$$

$$\dot{C}_2 = \lambda_{-\frac{1}{2},q} C_2 - U_{21} a^2 e^{2i\omega t_4} C_1, \tag{E.24}$$

where

$$U_{12} = K_{\mathbf{P}_2,\mathbf{K}_1,\mathbf{K}_1}^{(1)} = \alpha \left(\frac{9}{4} - q^2\right) + J_c q^2 \frac{3 - 4q^2}{1 + 4q^2}, \tag{E.25}$$

$$U_{21} = K_{\mathbf{P}_1,\mathbf{K}_1,\mathbf{K}_1}^{(3)} = \alpha \left(q^2 - \frac{3}{4}\right) + \beta \left(\frac{3}{2} - q^2\right). \tag{E.26}$$

The appropriately introduced growth rate solves the quadratic equation:

$$\lambda^2 - \lambda(S_1 + S_2) + S_1 S_2 - D = 0, \tag{E.27}$$

$$\tag{E.28}$$

with

$$S_1 = \lambda_{\frac{3}{2},q} + i\omega, \quad S_2 = \lambda_{-\frac{1}{2},q} - i\omega, \quad D = U_{12}U_{21}a^4. \tag{E.29}$$

Equation (E.27) remains valid with the same U_{12} and U_{21} for all the remaining patterns.

(iii) $p_2 = p_1 + 1 + \cos\theta$, $q_2 = q_1 + \sin\theta$, where

$$p_1(1 + \cos\theta) + q_1 \sin\theta + \cos\theta = 0. \tag{E.30}$$

This family belongs to *Class III* with $\mathbf{n} = \mathbf{K}_1$, $\mathbf{m} = \mathbf{K}_2$.

The amplitude equations are:

$$\dot{C}_1 = \lambda_{p_1 q_1} C_1 - iU_{12} a^2 e^{2i\omega t_4} C_2, \tag{E.31}$$

$$\dot{C}_2 = \lambda_{p_2 q_2} C_2 + iU_{21} a^2 e^{-2i\omega t_4} C_1, \tag{E.32}$$

where

$$U_{12} = K^{(3)}_{\mathbf{p}_2 \mathbf{K}_1 \mathbf{K}_2} = 2(\alpha - \beta)\mathbf{p}_1 \cdot \mathbf{p}_2 \cos\theta - \beta\left[p_1 - p_1^2 + p_2 - p_2^2\right], \tag{E.33}$$

$$U_{21} = K^{(1)}_{\mathbf{K}_1 \mathbf{K}_2 \mathbf{p}_1} = -2\alpha\left[\mathbf{p}_1 \cdot \mathbf{p}_2 \cos\theta + p_1 - p_1^2 + p_2 - p_2^2\right]$$
$$+ J_c(l_1^2 - 1)\left[q_1 \frac{p_2 \sin\theta - q_2 \cos\theta}{(1 + p_1)^2 + q_1^2} + q_2 \frac{p_1 \sin\theta - q_1 \cos\theta}{(p_2 - 1)^2 + q_2^2}\right] \tag{E.34}$$

with $l_1^2 = p_1^2 + q_1^2$. The growth rate is determined by (E.27) with

$$S_1 = \lambda_{p_1 q_1} - i\omega, \quad S_2 = \lambda_{p_2 q_2} + i\omega, \quad D = U_{12} U_{21} a^4. \tag{E.35}$$

For SSs and TSs, (E.31) and (E.32) are valid, but $-i$ and i have to be replaced with -1 there. For both types of rolls, this family is impossible as well as families (iv)–(vi).

(iv) $p_2 = p_1 + 1 - \cos\theta$, $q_2 = q_1 - \sin\theta$, where

$$(1 + p_1)(1 - \cos\theta) - q_1 \sin\theta = 0. \tag{E.36}$$

This family corresponds to *Class IV* with $\mathbf{n} = \mathbf{K}_1$, $\mathbf{m} = \mathbf{K}_2$ and with $\mathbf{n} = -\mathbf{K}_2$, $\mathbf{m} = -\mathbf{K}_1$.

The variables C_1 and C_2 are governed by the following equations:

$$\dot{C}_1 = \lambda_{p_1 q_1} C_1 + iU_{12} a^2 C_2, \tag{E.37}$$

$$\dot{C}_2 = \lambda_{p_2 q_2} C_2 - iU_{21} a^2 C_1 \tag{E.38}$$

with

$$U_{12} = K^{(2)}_{\mathbf{p}_2 \mathbf{K}_2 \mathbf{K}_1} - K^{(2)}_{\mathbf{p}_2, -\mathbf{K}_1, -\mathbf{K}_2} = (2\alpha - \beta)\left(p_1^2 - p_2^2\right)$$
$$+ J_c(l_2^2 - 1)\left[q_2 \frac{(q_1 \cos\theta - p_1 \sin\theta)}{(p_2 - 1)^2 + q_2^2} - q_1 \frac{q_2 \cos\theta - p_2 \sin\theta}{(p_1 + 1)^2 + q_1^2}\right], \tag{E.39}$$

$$U_{21} = K^{(2)}_{\mathbf{p}_1 \mathbf{K}_1 \mathbf{K}_2} - K^{(2)}_{\mathbf{p}_1, -\mathbf{K}_2, -\mathbf{K}_1} = (2\alpha - \beta)\left(p_1^2 - p_2^2\right)$$
$$+ J_c(l_1^2 - 1)\left[q_2 \frac{(q_1 \cos\theta - p_1 \sin\theta)}{(p_2 - 1)^2 + q_2^2} - q_1 \frac{q_2 \cos\theta - p_2 \sin\theta}{(p_1 + 1)^2 + q_1^2}\right]. \tag{E.40}$$

The growth rate is given by (E.27) with

$$S_1 = \lambda_{p_1 q_1}, \quad S_2 = \lambda_{p_2 q_2}, \quad D = U_{12} U_{21} a^4.$$

For SSs and TSs, (E.37) and (E.38) are the same with $+i$ and $-i$ replaced with -1 and redefinitions of U_{12} and U_{21}. For SSs $U_{12} = K^{(2)}_{\mathbf{p}_2\mathbf{K}_2\mathbf{K}_1} + K^{(2)}_{\mathbf{p}_2,-\mathbf{K}_1,-\mathbf{K}_2}$, for TSs $U_{12} = K^{(2)}_{\mathbf{p}_2\mathbf{K}_2\mathbf{K}_1}$; the same replacements are valid for U_{21}.

(v) $p_2 = 1 + \cos\theta - p_1$, $q_2 = \sin\theta - q_1$, where

$$\left(p_1 - \frac{1+\cos\theta}{2}\right)^2 + \left(q_1 - \frac{\sin\theta}{2}\right)^2 = \frac{1-\cos\theta}{2}. \tag{E.41}$$

This is an example of *Class V* with $\mathbf{n} = \mathbf{K}_1$, $\mathbf{m} = \mathbf{K}_2$.

The perturbations evolve according to

$$\dot{C}_1 = \lambda_{p_1 q_1} C_1 + iU_{12} a^2 e^{-2i\omega t_4} C_2^*, \tag{E.42}$$
$$\dot{C}_2 = \lambda_{p_2 q_2} C_2 + iU_{21} a^2 e^{-2i\omega t_4} C_1^*, \tag{E.43}$$

where

$$U_{12} = K^{(2)}_{\mathbf{K}_1\mathbf{K}_2\mathbf{p}_2} = 2\alpha\cos^2\theta + \beta\left[2p_1 p_2\cos\theta + (p_1 q_2 + p_2 q_1)\sin\theta\right] + $$
$$J_c(l_2^2 - 1)\left[q_1\frac{q_2\cos\theta - p_2\sin\theta}{(p_1 - 1)^2 + q_1^2} + q_2\frac{q_1\cos\theta - p_1\sin\theta}{(p_2 - 1)^2 + q_2^2}\right], \tag{E.44}$$

$$U_{21} = K^{(2)}_{\mathbf{K}_1\mathbf{K}_2\mathbf{p}_1} = 2\alpha\cos^2\theta + \beta\left[2p_1 p_2\cos\theta + (p_1 q_2 + p_2 q_1)\sin\theta\right] + $$
$$J_c(l_1^2 - 1)\left[q_2\frac{q_1\cos\theta - p_1\sin\theta}{(p_2 - 1)^2 + q_2^2} + q_1\frac{q_2\cos\theta - p_2\sin\theta}{(p_1 - 1)^2 + q_1^2}\right]. \tag{E.45}$$

$$\tag{E.46}$$

The growth rates are determined by (E.27) with

$$S_1 = \lambda_{p_1 q_1} + i\omega, \quad S_2 = \lambda_{p_2 q_2}^* - i\omega, \quad D = U_{12} U_{21}^* a^4.$$

For SSs and TSs, only $+i$ has to be replaced with -1 in (E.42) and (E.43) which is equivalent to redefinition of U_{12} and U_{21}, unimportant for D.

(vi) $\mathbf{p}_2 = -\mathbf{p}_1$ with

$$p_j^2 + q_j^2 = 1 \ (j = 1,2) \text{ or } p_1 = \cos(\phi + \theta/2), \ q_1 = \sin(\phi + \theta/2)$$

for arbitrary ϕ.

It belongs to *Class V* with $\mathbf{n} = -\mathbf{m} = \mathbf{K}_1$ and with $\mathbf{n} = -\mathbf{m} = \mathbf{K}_2$.

The perturbation amplitudes are governed by the equations

$$\dot{C}_1 = \lambda_{p_1 q_1} C_1 - U_{12} a^2 e^{-2i\omega t_4} C_2^*, \tag{E.47}$$
$$\dot{C}_2 = \lambda_{p_2 q_2} C_2 - U_{12} a^2 e^{-2i\omega t_4} C_1^*, \tag{E.48}$$

where

$$U_{12} = K^{(2)}_{\mathbf{K}_1,-\mathbf{K}_1,\mathbf{p}_2} - K^{(2)}_{\mathbf{K}_2,-\mathbf{K}_2,\mathbf{p}_2} = -2\beta\sin 2\phi \sin\theta, \tag{E.49}$$
$$\lambda_{p_1 q_1} = \lambda_{p_2 q_2} = \gamma - 2a^2\left(2\beta + (\beta + 2\alpha)(1 + \cos 2\phi\cos\theta)\right). \tag{E.50}$$

In fact, this perturbation has been already considered in Section 4.1.1.5; see (4.139) and (4.140). The only replacement is that the growth rate at small supercriticality, γ_2, is replaced with the entire growth rate, γ. This is unimportant because (i) the squared amplitude of the base solution is proportional to the same growth rate (γ_2 near the stability threshold and γ for finite-amplitude ARs) and (ii) both waves of perturbations have the same wavenumbers and, hence, the growth rates.

As a consequence, the stability conditions formulated in Section 4.1.1.5, (4.144) and (4.145), remain valid. The former one is guaranteed, if ARs-R is stable with respect to simple perturbations. The latter one can be rewritten as:

$$\left|(4\cos^2\theta - 3)\alpha - \beta - 2(2\alpha + \beta)\cos 2\phi \cos\theta\right|^2 - 4|\beta|^2 \sin^2 2\phi \sin^2\theta > 0. \quad (E.51)$$

For SSs $U_{12} = K^{(2)}_{\mathbf{K}_1,-\mathbf{K}_1,\mathbf{p}_2} + K^{(2)}_{\mathbf{K}_2,-\mathbf{K}_2,\mathbf{p}_2}$, whereas for TSs this resonant family does not exist.

(vii) $p_2 = 1 - p_1$, $q_2 = -q_1$;

$$\left(p_1 - \frac{1}{2}\right)^2 + q_1^2 = \frac{1}{4}. \quad (E.52)$$

This is a representative of *Class II* with $\mathbf{n} = \mathbf{K}_1$.

The amplitudes are governed by the equations

$$\dot{C}_1 = \lambda_{p_1 q_1} C_1 - U_{12} a e^{-i\omega t_4} C_2^*, \quad (E.53)$$
$$\dot{C}_2 = \lambda_{p_2 q_2} C_2 - U_{12} a e^{-i\omega t_4} C_1^*. \quad (E.54)$$

with $U_{12} = N^{(2)}_{\mathbf{K}_1 \mathbf{p}_2} = N^{(2)}_{\mathbf{K}_1 \mathbf{p}_1} = p_1(1 - p_1)\mu_1$.

The growth rate for this family satisfies (E.27) with

$$S_1 = \lambda_{p_1 q_1} - \frac{i\omega}{2}, \quad S_2 = \lambda^*_{p_2,-q_2} + \frac{i\omega}{2}, \quad D = |U_{12}|^2 a^2.$$

For the rest of rhombic patterns, the same amplitude equations are appropriate with U_{12} unchanged.

E.3 Resonant Perturbations for ARs

The analysis of resonant perturbations for ARs is cumbersome because the solution corresponding to ARs includes infinite sequence of base waves. Recall that each base wave and each pair of base waves produce several resonant families. This complexity is heavily grounded on the quadratic terms in amplitude equations (4.278); for $N^{(1,2)}_{\mathbf{k}_1 \mathbf{k}_2} = 0$, ARs comprise four base waves only, and the analysis of external perturbations becomes simpler. There are two reasons for vanishing of the quadratic terms, either due to the symmetry as for the buoyancy convection (see Section 4.1.1) or at large supercriticality (see Section 4.2.1.4 for the square lattice at large m_2).

In fact, most of the families of resonant perturbations can be considered on the base of Section E.2; one just has to set $\theta = \pi/2$ in the corresponding conditions and to exclude the quadratic interactions; the nomenclature of the resonant families is adopted from Section E.2. This leads to disappearance of the resonant family (vii) and simplification of (i); in contrast, the resonant family (iii) becomes more involved. For the sake of reader convenience, below we briefly list the corresponding families; if this is not mentioned specially, the perturbation is given by (E.1) and only $\mathbf{p}_{1,2}$ are presented.

(i) $\mathbf{p}_1 = K(1,q)$, $\mathbf{p}_2 = K(-1,q)$ with arbitrary q. The evolution of the amplitudes $C_{1,2}$ is governed by the dynamical system

$$\dot{C}_1 = \lambda_{1q}C_1 - U_{23}a^2C_2, \tag{E.55}$$
$$\dot{C}_2 = \lambda_{-1,q}C_2 - U_{23}a^2C_1, \tag{E.56}$$

which is the limiting case $B_1 = 0$, $B_2 = C_1$, $B_3 = C_2$ of (E.17)–(E.19) with the same U_{23}. The growth rate for this set of equations is determined by (E.27) with $S_{1,2} = \lambda_{\pm 1,q}$ and $D = U_{23}^2 a^4$, which provides

$$\lambda^{(1,2)} = \lambda_{1,q} \pm \left[2\alpha + (2 - q^2)\beta - q^2 J_c\right]a^2.$$

(ii) $\mathbf{p}_1 = K\left(\frac{3}{2},q\right)$, $\mathbf{p}_2 = K\left(-\frac{1}{2},q\right)$ for any real q. This family and its stability condition coincide exactly with family (ii) for ARs-R.

(iii) For this family three waves interact; the perturbation is presented as

$$h' = B_1 e^{i(\mathbf{q}_1 \cdot \mathbf{R} - i\Omega q^2 t_4)} + \left(B_2 e^{i\mathbf{q}_2 \cdot \mathbf{R}} + B_3 e^{i\mathbf{q}_3 \cdot \mathbf{R}}\right) e^{-i\Omega l^2 t_4},$$
$$\mathbf{q}_1 = K(-q,q), \quad \mathbf{q}_{2,3} = K(\pm 1 - q, \pm 1 + q), \tag{E.57}$$

where $l^2 = 2(1 + q^2)$. This family (similar to resonant family (iii) for ARs-R) realizes *Class III* with $\mathbf{n} = \mathbf{K}_1$, $\mathbf{m} = \mathbf{K}_2$ (interaction of the waves with the amplitudes B_1 and B_2) and with $\mathbf{n} = -\mathbf{K}_1$, $\mathbf{m} = -\mathbf{K}_2$ (interaction of the waves with the amplitudes B_1 and B_3). Thus, for the square lattice, the single wave with $\mathbf{p}_1 = \mathbf{q}_1$ gives rise to the resonant interaction with $\mathbf{q}_2 = \mathbf{q}_1 + \mathbf{K}_1 + \mathbf{K}_2$ and with $\mathbf{q}_3 = \mathbf{q}_1 - \mathbf{K}_1 - \mathbf{K}_2$. In contrast, for ARs-R those are two different interactions with their own $\mathbf{p}_1$.

The amplitudes of these three perturbation waves are governed by

$$\dot{B}_1 = \lambda_{-q,q}B_1 - iU_{12}a^2 e^{2i\omega t_4}(B_2 + B_3), \tag{E.58}$$
$$\dot{B}_2 = \lambda_{1-q,1+q}B_2 + iU_{21}a^2 e^{-2i\omega t_4}B_1, \tag{E.59}$$
$$\dot{B}_3 = \lambda_{-1-q,-1+q}B_3 + iU_{21}a^2 e^{-2i\omega t_4}B_1, \tag{E.60}$$

where

$$U_{12} = K^{(3)}_{\mathbf{q}_2 \mathbf{K}_1 \mathbf{K}_2} = K^{(3)}_{\mathbf{q}_3, -\mathbf{K}_1, -\mathbf{K}_2} = 2\beta q^2, \tag{E.61}$$

$$U_{21} = K^{(1)}_{\mathbf{q}_1 \mathbf{K}_1 \mathbf{K}_2} = K^{(1)}_{\mathbf{q}_1, -\mathbf{K}_1, -\mathbf{K}_2} = 4\alpha q^2 - 2J_c\frac{(2q^2 - 1)^2 q^2}{4q^4 + 1}. \tag{E.62}$$

There are two types of perturbations within this family: for the first one $B_2 + B_3 = 0$, its growth rate is given by $\lambda_{1-q,1+q} (= \lambda_{-1-q,-1+q})$; for the second one, $B_2 = B_3$ and the growth rate is determined by (E.27) with

$$S_1 = \lambda_{-q,q} - i\omega, \quad D = 2U_{12}U_{21}a^4, \quad S_2 = \lambda_{1-q,1+q} + i\omega. \qquad (E.63)$$

It is clear that the first type of perturbation is just a particular case of *simple* perturbation, whereas the equation for the growth rate for the second type coincides with that for family (iii) in Appendix E.2 at $\theta = \pi/2$ with the only distinction, D is twice larger for the square lattice, because two similar waves (with the amplitudes B_2 and B_3) interact with the B_1-wave.

The last three families are obvious limiting cases $\theta = \pi/2$ of the same cases for ARs-R. Since the stability conditions for square lattice are important and substantial simplification is possible, the corresponding reduced conditions are presented below.

(iv) $\mathbf{p}_1 = K(q-1,q)$, $\mathbf{p}_2 = K(q,q-1)$.
Substituting these values into (E.39) and (E.40), one readily obtains:

$$\lambda_{q-1,q} = \lambda_{q,q-1}, \quad U_{12} = U_{21} = (2\alpha - \beta)(1 - 2q).$$

Thus, the growth rates for this mode are given by

$$\lambda^{(1,2)} = \lambda_{q-1,q} \pm (2\alpha - \beta)a^2.$$

(v) $\mathbf{p}_1 = K(p,q)$, $\mathbf{p}_2 = K(1-p,1-q)$, where

$$\left(p - \frac{1}{2}\right)^2 + \left(q - \frac{1}{2}\right)^2 = \frac{1}{2}. \qquad (E.64)$$

For a square lattice, $U_{12} = U_{21} = \beta(p_1 - q_1)^2$ and, hence, $D = |\beta|^2(p_1 - q_1)^4 a^4$; $S_{1,2} = \lambda_{p_{1,2},q_{1,2}} \pm i\omega$; the growth rate is determined by (E.27).

(vi) $\mathbf{p}_2 = -\mathbf{p}_1 = K(p,q)$ with $p^2 + q^2 = 1$.
Similarly to ARs-R this family can be considered on the base of analysis of external perturbations near the stability threshold (Section 4.1.1.5). Therefore, the stability condition (4.146) (or (E.51) at $\theta = \pi/2$) is valid:

$$|3\alpha + \beta|^2 - 4|\beta|^2 > 0.$$

Also, any combination of the single-parameter families mentioned above should be considered separately. However, most of these perturbations have integer or half-integer values of p and q. The numerical simulations discussed in Section 4.2.1.4 show that all these perturbations decay. The only case which should be taken into account separately is

$$h' = b\exp\left[ik\left(-\frac{1}{2}X + q_1Y\right) - i\tilde{\Omega}\left(\frac{1}{4} + q_1^2\right)T\right]$$

$$+d\exp\left[ik\left(\frac{3}{2}X+q_1Y\right)-i\tilde{\Omega}\left(\frac{9}{4}+q_1^2\right)T\right]$$
$$+c\exp\left[ik\left(-\frac{1}{2}X+q_2Y\right)-i\tilde{\Omega}\left(\frac{1}{4}+q_2^2\right)T\right]$$
$$+f\exp\left[ik\left(\frac{3}{2}X+q_2Y\right)-i\tilde{\Omega}\left(\frac{9}{4}+q_2^2\right)T\right] \quad\quad\text{(E.65)}$$
$$q_{1,2}=\frac{1}{2}\pm\frac{1}{\sqrt{2}},$$

which is a combination of cases (ii) and (v) considered in Appendix C of [1].

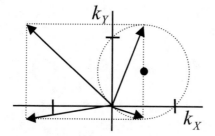

Fig. E.1 Wavevectors of four interacting resonant modes, (E.65)

The wavevectors of the resonant perturbations are shown in Figure E.1. A stability analysis not presented here shows that the critical value of $S+P$ for this mode varies with the Lewis number, but for any value of L, it always remains smaller than $26/3$. Thus, the limiting solution at large m_2 loses its stability when the inequality (4.308) is satisfied.

Reference

1. S. Shklyaev, A.A. Nepomnyashchy, A. Oron, SIAM J. Appl. Math. **73**, 2203 (2013)

Index

© Springer Science+Business Media, LLC 2017
S. Shklyaev, A. Nepomnyashchy, *Longwave Instabilities and Patterns in Fluids*,
Advances in Mathematical Fluid Mechanics,
https://doi.org/10.1007/978-1-4939-7590-7

Printed in the United States
By Bookmasters